Characteristics to Be Measured	Units and Approximate Metric-to-English Conversions*
Small liquid volume	1 milliliter is about 0.2 teaspoon (1 tsp is about 5 ml)
Medium liquid volume	1 liter is about 0.26 U.S. gallons (1 gal is about 3.8 l)
Large liquid volume	1 megaliter is about 0.8 acre feet (1 ac-ft is about 1.2 Ml)
River flow	1 liter per second is about 0.04 cubic feet per second (1 cfs is about 28 l/s) [1 cfs is about 0.65 million gallons per day (mgd)]
Speed	1 meter per second is about 2.2 miles per hour (1 mph is about 0.45 m/s)
Pressure	1 millibar is about 0.015 pounds per square inch (1 lb/sq in. is about 70 mb) (1 in. of mercury is about 34 mb)
Temperature	1 Celsius degree is about 1.8 Fahrenheit degrees, but the two scales do not have the same zero point; therefore $°C = (°F - 32) \div 1.8$ $°F = (°C \times 1.8) + 32$
Energy	1 joule is about 0.24 calories 1 kilojoule is about 0.95 BTU
	1 kilocalorie is about 4 British Thermal Units (one food calorie) (1 BTU is about 0.25 kcal)
Power	1 kilowatt is about 1.3 horsepower (1 HP is about 0.18 kcal/s)
Energy flow	1 langley is about 4.2 joules per square centimeter (1 ly is about 1 calorie per square centimeter)
	1 langley is about 3.7 British Thermal Units per square foot (1 BTU/ft^2 is about 0.27 ly)

*With abbreviations and approximate English-to-metric conversions

Physical Geography

Physical Geography

Philip Gersmehl
University of Minnesota

William Kammrath
Concordia College

Herbert Gross
Concordia College

1980 SAUNDERS COLLEGE
Philadelphia

Library of Congress Cataloging in Publication Data

Gersmehl, Philip.
Physical geography.

Includes index.
1. Physical geography. I. Kammrath, William, joint author. II. Gross, Herbert Henry, 1905– joint author. III. Title.
GB55.G35 910′.02 78-12212
ISBN 0-03-014476-0

Printed in the United States of America
012 039 98765432

PREFACE

This text presents the basic theoretical core of physical geography at the introductory level. The interrelationships between humans and their environment are stressed by placing contemporary environmental science in a geographical framework.

All too often texts present a host of details, examples, and applications, and the instructor is forced to spend class time abstracting the pertinent information for a particular group of students. Ours is a teaching book, deliberately organized with several levels of detail running concurrently, so that class time can be devoted to local examples, policy issues, slideshows, simulations, and field excursions. On the most fundamental level are the main text, illustrations, and the first few lines of each caption. This material provides basic grounding in ecological physical geography for students seeking a general understanding of natural environments. Definitions in the margins furnish a second level of detail and are intended primarily for students likely to continue studying the field since knowledge of these terms is assumed in advanced courses and textbooks. The extended figure captions and chapter appendixes provide a third level of information, often of a statistical or regional nature. Finally, reading lists, simulations, and student activities are included in the instructor's manual, which will be updated more frequently than the text.

We recommend reading the main text for the principal ideas (it is intentionally only about one-half as long as most of its competitors). Then, if necessary or desirable, reread the text with whatever supplementary information you find useful.

Special thanks go to

our wives for all kinds of assistance in writing, editing, typing, research, map drafting, and psychological counseling;

Barb VanDrasek, Carol Atchley, and Marge Rasmussen, for typing and manuscript assembly;

Bruce B. Adams, Bloomsburg State College; Robert Altschul, University of Arizona; Robert Burrill, University of Georgia; David Castillion, Southwest Missouri State University; George Dury, University of Wisconsin; Gary Elbow, Texas Tech University; Wallace Elton, University of Georgia; David Firman, Towson State University; Robert M. Hordon, Rutgers University; Robert Howard, California State University; Elias Johnson, Saint Cloud State College; William Johnson, University of Oklahoma; Donald Kiernau, Solano Community College; Edward Lyon, Ball State University; Elizabeth Marsh, Stockton State College; Thomas McDannold, Ventura College; Duncan Randall, University of North Carolina, Wilmington; Paul Rizza, Slippery Rock State College; Neil Salisbury, University of Iowa; George A. Schnell, College of New Paltz, SUNY; Walter Schroeder, University of Missouri; Judth Shulz, Old Dominion College; Donald Steila, East Carolina University; Thomas

Vale, University of Wisconsin; Harold A. Winter, Michigan State University; and C. E. Zimolzak, Glassboro State College, for critically reading the manuscript;

Rod Squires, Dwight Brown, Dick Skaggs, Ward Barrett, Marv Bartell, Clarence Drews, Bill Hussong, Bob Osinski, Vern Meentemeyer, James Shear, James Rahn, and other colleagues who allowed their ideas to rub off on us; and

Gail Dietrich, John Oppermann, Joe Mannion, Karen Van Buskirk, Kevin Hanron, and many other students who read parts of the manuscript and made appropriate suggestions.

Finally Philip Gersmehl and William Kammrath acknowledge their enormous debt to Herbert H. Gross, their undergraduate mentor, graduate advisor, faculty colleague, and close friend; he is thus ultimately responsible for much of the content of this book, although we are willing to admit that most of its errors are probably ours.

Minneapolis, Minnesota
River Forest, Illinois
September, 1979

Philip Gersmehl
William Kammrath
Herbert Gross

CONTENTS

Preface v

Prologue: Physical Geography 1

Appendix to Prologue Location 4

Chapter 1 *Energy Budgets: Power for the Earth Machine* 7

The Basic Energy Budget 8
Comparative Energy Budgets 11
Detailed Energy Budget 13
Energy Budget and Temperature Maps 17
Suburban Energy Budgets 19
Biological Energy Budgets 22
Energy Systems 25

Appendix to Chapter 1 Earth-Sun Relationships 26

Chapter 2 *Instability: Vertical Motion in the Atmosphere* 30

Changes in Rising Air 31
Principles of Air Stability 33
Evaluating Air Stability 35
Air Stability and Cloud Types 37
Air Stability and Atmosphere Pollution 39
The Inversion Type of Climate 43
The Instability Type of Climate 46

Appendix to Chapter 2 Meteorological Charts 48

Chapter 3 *Wind: Horizontal Motion in the Atmosphere* 50

Simple Convection Systems 51
Natural Convection Systems 53
A Grand Convection Model 54
Subtropical Highs 54
The Subsidence Type of Climate 56
Complex or Hybrid Climates 58
Hurricanes 60
Prevailing Winds 63
World Wind Patterns 64
Conclusion 65

Chapter 4 *Fronts: Battlefields in the Sky* 66

Synoptic Weather Maps 67
Air-Mass Identification 67
The Anatomy of Fronts 71
Surface Map Forecasting 76
Dynamic Forecasting 78
Analog Forecasting 80
Colloquial Forecasting 82
The Frontal Types of Climate 84
Conclusion 86

Appendix to Chapter 4 Common Weather Patterns 87

Chapter 5 *Climate: Expected Patterns of Weather* 92

Classification 93
A Genetic Classification of Climate 94
A Descriptive Classification of Climate 97
Translations 98
A Probability Classification for Farmers 99
A Recurrence Interval Classification for Engineers 101
A Duration Sum Classification for Furnaces 104
A Physiological Classification for Human Bodies 106
A Water Balance Classification for Crops 108
Conclusion 108

Appendix to Chapter 5 Climatic Graphs 110

Hints for Locating Places by Analyzing Their Climatic Graphs 112

Chapter 6 *Plate Tectonics: The Not-So-Solid Earth* 114

Internal Structure of the Earth 115
Evidence for Crustal Motion 116
Oceanic Effects of Crustal Motion 118
Continental Effects of Crustal Motion 121
Panagaea, The Ancient Supercontinent 122
Volcanic Features 123
Other Tectonic Features 126
Isostasy 128
Crustal Motion and World Climate 129

Appendix to Chapter 6 Geological Time 130

Chapter 7 *Weathered Rocks: Surface Earth Materials* 135

Rocks and Minerals 136
Frozen Liquids—The Igneous Rocks 137
The Four Fates of Igneous Rocks 140
Reincarnation—The Sedimentary Rocks 141
Rock Weathering 145
Climate and Weathering 147
Rock Composition and Weathering 149
Summary—The Geography of Weathering Products 150

Appendix to Chapter 7 Rock and Mineral Identification 151

Chapter 8 *Exchange Processes: Soil as a Storehouse* 157

Soil Texture and Storage Capacity 158
Exchange Processes 160
Leaching 161
pH and Base Saturation 162
Soil Moisture Tension 162
The Water Table 165
Soil Drainage 167
Capillarity and Desert Soils 168
Soil Structure 169
Conclusion 172

Appendix to Chapter 8 Plant Nutrients and Deficiency Symptoms 174

Chapter 9 *Decomposers: Natural Sanitary Engineers* **176**

Temperature 177
Moisture 179
Soil Texture 182
The C/N Ratio 183
Origins of Organic Matter 184
Benefits of Organic Matter in a Soil 186
Nitrogen Conversion 188
Conclusion 190

Chapter 10 *Soil: Reprocessed Dirt* **191**

Soil Series and Sequences 194
Soil Series and Crop Productivity 197
Land Capability Classes 198
Engineering Properties of Soils 202
The Seventh Approximation 205

Appendix to Chapter 10 Soil Orders 208

Chapter 11 *Tolerance and Succession: Life on Land* **231**

Tolerance Theory 232
Complex Environments and Organism Responses 236
Succession as Environmental Modification 238
Succession as Energy Optimization 241
Population Regulation 242
The Geography of Life 244
Carnivores, Energy, and Wilderness 246

Appendix to Chapter 11 Terrestrial Biomes 246

Chapter 12 *Biogeochemical Cycles: The Balance of Nature* **256**

The Nutrient Cycle as a Closed System 257
The Nutrient Cycle as an Open System 261

Harvests and Fertilizers 263
The Nutrient Cycle as an Externally Governed System 266
The Nutrient Cycle as an Internally Linked System 269
The Nutrient Cycle as a System Tending toward Equilibrium 270
Conclusion 271

Chapter 13 *Oxygen and Biochemical Demand: Life in Water* 272

Food and Oxygen in Water 273
Water Characteristics and Oxygen Supply 274
Oxygen Sag in a Flowing River 277
Oxygen Sags in Lakes and Rivers 278
Eutrophication 280
Pollution as Accelerated Eutrophication 282
Ocean Productivity 282
Upwellings and Ocean Currents 284
Estuaries, Shelves, and Salt Marshes 286

Appendix to Chapter 13 Aquatic Biomes 289

Chapter 14 *Water Budgets: Floods and Droughts* 295

The Water Budget 296
Refined Water Budgets 301
Groundwater Graphs 307
Stream Hydrographs 309
Regimes, Stages, and Frequencies 310
Flood-routing Graphs 312

Appendix to Chapter 14 Estimating Runoff, Erosion, and Sediment Yield 314

Chapter 15 *Erosion and Isostasy: Gravity at Work* 319

Principles of Landforming Energy 320
Downhill Movements and Gravity Components 321
The Divisions of a Mass Movement Slope 323

River Profiles and Graded Streams 324
Baselevel and the Cycle of Erosion 327
Isostasy 328
Spatial Differences in Landforming Energy 329
Temporal Variations of Landforming Energy 331
Climatic Control of Landforming Processes 332

Chapter 16 *Terrain Shapers: Landforming Processes* **333**

Terrain Shaped by Mass Movement 334
Terrain Shaped by Water—Pediments 339
Terrain Shaped by Water—Valleys 339
Terrain Shaped by Water—Plains 341
Terrain Shaped by Ice—Scour Zones 345
Terrain Shaped by Ice—Moraines 347
Terrain Shaped by Melting Ice—Outwash Deposits 349
Terrain Shaped by Wind—Pavements 350
Terrain Shaped by Wind—Sand Dunes 352
Terrain Shaped by Wind—Dust Caps 353
Terrain Shaped by Waves—Benches 356
Terrain Shaped by Waves—Beaches 358
Terrain Shaped by Waves—Bars 360
Terrain Shaped by Animals—Pits and Mounds 361
Conclusion 362

Appendix to Chapter 16 Morphogenetic Regions 363

Chapter 17 *Landforms: Structure, Process, and Time* **370**

Central Appalachia 371
The Wisconsin Glaciers 373
The Lower Forty and the High Plains 378
Canyonlands 381
Mount Rainier 384
Conclusion 386

Appendix to Chapter 17 Topographic Maps 387

Epilogue: Earth Systems **403**

Appendix to Epilogue Generalized Geographic Correlation of Major Environmental Process and Features 406

Index **409**

Physical Geography

PROLOGUE

Physical Geography

Geography is not a collection of facts; it's a way of looking at things.

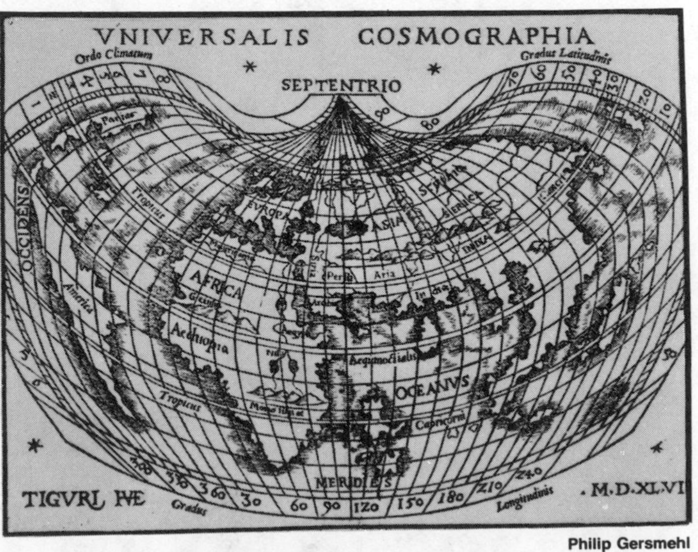

Philip Gersmehl

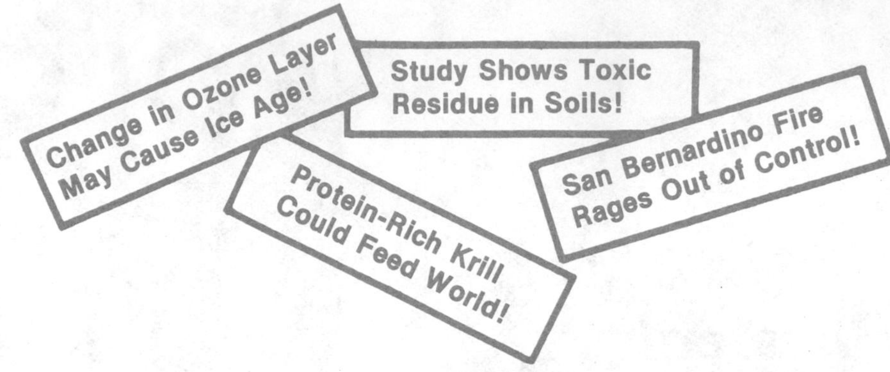

These recent headlines about resources, catastrophes, theories, technology, are all related to physical geography, the subject of this book. Physical geography is a study of the earth—what it is, what makes it function, how it developed into what it is, what kinds of changes are taking place, how things in nature are distributed, how these things affect each other, and how they affect people. Nature has many fascinating stories to tell. Every hill, every lake, every storm, every forest carries a message. Some messages provide pleasure when understood; others have an important bearing on the very existence of the observer and therefore should not be ignored.

Physical geographers depend on various sciences for much of the information they need (Fig. P-1). Earth sciences deal mainly with processes; they attempt to provide insight into how and why things happen in nature.

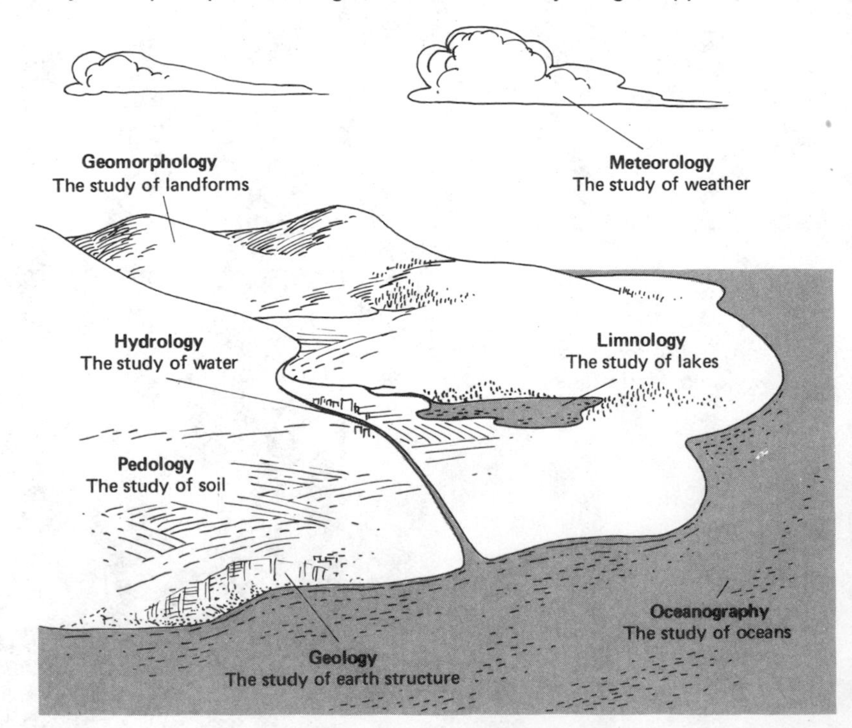

Physical geography is also concerned with processes, but to "how" and "why" it adds "where." The reader should sense in this book a pervasive conviction that the location of a place has much to do with the kinds of features and events that occur there. The belief that the world has a spatial as well as a causal logic is the peculiar bias of a geographer (Fig. P-2).

One obvious outcome of this emphasis on spatial differences is the realization that the earth contains many different environments, in which different processes operate, or in which the same processes have very different outcomes. For example, there are places in the world where deliberate forest fires are a sensible use of the physical environment. In fact, suppression of fires in those places can be very harmful, because the absence of fire leads to depletion of soil nutrients, deterioration of plant productivity, reduction of animal populations, and ultimately the development of catastrophic fires. "Fire environments" such as these are associated with certain types of vegetation that arise in response to particular patterns of weather, which, in turn, occur in predictable locations on the globe.

Many other natural catastrophes are also noted for their regular habits; they appear under particular combinations of circumstances in specific locations. Physical geographers insist that there is regularity in the spatial arrangements of most characteristics of the physical environment; there is a discoverable pattern to the distributions of areas of high mineral value, pollution risk, energy use, scenic beauty, agricultural productivity, or recreational potential.

Most of the hazards and resources of the physical world are subject to alteration by human actions. Whether the changes will be beneficial or harmful depends largely on the knowledge and values possessed by the humans who make the decisions. Most U.S. citizens have a very uneven understanding of environmental processes. Television forecasters impart a great deal of weather terminology and basic concepts, although most people rarely organize their weather knowledge into a comprehensive theory. Reports of catastrophic storms and earthquakes appear frequently in newspapers and other media, but consequences receive more coverage than causes. Natural terrain is not well understood; seen from a car window, the features are partly obscured by the land-forming work of house builders and highway engineers. Most urban dwellers take their water supply for granted, while many suburbanites garden a soil they do not really understand. A typical citizen of the United States knows far more about African lions and Antarctic whales than about the organisms beneath the grass of the front lawn.

Despite wide variations of background knowledge, citizens directly and indirectly make environmental decisions every day. Lawn fertilizers, political votes, grocery lists, vacation plans, and house furnishings all have effects on the physical environment, both in the local area and far away. The effects are different in different places; they form the subject of this book.

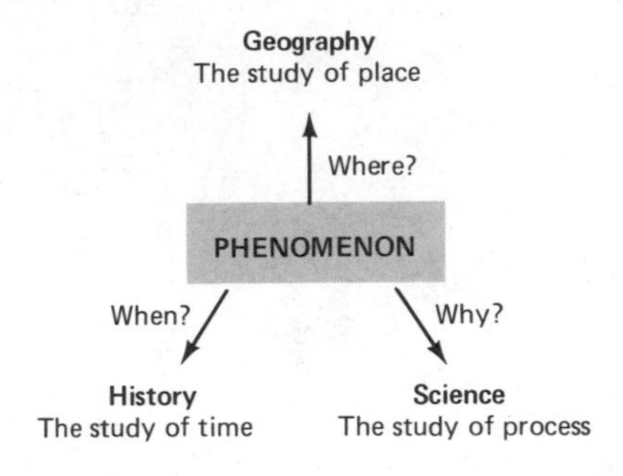

FIG. P-2. Kant's triangular division of knowledge. Each of the three approaches may look at the same phenomenon, but they organize their inquiry around different questions.

APPENDIX TO PROLOGUE
Location

A major article in the geographer's creed states that the location of a place profoundly influences its character. A geographic analysis of something begins with a description of where it is. The most straightforward expression of location relies on landmarks. A place is described in terms of its neighbors—"just beyond the mountain," "under the old oak tree," or "where the creek enters the river." Obviously, a landmark system fails as soon as there are two mountains or old oaks or creeks. Assigning proper names—"Lofty Mountain" or "Cripple Creek"—is at best a temporary solution, because the number of features to be named exceeds the number of easily remembered names. Landmark location therefore usually works only on local features or on very prominent global ones.

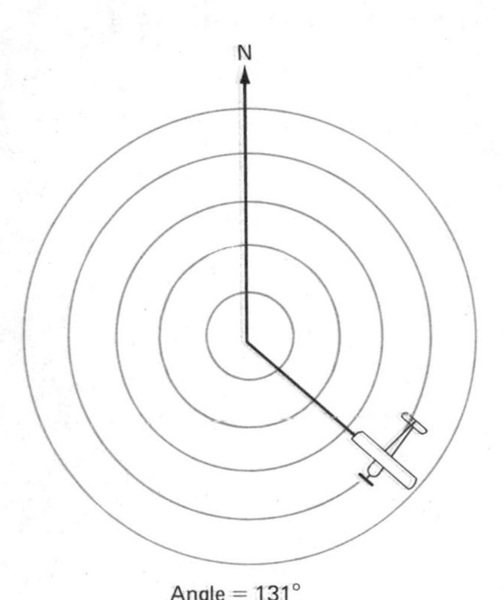

Agreement on a standard set of names for directions and distances allows people to use more sophisticated and more precise expressions of location, known to mathematicians as coordinate systems. Airports locate airplanes by means of polar coordinates that express position in terms of angles and lengths. The length represents the distance of the airplane from an arbitrary central point, usually the control tower; the angle is the direction of the plane relative to the central point, usually expressed in degrees clockwise from north. Since airport control towers are not mobile, the point-centered coordinate system is convenient and workable. However, trying to compute distance and directions from a noncentral point on polar coordinates is a complex and time-consuming chore. Furthermore, specifying the size of a given area is difficult with a polar system.

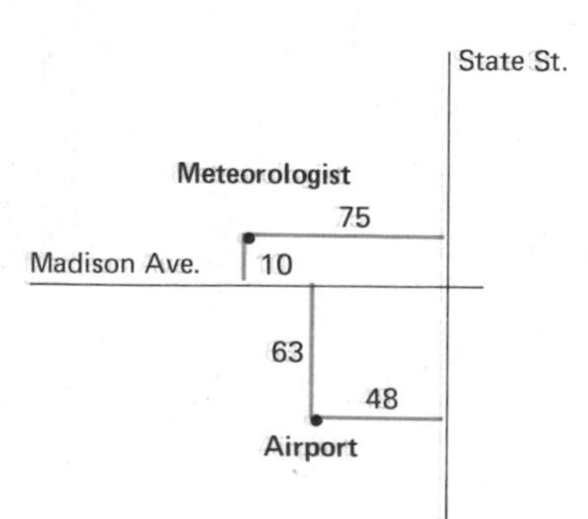

A coordinate system based on distances from two standard lines is more appropriate for general purposes. The street plan of Chicago is a good example of rectangular coordinates. State Street, running north and south near Lake Michigan, and Madison Avenue, an east-west road, are the reference lines for the grid system. A Chicago block is one-fifth of a kilometer or one-eighth of a mile long. A meteorologist lives 10 blocks north of Madison Avenue and 75 blocks west of State Street. The weather station at Midway Airport is 48 blocks west and 63 blocks south of the reference lines. When the directional words are translated into the plus and minus signs of a standard mathematical coordinate system, coordinates of the airport are −48, −63, and the meteorologist is located at −75, 10. The meteorologist would have to go 27 blocks east and 73 blocks south to get to the airport. The rectangular arrangement of streets makes the journey by land a 100-block ordeal.

Pythagoras and other mathematicians point out one of the advantages of a rectangular coordinate system. They can rise above the constraints of the street grid and calculate the distance along a direct path to the airport. Their formula says that the airport is $(27^2 + 73^2)^{1/2}$ or nearly 78 blocks as the the crow struggles through Chicago's air. Standard trigonometric tables furnish a quick determination of the exact direction. Calculation of area is also very easy on a uniform rectangular grid.

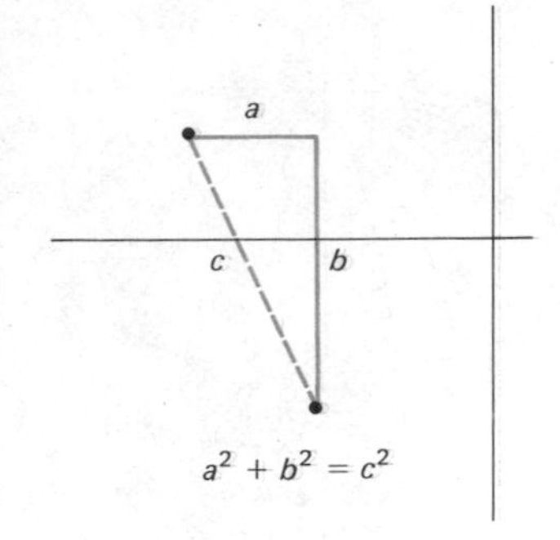

Polar and rectangular coordinate systems are products of plane geometry, whereas the earth is a big and nearly spherical ball. Applying a plane system to a curved surface involves some bending or adjustment. Allowances for the curvature of the earth are minor when mapping an area as small as Chicago. Some streets do not run exactly north and south, while others take small jogs every few blocks. A rectangular survey of an entire state encounters some serious problems. The United States Rectilinear Survey tried to divide much of the country into squares one mile on a side. A reasonably level part of Indiana or Kansas looks like a giant disjointed chessboard, as if different carpenters made different parts of the board and then brought them together and tried to make them fit. Some roads run due north for a dozen miles and then take sharp double bends as they pass onto a slightly offset part of the survey grid. Some of the squares are full-sized. Others may fall far short of the specified 640 acres. Still others are flanked by triangular patches of land where independent surveys met but did not fit. By the time the survey extended over half the continent, the problems of fitting a flat grid to a spherical earth became almost insurmountable.

Meridians

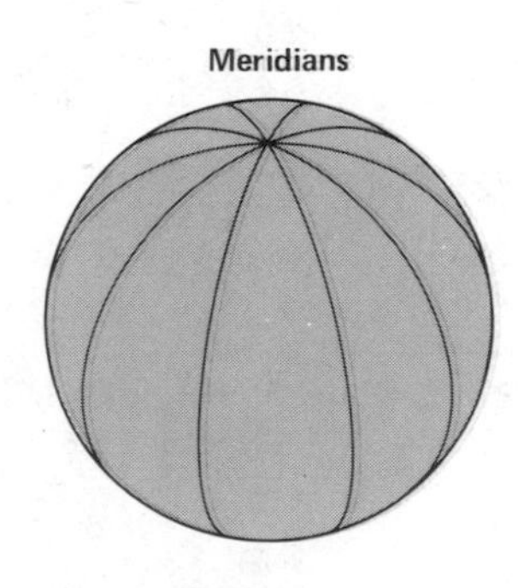

Fortunately, the earth itself provides the basis for its own system of expressing location. Its rotation on an axis furnishes two fixed points, the poles, upon which to build a coordinate system that is more appropriate for a globe than the plane systems. Meridians (lines of longitude) pass through both poles and run due north and south. They are numbered in degrees east and west of the Prime Meridian, which runs through Greenwich, a suburb of London, and was somewhat arbitrarily picked as the central meridian of the global coordinate system by a group of astronomers who lived there.

The equator, the standard east-west line for the global coordinate system, runs around the earth exactly halfway between the poles. Between the equator and the poles, other latitude lines are drawn parallel to the equator. The parallels (lines of latitude) are numbered in degrees north and south of the equator. The equator is the longest parallel. The earth is only half as far around along the 60th parallel as it is along the equator. The 90th parallel is only a point at the pole where all the meridians meet. The location of any place on the earth can be specified in terms of its position on the global coordinate system. New York (41° N and 74° W) is near the intersection of the 41st parallel north of the equator and the 74th meridian west of Greenwich.

Parallels

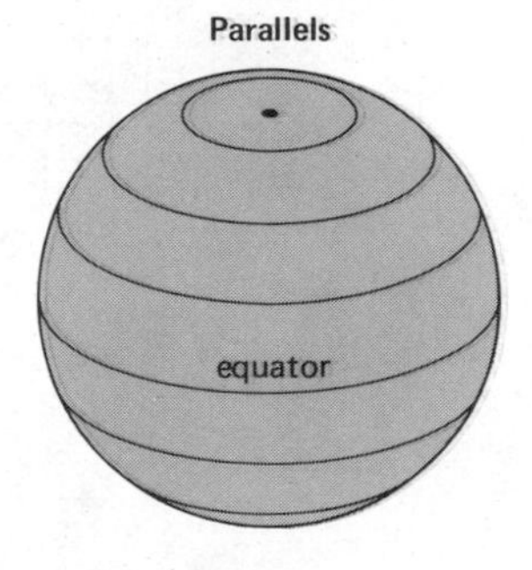

The earth grid near the equator resembles a rectangular coordinate system. Seen from above the poles, it appears to be a polar grid. The best way to become familiar with the characteristics of the global grid is to hold a three-dimensional globe in your hands and study it; most people have serious misconceptions about the world because they have seen it mainly through the distorted images of flat maps.

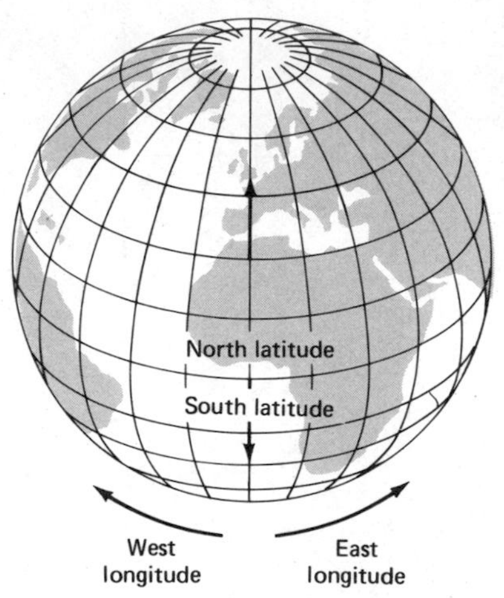

The problem of transferring coordinate systems from a spherical to a plane surface means that it is impossible to draw an accurate map of the earth on flat paper. All maps distort at least one of four characteristics of the earth's surface: direction, shape, area, or distance. A map with a particular purpose may have distortions of two of these characteristics in order to represent the other two accurately. For example, the map of wind directions in this book requires more faithfulness of direction than of area. Therefore it appears on a rectangular base map which preserves the traditional north-up, east-right, south-down, west-left orientation that most students expect. Directional accuracy has a price, though; global parallels get shorter with distance away from the equator, but the map parallels are equal in length. In effect, the global grid has been stretched to fit a rectangular coordinate system. Direction and shape survive the stretching, but area and distance are wildly distorted. Greenland appears to be half as big as South America, whereas it should be less than one-eighth the size of the southern continent.

Many atlas maps of vegetation patterns and geologic processes are drawn on a base map that sacrifices directional accuracy and puts a big gap between Greenland and Europe in order to get a better representation of the size of continents. Comparisons of maps on different bases will not be reliable unless the reader keeps the dissimilarities in the base maps in mind. To maintain easy comparability, this book uses the same map projection throughout.

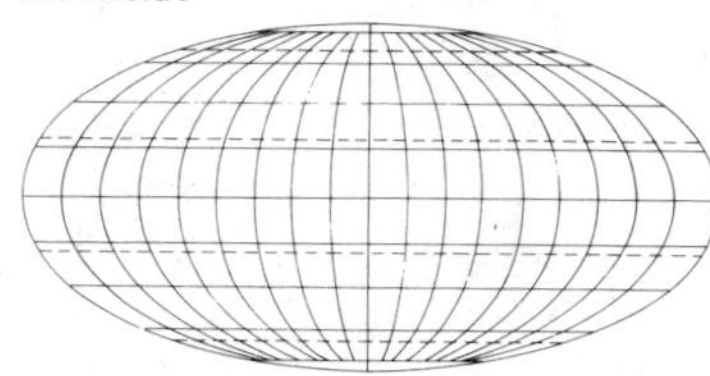

Mercator

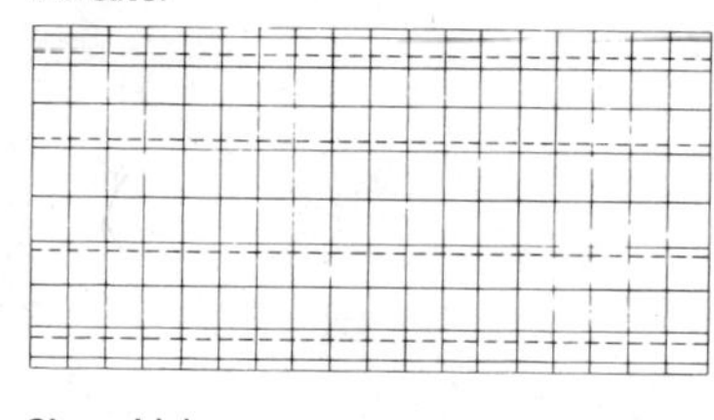

Sinusoidal

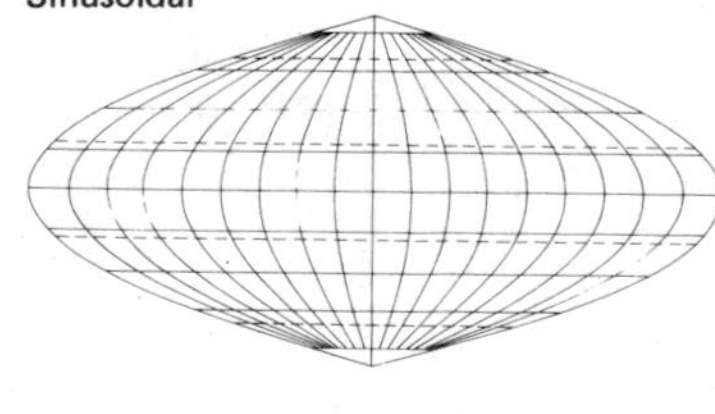

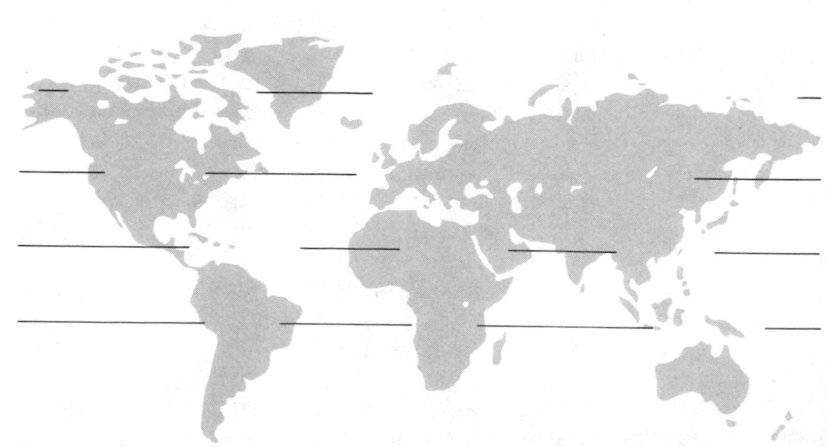

Chapter 1

Energy Budgets: Power for the Earth Machine

In a well-made universe the sun does far more than merely allow eyes to function.

Ted Hammond

energy ability to do work: to lift objects, raise temperatures, melt ice, evaporate water, form complex chemicals, and so on
(see Fig. 15-1)

Most of the energy available for use on the earth's surface comes from the sun. A waterfall may be yesterday's cloud, moisture lifted from the sea by the sun. A wood stove burns yesteryear's sunlight. The green plant of centuries ago is the coal of today. Even the wind blows not where it wills, but where the sun leads it, because wind gets its impetus from sun-caused differences in the heaviness of the air at various places.

Inequalities in the intensity and duration of sunshine have an important role in a variety of events: Nebraska droughts, Florida hurricanes, wildfires and mudslides on a California hillside, the formation of red soils in Alabama, the colorful spectacle of a Vermont autumn, the slow recovery from an Alaskan oil spill, the gouging of the Grand Canyon, the whispering wealth of an Iowa cornfield, and the flooding of suburban basements. Much of this book is an exploration of the many ways in which the sun influences these events, together with a look at the other entities assisting it and with some of the consequences for the occupants of the earth. The first phase of the exploration is an introduction to the study of energy, because it is extremely difficult to study other parts of the environment without understanding the distribution and conversion of energy from the sun (Fig. 1-1).

THE BASIC ENERGY BUDGET

latitude angular distance north or south of the equator
(see appendix to Prologue)

Sunlight strikes the earth in a geometrically regular pattern: on a given day, places at the same latitude always receive the same amount of energy (see Fig. 1-2 and the appendix to this chapter). The energy that comes to a place, however, can be spent in many different ways. Characteristics of the air and surface at a given place determine how the daily quota of energy will be divided among the various uses there. The division of sunlight begins

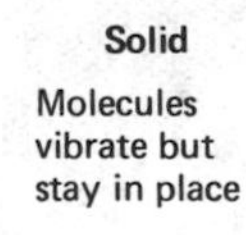

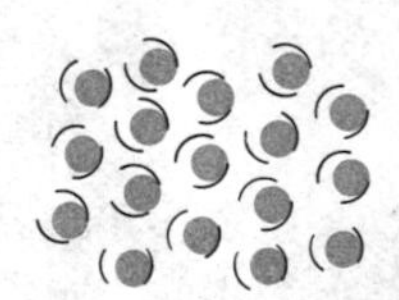

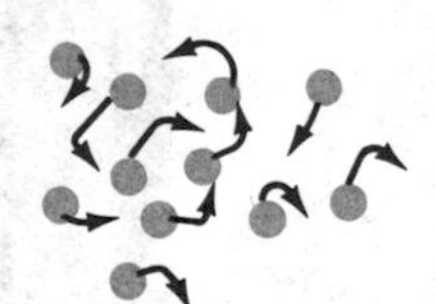

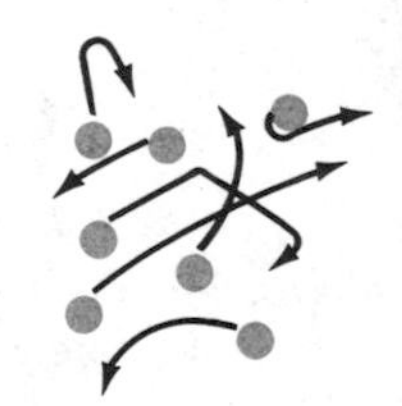

FIG. 1-1. An optional refresher course on the physics of heat energy. Temperature is a measure of molecular motion; the warmer a substance is, the more rapidly its molecules move. At low temperatures, molecular motion consists primarily of vibration within the confines of a rigid crystal structure. Each substance has its own rate of increase of molecular activity as energy is added, and each substance has its own critical temperature at which the motion of its molecules is so rapid that the material cannot remain in solid form. Molecules in a liquid are more mobile than are those in a solid, but they still are constrained by their neighbors. Additional heat causes molecules to move so rapidly that they escape the restraints of their neighbors entirely and fly off as a gas or vapor. Gases expand when they are heated because the molecules in a hot gas move around more rapidly than do those in a cool gas.

FIG. 1-2. Global radiation imbalances. Solar energy is reliably but unevenly distributed over the earth. The reason for the inequality lies not in the sun but in the shape of the earth (see Appendix to Chapter 1). The sun can shine directly down on only one part of the globe at a time. All other places on the sunlit half of the earth receive sunshine at an angle. Slanting rays take a long path through the atmosphere and spread themselves out over a large area, so that they deliver less energy to a unit of surface area than do vertical rays.

The pattern of incoming energy at any instant in time resembles a target centered in the place that receives sunlight from directly overhead. A place 5000 kilometers (3000 miles) in any direction from the favored spot gets only two-thirds as much energy from the sun. The spinning of the earth transforms the target pattern into a series of bands around the earth. All places at the same latitude are in the same energy income bracket; places at different latitudes receive different amounts of energy.

The exact amount of energy sent toward a particular place depends on two things: the latitude of the place and the seasonal position of the sun. These two geometric variables operate by altering the intensity of the noon sun and the length of the day. For the whole year the potential income of energy is greatest at the equator, decreases with distance from the equator, and reaches a minimum at the poles (see Fig. 1-9).

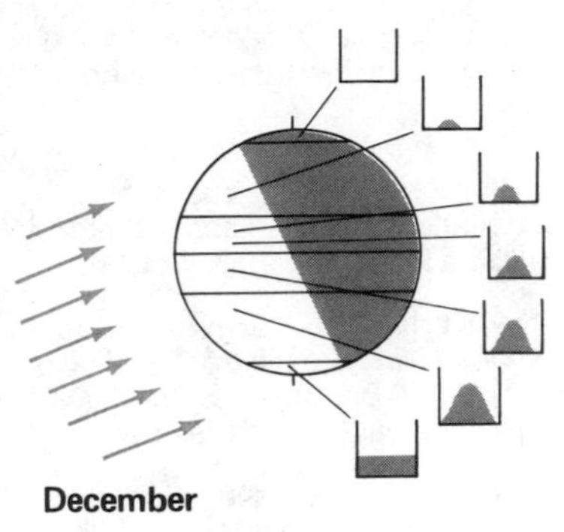

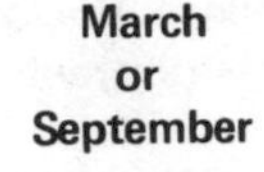

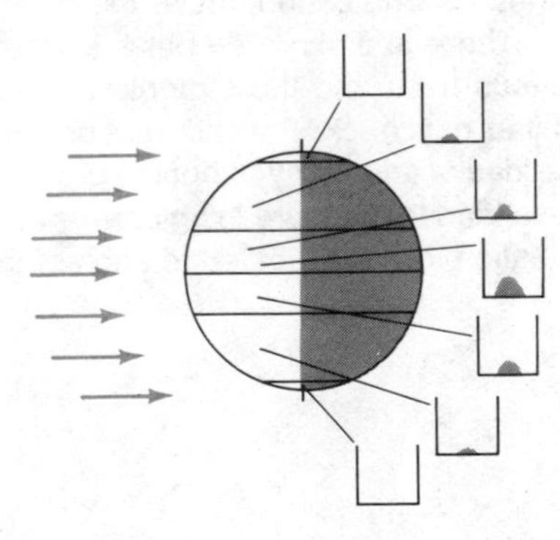

June

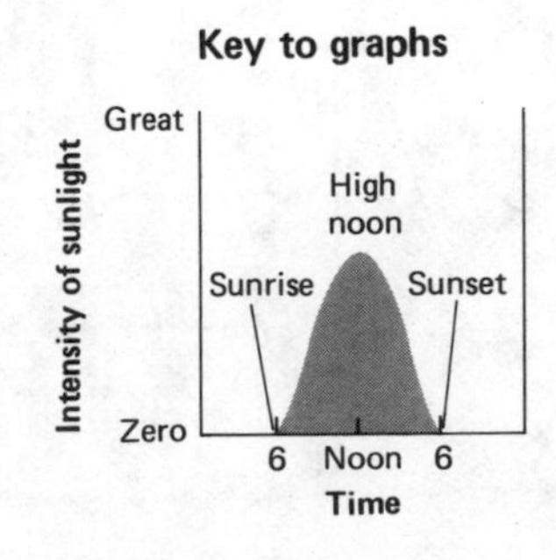

even before the rays reach the ground. Energy from the sun must first pass through a dirty window, the atmosphere. Atmospheric gases are essentially transparent to visible light, but suspended bits of liquid or solid material can absorb or reflect sunlight. A thick cloud may allow less than 10 percent of the sunlight hitting it to reach the earth's surface. Clouds generally behave like mirrors; they bounce sunlight off in a different direction rather than absorbing it. Reflected sunlight usually is permanently lost, unable to do any more work for the earth.

Other particles in the air are more possessive about sunlight. They absorb much of the energy and store it in the form of internal heat. Still other substances deflect the sunlight from a straight path, but it gets to the ground anyway. The blue color of the daytime sky is one consequence of indirect or scattered sunlight. Without the scattering effect, sky would be black and starry even when the sun was up. The amount of sunlight reflected, absorbed, or scattered depends on the angle of the sun's rays and the composition of the air on a particular day.

Some of the sun's energy passing through the atmosphere and striking the earth's surface is reflected; some is absorbed and converted to heat. Melting of ice or evaporation of water also takes some energy. Green plants make direct use of only a tiny fraction of the available sun energy, but their conversion of solar energy into plant material is obviously significant to all creatures who cannot make their own food.

There is a similarity between the process of allocating solar energy to various uses and the struggle to pay for food, rent, and clothing out of a limited paycheck. For this reason, environmental scientists frequently use the idea of an energy budget to organize the study of energy. Many scientists have tried to measure the global energy budget, the average amount of solar radiation that is allocated to each of the various uses on earth (Fig. 1-3). An

heat molecular motion; warm molecules move more rapidly than cool ones

diffuse insolation sun energy that is scattered by the atmosphere but still reaches the earth's surface

latent heat energy needed to change water from ice to liquid (about 80 calories per gram) or from liquid to vapor (about 600 calories per gram)

photosynthesis sun-assisted conversion of water and carbon dioxide into chemical energy by green plants (see Table 13-1)

energy budget allocation of solar energy to various uses: reflection, surface heating, evaporation, photosynthesis, and so on

FIG. 1-3. Global energy budget calculated by different scientists.

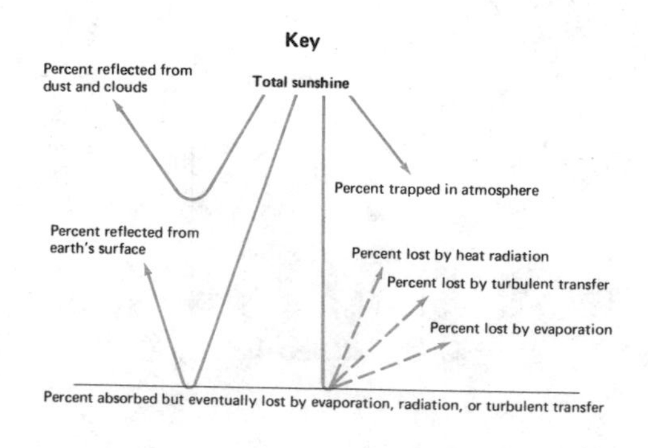

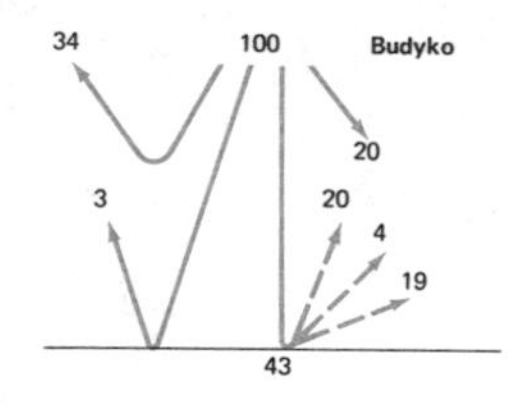

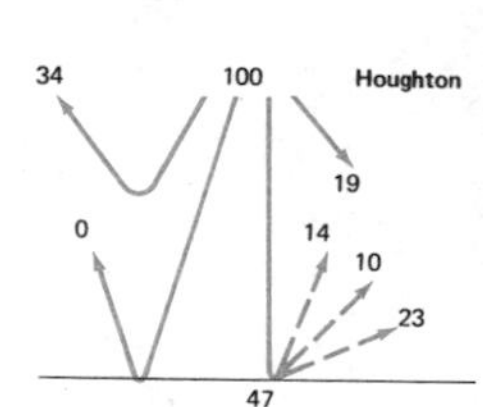

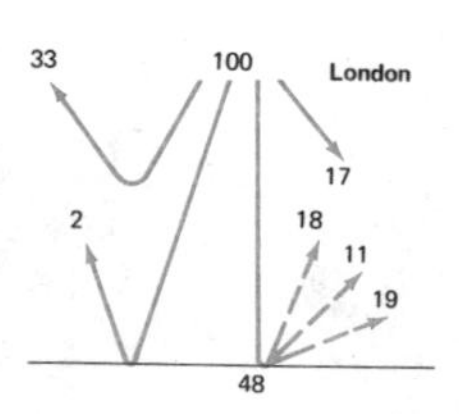

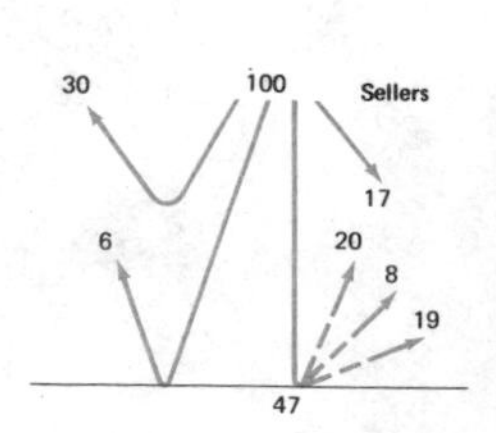

average energy budget, however, may be misleading precisely because it is an average. Different locations have different energy incomes and different ways of spending their share of sunlight. Even in the same place, energy income and expenditures usually change from season to season.

Despite some important differences among places, every location has the same basic obligation: each must spend its entire daily income of solar energy in some manner. Whatever is not blocked by the atmosphere or reflected from the surface must be used to evaporate water, melt ice, heat the environment, or make plant material. It is quite difficult to store energy without using it to do something. By its very nature, energy is subject to a variety of "tax laws" that compel energy-rich areas to transfer some of their wealth to poorer areas. This flow of energy from surplus areas to deficit areas provides the force that starts and maintains most of the environmental systems described throughout this book.

COMPARATIVE ENERGY BUDGETS

A map of world temperatures illustrates the consequences of dissimilar energy budgets. Record high temperatures invariably occur in desert regions. A desert and a seacoast at the same latitude have exactly the same potential income of solar energy. However, the energy budgets of the two places are very different, and the temperatures of the places reflect the dissimilarity. The desert may spend as much as one-fourth of its energy income in reflection from the bare rock and soil. Except at sunrise and sunset, when the sun is very close to the horizon, the ocean absorbs nearly all of the sunlight that strikes the surface. After the loss by reflection is deducted, the desert has 75 percent of its energy paycheck left over, whereas the ocean has 95 percent. In the ocean, however, evaporation may consume as much as 75 percent of the available energy. The dry desert uses very little of its energy to evaporate water or to produce green plant material. When losses by evaporation and reflection are subtracted from incoming solar energy, the energy available for absorption by the desert is nearly four times as great as that of the ocean (Fig. 1-4).

albedo percent of incoming light that a surface reflects:

surface	albedo
mirror	90–100
snow	40– 80
sand	20– 40
rock	10– 20
plants	5– 15
water	2– 10

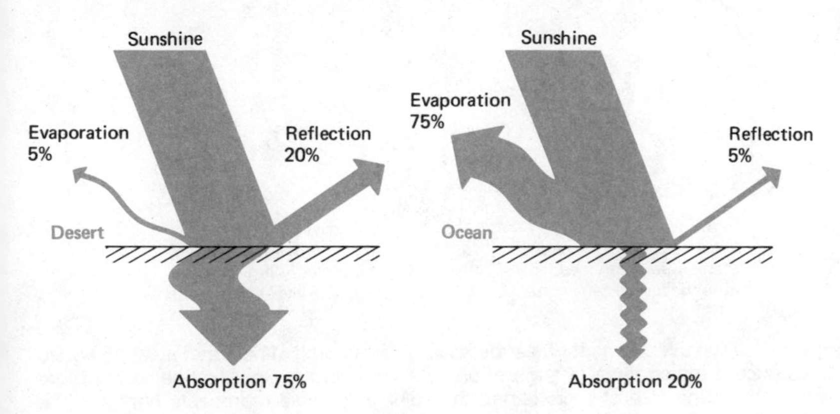

FIG. 1-4. Surface energy budgets. Different surfaces allocate the incoming energy in different ways.

calorie (cal) energy needed to raise the temperature of one gram of water one Celsius degree

specific heat energy needed to raise the temperature of one gram of a substance one Celsius degree

radiation outward flow of energy from any object that has internal heat

thermal equilibrium state of constant temperature; heat inflow equals heat outflow

This ratio of four to one does not mean that the desert surface is four times as hot as the ocean surface. The relationship between energy and temperature is not that simple. Different materials react in different ways to a surplus of energy. By definition, one calorie of energy will raise the temperature of one gram of water one Celsius degree. Soil has a different heat capacity: it takes only one-fifth as much energy to produce a similar temperature increase in the same quantity of dry soil (Fig. 1-5). Moreover, sunlight spreads through a thick layer of water but concentrates its heating effect on a thin surface layer of soil. Compared with dry soil, therefore, water can take in more energy without getting as hot.

As the desert and ocean become warmer, they are confronted with another inescapable fact: permanent energy storage is nearly impossible. Any object that contains heat always loses energy by radiation. Outgoing radiation from material things is like a tax on wealth: anything possessing heat energy is forced to pay out an amount that is determined mainly by the temperature of the object (Fig. 1-6). Nearly everything in the universe is radiating energy at all times. A comparatively cool surface receives more energy from its neighbors than it radiates away and will get warmer. The warmer the surface becomes, however, the more rapidly it must radiate energy away. Eventually, it should get hot enough so that outgoing radiation exactly balances incoming energy. The desert spends most of its income to heat the surface, and therefore it must reach a higher temperature than the ocean in order to radiate away as much energy as it absorbs. The temperature of either surface is ultimately determined by the balance between incoming and outgoing energy.

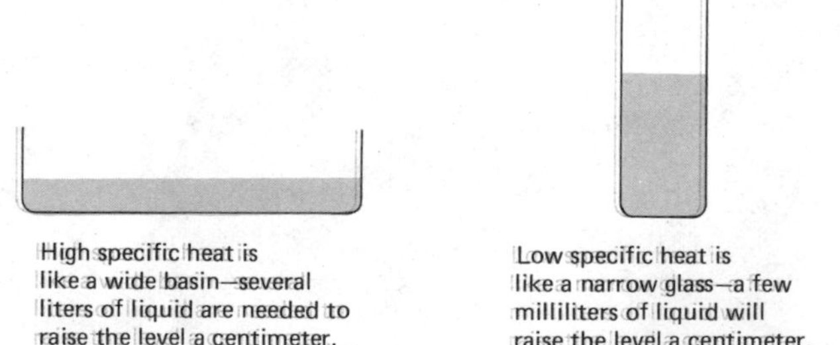

FIG. 1-5. The specific heat of a substance is the amount of heat that must be added to produce a temperature change of one Celsius degree. Specific heat is therefore analogous to the size of a container; the volume of liquid represents heat and the depth of liquid is equivalent to temperature.

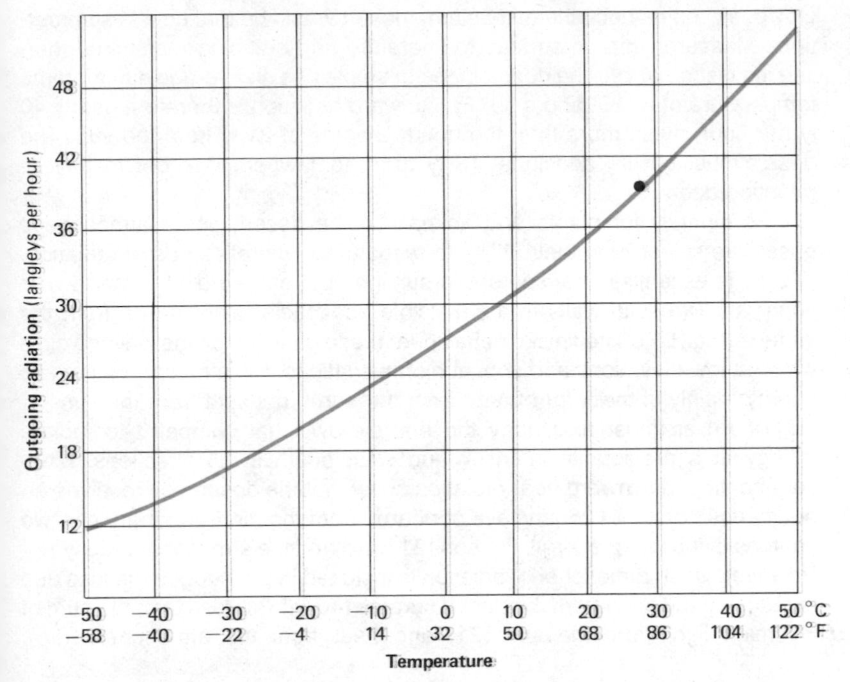

FIG. 1-6. Energy losses from a surface. A hot object is characterized by rapid molecular motion and must give up more energy by radiation than a cool object. The dot shows the temperature (27°C) and outgoing radiation (40 ly/hr) of the desert mentioned in the text.

Stefan-Boltzmann Law objects radiate energy at an intensity proportional to the fourth power of their Kelvin temperature (°K = °C + 273)

DETAILED ENERGY BUDGET

A more detailed version of the energy budget of the desert should clarify the principle of energy balance. On a normal summer day the sun sends about 960 langleys of energy toward a typical desert in the middle latitudes. (One langley is equal to one calorie of energy coming into or going out of one square centimeter of surface area, and it is certainly easier to write ly rather than cal/cm^2.) About one-fourth of the solar income of the desert is reflected or absorbed into the atmosphere. Thus the surface has an energy income of about 720 ly. Reflection from the ground takes 25 percent of the surface income, or 180 ly. Evaporation takes another 5 percent, or 36 ly. The remaining energy after these losses are deducted is a bit more than 500 ly per day. The average income over a 24-hour period therefore is about 21 ly per hour.

langley (ly) energy flow of one calorie per square centimeter

At this point, the idea of a radiation balance runs into trouble. The temperature of the surface would have to be −15°C (+5°F) in order for outgoing radiation to balance the average income of 21 ly per hour (Fig. 1-6).

Obviously, no respectable mid-latitude desert would be that cold in summertime. Measurements of surface temperature and outgoing radiation underline the discrepancy. The desert in the example has an average summertime temperature of 27° C (about 80° F). Outward radiation is therefore nearly 40 ly per hour, much more than the hourly income of 21 ly from the sun. The desert must get an additional 19 ly from somewhere in order to pay its radiation debt.

greenhouse effect atmosphere allows shortwave (visible) radiation from the sun to pass through unhindered, but it prevents the escape of longwave (heat) radiation from the earth

Wien's Law objects radiate energy at a wavelength equal to 2940 divided by their Kelvin temperature

counter radiation downward radiation from the atmosphere

Fortunately for plants and animals in the desert, some atmospheric gases are selective in their ability to respond to different kinds of radiation. The air is essentially transparent to sunlight, but carbon dioxide and water vapor are like solid walls to the invisible heat radiation given off from the surface (Fig. 1-7). The finicky behavior of these gases explains why the air is warm at low elevations and cool at high elevations: the atmosphere gets its energy mainly in the form of heat from the earth, not light from the sun.

Warm air must also obey the energy laws that compel it to radiate energy at a predictable intensity and wavelength in all directions. Measurements of downward heat radiation show that the desert soil receives an hourly dose of about 25 langleys of energy from the air. Thus there are two sources of incoming energy: the sun (21 ly/hr) and the atmosphere (25 ly/hr). However, when atmospheric radiation is included in the budget, the loss due to outgoing radiation from the desert surface (40 ly/hr) is less than the sum of incoming light from the sun (21) and heat from the atmosphere (25).

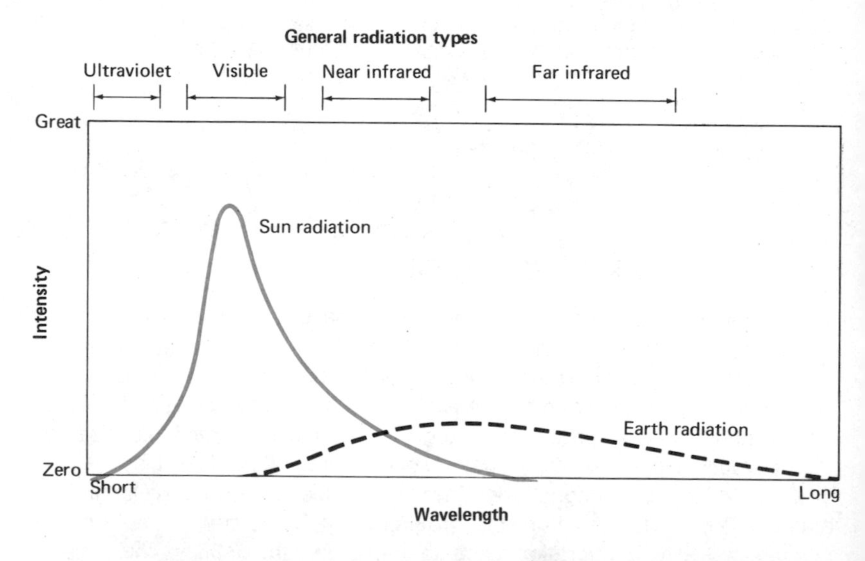

FIG. 1-7. Characteristics of radiation at different temperatures. Both the intensity of radiation and the wavelength of radiation depend on the temperature of the radiating body. Hot bodies radiate at high intensity and short wavelength. This diagram portrays the relationships between intensity and wavelength for the radiation given off by the sun, with a surface temperature of nearly 6000° C, and the earth, with an average surface temperature of about 15° C.

Moreover, the balance between incoming and outgoing energy must be accomplished by some other means, because all methods of energy exchange by radiation have been measured and accounted for.

Some heat possibly moves down into the soil. In most places the flow of energy between the surface and the deep underground is upward and very small, on the order of 1 ly per day or less. Internal exchanges of heat are quite significant in the oceans and in a few other special environments, such as hot springs or volcanos. Because the desert in this example is neither watery nor volcanic, the movement of energy beneath the surface has little effect on the energy balance of the surface.

geothermal energy upward movement of energy from the hot interior of the earth

Shimmering air and miniature whirlwinds above a hot surface suggest another possible means of heat transfer. Desert soil is often far warmer than air at shoulder height. A warm earth surface heats the bottom layers of the air. The warmed air is light, and its buoyancy causes it to rise and be replaced by cooler air, which is heated in turn. It is difficult to measure the amount of heat actually carried away from the earth's surface by air movement. Strong winds from other regions can take heat from a surface or add energy to it. The direction and rate of energy transfer depends on the temperature of the air and the speed of the wind. Even a weak desert wind can move as much as 12 ly per hour during the afternoon and evening. The ground and the air near it are cool at night, and cool air tends to be calm, so that the turbulent transfer of heat to the atmosphere at night is practically zero. The transfer of heat to the desert air thus averages about 6 ly per hour.

convection rising of warm and sinking of cool material
(see Fig. 3-1)

turbulent transfer removal of heat from a surface by convection

conduction movement of energy from molecule to molecule

The desert is now in energy balance: the sum of all daily energy flows by shortwave and longwave radiation, reflection, evaporation, conduction, and turbulent transfer to the atmosphere is zero. Income and expenditures are equal, and therefore the temperature does not change (Table 1-1).

thermal equilibrium temperature at which incoming and outgoing energy flows are equal

TABLE 1-1 ENERGY BALANCE OF THE SAMPLE DESERT[a]

Income		Outgo	
By shortwave radiation from the sun	720	By reflection from ground	180
		By evaporation of water	36
By longwave radiation from the atmosphere	600	By radiation from ground	960
		By turbulent transfer	144
Total in	1320	Total out	1320

[a]The figures represent the total energy flows in langleys for a 24-hour period.

Suppose that the surface temperature were higher than 27° C, while the incoming energy remained the same. Under these conditions, turbulent transfer and outgoing radiation would be speeded up. The temperature must then decrease because heat losses would be greater than income. If surface temperature were lower than 27° C, outgoing radiation and turbulent transfer would take less heat away than at 27° C, but the income from the sun would still be 21 ly per hour, and the temperature would rise. A decrease in the income from the sun (for example, on a cloudy day or in late afternoon) would be followed by a decline in surface temperature because outgoing radiation and turbulent transfer would remove heat faster than the sun replaces it (Fig. 1-8).

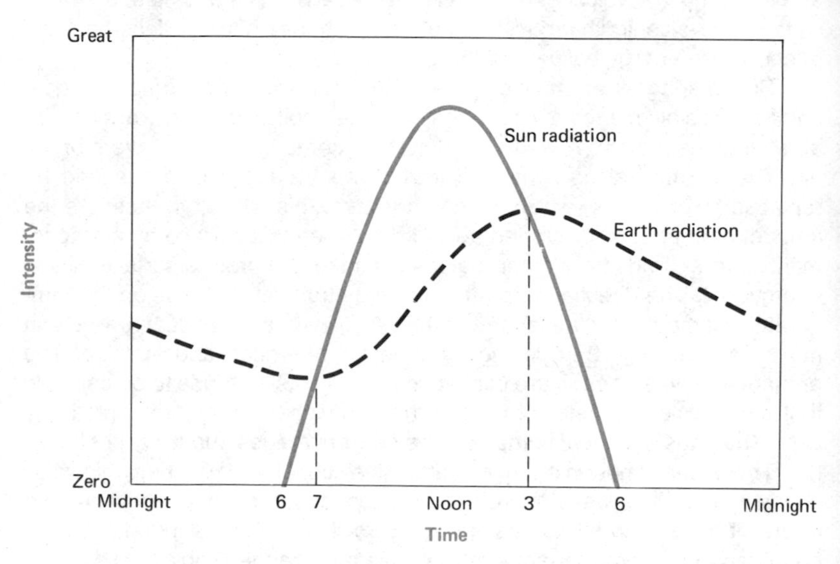

FIG. 1-8. Incoming and outgoing radiation during a 24-hour day in a desert. Energy comes in during only half of the day. The flow of energy from the earth is proportional to the temperature of the surface, so that the dotted line could represent either radiation or temperature. From about 7 a.m. until about 3 p.m., incoming solar radiation is greater than losses of energy from the earth, and therefore the temperature of the surface rises. The other 16 hours of the day are marked by net outward flow of energy and decreasing surface temperatures.

In summary, two major ideas emerge from this discussion of a detailed energy budget:

feedback loop mechanism for sending information about an effect back to change the causes; a thermostat is a negative feedback device that turns a furnace down if room temperature increases; a microphone in front of its own loudspeaker is a positive feedback device that makes louder and louder noises until something breaks

1. Outgoing radiation, evaporation, and turbulent transfer are consequences of temperature; they serve to offset changes in energy budgets and to bring energy exchanges back into balance. Almost all parts of the environment have similar procedures to compensate for changes. An appreciation of the role of this kind of natural regulator is an essential prerequisite for understanding the way the environment works.

2. The heat conveyed to the atmosphere by radiation, evaporation, and turbulent transfer is the power that drives most atmospheric processes. The atmosphere above the desert has a positive radiation balance. It receives about 27 ly of energy from the sun and the earth, but its outgoing radiation is only 25 ly per hour. The remaining 2 ly lift the air and push it into some other region where the local radiation balance is negative (Fig. 1-9). Thus the energy budget of a place provides a means for the conversion of solar energy into wind and other weather processes.

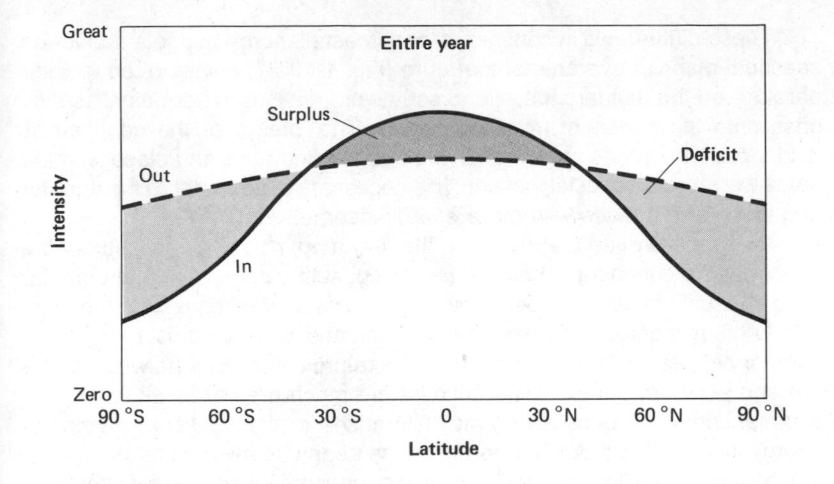

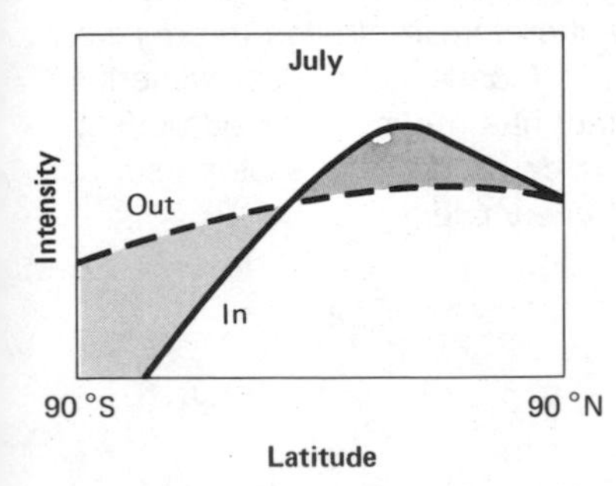

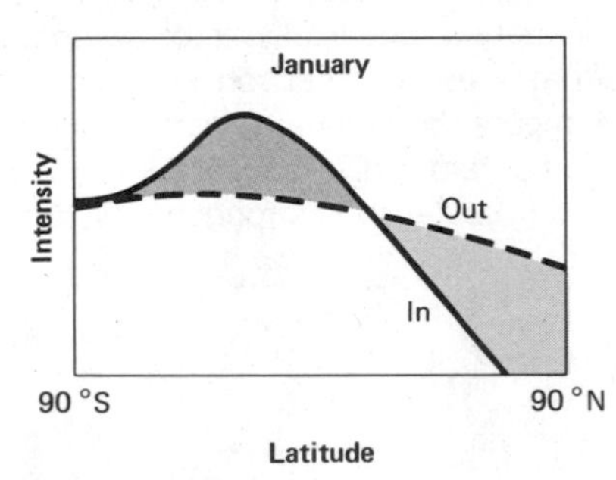

FIG. 1-9. Radiation balance at various latitudes. Air and ocean currents are the means that the earth uses to transfer heat from areas of surplus to places with deficit.

ENERGY BUDGET AND TEMPERATURE MAPS

The refined energy budget is a useful tool because it can explain most of the observed patterns of temperature. Northern Canada usually is the coldest part of North America because it receives little energy from the sun. Places closer to the equator have a greater energy income and are understandably warmer than Canada.

South Dakota and coastal Oregon have the same potential income of energy, but their temperatures are quite different. The center of a continent is usually dry, and dryness has three consequences for energy budgets. The dry Dakota soil has a low specific heat and therefore responds rapidly to an energy imbalance. The dry Dakota air has few clouds to reflect sunlight and little water vapor to trap heat radiation from the earth and to send it back to the surface. The dry Dakota land has few trees to evaporate water in summer or to stand above the snow and absorb sunlight in winter. By contrast, the Oregon coast swaps energy with its humid atmosphere and the vast storehouse of heat in the ocean. As a consequence, the temperature does not fluctuate nearly as much in Oregon as it does in South Dakota.

specific heat energy needed to raise the temperature of one gram of a substance one Celsius degree (see Fig. 1-5)

transpiration evaporation of water from leaf openings of plants

maritime effect moderation of temperature extremes by nearby bodies of water

isoline line that has some arbitrary value on a map and separates areas with greater values from those with less than the line value

isoline	characteristic
isotherm	temperature
isobar	pressure
isohyet	rainfall
isotach	wind speed
isohypse	elevation
isochrone	time
isopach	thickness

temperature gradient change in temperature across an arbitrary distance

These differences in continental and coastal energy budgets appear on seasonal maps of average temperature (Fig. 1-10). The lines of equal temperature on the January map bend southward, toward the equator, as they pass onto the continent from the ocean. The interior of the continent is colder than the edges at any given latitude. To find an inland place with the same warmth as a coastal location, it is necessary to go closer to the equator and to look for a place with more energy income.

An opposite pattern appears on the July map. At any given latitude, the interior of the continent is hotter than the coast in summer, so that the map lines of equal temperature bend toward the pole as they pass onto the land.

Another significant difference between the two maps is the greater number of lines on the winter map. Temperature differences between northern and southern parts of the continent are much greater in winter than in summer, primarily because of the differences in the seasonal pattern of energy income. Northern Canada receives nearly as much solar energy as the equator does in July, but the energy income of northern Canada in January is practically zero. Temperature therefore fluctuates more widely from season to season in Canada than in Florida. The great wintertime temperature difference across the continent has a number of side effects. Temperature differences are the driving force of the wind system, so that wind circulation is stronger in winter than in summer.

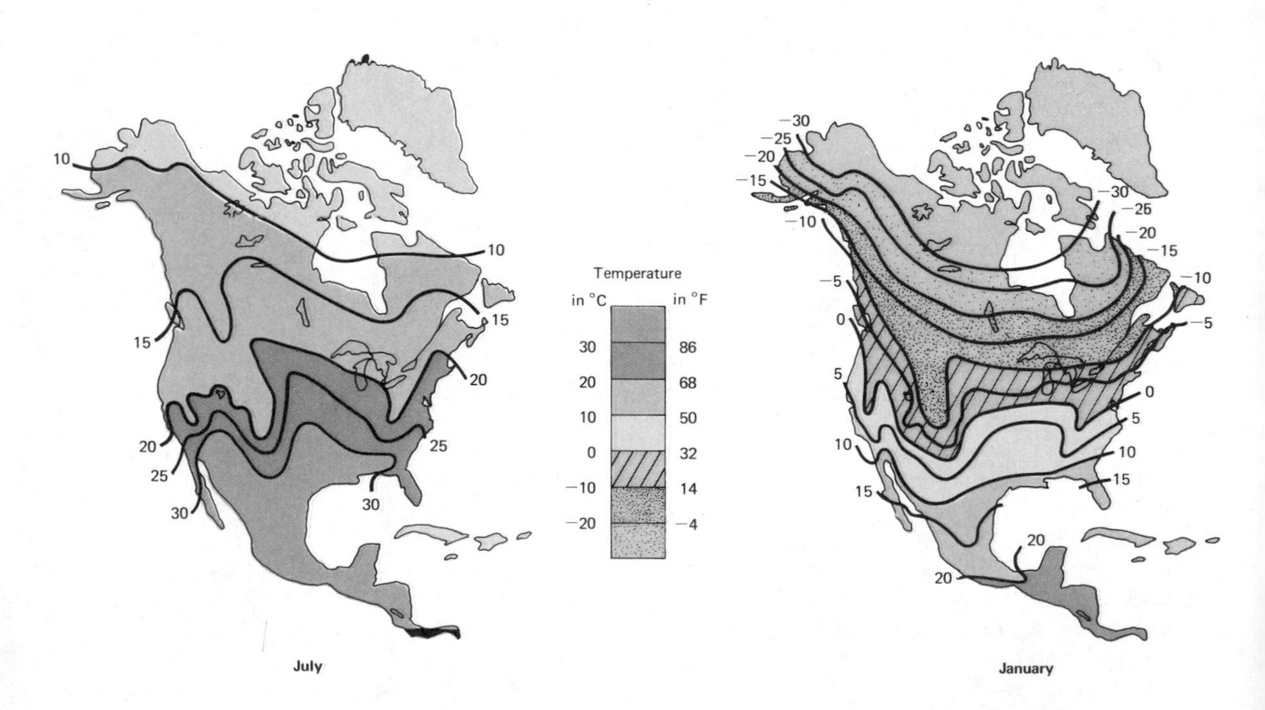

FIG. 1-10. Isoline maps of average temperatures in North America.

Temperatures vary less from season to season in the eastern forest than in the desert areas of the west. This difference is not as striking as that between coast and inland, but the cause is basically the same. Forests occupy rainy places, and their temperatures are moderated by clouds, humid air, waterlogged soil, and evaporation of water from the vegetation.

Mountain ranges appear as areas of low temperature on both maps. A mountaintop starts with a slightly greater energy income than a lowland, because sunlight does not have to pass through very much air on its way to the surface. However, outgoing radiation from a mountaintop also encounters little resistance on its way to outer space. Air normally absorbs heat energy from the earth and radiates it back down toward the surface. A reduction in the thickness of the air allows more of the earth's radiation to escape into outer space. The resulting loss of energy keeps mountaintops cooler than lowlands (Fig. 1-11).

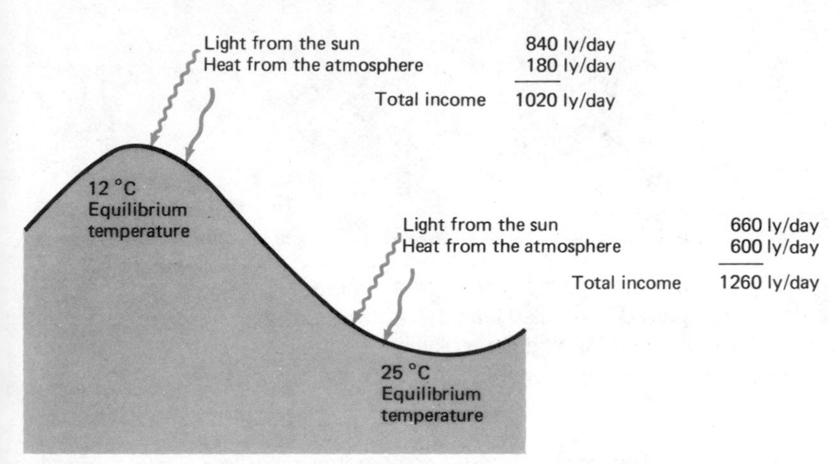

FIG. 1-11. Energy balance of a mountaintop and a valley. The valley has a higher total income and therefore needs a higher temperature in order to lose the extra energy by radiation, evaporation, and turbulent transfer.

SUBURBAN ENERGY BUDGETS

The principles of energy budgeting can also help interpret or predict the energy consequences of changes in the use of land in a particular place. As an illustration of applied energy budgeting, consider the following catalog of contemporary suburban construction malpractices:

Vegetation removal prior to building. A single large tree in Atlanta may transpire 200–300 liters of water (50–80 gallons) each day. The daily water loss by transpiration also removes about 120 to 180 million calories of energy from the local area. Rainwater in a treeless urban area goes mainly into sewers, where it is lost from the local environment. The energy formerly used for evaporation of water still comes into the building site and insists on being used for some purpose. In effect, the tree was a natural air-

transpiration evaporation of water from leaf openings of plants

FIG. 1-12. Typical United States residential subdivision under construction. Many builders prefer a deforested site because they see only the extra cost of working around trees. (Courtesy of C. J. DeLand of Soil Conservation Service.)

conditioner with the cooling power of a medium-sized window unit. The eulogy for departed trees also includes an account of their value as shade, noise insulation, wind protection, pollution absorption, and beauty. Unfortunately, many builders and buyers are able to see only the aesthetic benefit and consequently put an unreasonably low value on trees (Fig. 1-12).

albedo reflectivity of a surface (see p. 11)

Dark roofs. The energy budget of a roof is easy to calculate. Nearly every ray of incoming sunlight that is not reflected must go to heat the surface. Dark green and blue roofs absorb significant amounts of heat during the summer and put an unnecessary load on air-conditioning systems. Winter may bring a reflective blanket of snow that cancels the energy-absorbing advantage of a dark roof. The heating season in a snow-free place is shorter than the air-conditioning season anyway, so that the choice of a roof color as a heating aid makes little sense in most areas.

specific surface amount of exterior surface area per unit of volume (see Fig. 8-3)

"Ranch-style" architecture. A one-story house has more exterior surface than a two-story house with the same amount of interior space (Fig. 1-13). If the inside of the building is warmer or cooler than the outside, heat will pass through every square centimeter of exterior surface. The large surface area of a one-story house makes it expensive to insulate and difficult to keep heated or cooled. Ironically, when it comes to adding heat to an environment, air conditioners are just as bad as furnaces, because both burn fuels somewhere in the house or at a generating station. Furthermore, neither use of energy is very efficient; several units of fuel must be burned to obtain one unit of heating or cooling.

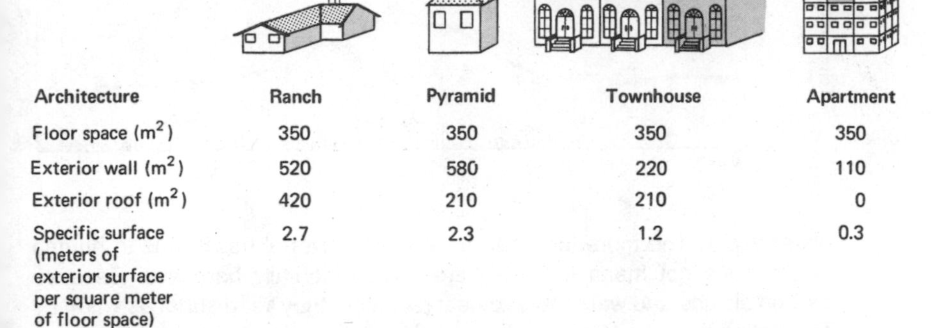

Architecture	Ranch	Pyramid	Townhouse	Apartment
Floor space (m^2)	350	350	350	350
Exterior wall (m^2)	520	580	220	110
Exterior roof (m^2)	420	210	210	0
Specific surface (meters of exterior surface per square meter of floor space)	2.7	2.3	1.2	0.3

FIG. 1-3. Comparisons of surface area and floor space for typical dwellings.

Randomly oriented floor plans and roof lines. Each day the sun makes a predictable journey across the sky. Likewise, many activities in different parts of the house occur at regular times of the day. Properly arranged rooms would allow the sun to provide light or heat for some domestic activities. For example, rooms facing east are brighter in the morning and cooler in the evening than rooms facing west. In most climates, this combination of features is better suited for bedrooms than for living rooms. The sun gives heat and light on the west side of the house in the evening, when most living or family rooms are in use. It is also possible to design roof overhangs and windows to aid the occupants by blocking sunlight or by letting it in, as desired.

Dark streets and driveways. Asphalt absorbs more sunlight than does light-colored concrete. Both surfaces shed water quickly and therefore lose the opportunity to cool themselves by evaporation. The summertime result is high temperatures on both surfaces and excess heat in the local environment. In winter, dark surfaces warm quickly during the short sunlit days and cool rapidly at night. Expansion and contraction during this daily temperature cycle often cause cracking and deterioration of the pavement.

Wide spacing of houses. Great distances to schools, shops, and places of work necessitate some form of private mechanized transportation for nearly every trip from a typical suburban home. The resulting use of fuel has at least three general effects on the local energy budget. First, automobiles add heat directly to the environment. The internal combustion engine is notoriously inefficient; it adds 5 to 10 calories of heat to its surroundings for every one it manages to convert into forward motion, and even that 1 calorie eventually becomes heat because of friction with the pavement or the brake pads. Second, auto exhaust alters the transparency of the air above urban areas and thus reduces incoming solar energy. Third, even perfect combustion of fuel in a "pollution-free" car has two by-products—carbon dioxide and

photochemical smog blue-gray haze formed when sunlight reacts with unburned hydrocarbon fuels and nitrogen oxides from internal combustion engines

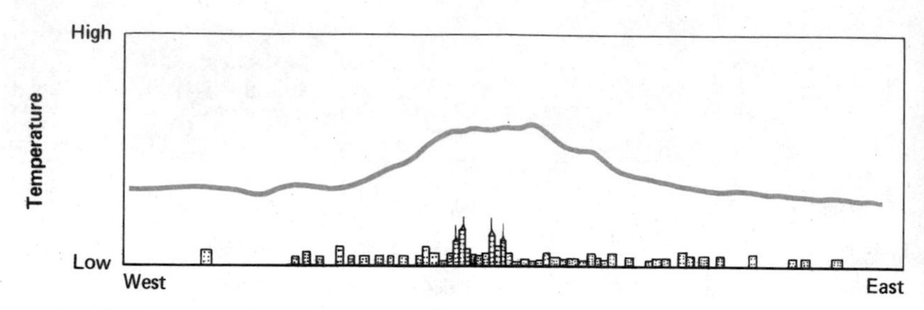

FIG. 1-14. Temperature profile across a typical urban area.

water vapor. The mere fact that these gases are not hazardous to human health does not mean that they are environmentally harmless. Because carbon dioxide and water absorb heat radiation, they keep surfaces warm by preventing the loss of energy by outgoing radiation.

The mere existence of a city therefore has a profound effect on the local energy budget. Income of energy from the sun is reduced slightly, but the city produces significant quantities of energy internally and interferes with many of the mechanisms for energy loss. Thus a typical city is measurably warmer than surrounding countryside (Fig. 1-14). The relatively high temperature of an urban area inevitably has side effects: decreases in air stability, alterations in rainfall patterns, changes in wind and humidity, concentrations of pollution, and so on. Scientists are just beginning to discover the more subtle urban effects on climate.

heat island zone of high temperature around an urban area; altered wind pattern over a city gives rise to a related feature called a dust dome

BIOLOGICAL ENERGY BUDGETS

The share of solar energy that is used to make plant tissue seems pitifully small (Fig. 1-15). Few green plants store more than 1 percent of the sunlight striking them. Nevertheless, photosynthetic use of energy is

FIG. 1-15. Energy budget of a typical leaf (not every leaf).

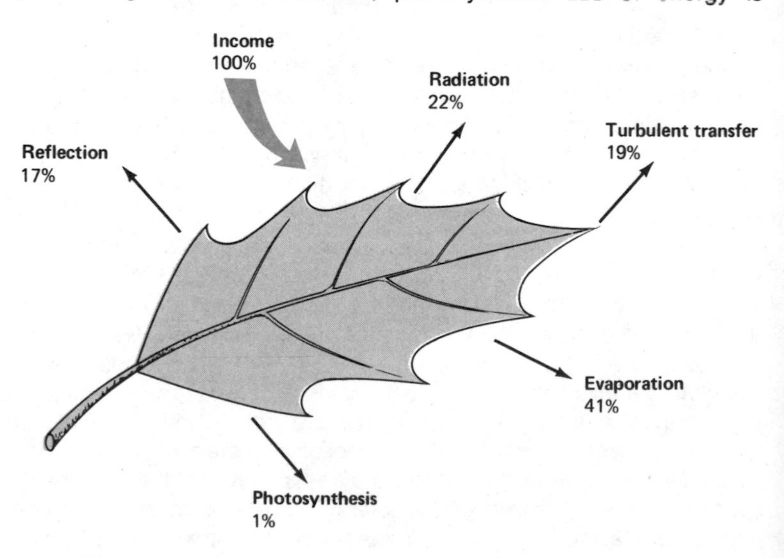

significant for a variety of reasons. Without green plants, animals would have nothing to eat. In a more abstract sense, photosynthesis is a temporary eddy in the continual flow of energy from hot to cold. Plants accumulate energy in chemical form in their tissues, thus making it available to provide power for other biological processes. Moreover, plants stand at a key environmental intersection, where they direct the flows of energy, water, and mineral substances. Their strategic location multiplies the importance of plants far beyond the 1 percent of available energy they represent.

entropy state of energy dispersal; all physical processes tend to increase entropy by taking heat from warm places and giving it to cold places

ecosystem inanimate environment and its living occupants, held together by exchanges of energy, water, and minerals
(see Fig. 12-4)

There is little reason to expect any great increase in the efficiency of photosynthetic use of energy. Plants are engaged in a competitive struggle from the beginning of their lives. Any plant species that developed a significantly better way to use sunlight would be able to grow taller or stronger than its neighbors and eventually deprive them of life-giving sunlight. Through the ages, plants have had more than enough time to experiment with genetic combinations in an effort to boost their efficiency. Their apparent failure to improve on the low average seems to suggest that green-plant chemistry may, indeed, have an efficiency limit of a few percent.

primary production amount of plant material produced

From an energy viewpoint, the headline-grabbing "miracle" crop breeds are misnamed. The spectacularly high efficiency of some experimental crops is based on incomplete energy accounting. The new varieties usually require heavy energy subsidies in the form of fertilizers, soil treatments, and pest control. The advantage of the new varieties over their wild ancestors lies not in their efficiency but rather in the fact that a large fraction of their mass is harvestable and usable.

primary efficiency percent of solar energy converted to plant material

Animals seem to be more efficient than plants in using energy, but only because plants have already built many of the complex chemicals that animals need to sustain life. (Fig. 1-16). Animals obtain energy in a highly refined chemical form, but they lose energy by radiation from their bodies, evaporation of water, and friction with their surroundings as they move about. As a result, only about 10 percent of the energy stored in plant tissue actually becomes animal tissue. Efficiency declines in harsh environments; a

herbivore animal that eats plants

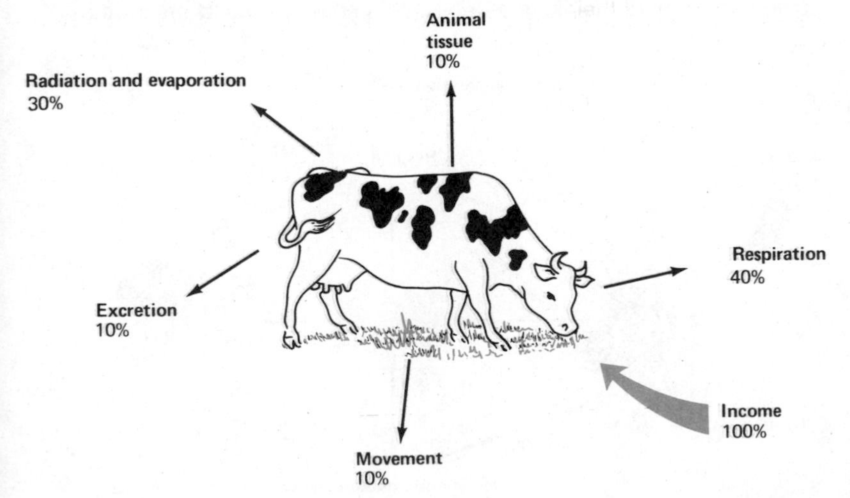

FIG. 1-16. Energy budget of a typical animal (not every animal).

desert animal uses much energy for cooling and for transportation over large barren areas in search of food. On the positive side, plant-to-animal energy efficiencies can go beyond 40 percent in some controlled situations. Meat producers confine and practically hand-feed their cattle and chickens in order to keep them from wasting energy by doing anything except gaining weight. The growth of these animals is aided by expenditures of energy to harvest, transport, dry, store, grind, and enrich their food. Indeed, the fences that keep the animals from straying and the roofs that protect them from the weather also represent expenditures of energy. Adding these energy subsidies to the food input yields a true energy efficiency of 10 percent or less.

carnivore meat-eating animal

The food of a meat-eating animal is chemically very similar to its own body. Prey animals, however, are often mobile and must be caught before they can be eaten. Getting high quality food but with great effort means that meat eaters ultimately have about the same net efficiency as plant eaters: about 10 percent or less.

trophic levels steps along a food chain

organism	trophic level
plant	primary
herbivore	secondary
carnivore	third, fourth, and so on
top carnivore	highest

Additional links in the food chain have essentially the same efficiency (Fig. 1-17). A huge initial input of energy shrinks quickly to insignificance when 90 percent of the available energy is lost at each step along a sequence. Most natural food chains have only three or four links. Long chains are possible only under very favorable conditions, such as a protected ocean area occupied by a large fish that can afford to wait for days between big meals.

An examination of biological energy budgets shows that more food energy is available to plant eaters than to animal eaters, because plant eaters are closer to the original source of energy—the sun. In theory, a vegetarian society should be able to support a larger population in the same area than could a meat-eating culture. In practice, the entire energy budget of the food system must be considered. Gathering and processing plant foods may be less efficient than allowing animals to gather the food and then eating the animals. Carnivorous animals as food or pets, however, are clearly environmental luxuries that may not be tolerable if humans insist on putting as many of their kind as possible on the surface of the earth.

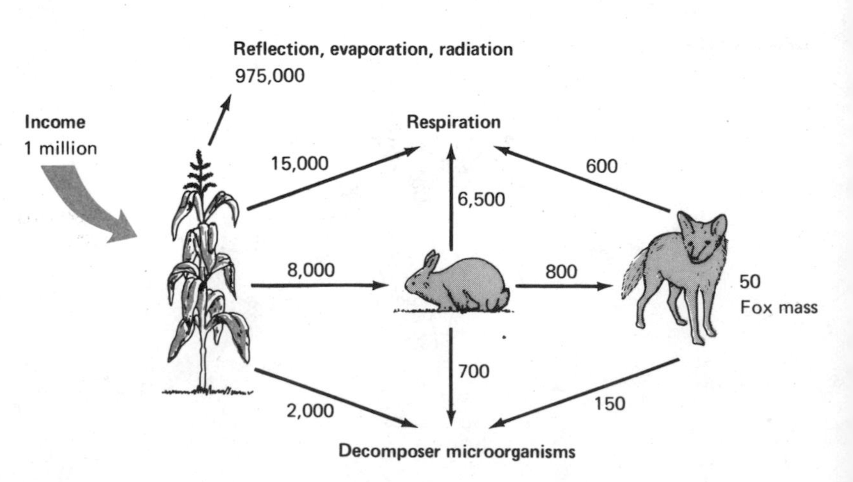

FIG. 1-17. Energy budget along a typical food chain (not every food chain).

ENERGY SYSTEMS

All parts of the environmental system are energy spreaders. Solar radiation is simply a means of taking energy from the sun and giving it to cooler places. Most activity on the earth's surface depends on a continual input of energy from the sun. When the sun ceases to add heat, the earth will cool rather quickly by radiating away the energy it has stored. Other earth processes, deprived of their energy source, will also run down.

maximum entropy ultimate state of maximum energy scattering; no one place has more energy than another

A one-paragraph excursion into the realm of astronomy is probably sufficient for an introduction to physical geography. It is true that a comprehensive analysis of the earth should include a study of the sun and other outside forces affecting the earth. By logical extension, only the universe as a whole is big enough to be unaffected by outside forces. The universe is therefore the only system that could be truly predictable if understood, but it is so big and complex that it simply cannot be studied in its entirety. Thus a student of the earth is advised to draw the boundaries of investigation at the outer limits of the atmosphere. The system enclosed by these mental boundaries is admittedly an incomplete one. Worse yet, it can never be completely predictable, because changes in the external forces can alter the earth's system without warning. Nevertheless, some useful ideas are gained from realizing that the physical geographer sees only a portion of the total picture.

closed system collection of objects and relationships that operates without being affected by outside forces or events; the universe is the only truly closed system, but other systems may have different degrees of openness

open system system that is affected by outside forces; to someone studying an open system, these external features are treated as sources or destinations for inputs and outputs of energy, materials, or information (see Chapter 12)

The most important of these ideas is interrelatedness. Every system described in this book is made of smaller systems and is also a part of a hierarchy of bigger ones. The energy system of the earth is driven by outside forces and governed by a complex web of internal and external connections. Within this large system are smaller systems that interact in different ways: the wind system, the carbon cycle, soil-forming processes, biological food chains, and so on.

The complexity of the energy system may mean that the answer to one question raises five more questions, but environmental scientists are rarely frustrated by the incompleteness of their knowledge. Acting on the basis of a preliminary answer while investigating the next question is preferable to acting on the basis of complete ignorance.

The openness, interrelatedness, and energy-dependence of environmental systems are extremely useful preliminary conclusions. *Homo myopicus,* the shortsighted human, often prefers to treat environmental systems as separate and unconnected. If people need houses, they cut down trees; if the houses are then too warm, they invent air-conditioners. If the resulting increase of urban temperature provokes violent storms, people talk about "defusing" them by cloud seeding. If the inevitable change in wind flow affects crop production, a clamor arises for another kind of environmental management. Each action postpones and possibly exacerbates an eventual reckoning with the energy budget. On the other hand, simply knowing that trees are linked to the temperature system could induce humans to look for ways to build houses that are more efficient in their use of energy.

APPENDIX TO CHAPTER 1
Earth-Sun Relationships

If the sun were a volleyball on the pitcher's mound of a typical baseball field, the earth would be a speck the size of a pinhead on home plate. A year is a home run, and it takes the speck 365 days, 6 hours, 9 minutes, and 10 seconds to touch all of the bases.

The speck does not stand vertically; it resembles a baseball runner who goes around the bases but is always leaning slightly toward right field on the way around. The earth tilts about 23.45° from "upright" as it circles the sun, and the seasons are consequences of the tilt of the earth (Fig. 1A-1). The United States has summer when the northern part of the earth is leaning toward the sun. Half a year later, the earth is half way around its orbit, the northern part leans away from the sun, and the United States experiences winter. Most globes have an analemma or ellipse marked on the globe to show the latitude of direct overhead sunshine at different times of the year (Fig. 1A-2). The graph can be used to determine the angle of the noon sun for any place on earth on any day of the year (Fig. 1A-3).

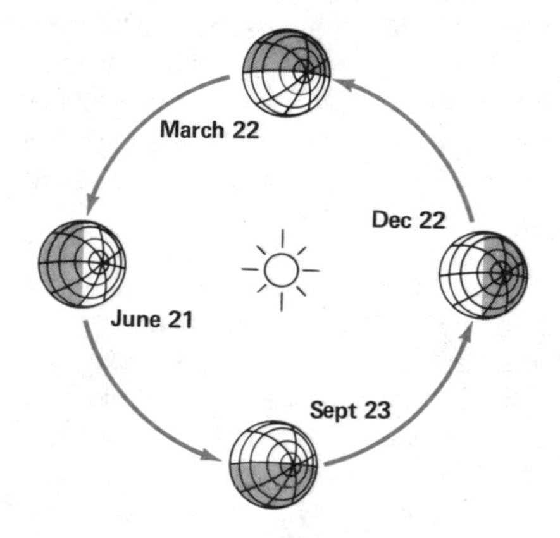

FIG. 1A-1. Seasons as consequences of the constant tilt of the earth. (Scale severely distorted.)

The angle of the noon sun is directly related to the maximum intensity of incoming solar energy on a particular day. In turn, the maximum intensity and the duration of daylight can be combined to provide an estimate of total incoming energy. The remaining three graphs in this appendix give the annual patterns of noon sun angle, duration of daylight, and total solar energy per day at different latitudes. (Figs. 1A-4, 1A-5, and 1A-6). The values are read from the graphs by finding the intersection of the lines for the selected latitude and date. For example, the dots on the graphs show that Los Angeles (34°N) on February 15 has a noon sun angle of 47° down from overhead, a daylength of 11+ hours, and a total incoming energy of 550 ly.

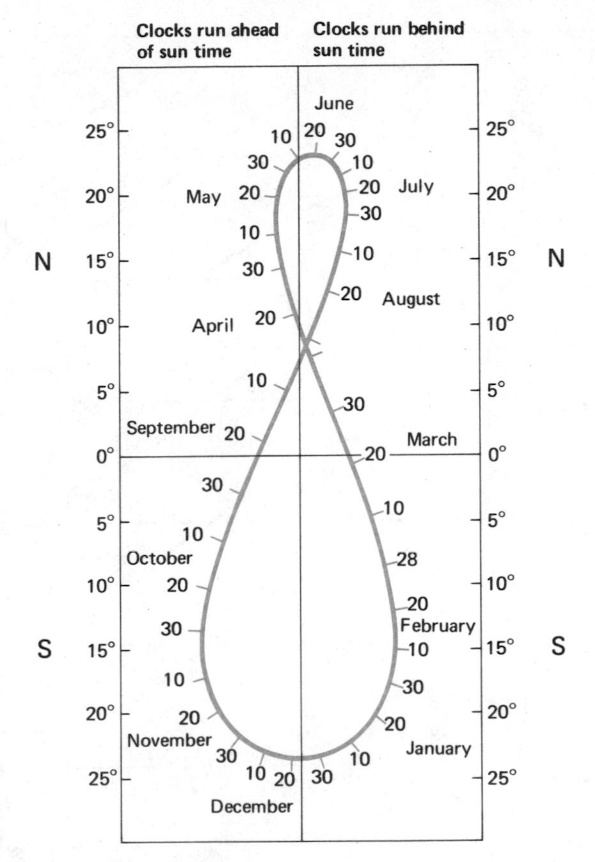

FIG. 1A-2. An analemma.

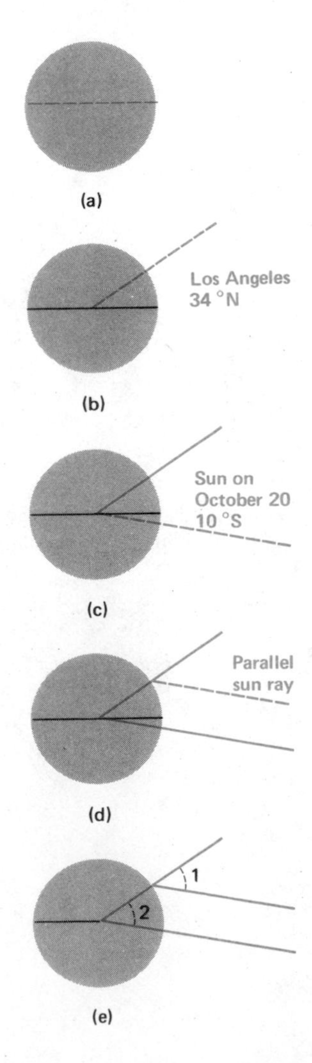

FIG. 1A-3. Determining the noon sun angle.
(a) Draw a circle with a horizontal line that bisects the circle and represents the equator.
(b) Latitude is defined as angular distance north or south of the equator. Show the latitude of the selected place by a line with the appropriate angle north or south of the equator.
(c) The analemma indicates the latitude of the overhead sun on the given date. Show the latitude of the direct noon sun by another line with the appropriate angle north or south of the equator.
(d) Rays from the sun arrive parallel to one another. Draw another line that is parallel to the direct noon sun line but strikes the selected place.
(e) Euclid says that when parallels are cut by a transverse, corresponding angles are equal. Therefore angle 1 (the angle of the sun down from overhead at noon at the selected place) equals angle 2 (the latitudinal difference between the selected place and the direct noon sun).

FIG. 1A-4. Zenith angle (angle down from directly overhead) of the noon sun at different latitudes and months.

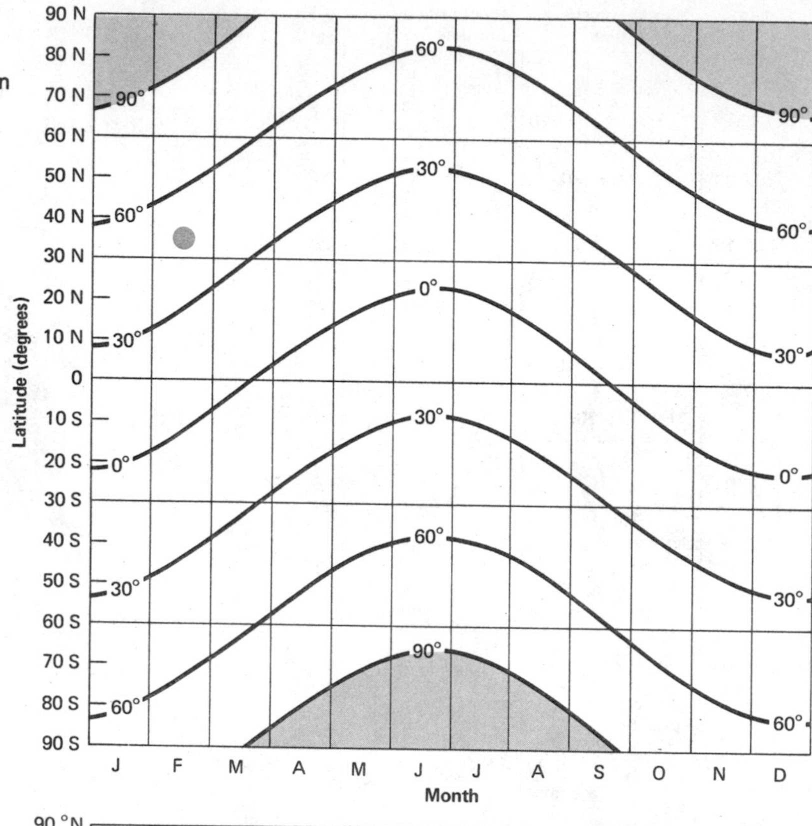

FIG. 1A-5. Daylength in hours at different latitudes and months.

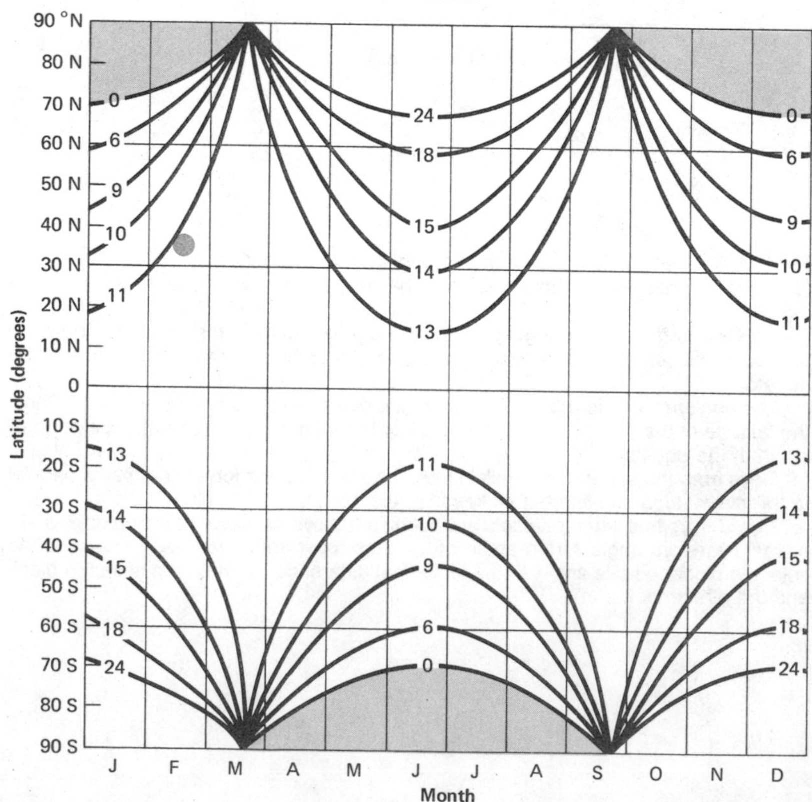

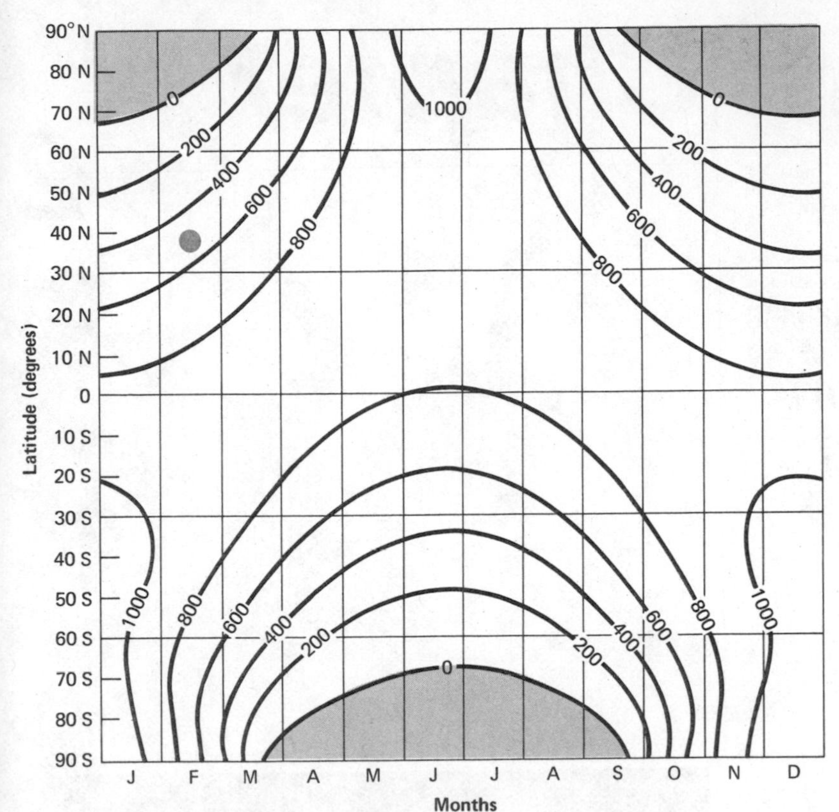

FIG. 1A-6. Total solar radiation in langleys per day at different latitudes and months.

TABLE 1A-1 MEASUREMENTS AND OTHER VITAL STATISTICS

	Kilometers	Miles
Sun diameter	1,394,000	866,000
Earth diameter		
equator	12,757	7,927
poles	12,714	7,900
Orbit radius	150,000,000	92,900,000
Orbit speed	107,000	66,600
Earth circumference	40,075	24,902
Rotation speed per hour		
at equator	1,670	1,038
at 10 °N or S	1,645	1,022
at 20 °N or S	1,570	975
at 30 °N or S	1,447	899
at 40 °N or S	1,281	795
at 50 °N or S	1,075	668
at 60 °N or S	837	520
at 70 °N or S	573	356
at 80 °N or S	291	181
at 90 °N or S	0	0

Chapter 2

INSTABILITY: Vertical Motion in the Atmosphere

So foul a sky clears not without a storm.

Philip Gersmehl

The noon sun beats down from nearly overhead at the equator. Sudden and often intense showers are common in the afternoon, although they usually end by early evening. It is a familiar atmospheric drama, performed with the indifference of a cast that has played the roles countless times before. However, the play is bigger than its equatorial stage. The sun and rain start a global series of events that will parch the sands of the Sahara, give the Irish yet another day of fog, and guide a storm across the plains of Oklahoma. The entire world system of winds and ocean currents behaves like a planetary Robin Hood with one simple goal: to equalize the temperature of the earth by taking heat from the equator and giving it to the poles. As with any great transaction, the means are considerably more complicated than the end would first imply. At the same time, the sun continually thwarts the process by contributing more energy to the already wealthy equatorial areas.

The equatorial surplus of energy appears first at the earth's surface. Soil and plant temperatures rise and water evaporates. The surface soil dries out as the day wears on. Evaporation therefore becomes less and less significant, and a greater fraction of the incoming energy is used to heat the soil at noon than in the morning. The warm surface, in turn, gives heat to the atmosphere. Air near the earth's surface usually is warmer than air a few hundred meters above the ground, because the atmosphere absorbs heat from the earth more efficiently than energy from the sun.

lapse difference in air temperature at different elevations; on a typical day, the lapse in the lower atmosphere is 0.6° C/100 m, the air 1 km above the ground is 6° cooler than air at the surface

The difference in temperature at different heights above the ground is important information for a weather forecaster. Knowledge of upper-air temperature helps the forecaster evaluate the stability of the atmosphere. The basic idea of air stability is simple: warm air is lighter than cold air and therefore tends to rise above it. The significance of air stability is also simple: unstable air is an essential ingredient in all thunderstorms, tornadoes, hailstorms, and most other violent weather catastrophes. Stable air also has undesirable effects, because very stable air is an important component of air pollution emergencies, unseasonable frost, heat or cold waves, drought, and many of the more insidious weather catastrophes.

air stability tendency for air to resist vertical motion

CHANGES IN RISING AIR

The evaluation of air stability starts with the realization that the atmosphere consists of gas under the influence of gravity. Because air is a gas, it can expand to fill any volume into which it is put. However, air is also under the influence of gravity, and gravity exerts a constant pull down toward the surface of the earth. The outermost edge of the atmosphere is hundreds of kilometers above the ground, but the lowest 5.5 km (18,000 ft) contains nearly half of the total mass of air.

At sea level an adult of average size supports a column of air weighing more than a ton. Weather scientists usually express this weight or pressure of air in millibars. Sea level air pressure fluctuates from time to time, but it is rarely more than 1040 or less than 960 mb. Air pressure on a mountaintop is lower than at sea level. The decrease in air pressure with increasing height

millibar (mb) one thousandth of a bar, the standard unit of push or pull; the word comes from the Greek *baros* (heavy), as does barometer, the name of the instrument used to measure air pressure

follows a fairly regular pattern. As a general rule, the pressure of the air 1 km above a particular place will be about seven-eighths of the pressure at that place.

density mass per unit of volume

The actual rate of pressure decline depends on the constant force of gravity and the changing density of the air. The density of air, in turn, is influenced by pressure, temperature, and moisture content. The effects of these three variables are easy to see when each one is considered alone. Air under high pressure is heavier than the same volume of air under low pressure. Cold air weighs more than warm air. Dry air is heavier than humid air. A combination of these three factors, however, is difficult to evaluate. Most weather forecasters use a table or graph to assist them in comparing the densities of different kinds of air.

Dense air from near the earth's surface, if forced to ascend, will rise above some of the atmosphere. As the amount of air above it decreases, the rising air responds to the decrease in pressure by expanding and pushing out on the thinner air around it until its density is equal to that of its surroundings. Outward shoving, however, takes energy, and the only source of energy for a typical mass of air is its own internal heat. In effect, rising air must trade heat for space; its temperature must go down as it expands. The contents of spray cans or pressurized fire extinguishers get very cold when they are released from confinement and allowed to expand. Airline companies thoughtfully provide oxygen masks for passengers on a jet plane, but the masks offer no protection from another consequence of an accident at high elevation. If the pressurized air escaped through a hole in the cabin, the temperature in the plane would drop 50 to 100 degrees. The decline in temperature of rising and expanding air follows a complex mathematical equation, but for the lowest layers of the air a simple rule is reasonably accurate: rising dry air will lose about one Celsius degree for every hundred meters it goes up (5.5 Fahrenheit degrees per thousand feet).

adiabatic process change that takes place without the addition of external heat

dry adiabatic rate of cooling decline of temperature as unsaturated air rises: 1C°/100 m

The inclusion of the word "dry" in this rule guards against the major complication in the process. Rising air cools, and cool air cannot hold as much water vapor as warm air (Fig. 2-1). Every cloud in the sky is a billboard proclaiming that the air at that level has become too cold to hold all of the water vapor that was in it. Every raindrop asserts that air somewhere has been cooled below the critical temperature at which it could barely hold its burden of evaporated water.

dew point temperature at which a mass of air would be saturated, unable to hold any more water than it already has

A temperature decrease is needed for clouds to form, but the interaction between temperature and clouds is not a one-way process. The formation of clouds and rain also affects the temperature of the air. Chapter 1 emphasized that water absorbs energy when it changes from a liquid to a gas. Water vapor is forced to release that energy when it condenses (changes back into a liquid—Figures 1-1 and 8-6 add more detail). The heat given off by condensing water offsets some of the cooling due to expansion; air that is producing clouds or rain cannot cool as quickly as rising dry air. The mathematical or verbal expression of the rate of cooling of waterlogged air is very complex, because both pressure and temperature are involved. Nevertheless, knowledge of the probable loss of temperature of rising air is critical in determining the stability of the air.

latent heat of vaporization energy needed to evaporate a liquid; the latent heat for water is about 600 calories per gram

wet adiabatic rate of cooling decline of temperature as saturated air rises: roughly 0.5C°/100 m

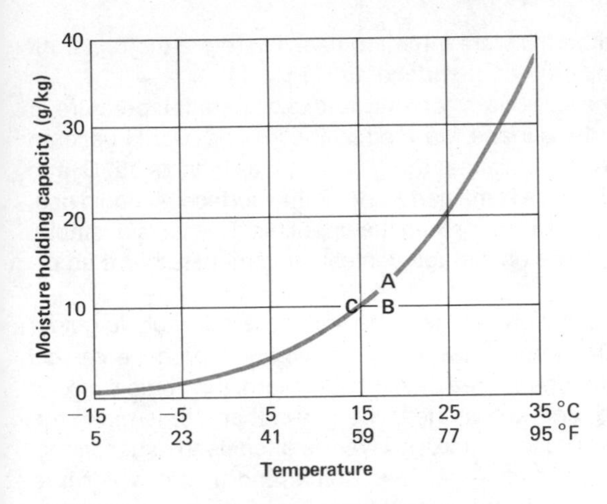

FIG. 2-1. This graph shows the maximum amount of moisture that the air is able to hold at different temperatures. Rapid molecular activity at high temperatures enables more water to remain in the vapor state in hot air than in cold air.

The relative humidity of a mass of air is defined as the amount of moisture actually in the air divided by the maximum amount that the air would be able to hold at its temperature. A kilogram of air at sea level is a mass about the size of the trunk of a mid-sized automobile. At 18° C, it can hold as much as 13 g of water vapor (**A**). If a kilogram of 18° air has only 10 g of water in it (**B**), its relative humidity is about 75 percent (10 is 75 percent of 13). The relative humidity of this air would be 100 percent if it were cooled to 14° C (**C**). Lowering its temperature below its dew point (the temperature at which it is saturated) forces some of the water to condense.

PRINCIPLES OF AIR STABILITY

Warm air tends to rise above cold air. This elementary principle is the key idea of air stability, but it is complicated by the changes that occur when air moves up or down. Suppose that air at the earth's surface on a particular day had a temperature of 25° C, while air 2 km above the ground had a temperature of only 12° C. A simple comparison of temperatures leads to the conclusion that the warm air near the surface should rise above the cold air aloft, yet observation shows that it does not.

The key to this puzzle is the inevitable change of temperature in rising air. Imagine that some outside force lifted an armful of air from the layer near the surface. The temperature of the lifted air would decline as the air ascended and the pressure on it decreased. A loss of one Celsius degree for every hundred meters of a 2000-meter ascent gives the rising air a final temperature of 5° C. Meanwhile, the air already 2 km up still has a temperature of 12° C. Cold air is heavier than warm air (if both have the same pressure and moisture content), so that the lifted mass of air would have a tendency to sink back toward the ground. Actually, an armful of surface air under these conditions is unlikely to even begin to rise, which is precisely the point. The air in this example is stable. The air above is warmer and lighter

stable air air that resists vertical motion

than the air below would be if it were forced to rise. Therefore the upper air tends to resist vertical movement of surface air (Fig. 2-2).

Now suppose that the sun beams brightly and raises the temperature of the ground. The air near the surface, warmed by the ground, would become lighter. If the temperature of the air near the ground got as high as 35° C, and if the air 2 km above the surface remained at 12° C, the surface air could rise, cool as it expands, and still be warmer than the upper air. This air is unstable: its temperature characteristics encourage vertical movement rather than resisting it (Fig. 2-3).

unstable air air that tends to move vertically

In fact, the temperature of the air near the surface is not likely to get as high as 35° C, because the very instability of the air would produce vertical motion as soon as the temperature reached 32° C. In turn, the vertical mixing would carry heat upward, cool the air near the ground, and thus make the atmosphere stable again. The term "mixing layer" is often used to describe a layer of air if it is on the borderline between stable and unstable. Air in a mixing layer can move up or down without resistance, so that any disturbance can lead to vertical mixing.

turbulent transfer removal of heat from a surface by air currents (see p. 15)

mixing layer layer of air characterized by frequent vertical movements

FIG. 2-2. Air stability as a tenancy arrangement. Gravity acts as a landlord who prefers the coolest possible tenants in the ground-floor apartments. If there is not enough temperature difference between downstairs and upstairs air, gravity does not gain anything by trying to evict the occupants of the first floor.

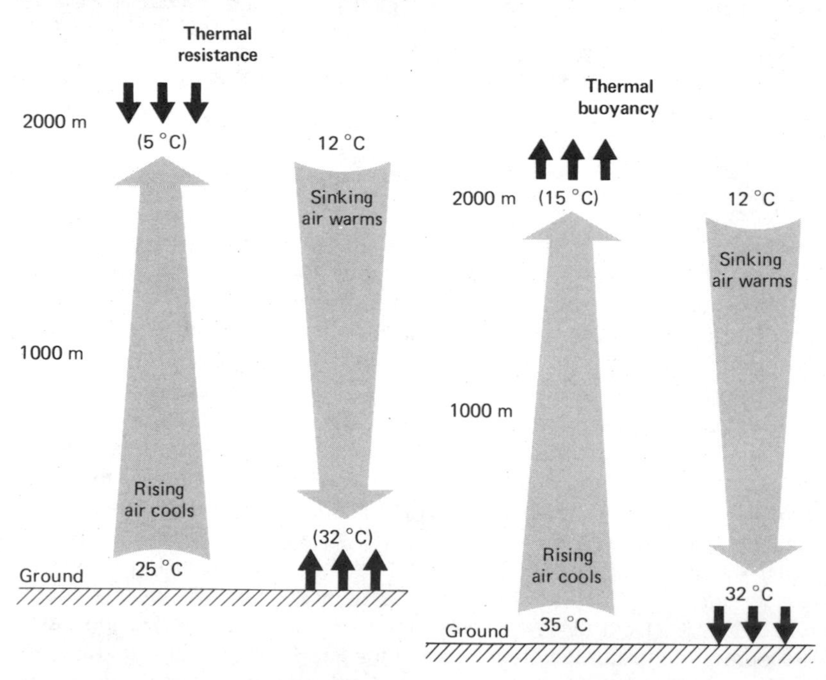

FIG. 2-3. Instability and the uppity tenant. If air in the ground-floor apartment gets too hot and lightheaded, gravity forces the lower air to rise by pulling cool and heavy air down from upstairs.

FIG. 2-4. A weather balloon with its instrument pod. The instruments measure the temperature, humidity, and pressure of the air around the balloon. A small radio transmitter sends the data down to the surface at regular intervals. At the same time, ground instruments follow the ascent of the balloon and gather information about upper-air wind speeds and directions. (Courtesy of the Bureau of Reclamation.)

EVALUATING AIR STABILITY

An evaluation of the stability of the air is essential in assessing the risk of air pollution, making an accurate weather forecast, or trying to fly a kite. The 1 C°/100 m rule of thumb given above is accurate enough for kite flyers, but forecasters, pilots, and pollution control officers need a more precise measure of air stability. Fortunately for students, scientists have put an end to much of the drudgery of calculating and comparing temperatures at different elevations by devising some reasonably straightforward graphic tests of air stability.

A weather analyst who wishes to evaluate the stability of the atmosphere first sends up a weather balloon (Fig. 2-4). As the balloon rises, its instruments transmit information on the temperature, dew point, and pressure of the different layers of the air. On the graph, the forecaster plots the temperature of the air at each pressure level, connects the dots with a line,

balloon sounding series of temperature and pressure measurements sent down by the instruments on a rising weather balloon

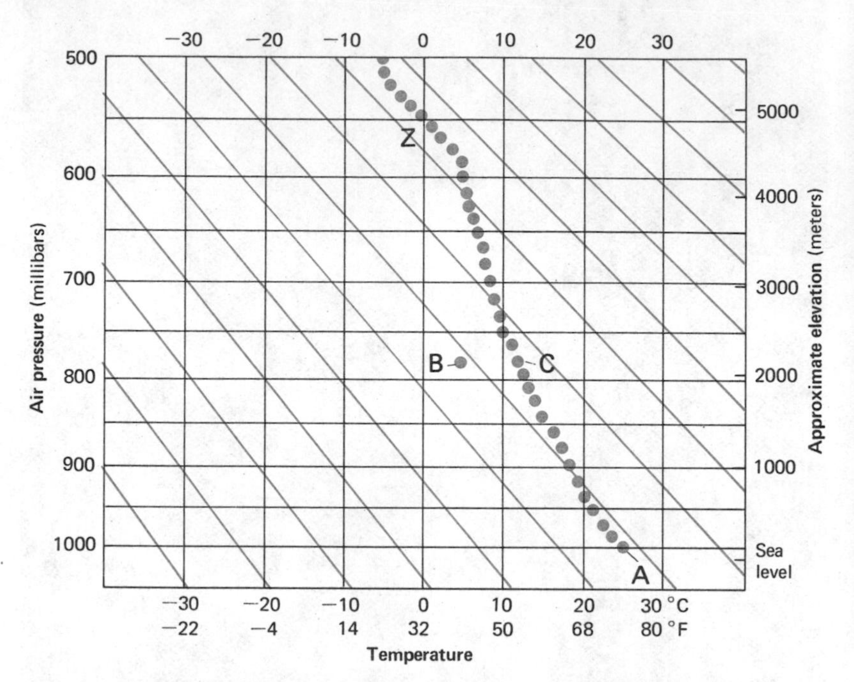

FIG. 2-5. A dry adiabatic chart, a graph designed with slanting lines to show the temperature and pressure changes that occur when a mass of dry air moves up or down in the atmosphere. Point **A** indicates a mass of air with a temperature of 25° C and a pressure of 1000 mb (the air near the ground in the example given in the text). If this air were lifted to where the pressure is only 780 mb (about 2 km up), it would expand and cool. On the graph, the air would "move" parallel to the slanting lines until it reached a temperature of 5° C at the 780 mb level (point **B**).

The dotted line shows the actual air profile described in the text. The balloon instruments report that the actual air temperature at 780 mb is 12° C (point **C**). If air from the surface were lifted to that height, it would be 7° C colder than the air around it, and therefore it would tend to settle back toward the surface.

and compares the steepness of the "air profile" with the slope of the slanting lines (Fig. 2-5). The graph is designed so that the slanting lines indicate the slope of a profile of air that is just barely stable. The illustrated profile stands more vertically than the slanting lines and therefore is safely on the stable side of the critical slope.

Airplane pilots would be particularly interested in knowing about the layer marked **Z** on Figure 2-5, where the air pressure is only 550 mb. The air profile at that level slopes as steeply as the slanting lines on the chart; thus a small mass of air that began to rise would not be pushed back down by warmer surrounding air when it got into a higher layer. Likewise, sinking air would not be pushed up by cooler air around it. In such an unstable layer of air, seemingly unconnected and unpredictable upward and downward movements of air are common. Obviously, it is not good for passenger comfort or company profits when a mass of sinking air pushes a plane wing down while the tail is lifted by an updraft; pilots avoid unstable layers of air whenever possible.

clear-air turbulence vertical motions caused by differences in horizontal wind speed in different layers of air

AIR STABILITY AND CLOUD TYPES

The shape of a cloud allows an observer without a balloon to judge the stability of the air at high elevations. Scattered puffy clouds indicate areas where columns of air are rising and cooling sufficiently to cause water vapor to condense (Fig. 2-6). The spaces between the clouds are just as significant as the shape of the clouds, for they indicate that the rising air is scattered, not universal. Isolated updrafts of air are most common on sunny days, when the ground surface is considerably warmer than the air above. Surplus heat from the ground warms the lower air and makes it less stable and more likely to rise. The more unstable a layer of air is, the higher an updraft can go and the more prominent the clouds become. Cumulus clouds that are taller than they are wide are symptoms of very unstable air (Fig. 2-7). The higher air

cumulus clouds isolated clouds without great horizontal or vertical extent

cumulonimbus clouds clouds with great vertical extent; precipitation from such clouds is usually intense

thunderhead cumulonimbus cloud that has enough vertical air movement to form electrical charges, lightning, and thunder

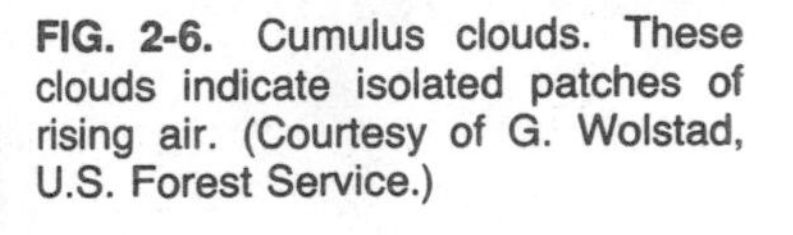

FIG. 2-6. Cumulus clouds. These clouds indicate isolated patches of rising air. (Courtesy of G. Wolstad, U.S. Forest Service.)

FIG. 2-7 Cumulonimbus clouds. These storm clouds are symptoms of strong upward movements of very unstable air. Cumulonimbus clouds may occasionally be as high as 15 km. (Courtesy of E. E. Hertzog, Bureau of Reclamation.)

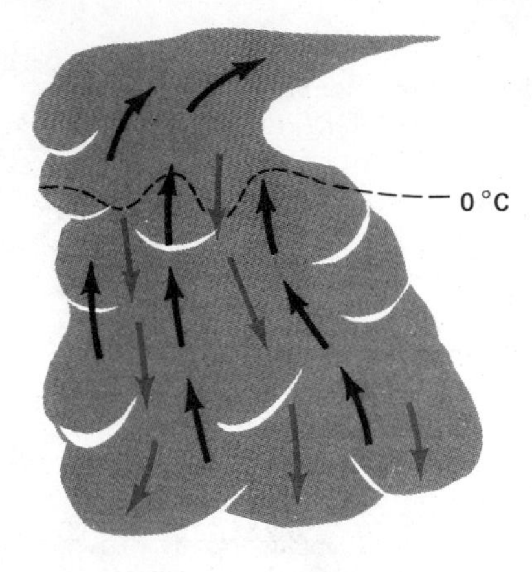

FIG. 2-8. A sketch of a typical hailstorm. Updrafts (black arrows) alternate with areas of sinking air (arrows). Rapidly rising air pushes water droplets up to great heights, where the temperature is below zero and the water droplets freeze. A given hailstone may pass through many cycles of lifting, growing, freezing, and falling, before it finally drops from the cloud to the earth.

rises, the more it cools and the less water it can hold. Lofty clouds in extremely unstable air can produce torrential rain, hail, or even tornadoes (Fig. 2-8).

A flat cloud, by contrast, is a symptom of stable air (Fig. 2-9). The temperature characteristics of stable air resist vertical movement; the air goes upward rather reluctantly if something tries to force it to rise. Despite its inability to rise by itself, stable air may meet with situations that result in upward movement. A strong steady wind may push a mass of stable air up the slope of a mountain. Cold dense air is an invisible but real obstruction that forces warm air to rise above it, even though the warm air may be basically stable. Rain or snow may fall from the stratus clouds, but the innate stability of the air makes thunder, hail, or torrential rain almost impossible.

orographic rain rain caused when air is pushed up a hillside

stratus clouds clouds with great horizontal extent

FIG. 2-9. Stratus clouds in a mountain valley. These flat clouds indicate stable air. A detailed analysis of clouds will recognize a variety of subtypes, usually named on the basis of altitude or precipitation behavior. For example, altostratus or altocumulus clouds are between 3 and 8 km above the ground; cirrostratus, cirrocumulus, and cirrus clouds are icy clouds more than 8 km up. Nimbostratus clouds deliver drizzle, rain, or snow.

AIR STABILITY AND ATMOSPHERE POLLUTION

The same basic principles of air stability and cloud shape can be helpful in analyzing man-made clouds of waste materials. The air stability aspect of air pollution has five dimensions:

Diurnal patterns of air stability. During a typical 24-hour day the atmosphere is usually least stable in midafternoon, when temperatures near the ground are probably as high as they will get and the air near the ground is most likely to rise. The afternoon instability, however, may characterize only the lowest layers of air. Stable air at high elevation may resist the upward movement of warmer air above a particular level. Pollutants therefore get trapped in a shallow layer of air immediately below the stable layer. After sunset, the earth loses heat, the lower air also cools, and the atmosphere becomes more stable. Rising drafts of warm air no longer carry air pollutants upward, and the contaminants begin to concentrate in precisely those layers of the air which city dwellers breathe. At the same time, the afternoon smoke and dust, no longer held up by rising air, may begin to settle toward the ground. Even though pollutants may enter the air more rapidly during the day than at night, the daytime instability dilutes them in a large volume of air. The concentration of pollutants in each breath of air is often greater at night, when the air is more stable. It may be no mere coincidence that hospital emergency rooms receive more gasping calls for help from respiratory patients between midnight and dawn than at other times of the day. Understandably, some industrialists would like to measure air pollution during the afternoon, but the suffering public would prefer a mandatory 24-hour monitoring system.

upper-air inversion exceptionally stable layer of air high above the ground; warm air overlies cold in this layer, the reverse of normal decline in temperature with increasing height

fumigation concentration of pollutants in a small area by wind activity

Day-to-day changes in air stability. Some of the vertical temperature characteristics of the air over a particular place are caused locally. Bright sunshine on an asphalt parking lot warms the lower air and decreases its stability. A leaden gray cloud has the opposite effect because it keeps the surface cool and thus promotes stability. The stability of the air on a particular day also reflects previous events in other areas. Air that was over a cool ocean last week will be more humid and stable than air from a hot desert. A steady wind from the northwest carries air from North Dakota to St. Louis in a few days. The surface between Dakota and Missouri will change some of the characteristics of the incoming air. If the ground between North Dakota and St. Louis is warm, the air will lose stability as it moves southward. If snow covers a frozen soil, air moving over it will stay cool and stable. The amount of change depends on the speed of the wind and the temperature difference between the air and the ground. Nevertheless, the fact that the air came from Dakota and not Arizona or the Gulf of Mexico has certain implications for the steel mills and emphysema victims of East St. Louis.

destabilization reduction in the stability of air as it passes over warm surfaces

emphysema smoke-aggravated lung disease

FIG. 2-10. Frequency of stable air in the United States.

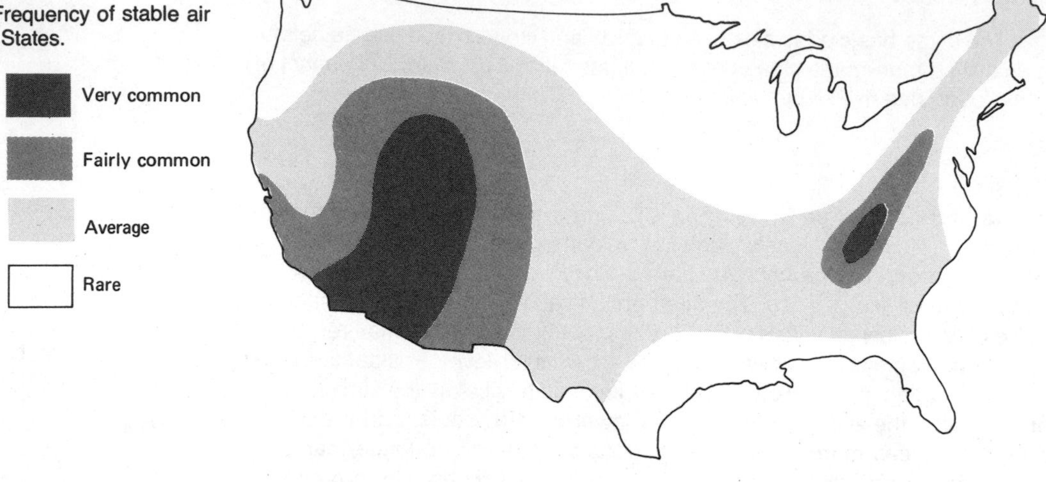

Spatial patterns of air instability. It is possible to map the stability of the atmosphere across an entire continent on a particular day. Indeed, a map of stability is essential for a comfortable long-distance airplane flight. A large number of daily maps can be combined to produce a composite map of the frequency of stable air across the United States (Fig. 2-10). Pittsburgh, Denver, and Los Angeles appear as exceptionally poor places for the coexistence of smokestacks and people. The average stability of the air at those cities is not conducive to efficient dilution of pollution (Fig. 2-11). The "pollutability" of the local environment has only recently become a significant consideration for those who make decisions about factory locations. In former times the inhabitants of an area were handed a take-it-or-move ultimatum.

Seasonal differences in air stability. Many regions of the United States have distinct seasonal differences in the stability of the air. For example, a mountain valley may be a winter resting place for cold air whose stability is undisturbed by a sun that does not climb high enough in the sky to peer over the nearby ridges. In the summer the sun rises high in the sky, beams down through the thin air, warms the valley, and makes the air predictably less stable than it was in winter. Late afternoon thunderstorms are regular summertime events in the Rocky Mountains of the United States and Canada. Seasonal differences in average air stability imply seasonal changes in the ability of the atmosphere to dilute pollutants.

valley inversion mass of cold air settling in a low place and resisting any movement

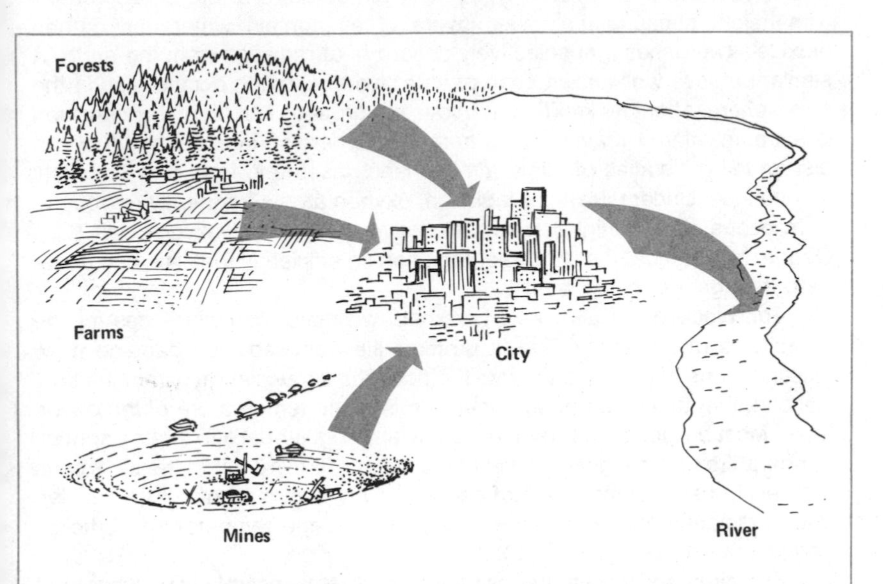

FIG. 2-11. A supplementary essay on "dilution as a solution to pollution."

Heaving garbage out of a window makes more sense to the thrower if there is a fast-moving river rather than a motionless lawn next to the house. Someone who plans to use some part of the environment as a sewer would be wise to check first whether that part can move. The ancient and much maligned system of sending wastes elsewhere still has a certain amount of validity. A modern city is an efficient means of bringing products together from distant places. Food, fuels, raw materials, even water, come in from producing areas that may be far from the city. Unfortunately, the distribution system is not nearly as well-developed as the concentration system. Cities tend to gather products from a large area and then deposit the waste in a single river or landfill or smokestack. Pollution is often a problem of quantity, not quality; the environment is overloaded, not poisoned.

The authors of this book have no objection to making paper out of a forest, printing this book, and then burning the paper when the book is obsolete, *provided* that the trees are cut no faster than they grow, the ink is not permanent, and the burning is done in a proper manner—at the right time of day and with a smokestack designed to put the smoke into the right kind of air. A protest is in order when too many trees are cut, the forest ecosystem is stressed, the ink is a persistent or toxic chemical, or a short smokestack deposits the ashes from 20 square miles of trees onto six city blocks. The writers also object to simplistic solutions that do not solve basic problems. Some incinerators need paper in order to maintain a proper temperature for the combustion of other materials. A newspaper-recycling program may help make some people aware of the solid waste problem, but it does not solve it. A municipal pyrolysis plant that converts solid waste into energy and fertilizer may make more sense than recycling in many towns, but the development of a rational system of "waste disposal" requires knowledge as well as the urge to "save the ecology." After all, ecology is a science, not a perishable commodity.

The upper limits of instability. The temperature of the air decreases with height because the lower layers of air contain water and carbon dioxide—two gases that selectively absorb heat radiation from the earth. A similar but less well-known case of selective absorption occurs high in the atmosphere, about 30 km (20 mi) above the ground. The air at this elevation is extremely thin. Ultraviolet rays from the sun are so intense that they can disrupt the molecules of some atmospheric gases and form new chemical compounds. Under ultraviolet radiation, oxygen atoms frequently regroup to form ozone (O_3), a highly reactive relative of "normal" oxygen gas (O_2). Ozone, in turn, absorbs ultraviolet radiation so efficiently that little can penetrate through the ozone layer.

ultraviolet radiation radiation from hot sources; the rays have short wavelengths and high intensity

The blockage of ultraviolet rays has two major consequences for the inhabitants of the earth. First, it protects life from radiation damage. Few existing forms of life could tolerate the powerful ultraviolet rays from the sun. Second, the absorption of sun energy raises the temperature of the ozone layer. Most balloon measurements show an irregular decline in atmospheric temperature up through the lowest 10 or 15 km (6 to 10 mi) of the air. Above that level, the warming effect of ozone activity begins to offset the cooling due to distance above the earth's surface, and the temperature of the air begins to increase again (Fig. 2-12).

troposphere layer of occasionally unstable air near the ground; air temperatures in this layer usually decrease with height

tropopause upper boundary of the troposphere, about 10 km above the ground at the poles and 20 km over the equator.

stratosphere layer of stable air above the tropopause; air temperatures in this layer increase with height

The air in and above the cold layer is extremely stable. The warmth of the ozone layer above puts an effective lid on the lower atmosphere. Even a strong updraft of air from the lower atmosphere cannot penetrate the stable layer very far before it becomes colder than its surroundings and loses its impulse to continue upward. The hailstorm cloud in Figure 2-8 has the characteristic flat top that develops when uprushing air reaches the stable layer, stops rising, and begins to spread out sideways.

The stability of the air above thunderclouds is both a benefit and a potential threat to humans. On the one hand, the stable air is an aviator's dream, clear and often free of turbulence. On the other hand, there is no vertical mixing of the air and no snow or rain to wash pollutants out of the air. Any contamination of the air at that elevation may persist for months or even years.

residency time amount of time a particle typically remains in a place; average residency times are as follows:

place	time
lower troposphere	weeks
upper troposphere	months
lower stratosphere	years
upper stratosphere	decades

Weather observers in recent years have noted a general worldwide increase in the amount of very high clouds. They think that high clouds are blocking some of the incoming sunshine and thus cooling the earth's surface. Some scientists believe that the cloudiness is a consequence of pollution of the upper air. Other scientists doubt the importance of pollution and point to a general increase in volcanic activity, while still others question the accuracy of the temperature and cloud measurements. It may take decades for scholars to find the reason for the apparent increase in cloudiness, because the upper atmosphere takes so long to react to changes. The consequences of human activity in the upper layers of the atmosphere may not be known in this century, a thought worth considering before embarking on any widespread program of upper-air use.

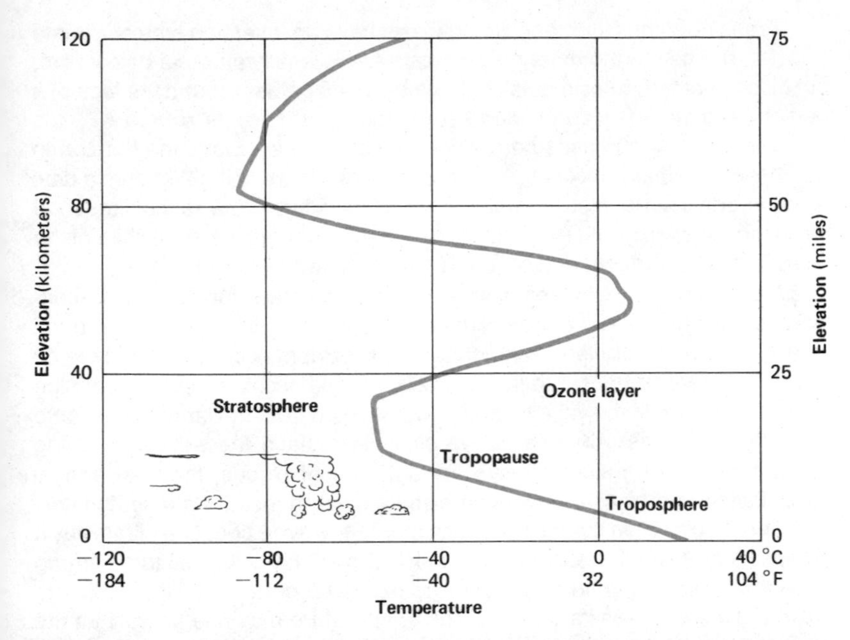

FIG. 2-12. A typical vertical profile of atmospheric temperature. The air near the ground consists mainly of nitrogen, oxygen, argon, carbon dioxide, and water vapor. It is warmed primarily by heat radiation from the earth, and therefore its temperature decreases with height. Air at high elevations contains more ozone and hydrogen and less oxygen and carbon dioxide. This layer of air is quite warm because ozone absorbs ultraviolet radiation from the sun. Between the ground layer and the ozone layer is a cold layer that receives little usable radiation from either earth or sun. The cold air and the warm ozone layer above it are exceptionally stable; they resist vertical movements very effectively. Rain or snow clouds therefore are limited to the lowest layer of air—the troposphere.

INVERSION TYPE OF CLIMATE

People accustomed to the varied fare of mid-latitude weather often associate the word "weather" with the idea of change. To them, therefore, it seems strange, almost alien, when a single atmospheric process dominates in a region, monotony of weather is the rule, and changeability of weather is really quite unusual. This brief introduction to the physical geography of the earth organizes the world pattern of climate into a set of four distinct types of climate, only one of which is characterized by significant changeability. Each of these climatic types is the consequence of atmospheric processes that operate in a specific part of the world. Two of the types are caused by large-scale air motion and are explained in Chapters 3 and 4. The other two are direct consequences of the average stability of the air in particular temperature environments.

climatic type long-term pattern of weather caused by the major atmospheric processes in a particular region

boundary layer layer of modified air near a surface

albedo reflectivity of a surface
(see p. 11)

inversion warm air overlying cool air; a reversal of the normal air temperature lapse

wind chill measure of the combined cooling effect of low temperature and wind
(see Fig. 5-9)

polar anticyclone high-pressure area over the poles

Areas near the North and South Poles have an inversion type of climate (Fig. 2-13). The thermometer may indicate an air temperature far below zero, but it does not feel so cold outside. Breath clouds gather around the face of a person, and the absence of wind keeps body heat near its source.

The summer sun rises barely one-fifth of the way above the horizon to overhead. It partially makes up for its low intensity by shining 24 hours a day, but the earth's surface still remains frozen. The white snow reflects much of the incoming energy and is too deep to melt away during the sunny half of the year. The cold surface, in turn, cools the air near the ground.

Instruments on weather balloons usually indicate that the air is considerably warmer a few hundred meters above the surface than at ground level (Fig. 2-14). Turbulence and gusty winds seldom occur when surface air is much colder and denser than the air above. Heavy cold air slides down the hillsides, gathers in the valleys, and flows outward from the landmasses onto the nearby oceans. Air from above sinks to replace the outward moving surface air. Sinking air is always dry and free of clouds; the total annual precipitation in regions of inversion climate is often less than a centimeter.

The precipitation seems like much more, however, because of the way it arrives. A drop in air pressure signals the arrival of an external force strong enough to disrupt the normal pattern of calm and cold. The wind begins to blow and the air loses its stability. The temperature may rise sharply as the surface layer of dense cold air is pushed away. During the storm the wind picks up snow from the ground and blows it around, making it difficult to separate new-falling snow from old snow that is being redistributed.

After the storm the new surface air slowly cools and becomes more stable (Fig. 2-15). Stable air resists motion and thus produces the calm that allows the air to cool further and to become still more dense. The mutual reinforcement of cold and stability continues, perhaps for weeks and months, until another outside disturbance appears with enough strength to penetrate the stable air mass and disrupt the self-imposed calm.

The eastern highlands of Antarctica and the northern part of ice-covered Greenland are the inner bastions of the inversion type of climate. Their high elevations aid their polar location in protecting them from invasion. Even the poles have more weather disturbances than these mountain areas near the poles. The farther a place is from the core of the inversion type of climate, the more frequently its calm weather is disrupted.

Wet air from the oceans often invades the margins of the polar regions. This air cools rapidly as it passes over the icy land surface. Its stability increases, fog is common, and rain or snow falls gently but persistently, often accumulating considerable quantities after seemingly endless days of leaden skies. The humid coasts are transitional to the inversion climate rather than typical of it.

The polar climate controls only a small area during the summer half of the year. Summer sunshine promotes air instability by warming the ground and making the lower air more buoyant. The inversion climate shrinks into the inner reaches of the permanently frozen regions. In the long dark winter, however, the inversion climate reaches far toward the equator and dominates much of Eurasia and North America.

Up here, the words "day" and "night" are used to describe halves of the year.

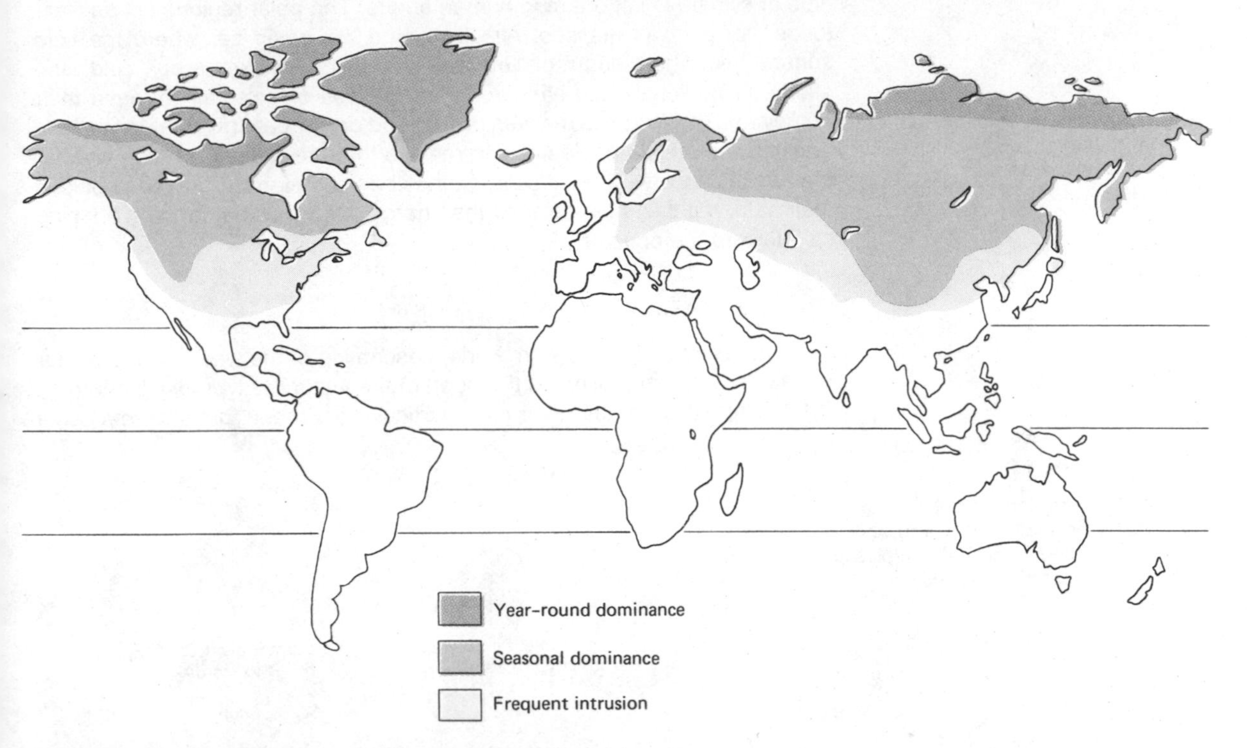

FIG. 2-13. The domain of the inversion type of climate, the calm and clear consequence of a cold surface. Antarctica is not shown on this map because of the nature of the map projection, but it is nearly all dominated year-round by inversion climate.

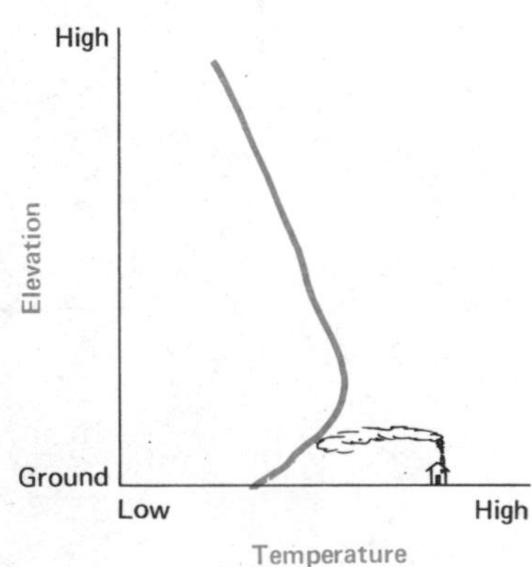

FIG. 2-14. A sample vertical profile of the air over a station in the inversion type of climate. Exceptionally cold air near the surface and warmer air aloft make vertical motion of the air very difficult.

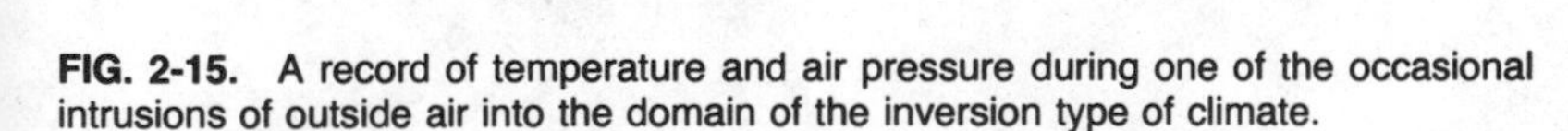

FIG. 2-15. A record of temperature and air pressure during one of the occasional intrusions of outside air into the domain of the inversion type of climate.

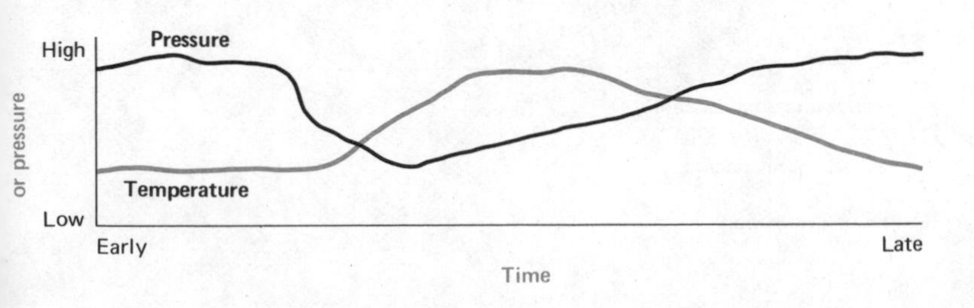

Few people venture into the forbidding polar regions, yet the inversion type of climate is important in human affairs. The polar regions act as "factories" for cold air masses. Air sinks onto the polar ice, where the cold surface takes heat and moisture from the air and makes it dry, cold, and stable. Occasional disturbances cause masses of polar air to leave their homeland. The expelled masses of cold and dry air can travel far toward the equator and affect people living in regions that are ordinarily much warmer than the poles. Invasions of polar air may be uncomfortable for the people in their path, but they help balance the energy budget of the earth by bringing cold toward the equator.

air mass large mass of air with essentially uniform temperature and moisture characteristics
(see p. 67)

THE INSTABILITY TYPE OF CLIMATE

This chapter opened with a brief description of the weather at a typical equatorial weather station, in the heart of the instability type of climate (Fig. 2-16). Rain may not fall there on a particular afternoon, but showers occur

FIG. 2-16. The domain of the instability type of climate, the warm and showery consequence of surplus energy at the earth's surface.

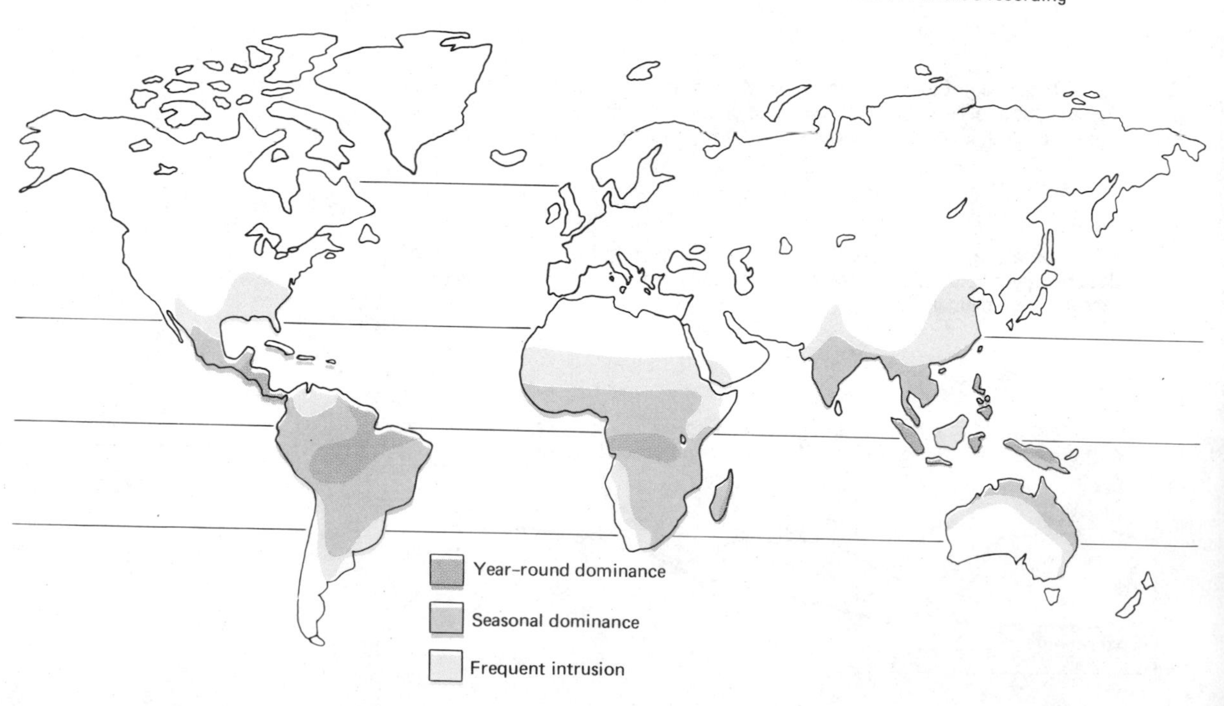

between noon and sunset on 320 days of the year. The hot overhead noon sun causes the afternoon storms. During the day, energy from the sun makes the air near the ground progressively warmer, lighter, and less stable. By midafternoon the lower air begins to rise. The rising air cools and reaches the critical temperature where it must give up moisture. Clouds and showery rains are inevitable consequences of rapidly rising air.

The rise-and-rain process is reinforced by masses of air that come in from north and south of the equator to replace the rising air. Average air pressure in the instability zone is low because the air is usually warm and light. In turn, the low pressure lures air inward from areas of higher pressure both south and north of the instability zone. Clouds and rain may also occur when incoming air masses collide.

intertropical convergence zone (ITC or ITCZ) area of low air pressure and converging winds near the thermal equator

Atmospheric instability and low air pressure are not absolutely uniform throughout the equatorial regions. Distinct surface features, such as mountain ranges or bodies of water, have different energy budgets that lead to irregularities in the pattern of equatorial instability. Variations in the flow of wind in the upper atmosphere produce migratory zones of exceptionally low air pressure. The convergence and rising of surface air is stronger and the resulting precipitation heavier when air pressure in the center is lower than usual. The weather map of the equator for a given day usually contains several distinct zones of instability and air mass convergence, separated by areas where the weather may be calm and settled.

easterly wave moving zone of low pressure that travels to the west in the ITCZ

Despite minor day-to-day variations, most areas under the influence of the instability type of climate have high temperature and abundant rainfall in every month of the year. In fact, the temperature and humidity differences between day and night are usually much greater than the variations from month to month. The lack of seasonality is one of the most striking features of the climate near the equator.

Even more important, the instability of the equatorial atmosphere is the first cog in the entire global wind machine. The next chapters carry the story further by tracing the consequences of a cornerstone concept: air movements on the earth have their ultimate origin in the surplus of heat at the equator.

APPENDIX TO CHAPTER 2

Meteorological Charts

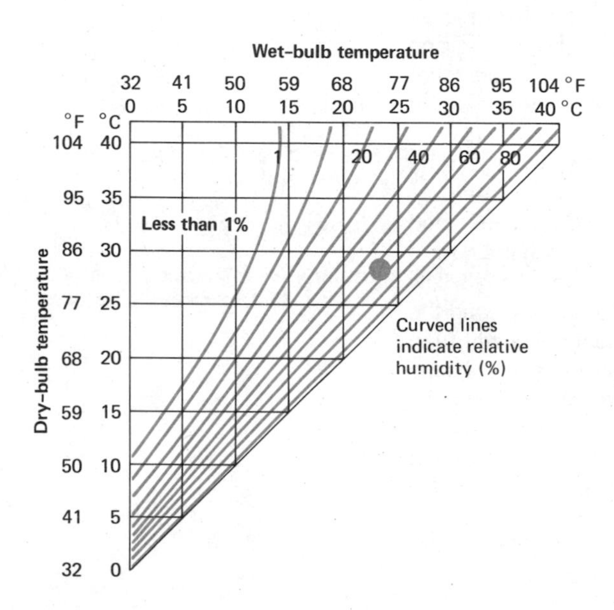

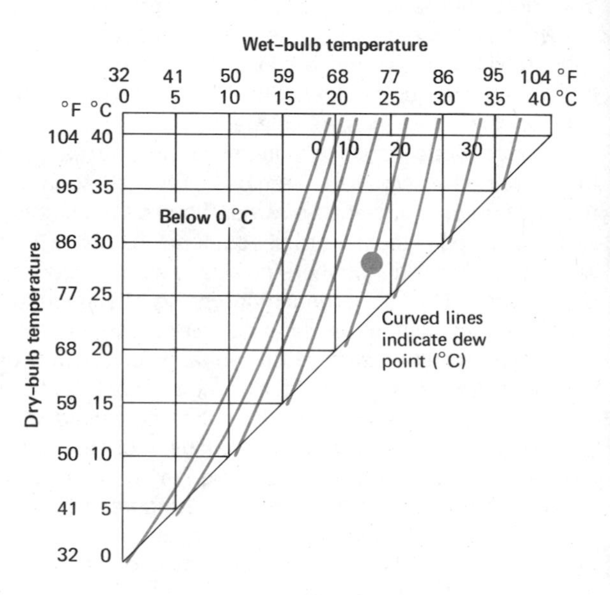

PSYCHROMETRIC GRAPHS

The ability of air to hold moisture depends on temperature. Relative humidity is the amount of moisture in the air expressed as a percent of maximum amount the air could hold at its temperature. Dew point is the temperature at which the actual amount of moisture in the air would be all that the air could hold. Both dew point and relative humidity are usually determined with a psychrometer and tables or graphs. A psychrometer consists of two thermometers, one dry and one moistened with water. Moisture evaporates from the wet thermometer, and evaporation takes energy; therefore the temperature shown by the wet thermometer decreases. Since the amount of evaporation depends on the dryness of the air, the temperature of the wet bulb is an indirect but accurate indication of the moisture already in the air. The graphs are read by finding the intersection of the appropriate dry-bulb and wet-bulb temperature lines and by reading the dew point or relative humidity on the nearest curving lines. For example, air with a dry-bulb temperature of 28° C (82° F) and a wet-bulb temperature of 23° C (73° F) would have a relative humidity of 65 percent and a new point of 20° C, as shown by the dots on the charts.

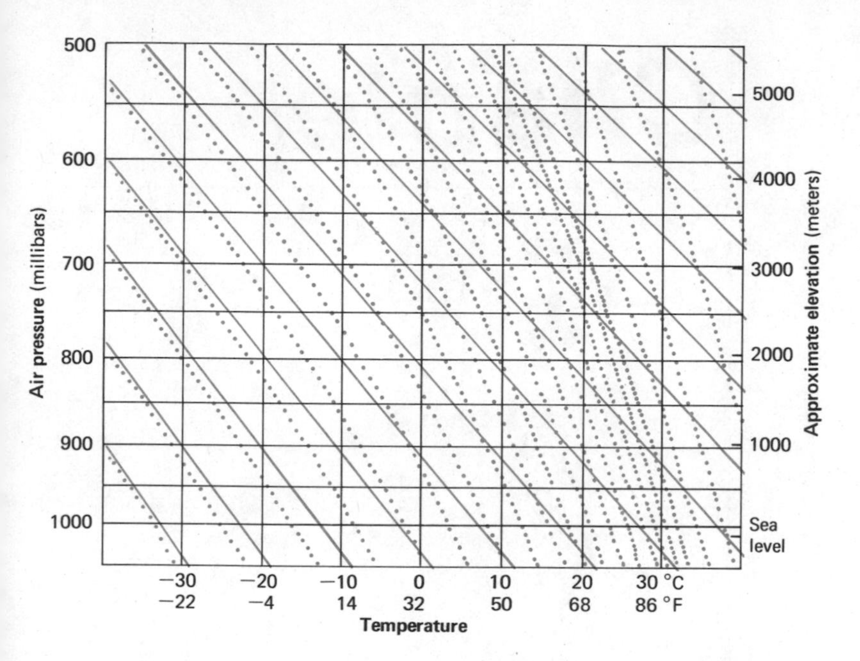

PSEUDO-ADIABATIC CHART

The vertical lines indicate air temperature. The horizontal lines represent air pressure. The solid slanting lines are the stability test for unsaturated air. The curving dotted lines are the stability test for air that is saturated with water and therefore will precipitate when lifted. To test the stability of the air on a particular day, send up a weather balloon and use the vertical and horizontal lines as guides to plot the balloon measurements of temperature and pressure on the graph; then, connect the resulting dots to form an air profile.

1. If the profile stands upright or leans to the right, the air is stable and resists vertical movements.
2. The more the profile leans to the left, the less stable the air is.
3. If the profile leans as steeply as the solid slanting lines, the air is unstable and apt to rise all by itself.
4. If air contains all the water it can hold, it is unstable if the profile leans as steeply as the dotted lines, not the solid slanting lines.
5. Different layers of the atmosphere may have different degrees of stability on a particular day.

Chapter 3

Wind: Horizontal Motion in the Atmosphere

East brings the sun, the source of light
South brings the summer, the wind of
growing
West brings the whirlwind, thunder and dust
North brings the cold wind, a time of dying.

U.S. Forest Service

Winds are horizontal movements of air across the earth's surface; they have a bewildering variety of speeds, directions, reliabilities, persistencies, and effects. Many of the basic observations of wind patterns were made centuries ago, when the wind was a major source of power for industry and transportation. Its importance as an energy source has declined sharply, but it is hard to overestimate the significance of wind as an element of weather and influence on climate. Despite the great variety of winds, the major principle of the previous chapter holds true: winds have their ultimate origin in the unequal amounts of energy the sun gives to different places on the earth. The global wind system begins with the surplus of heat near the equator.

SIMPLE CONVECTION SYSTEMS

Air over a warm surface becomes unstable and tends to rise. A cool surface makes air dense, stable, and likely to sink toward the surface. Vertical motions, when isolated, are self-eliminating. They cannot persist for long without some kind of compensating horizontal movement. Something must bring additional air to a warm surface in order for the heating and rising to continue. The concept of local air pressure helps connect vertical motions into a circulation system that provides a mechanism for the needed resupply of air.

As an illustration of local air pressure, consider a closed room with a heater on the floor near one wall (Fig. 3-1). Over the heater, air is warmed, becomes less dense, and rises. While the density of the air around the heater decreases, the air near the ceiling becomes more crowded. A barometer held near the ceiling records the crowding above the heater as an increase in the local air pressure.

barometer instrument to measure air pressure

Meanwhile, at the cold end of the room, air is losing more heat than it receives from its surroundings. As it cools, the air becomes more dense and sinks toward the floor. The local air pressure near the ceiling decreases as the air cools and settles.

The combination of crowding high above the heater and sinking at the other end of the room forces some of the air to leave the heated end and migrate along the ceiling to the less crowded end. The ceiling "wind" adds to the total mass of air on the cool side of the room. A barometer near the floor records an increase in the total weight of the air above it, and the increase in

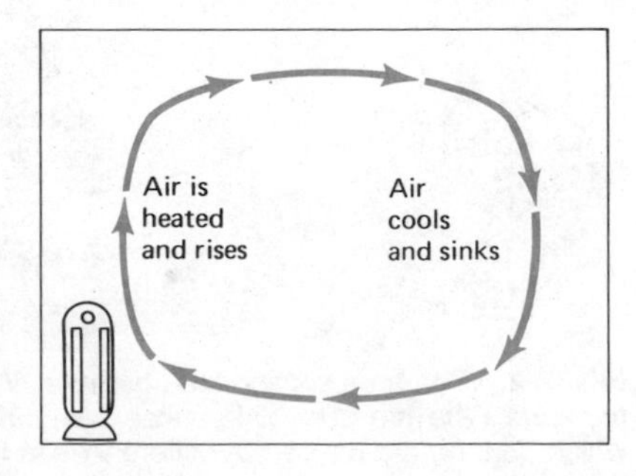

FIG. 3-1. A closed convection system. The cycle consists of two vertical air currents (one rising above the heater and one sinking at the other end of the room) and two horizontal winds (one going away from the heater along the ceiling and one going toward the heater along the floor).

air pressure forces some of the air near the floor to move. The result is a "wind" moving along the floor toward the heater and replacing the air that is rising in the warm end of the room.

convection vertical movement of air or liquid due to difference in temperature

The circulation of air persists as long as heat is added at one end of the room and removed at the other. The convection system has been outlined as a sequence of events, but it is actually a self-starting combination of four different air movements. The events take place simultaneously, not in the sequence that is needed to explain the process on paper. Many natural processes take place all at once, not one at a time in order. Simultaneity may make it difficult to find the ultimate cause, but an understanding of the reason behind a particular process makes it safer to tamper with the machinery. It would be useless to try to stop the convection in the room by putting a low barrier across the path of the floor breezes. The floor wind is not an isolated event; it is part of a system, and since it is ultimately caused by heat at one end of the room, the way to stop the wind is to remove the source of heat.

NATURAL CONVECTION SYSTEMS

sea breeze daytime wind from water onto heated land

Numerous outdoor situations show that solid walls around a room are not necessary for the existence of a convection system. On warm afternoons beaches frequently enjoy cool breezes that blow from the water. The sea

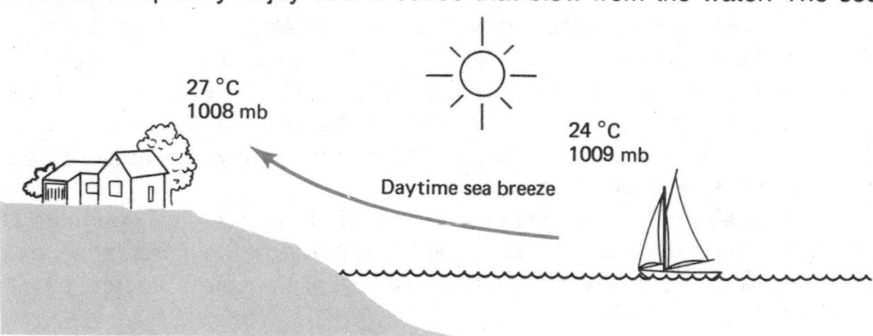

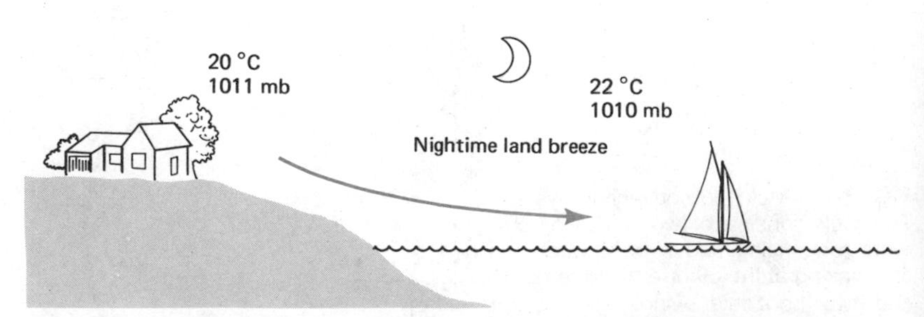

FIG. 3-2. Sea breezes and land breezes. Winds are responses to differences in air pressure. The pressure differences are consequences of temperature differences, which, in turn, are caused by differences in the energy budgets of land and water.

breezes are part of a larger circulation that involves rising currents of air above the warm land and sinking air over the cool water. Onshore winds near the ground are compensated by air moving from land to sea at higher elevations. The balance may not be apparent locally, because the offshore flow may occur through several intermediate steps in other places. The unequal temperature of land and water causes convection, but without walls the atmosphere has more freedom to work out the details of the system.

Nighttime reverses the relative temperatures of water and land, and the temperature change leads to a reversal of local air pressure. The winds respond to the altered pressure readings by blowing away from the land at night (Fig. 3-2).

land breeze nighttime wind blowing from cold land toward warm water

Seasonal maps of world air pressure reveal similar reversals of temperature and pressure over the continents and oceans in many parts of the world. The summer sun raises the temperature of the land more rapidly than it can warm the water. Air pressure over the land usually is lower in summer than in winter. Summertime winds blow toward the land from the sea, while wintertime air is more likely to move away from the continents (Fig. 3-3). Because of the immense size of the circulation system, the seasonal reversals are not very regular. Local exceptions are common, and there seem to be many times when the general rule does not apply.

monsoon seasonal reversal of wind direction between continent and ocean; the South Asian or Indian monsoon is most famous

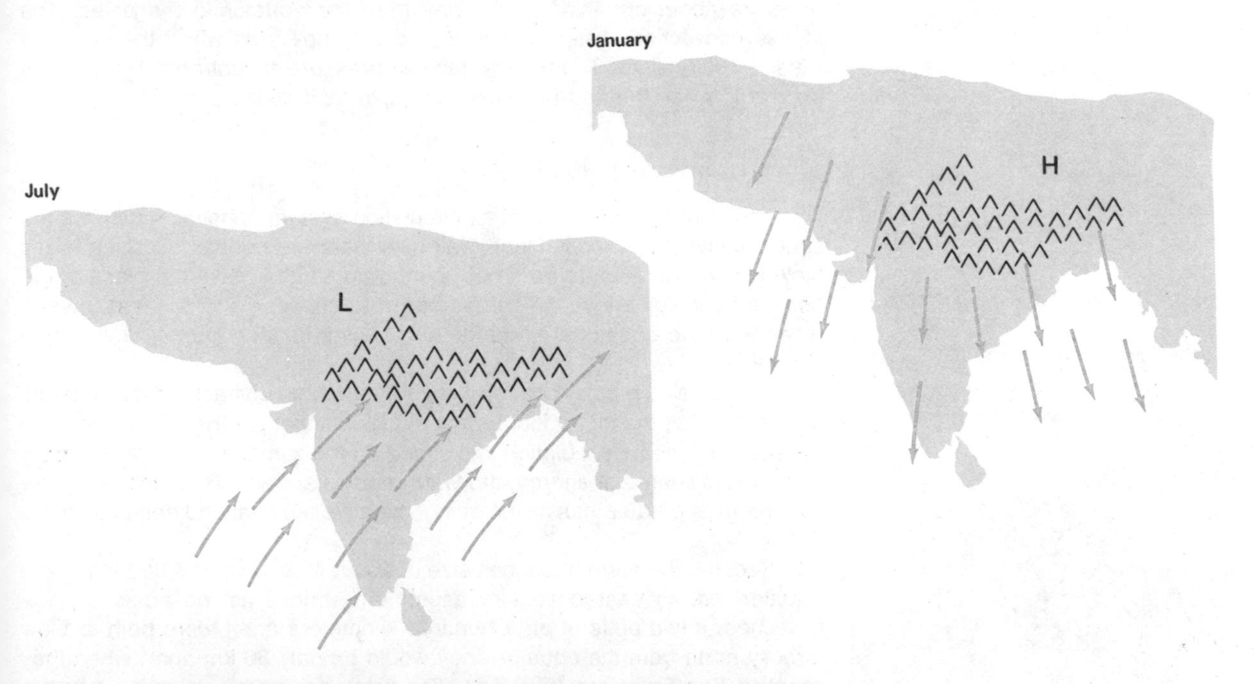

FIG. 3-3. Seasonal reversal of wind direction over India. This reversal is the famous monsoon, the bringer of summertime rain. The classic explanation of the monsoon emphasizes the seasonal reversal of temperature and pressure over the continent and the adjacent ocean. Modern studies assign considerable importance to several dynamic characteristics of the upper atmosphere, particularly the sudden shift of the prevailing winds from south to north of the Himalaya Mountains in summer.

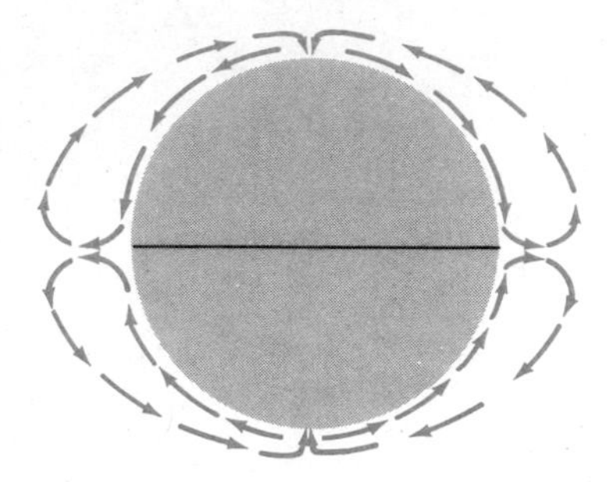

FIG. 3-4. Hypothetical convection system on a smooth nonrotating globe. Air would rise at the equator, sink onto the poles, and flow in horizontal currents between the warm equator and cold poles.

instability climate warm and showery weather near the equator, caused by rising unstable air.
(see p. 46)

tropic 23½° line of latitude

subtropical near the 30° latitude lines

A GRAND CONVECTION MODEL

There would be no wind on the earth if it were a smooth, nonrotating, uniformly heated sphere. Adding more heat to the equator of a stationary ball yields a simple but grand convection system involving the entire atmosphere. The surplus of energy at the equator induces upward air movements. The rising equatorial air moves high above the ground toward the poles. As it travels away from its warm "birthplace," the air loses heat, sinks toward the surface, and eventually becomes a surface wind blowing back toward the equator (Fig. 3-4). Surface air pressure is high at the poles and low near the equator. The pressure of the air far above the ground, while very low by surface standards, is greater above the equator than over the poles.

Actual patterns of atmospheric pressure and wind on the earth are compatible with some of the conclusions of this general model. The equator does have lower pressure near the ground and higher pressure aloft than do the poles. High-elevation winds tend to blow poleward, but there is no continuous surface wind blowing from the poles toward the equator. Furthermore, air pressure on the surface does not increase evenly from the equatorial low to the polar highs. Rather, there are striking zones of high surface air pressure about one-third of the way from the equator to the poles. The simple convection system becomes more complicated when the model is made more realistic. The high surface air pressure in subtropical latitudes is inevitable when the earth is allowed to spin as it does.

SUBTROPICAL HIGHS

convergence coming together

divergence spreading apart

The first part of the global air circulation system behaves much like the simple convection model. Heated air rises near the equator. Surface winds blow toward the equator from both north and south to take the place of the rising air. Air high above the ground becomes crowded and pushes outward, away from the equatorial updrafts, and begins its long journey toward the cold poles.

On its way the poleward wind encounters three obstacles that hinder its progress. First, the air no longer has a thick blanket of other air above it, so it loses much heat by radiation into space. At the same time, it is farther from its principal source of energy, the warm earth's surface. Because its outgoing energy is greater than its incoming, the air loses heat and becomes more dense.

Second, the earth has a belt size of about 40,000 km (25,000 mi) at the equator, but its east-to-west measurement shrinks as the poles are approached. If two puffs of air a hundred kilometers apart were both to blow directly north from the equator, they would be only 80 km apart when they reached San Francisco (Fig. 3-5). The rising equatorial air, when it heads toward the poles, is squeezed together in a similar manner.

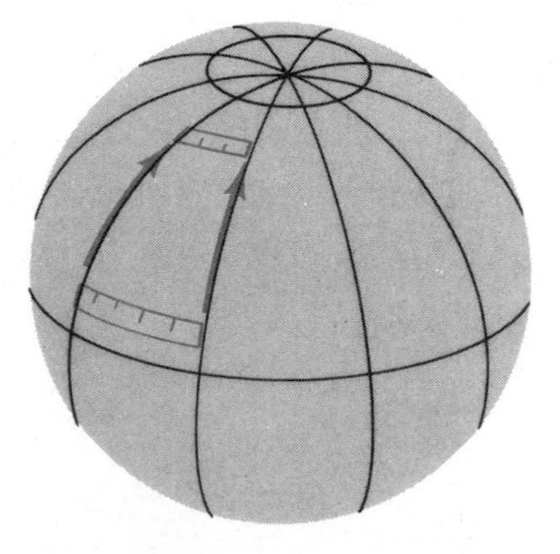

FIG. 3-5. Convergence of air traveling poleward on the globe.

FIG. 3-6. The Coriolis effect. The behavior of air as it moves from an area of high pressure to one of low pressure would be much easier to understand were it not for the so-called Coriolis force. This "force" makes sense only if one keeps in mind that it is not really a force at all; rather, it is a tribute to the limited vision of earthbound observers.

To visualize the Coriolis effect, imagine a Super Archer standing on the North Pole and aiming at a target 500 km (300 mi) to the south (What other direction is there from the North Pole?). Suppose this Archer shoots an arrow with a speed of 500 km/hr. By the time the arrow arrives at the target, one hour later, the target itself will have moved about 130 km (80 mi) to the east. The Archer must "lead" the target, or the arrow will land to the west of the target, because the target is fastened to a moving earth.

An observer in space would see the arrow traveling in a straight line, but to the Polar Bowperson, the arrow appears to veer to the right of its intended path, because the Archer is slowly being turned around, once every 24 hours, as the earth rotates. The basic principle holds regardless of where the Archer stands or where the arrow is aimed: objects in motion in the Northern Hemisphere always appear to be deflected to the right when viewed by someone on the surface of the earth. (In the Southern Hemisphere, deflection is always to the left of the intended path.) It may be difficult to visualize why the deflection should always be to the right in the Northern Hemisphere, even when the arrow is shot in another direction besides south. The key idea is simple: an arrow is already in motion even before it is shot. A released arrow continues its own motion as well as the one the Archer wishes to give it. On a spherical earth the combination of both kinds of motion inevitably produces an apparent deflection to the right in the Northern Hemisphere. (Fig. 3-14 presents an alternative way of visualizing the process.)

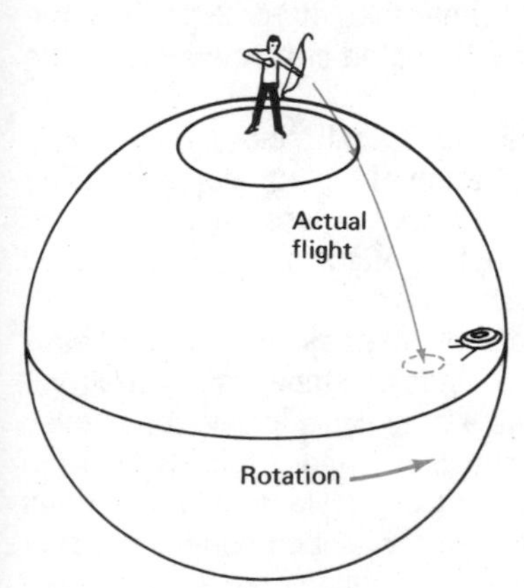

Arrow travels in a straight line, but target moves while arrow is in flight

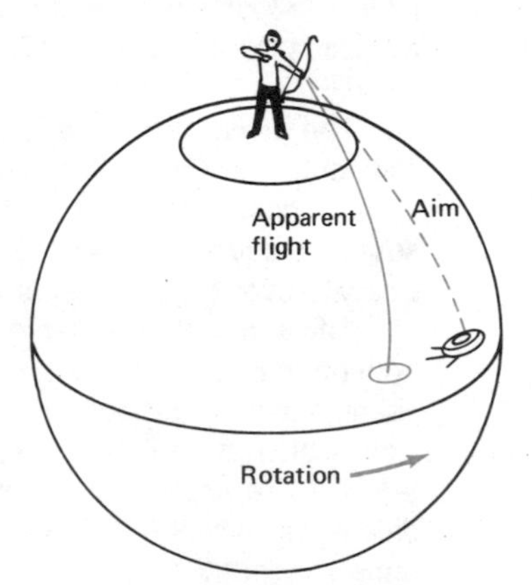

Arrow appears to follow a curved path

Third, a moving object on the spinning earth inevitably deviates from an apparent straight line (Fig. 3-6). The deflection is insignificant near the equator, but the tendency to detour from a straight path becomes stronger as the high-altitude winds move toward the poles. Air traveling north or south from the equatorial uplift usually finds itself moving eastward instead of poleward long before it gets even halfway to the poles.

Coriolis effect apparent deflection of objects in motion on the surface of the rotating earth; the deflection is to the right in the Northern Hemisphere, to the left in the Southern Hemisphere

subtropical highs zones of high air pressure roughly centered on the 30° parallels of latitude in both hemispheres

The combination of cooling, compression, and deflection means that the poleward journey of most of the high-flying equatorial air has little chance of successful conclusion. Some of the air begins to sink toward the earth's surface shortly after it leaves the ground. Most of the subsidence occurs about 30 degrees of latitude away from where the air rose, although persistent upper-air currents may continue farther toward the poles.

Laboratory studies of the behavior of unequally heated material on a rotating surface showed that the circulation pattern depends on three things: the nature of the circulating fluid, the temperature difference, and the speed of rotation of the surface. The circulation of winds on the earth would be slower if the air were thicker than it is. A smaller temperature difference between equator and pole would weaken the subsidence and move it toward the pole. A more rapid rate of earth spin would make the deflection force stronger, and the subsidence would occur closer to the equator than it does now. The location of the subsidence at 30° rather than 40° or 20° is thus the fortuitous product of a number of otherwise unrelated characteristics of the earth.

The great atmospheric dropout in the subtropical regions produces a distinct band of high air pressure about 30° south of the equatorial low. The zone of high pressure in the Northern Hemisphere is less regular than its southern counterpart because of the disruptive effect of different energy budgets over the continents and oceans.

The subtropical high-pressure areas are the sources for winds that blow both north and south along the ground. The winds that blow from a subtropical high toward the equatorial low complete a neat cycle, called the Hadley Cell, which superficially resembles a convection system (Fig. 3-7). High temperature at the settling end of the Hadley Cell, however, is an obvious difference between the tropical circulation and the closed convection systems described earlier. The high temperature is unusual, because it is associated with sinking air and high surface air pressure. In addition to demonstrating that simple convection models are not enough to explain the global wind system, these warm subtropical highs, with their settling air, are responsible for a distinct and important climatic type.

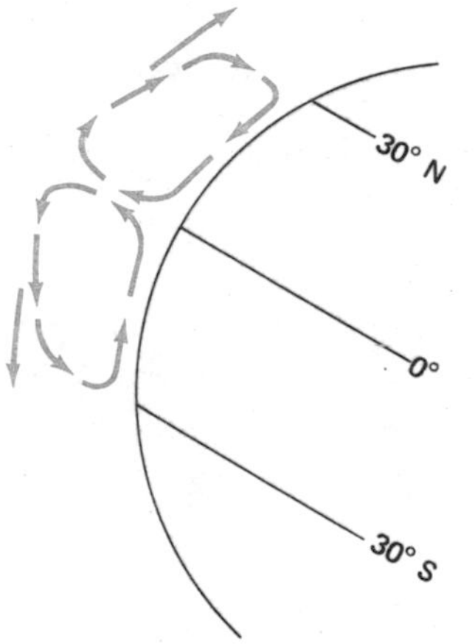

30° N
0°
30° S

FIG. 3-7. The Hadley Cell circulation of the tropical atmosphere. The movement of air is powered by the surplus of heat near the equatorial surface, but the poleward air currents are forced downward by dynamic processes, not thermal weight.

THE SUBSIDENCE TYPE OF CLIMATE

The weather in central Arabia does not make sense to a person who has just learned about atmospheric instability. From nearly overhead in a cloudless sky, the noon sun blazes down upon the shimmering rock. Aside from a few widely scattered puffs of dust, the air is still. Sparse and thorny shrubs endure the heat in silence. The expected upward rush of superheated surface air simply does not occur. Winds may blow from different directions, the air may be dusty or clear, thin clouds may form at different heights, the temperature may change dramatically as the hours and days pass, even fog may appear, but the stark aridity of the subtropical desert remains. In the heart of the subsidence type of climate, decades may pass between rains (Fig. 3-8).

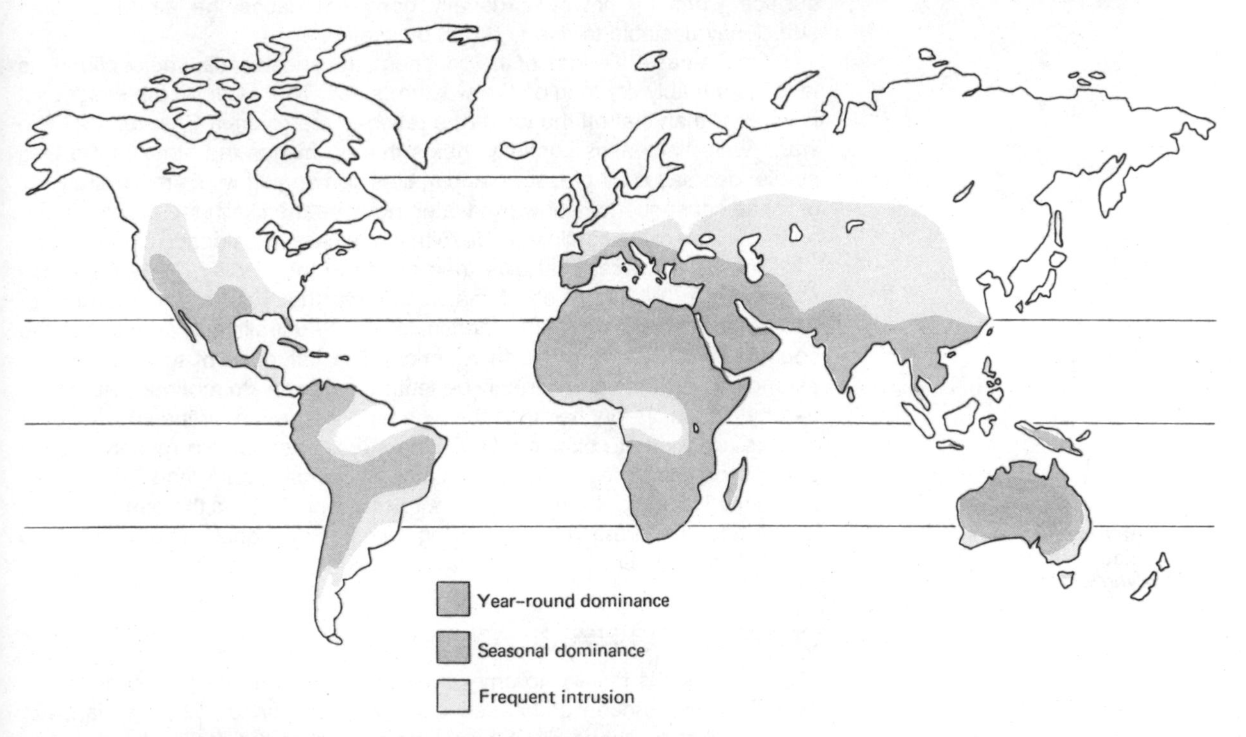

FIG. 3-8. The domain of the subsidence type of climate, the hot and dry consequence of sinking air.

The reason for the prolonged drought is found in the reliability of the rains in the nearby instability climate. The regular upward movement of unstable equatorial air drives the Hadley Cell circulation. Consequently, air is generally being pushed downward in the region of the subtropical high. Clouds are unlikely because the air is compressed, becoming warmer as it sinks. Compression also makes the air more stable and less likely to rise by itself. Meanwhile, additional air from the equator has moved in aloft, making it more difficult for the sinking air to turn around and to rise again. Finally, the outward spread of dry sinking air hinders the penetration of humid air from the oceans. The combination of warmth, dryness, stability, and outward wind makes rain almost impossible.

Hadley Cell circulation of winds, rising at the equator, moving to about 30°N or S, subsiding, and blowing back toward the equator along the ground

adiabatic warming increase of temperature of air as it sinks (see p. 34)

On the west side of a continent, the subtropical desert may extend right to the ocean shore. The low temperature of the water off the west coast is one key to the strength of the subsidence there. Cold water cools the air near the ground and makes the atmosphere stable. Complete saturation of the lowest layers of the air may produce a dense fog above the cold ocean. The fog offshore gives the illusion that rainfall is possible at any time, but the promise is rarely fulfilled. Even if the wind blows onshore, the warm land quickly heats the air, and the fog dissipates as the air temperature rises. The

upwelling rising of cold ocean-bottom water to the surface; most common off the west coasts of the continents in subtropical latitudes (see Fig. 13-10)

fog stratus cloud near the ground

surface warming, however, usually does not cause the air to become sufficiently unstable to rise and rain by itself.

On the eastern edges of the continent, the subsidence type of climate is less dependably dry than on the western sides. The water off the east coast is warmer than that off the west (the reasons are given in Chapter 13). The warm water increases humidity, heats the air, makes the atmosphere less stable, decreases air pressure, and makes converging winds more likely. All of these consequences of warm water increase the likelihood of rain on the eastern sides of the continents, despite the general subsidence of upper air.

The list of places that owe their aridity to the Hadley Cell subsidence reads like a "Who's Who" of major deserts: the Sahara and Kalahari of Africa, the Atacama of South America, much of Australia and Arabia, and the southwestern deserts of North America. The influence of the subsidence extends beyond the area directly beneath the downward motion of air. Many nearby places are dry because the high air pressure under the subsiding air causes dry winds to blow outward. The subsidence is also responsible for dry seasons in places such as southern California, Spain, and Nigeria, because the Hadley Cell circulation is not fixed in position on the earth. Rather, it migrates during the year, producing complex seasonal climates near the edges of its domain.

rainshadow dry area on the leeward side of mountains in an area of prevailing winds

COMPLEX OR HYBRID CLIMATES

The scene is familiar to any viewer of midnight movie reruns on television. The fever-ridden ambassador lies in a darkened room. A giant fan circles feebly overhead while potted palms wilt in the stifling heat. A few faithful friends wait for the monsoon to come and ease the fever. Immediately after the last commercial break, the cooling rains come, as inevitable as the march of the seasons.

thermal equator line around the earth following the area of greatest temperature at a given time

And well they should, because the sun drags a complex chain of related events along with it on its annual trek from 23+°N of the equator to 23+°S. The most direct consequence of the north-south movement of the zone of maximum incoming energy is a shift of the area of greatest surface heat. The motion of the "heat equator" lags a month or two behind the solar cycle, for the same reason that noon is not the warmest time of the day: the energy surplus continues beyond the time of maximum incoming radiation. The heat equator reaches its most northerly position in early August. Over the continents, the heat equator can shift as much as 15 or 20 degrees of latitude away from the global equator. Oceans have less responsive energy budgets than landmasses; over oceans the heat equator rarely wanders more than a few degrees from its yearly average position.

ITC (Intertropical Convergence) the zone of converging winds over the thermal low-pressure area near the equator

The belt of unstable air and low pressure obediently follows the heat equator north and south. The Hadley Cell circulation of the tropics therefore does not always force air downward precisely over the 30° lines of latitude. Rather, air may rise about 15°N in late August, move north and south at high elevations, and subside 40°N and 20°S. In February, the Hadley Cells consist of updrafts a few degrees south of the equator and a subsidence areas about 20°N and 35°S.

One result of the massive annual migration of temperature, pressure, and winds is a seasonal shift in the positions of the major types of climate. For instance, even though the rainy zone of unstable air extends over some 25 degrees of latitude at any one time, only a few degrees on either side of the equator are dominated by instability throughout the year (Fig. 3-9). The town of Kisangani lies nearly on the equator; it has high temperatures and abundant rainfall in every month. Mongalla (5°N) experiences unstable air and heavy rainfall for only three-quarters of the year. In December and January, when the sun is farthest south of the equator, the instability and subsidence zones are also south of their average positions, and Mongalla has a dry "winter" under the influence of subsiding air. The year is divided almost evenly between instability and subsidence climates at Gambela (8°N). Still farther north, the balance of dominance shifts to the subsidence type; at Khartoum (16°N) only two summer months show an appreciable influence of unstable air and converging winds. At Aswan (24°N), the reign of subsiding air is essentially unbroken throughout the year. The climates at Kisangani and Aswan may be described as "simple" or "pure" because all seasons are under the control of a single major climatic type. The other places have "complex" climates, seasonally affected by two or more different climatic regimes as the pressure and wind system of the earth shifts in position.

pure climate climate under the control of a single major atmospheric process

complex or hybrid climate climate seasonally under the influence of two atmospheric processes

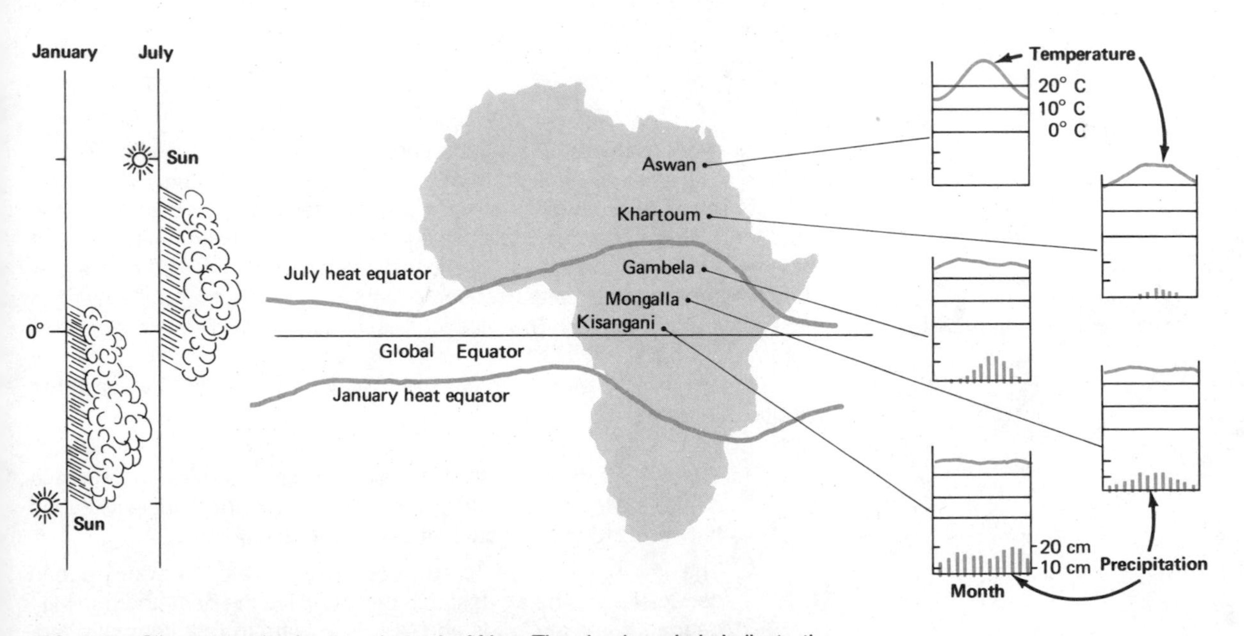

FIG. 3-9. Climate graphs for five places in Africa. The cloud symbols indicate the latitudes generally affected by the unstable air of the intertropical convergence during January and July. The curved lines on the graphs represent temperature; the vertical bars depict monthly precipitation from January through December.

In summary, the complex climate of a place about 15°N or 15°S is the result of two kinds of north-south motion in the atmosphere:

1. The high-elevation part of the Hadley circulation. These winds in the upper atmosphere carry air away from the zone of equatorial instability and push it down onto the subtropical regions. The subsiding air produces clear and dry conditions on the surface.
2. The seasonal shift of both instability and subsidence climates as they "follow the sun" on its annual migration from Tropic to Tropic.

Mediterranean climate summer-dry and winter-wet hybrid just to the poleward side of the subsidence climate

Other complex climates, similar in principle, occur in other regions of the world. Southern California is on the poleward side of the subtropical high. It has a mild winter with variable weather and frequent rains, caused by processes that will be described in the next chapter. Summer, however, is hot and dry, parched by subsiding air from the Hadley Cell.

There is another complex climate near the poles, where areas experience the cold and calm weather of inversion climate in the winter, and variable but generally warm conditions in summertime.

The key feature of all complex climates is alternation, not blending. The characteristics of a complex climate are not halfway between those of the basic types it lies between. The instability/subsidence hybrid is not partly humid all of the time. Rather, the climate is seasonal, with part of the year dominated by a wet type of climate, and the remainder under the influence of a dry one.

HURRICANES

Tropical depression, tropical storm, hurricane (or typhoon) degrees of intensification of a low-pressure area in tropical waters; the three stages are marked by an increase in wind speed and storm fury

Every year, some of the islands and coasts around the Gulf of Mexico are battered by violent circular storms, complete with high waves, damaging winds, torrential rains, and that strange calm center known as the "eye of the hurricane." Newspapers and television have made the destructive nature of hurricanes known to nearly all Americans. Most people are also aware that Gulf hurricanes have a limited season of activity, with the majority arriving in September and October. The distinct seasonality of hurricanes is a good clue to their origin.

A tropical storm with a circular wind flow can form only when three conditions are met:

1. There must be a center of low air pressure. Lows are common in the instability zone near the equator and in the middle latitudes between the subtropical highs and the poles.
2. The low must be centered above warm water. A warm ocean surface keeps the air unstable and supplies the heat and humidity needed to make clouds and rain. The warmth requirement effectively bars the mid-latitude oceans from the International Hurricane Builders Union. The cool ocean currents off the west coasts of the continents also fail to qualify for membership.

3. The low-pressure area over warm water must be at least a few degrees of latitude away from the equator. Hurricanes cannot form in the normal range of the instability climate, because the deflective "force" described in Figure 3-6 is not strong enough near the equator to bend the winds into a circular storm.

Coriolis effect tendency for objects in motion on the earth to be deflected from their intended path (see Fig. 3-6)

The September hurricane thus has its roots in June, when the sun is farthest north of the equator. The direct rays of the summer sun slowly raise the temperature of the water. Warmer water makes the air more humid and less stable. By late August, the instability climate reaches its northernmost extent over the oceans. At this time, an occasional area of especially low pressure may form (for reasons discussed on p. 79). A low-pressure area usually dies young if it is born close to the equator. Winds from all sides blow inward, filling the low with air and thus eliminating it. A low that develops farther from the equator is more persistent because incoming winds are deflected from their intended paths and form a spiral pattern around the center of low pressure (Fig. 3-10).

The motion of the wind at high elevations, where there is no friction, is almost a perfect circle around the low. Circular movement does not add much air to the center, and therefore the low pressure area can survive for a long period of time. The paths of the winds at the surface, however, cannot be perfect circles. The winds are affected by friction, so that they blow slightly inward as they go around the central "eye" of the storm (Fig. 3-11).

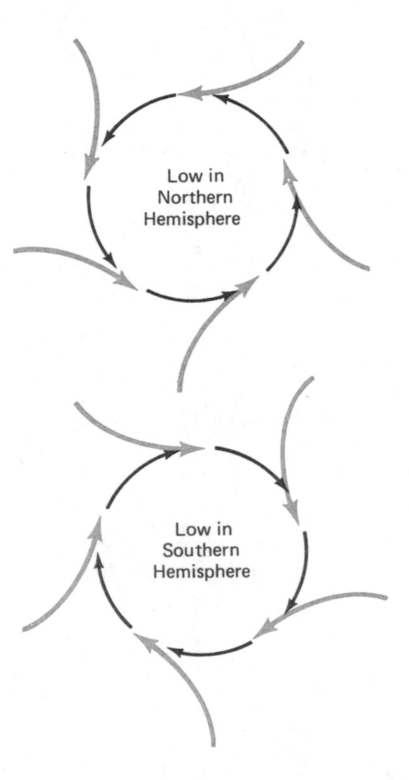

FIG. 3-10. Circulation of air around a low pressure area. Air moving in toward a center of low pressure anywhere except on the equator is deflected from a straight inward path and becomes a rotating mass of air. In the Northern Hemisphere the incoming air is deflected to the right and the cyclone rotates counterclockwise. In the Southern Hemisphere the deflection is to the left and the rotation is clockwise.

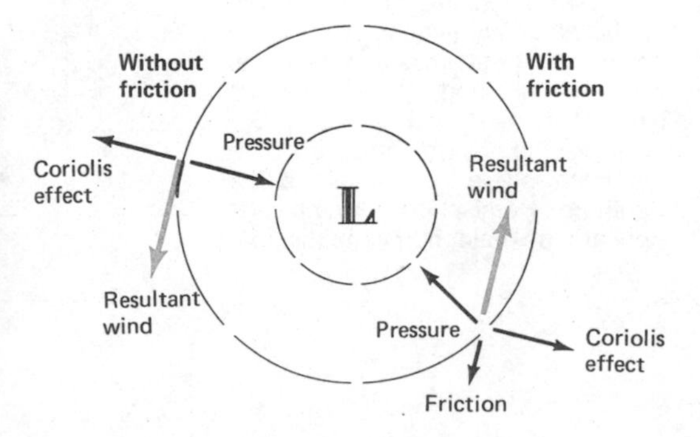

FIG. 3-11. The balance of forces around a low pressure area in the Northern Hemisphere. The Coriolis deflection (aided by inertia) pulls outward, whereas low pressure at the center draws air inward. Air moving toward the center goes "down" the pressure "hill" and will speed up. At greater speeds, however, the Coriolis deflection becomes stronger and the air tends to turn outward. If air moves outward, on the other hand, it must go "uphill" and therefore will lose speed. The decrease in speed weakens the deflective forces and allows the air to turn back toward the center. The result is a balance of forces that produces perfectly circular motion, provided there is no friction. Friction with the ground slows the air, weakens the deflective forces, allows the pressure forces to win, and thus produces an inward spiral of air near the surface.

The convergence of winds from all sides forces some of the surface air to rise and pushes upward on the entire rotating storm. The upward push turns the windstorm into a rainstorm, because the air is already unstable and almost saturated with water, and a slight lifting causes the air to lose moisture. The condensation of moisture adds heat to the storm, makes the air still more unstable, decreases the air pressure in the center, and drives the winds even faster.

latent heat heat released when water vapor condenses (see p. 10)

Like the convection system described at the beginning of this chapter, the events of a hurricane are lengthy to describe, but they occur simultaneously. Air is warmed, filled with moisture, drawn toward an area of low pressure, deflected from its path, forced to rise, and emptied of moisture, all at the same time. The process continues as long as the surface supplies heat and moisture to the air and the winds continue to blow in spiral paths. If the storm strays too close to the equator, it loses its spiral arrangement and develops a fatal case of disorganization. If a hurricane passes over dry land, it dies of moisture starvation. If it crosses too many mountainous islands or forested areas, friction with the rough surface interferes with the circular winds and weakens the storm. If a hurricane wanders too far from the equator and passes over cold surfaces, it slowly loses its energy, weakens, and becomes more stable, doomed to spend the rest of its days in the less glamorous career of a mid-latitude storm. On occasion, however, it may join forces with an existing storm and produce torrential rains and severe floods.

ventilation removal of air by upper-atmosphere winds, a process that makes a hurricane more intense

cyclone large storm organized around a low-pressure area in the middle latitudes (*not* a tornado)

In summary, hurricanes form under a particular set of conditions and may move in virtually any direction. Despite the apparent freedom associated with size and strength, hurricanes do not last long when they leave their preferred environment. Therefore the majority of hurricanes follow a few well-known paths and focus their destructive action on a relatively small fraction of the world (Fig. 3-12). The hurricanes that remain active for a long time generally migrate westward, urged toward the setting sun by a prevailing wind, another of the many effects of the Hadley subsidence.

hurricane track typical path of a hurricane toward the east coast of a continent

FIG. 3-12. Typical hurricane tracks on the earth. Hurricanes are virtually absent on west coasts, where the water is cold and prevailing winds blow offshore. September and October are the peak months for Northern Hemisphere hurricanes. In the Southern Hemisphere, March and April are the major hurricane months.

PREVAILING WINDS

Much of the air that sinks toward the ground in the subsidence areas must eventually make its way back toward the equator. The return trip is not a direct path, however, because the surface of the spinning earth is moving at a higher speed at the equator than at 30° N or S. Like a slow auto trying to merge into a faster lane on the freeway, a wind going toward the equator from the subtropics lags behind the surface beneath it. To an observer on the ground, winds that blow toward the equator appear to come from the northeast and southeast (Fig. 3-13). The easterly flow of air a short distance from the equator is the reason for the generally westward movement of hurricanes. (Winds are named for the direction *from* which they blow. A northeasterly wind therefore blows toward the southwest. The possibilities for confusion are obvious, and confusion about the wind on a sailboat could be quite disastrous, so sailors long ago adopted the arbitrary but rigid rule of naming winds for their origins.)

The Hadley Cell circulation is immense and persistent, a reliable driving force for the easterly winds between the subsidence and instability zones. In the days of sailing ships the tropical easterlies had a widespread reputation as safe, steady, and profitable, at least for the eight or nine months of the year when they were free of hurricanes. Christopher Columbus was undoubtedly aware of the easterlies. During the years before his famous expedition of 1492, he had traveled, observed, and inquired about the weather all along the west coasts of Africa and Europe. At 15°N, people reported that the winds were easterly: reliable, predictable, but somehow not very encouraging to someone who pictures an abrupt end to the world just beyond the western horizon.

trade wind wind blowing generally toward the west near, but rarely on, the equator

The information from England was more reassuring. At that latitude (more than 50°N), the prevailing winds blow toward the east (Fig. 3-14). The

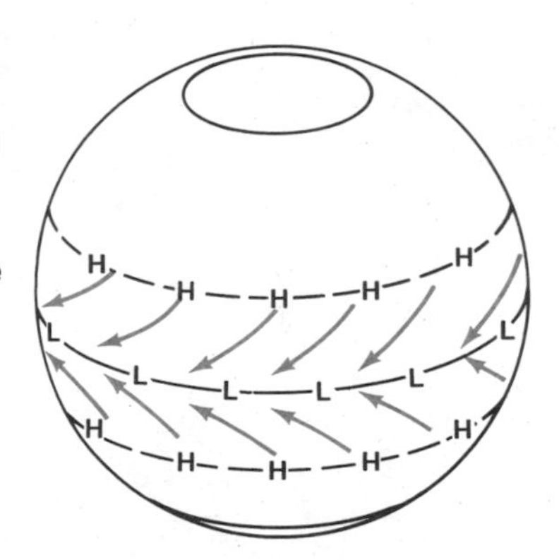

FIG. 3-13. The origin of the trade winds. Air blows from both north and south toward the band of low pressure near the equator. The rotation of the earth deflects winds toward the west and turns them into easterly winds.

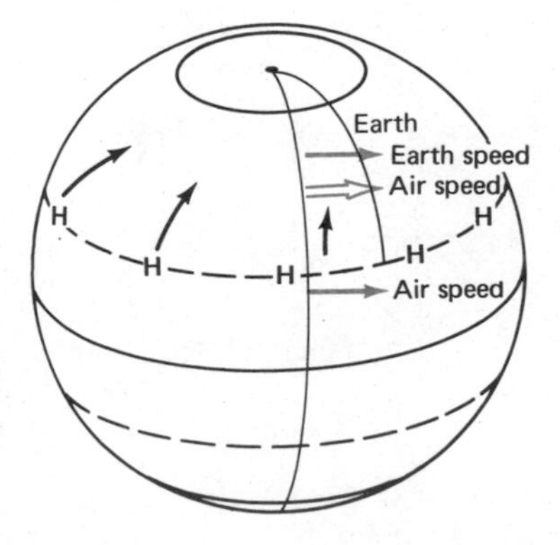

FIG. 3-14. The origin of mid-latitude westerly winds. Air at 30°N is moving from west to east at a rate of 1500 km per hour as the earth spins on its axis. When this air moves toward the pole, it goes over parts of the earth that are moving more slowly, and the air tends to outrun the earth beneath it. To an observer on the earth surface, it appears that the air is blowing from the west. Friction hinders the westerly wind near the ground. At high elevations, however, the wind may be traveling as much as 300 km per hour (200 mi/hr) faster than the ground below. These zones of exceptionally fast winds are known as jet streams.

westerly wind blowing to the east; most commonly between 40 and 50 degrees north and south of the equator

mid-latitude westerlies are less regular than the tropical easterlies, but at least they sometimes blow in the right direction for homebound European ships. Columbus probably reasoned that even if the easterlies did carry his ship to an edge of the world, the sailors could turn northward, go along the edge until they got to the 45th parallel of latitude, and then hitch a ride home with the favorable westerlies.

What Columbus apparently did not predict was the seasonal shift of the Hadley Cell and its associated surface winds. By October the sun is in the Southern Hemisphere. The equatorial zone of high temperature, low pressure, and rising air also moves to the south. The subtropical high-pressure area follows its instability parent and pushes the easterlies to the south. A sailing ship heading due west toward Florida in late summer would find itself deserted in the calm of the subsidence when the easterly winds moved south toward the equator. Fortunately (from a narrow European point of view), Columbus recognized the nature of the calm. He ordered a change of course toward the south in order to find the easterly winds and resume the westward journey to the Americas. Unfortunately (from an American Indian point of view), his ships entered the Gulf of Mexico without encountering a hurricane, a regional hazard unknown to Europeans.

WORLD WIND PATTERNS

A convenient way to summarize this chapter is to take a mental cruise south from Alaska, past Hawaii and the equator and on toward Antarctica. The nature of the world wind machine makes a journey directly to the south a long and difficult task in a sailboat. The ship must have another source of power to maintain its schedule, but this chapter is about winds, so the ship is instructed to pause every 800 kilometers to permit some research on the prevailing winds.

The results of the investigation fit neatly into a single graph of seasonal wind roses (Fig. 3-15). Four broad zones of prevailing winds are evident on the summary graph. Winds blow generally from west to east about 50 degrees from the equator and in the opposite direction near the 20-degree parallels. The equator and 30-degree lines generally experience calms or gentle breezes.

The seasonal shift of the wind zones is also apparent. A place at 40°N is buffeted by strong westerly winds in January. Then, when the sun drags the wind system into its northernmost position in July, the same place may experience many days of subsiding air and utter calm. A place about 35°S has the same pattern of winter westerlies and summer calms, but the seasons are reversed in the southern hemisphere and the calm summer arrives in January.

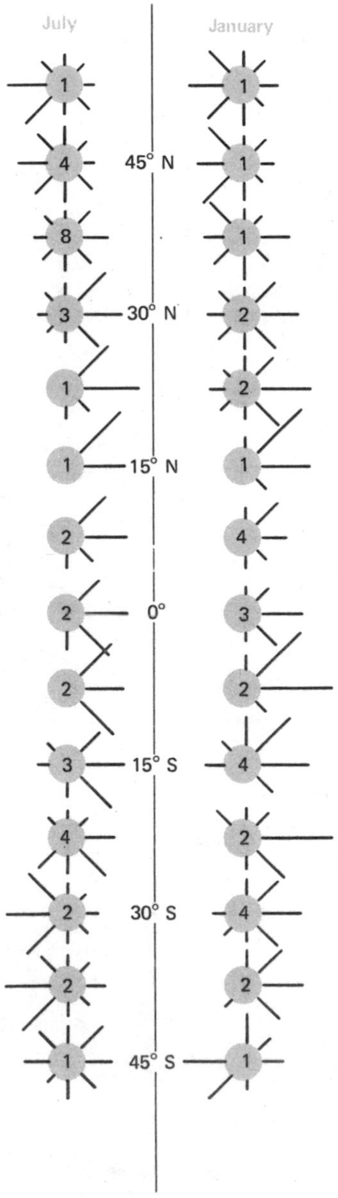

FIG. 3-15. Wind roses on the 150th meridian west of Greenwich. The length of a line pointing east from the center of a wind rose represents the number of days the surface wind blows from the east. The figure inside the circle represents the number of calm days at the measurement station.

The graph of wind roses adds a third fact about the world wind pattern, and the new data should make a person pause before confidently looking to the east for tomorrow's weather in Hawaii. Practically every wind rose on the chart proclaims that the word "prevailing" means "much of the time," not "all of the time." The world weather machine is big and strong and reliable, but it is not monotonously precise.

wind rose graph showing the fraction of time the wind blows from each major direction

CONCLUSION

Even the awesome dryness of the Mojave Desert can be broken by an occasional rainfall. It requires only the chance combination of several events, each of which is fairly unlikely by itself. A mass of air from the Gulf or Pacific Ocean may drift inward under the influence of a particular set of upper-air conditions. This air will become quite dry and warm by the time it climbs the mountains and crosses the desert. Nevertheless, it retains a faint but significant hint of its oceanic origin, in the form of relatively low temperatures and moist air in its upper layers. The second requirement for an Arizona storm is a weakening of the subsidence. Perhaps the instability storms were weaker than usual a few days ago over the central Pacific. The weaker equatorial storms put less air into the Hadley circulation, which means less air subsiding over Arizona. Finally, the Arizona soil must be warmed by a few days of unobstructed sunshine. At this point, the three factors begin to cooperate. The unstable lower air rises, and the upper air and weak subsidence do not resist the upward motion as well as normal. The result is a typical desert thunderstorm, usually brief but very intense. Half of the annual total of rain may appear in a single convectional storm that drenches one valley while leaving an adjacent one dry.

Hadley Cell tropical air circulation (see Fig. 3-7)

convectional storm showery rain caused by rapid rise of unstable surface air
(see Fig. 2-8)

The occasional storms are noteworthy precisely because they are so unusual. Conditions over southern Arizona usually consist of subsiding air, high air pressure, and little likelihood of invasion by humid air from the oceans. In a similar manner, the prevailing winds are largely responsible for the general weather conditions over many parts of the earth. Even in those areas where winds seem irregular, the horizontal movements of air still respond to the same forces and work toward the same goal. Winds on the earth are driven by the temperature inequalities that they seem to be trying to eliminate.

Chapter 4

Fronts: Battlefields in the Sky

Believe it or not, there's a pattern to the weather here.

Leo Ainsworth, National Severe Storms Laboratory

The atmosphere above the middle of the United States is a gigantic battleground. Cold and dry air from northern Canada fights for territory against warm and moist air masses from the Gulf of Mexico. Each side has allies from the Atlantic and Pacific Oceans which occasionally join the fray.

A typical mass of Canadian or Gulf air may be thousands of kilometers in diameter and tens of kilometers thick. Conditions inside an air mass resemble the weather where it originated. Weather analysts, using terminology developed during World War I, refer to the areas of conflict between air masses as fronts. Atmospheric fireworks such as tornadoes or blizzards occur along fronts; air-mass collisions produce much of the precipitation that falls in the middle latitudes.

air mass vast amount of air with horizontal uniformity of temperature and moisture

front zone of contact between unlike air masses

The key to successful weather forecasting in the middle latitudes is the ability to predict movements and future characteristics of air masses. The atmospheric battleground, however, is immense and complex. Forecasters are like reporters who receive scattered pieces of information from different parts of the battlefield and then must organize them into a pattern. It is not surprising that several different methods of forecasting have appeared, each one viewing the atmospheric battle from an individual perspective. This chapter will examine four different ways of forecasting, each able to shed its own kind of light on weather in the middle latitudes.

SYNOPTIC WEATHER MAPS

Trained observers note the local weather conditions every six hours at several hundred places in North America and the nearby oceans. Then they translate their observations into a standard numerical code and send the coded message to the national forecasting office. At the central office, analysts decode the hundreds of incoming messages and plot the information on a master weather map (Fig. 4-1).

synoptic map map of weather observations taken simultaneously at many different places

station model pictorial display of weather information on a synoptic map

The first step in analysis of a daily weather map is a search for general patterns among the observations. The search often begins with an attempt to determine what kind of air is over various parts of the country. On some days the battle lines are clearly drawn, with well-organized masses of air facing each other across sharply defined fronts. On other days the map is chaotic and patterns are difficult to discern.

Even on the easy days the analysts will always be somewhat uncertain about the pattern. Weather observers are competent but few in number, with hundreds of kilometers between some of the stations. Furthermore, the observations are already an hour or so old by the time they are ready for analysis. Radar images and information from satellites can assist the analyst in locating major storms and cloud groups, but the new techniques are not very helpful in updating maps of wind, dew point, or air pressure.

AIR-MASS IDENTIFICATION

In making their preliminary analysis, forecasters rely on a variety of clues to guide them in identifying air-mass territories.

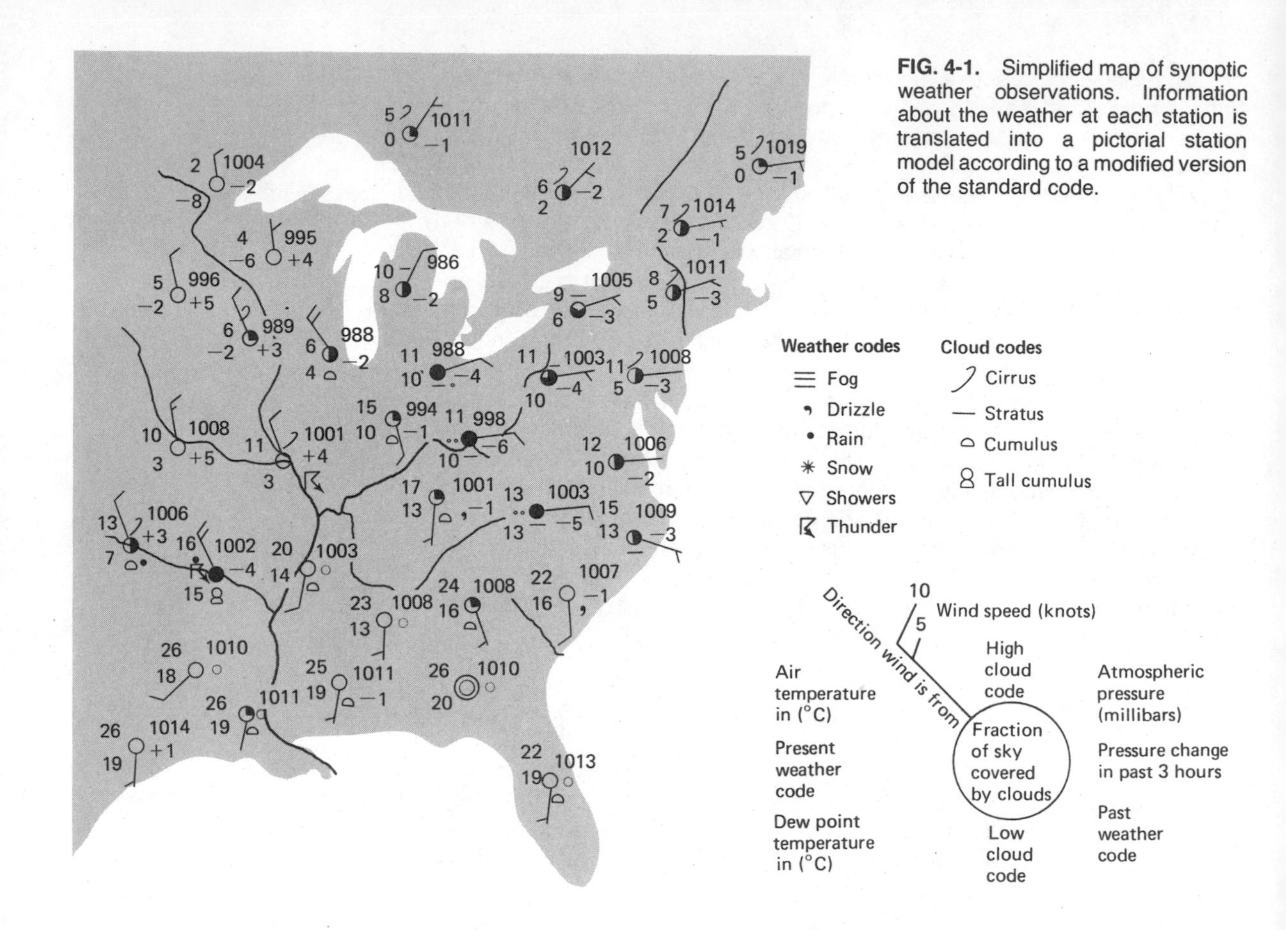

FIG. 4-1. Simplified map of synoptic weather observations. Information about the weather at each station is translated into a pictorial station model according to a modified version of the standard code.

pressure gradient difference in air pressure across an arbitrary unit of distance; the slope of an imaginary pressure surface

Air pressure. Air masses come together only in places with low air pressure. High air pressure pushes air outwards, away from the center of the high. Like marbles on a waterbed, air masses move away from high and toward low areas. The lower the air pressure, the stronger the inward pull and the more violent the interaction between air masses may be.

Coriolis effect deflection of moving objects from a straight line on the earth's surface
(see Fig. 3-6)

Wind direction. Drawn inward by low air pressure, winds usually blow from roughly opposite directions toward a front. A true head-on collision is rare, however, because local obstructions, the rotation of the earth, and other movements of air masses usually deflect winds from directly opposing paths.

Air temperature. An air mass that comes from the plains of northern Canada will be colder than one from the Mexican desert. Atlantic air masses are cooler than those from the Gulf of Mexico. The greater the temperature difference on opposite sides of a front, the more spectacular the effects along the front are likely to be.

Dew-point depression. The dew-point depression is the difference between the measured temperature of an air mass and the calculated temperature at which it would begin to form clouds and possibly rain or snow. Air with an oceanic background usually has a smaller dew-point depression than continental air; oceanic air therefore needs only a small amount of uplift and cooling to produce clouds.

dew point temperature at which a mass of air cannot hold any more water than it already has; air is saturated at its dew point; its relative humidity is 100 percent
(see Fig. 2A-1)

Stability. Observers at about half of the weather stations send balloons up to measure the characteristics of air at high elevation. The resulting vertical profile of an air mass offers one more clue to its origin (Fig. 4-2).

atmospheric stability likelihood of vertical movement of air; stable air tends to stay in place
(see p. 33)

FIG. 4-2. Balloon profiles and other characteristics of five typical air masses over North America. Which air mass is the least stable (see Appendix to Chapter 2)? Which one could deliver the most rain?

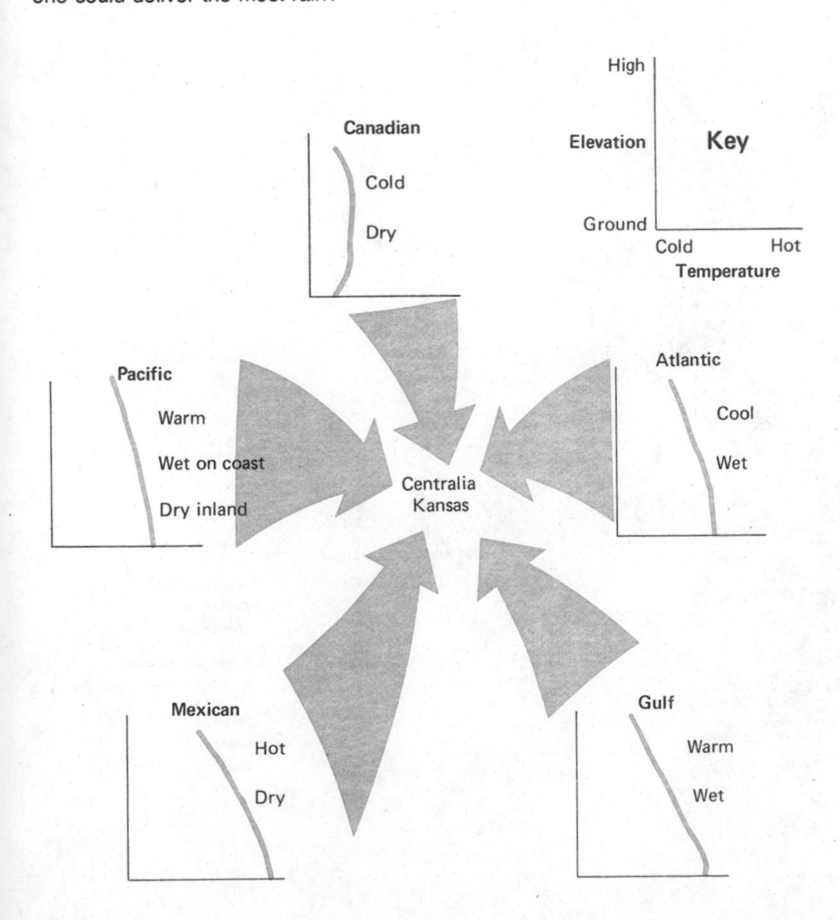

Cloud cover. Some air is forced to rise whenever air masses come together. If the rising air cools below its dew point, clouds appear. Clouds may form without an air-mass collision and fronts are not necessarily cloudy, but clouds and fronts are more often together than alone.

The weather analyst correlates these six kinds of information and tries to determine which areas are under the control of various air masses on a given day (Fig. 4-3). Once the fronts and air masses are mapped, the next step is to note the weather associated with each front.

FIG. 4-3. Synoptic map from Fig. 4-1 with the major air masses identified and the fronts located. Warm air is "winning" on the east side of the "low" over Lake Michigan while cold air advances on the west side, because the Coriolis effect causes winds from both the north and south to deviate to the right of a direct path toward the center of the low.

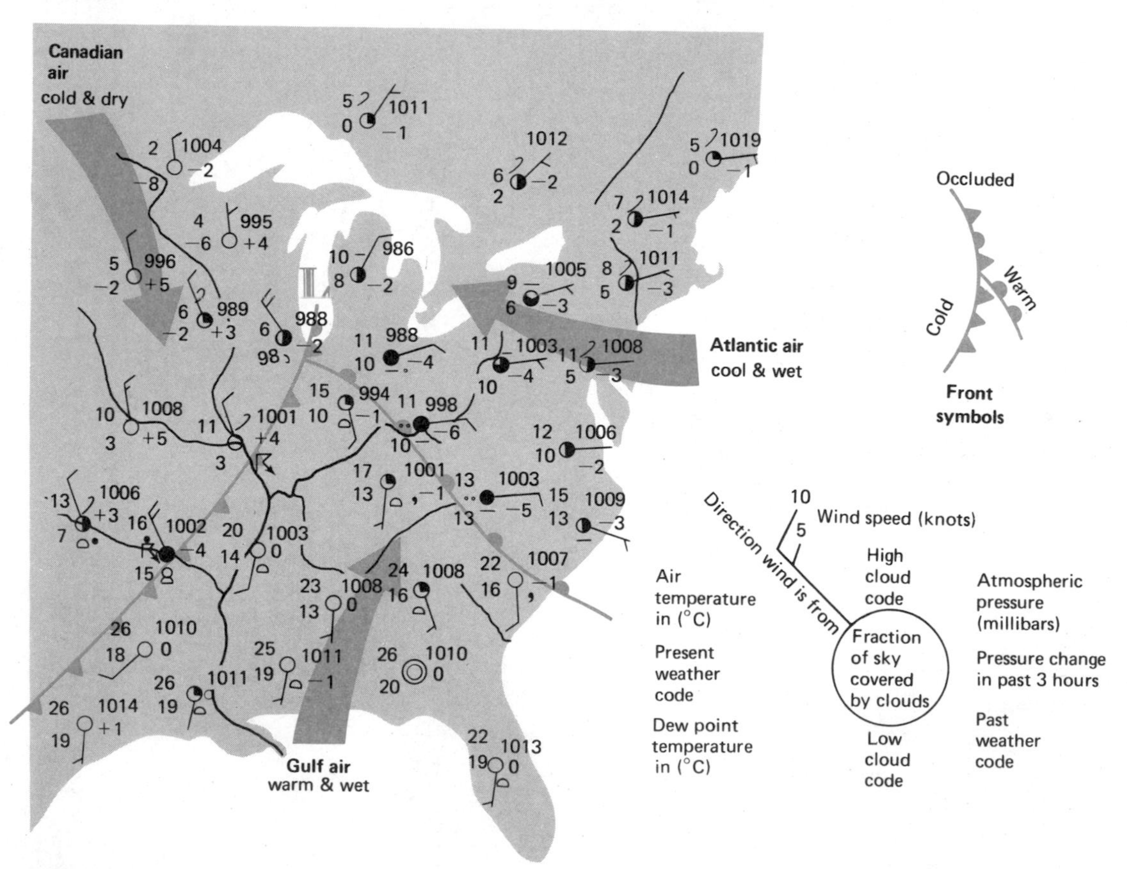

THE ANATOMY OF FRONTS

Fronts form when air high in the atmosphere begins to spread apart. Surface air rises into the emptying space above. Air from other regions then blows inward along the surface to take the place of rising air. The inward flow draws air masses with different temperature and moisture characteristics together (Fig. 4-4).

divergence spreading apart of air; upper air divergence is necessary for fronts to form near the ground

The line of contact seldom remains vertical when warm and cold air masses are pulled together. Cold air uses its weight to claim the space near the ground and to push the lighter warm air upward. The result is a slanting zone of air-mass contact that finally reaches the upper atmosphere hundreds of kilometers away from where it touches the ground. The steepness of the slanting front depends on how fast the air masses are moving and which side is "winning." Most fronts are named for the kind of air that is advancing into the territory of the other air mass along the front. Each type of front has characteristic features and directions of movement.

density weight per unit volume of a material; cold air is more dense than warm; dry air is more dense than humid air

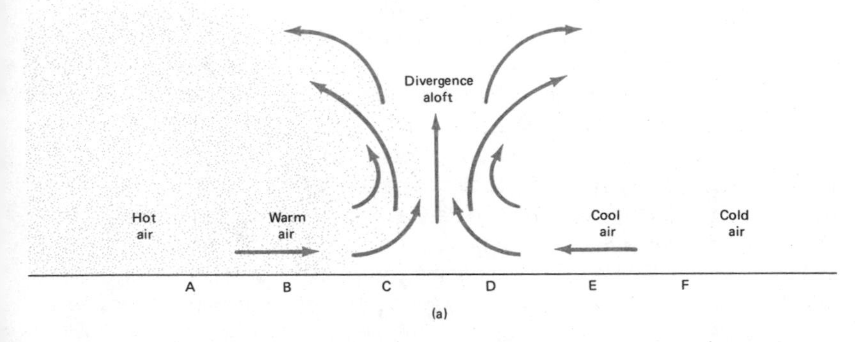

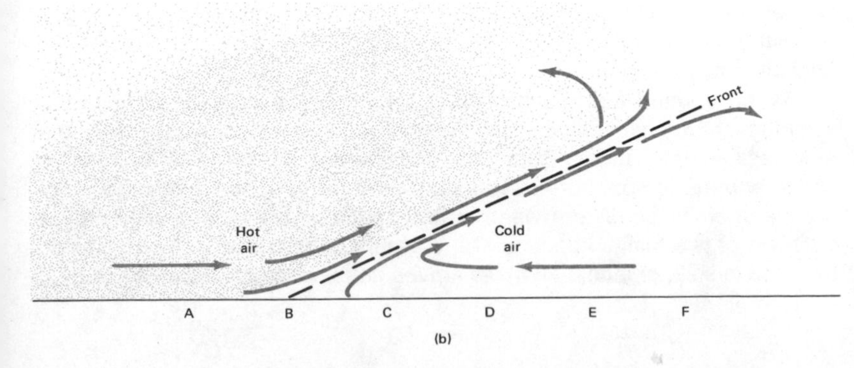

FIG. 4-4. Birth of a front. (a) Air is drawn upward into the upper-air circulation. Surface air masses with different origins come together in the vacated space. (b) The resulting area of abrupt temperature difference is the front. On a map, the front would appear at B, whereas in three-dimensional reality, the effects of the front—clouds, wind, perhaps rain or snow—can be felt at all places from B through E (see Fig. 4-5).

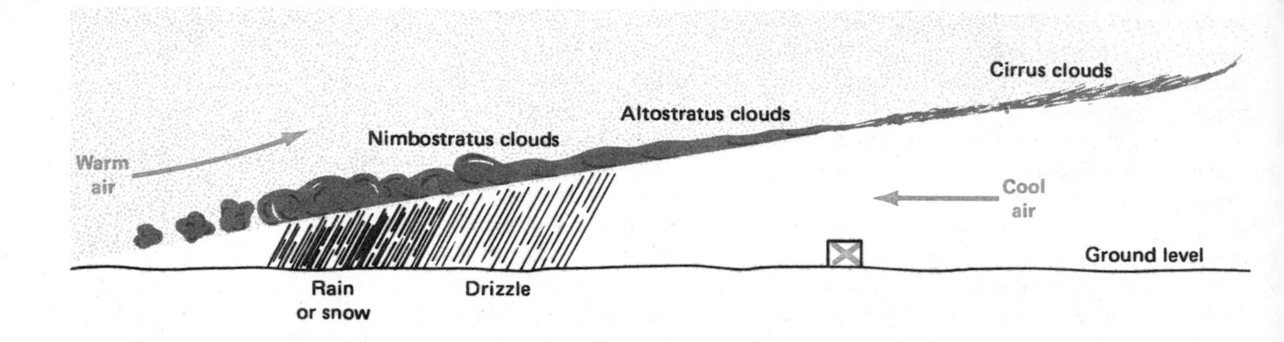

FIG. 4-5. A side view of a typical warm front. Friction with the ground slows the air near the surface, while the upper air advances unhindered. Therefore the front slopes more gently than it would if it were not moving. Place X can expect drizzle and then rain or snow in the future, because the front moves from left to right.

Warm fronts. Warm air masses usually win territory by gently ascending over cold air (Fig. 4-5). Thin wispy clouds often foretell the arrival of a warm front for an observer on the ground in eastern North America. The icy streamers are cirrus clouds that appear where warm air has risen so high that it has cooled below the freezing point.

cirrus clouds thin icy clouds at very high elevation; such clouds often are forerunners of approaching warm fronts

The cloud cover gets lower and thicker as the warm air advances. Air pressure decreases, partly because the column of air above the barometer now contains a thinner layer of cold dense air beneath a progressively greater amount of light warm air. Temperatures become more moderate as low clouds block sunlight and trap outgoing radiation. Drizzle and then steady rain are common near warm fronts in spring and fall. Warm fronts may bring unstable air and thundershowers in summer, whereas winter warm fronts often deliver deceptively gentle snowfalls that can accumulate to considerable depths and have a great effect on the local economy. The most dangerous warm front brings rain that freezes to form sleet as it falls through the cold air below the warm clouds.

stratus clouds low flat clouds that often produce drizzle or rain (see Fig. 2-9)

surface front place where sloping front between air masses touches the ground

When a warm front passes an observer on the ground, many of the symptoms of atmospheric conflict cease. The air pressure stops falling and may begin to rise. The sky clears, the surface wind shifts and blows from the south, and the temperature rises as the area falls under the domination of the warm air mass. Warm weather and southerly breezes usually prevail until one of two things happens: the pressure pattern changes and causes the wind to shift, or a cold air mass arrives and pushes the warm air away.

Cold fronts. Cold air masses usually win territory in a more brutal fashion, violently thrusting warm air up and out of the way. Rapidly advancing cold air creates a steep and abrupt front that arrives with little warning and passes more quickly than does a warm front (Fig. 4-6). The lack of warning is a problem for people on the ground, because cold fronts can produce heavy precipitation, thunderstorms, hail, or tornadoes, if the warm air is unstable enough. The storm activity may even extend ahead of a cold front, in the form of a squall line of thundershowers. These forerunners of the front may be even more violent than the storms directly on the front.

cumulonimbus clouds tall thunder clouds usually frozen at the top and often the sources of torrential rain or hail
(see Fig. 2-7)

Despite their fearsome reputation, cold fronts are not necessarily sudden and spectacular. A slow-moving cold front may have a gentle slope that resembles a warm front with its gearshift in reverse. If the warm air is stable, the clouds along a cold front will not be very tall and the precipitation will not be intense. If the warm air is not very humid, a cold front may be merely a temperature drop and wind shift, with few clouds and no precipitation.

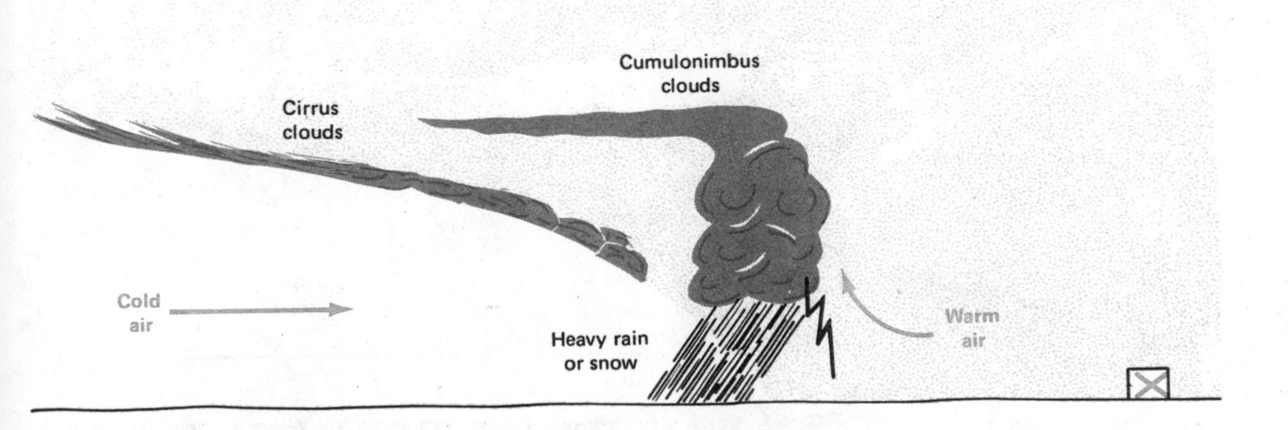

FIG. 4-6. A side view of a typical cold front. Friction with the ground slows the lower layers of the air, while the upper air advances unhindered. Therefore the front slopes more steeply than it would if it were not moving. The front moves from left to right, so that place X can expect storms followed by lower temperatures in the future.

Occluded fronts. Characteristically, North American cold fronts move more rapidly than warm fronts. A fast cold front from the northwest may overtake a warm front moving northeast. The result is an occluded front, a three-dimensional complex involving three different air masses (Fig. 4-7). To an observer on the ground, such an occlusion resembles a warm front that is followed by cold instead of warm air. Occluded fronts characterize the last stages of typical mid-latitude cyclonic storms (Fig. 4-8).

occlusion complex front caused when a cold front overtakes a warm front

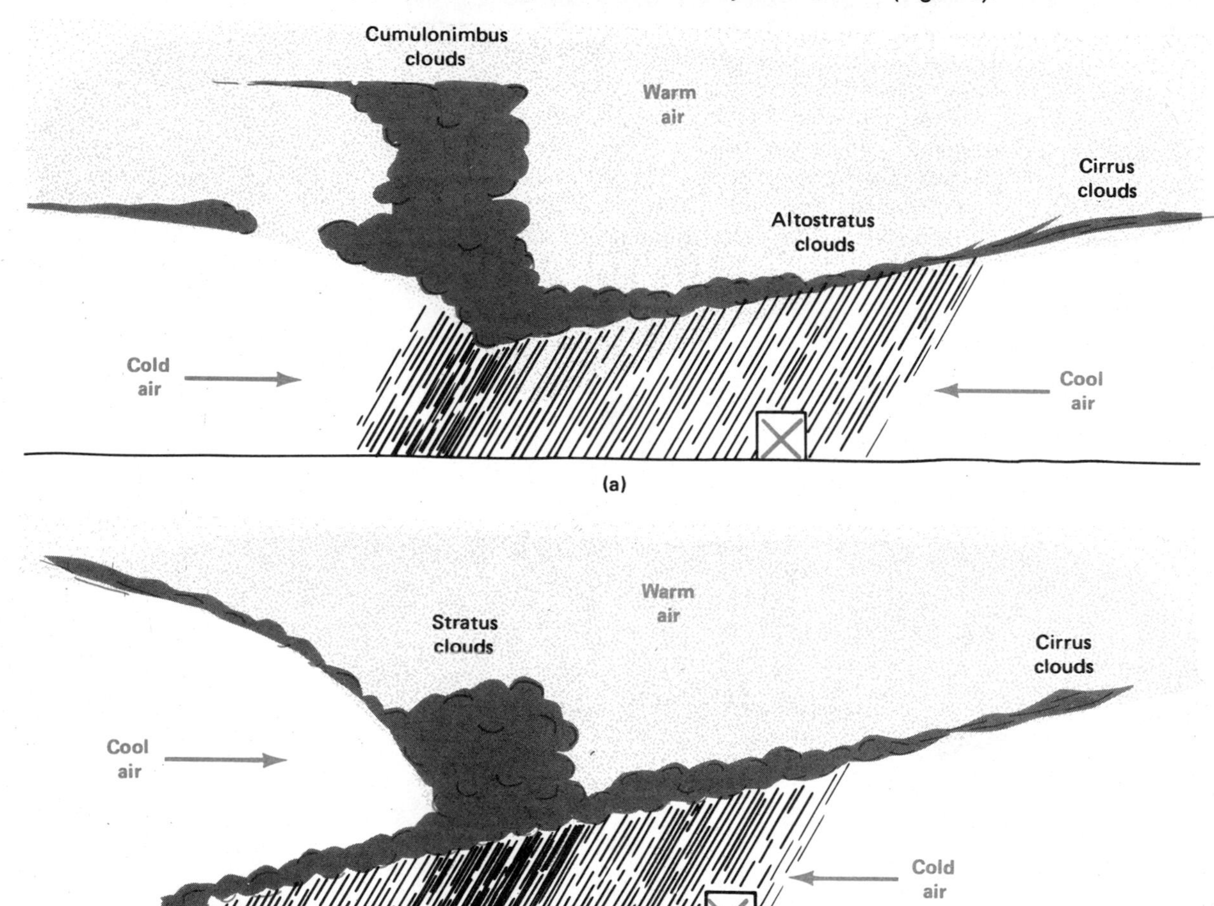

FIG. 4-7. (a) A side view of a typical occluded front. Warm air from the south is trapped between cold air from the northwest and cool air from the east. The warm air is the lightest of the three and rises above the other two. The cold air also forces the cool air upwards. The entire occlusion moves toward the right, so that place X can expect intense rain and then cold weather in the future. (b) A side view of a different kind of occlusion, one often seen near the west coast. Warm air is still trapped above a collision between cold and cool air, but the cold front rides up over the extremely cold air ahead of the warm front. The entire occlusion moves toward the right, so that place X can expect intense rain, then gentle rain again, and finally a rise in temperature.

cyclone traveling area of low pressure in the middle latitudes, often but not always a spiral storm with three or more air-mass fronts; some people also use the word to mean tornado

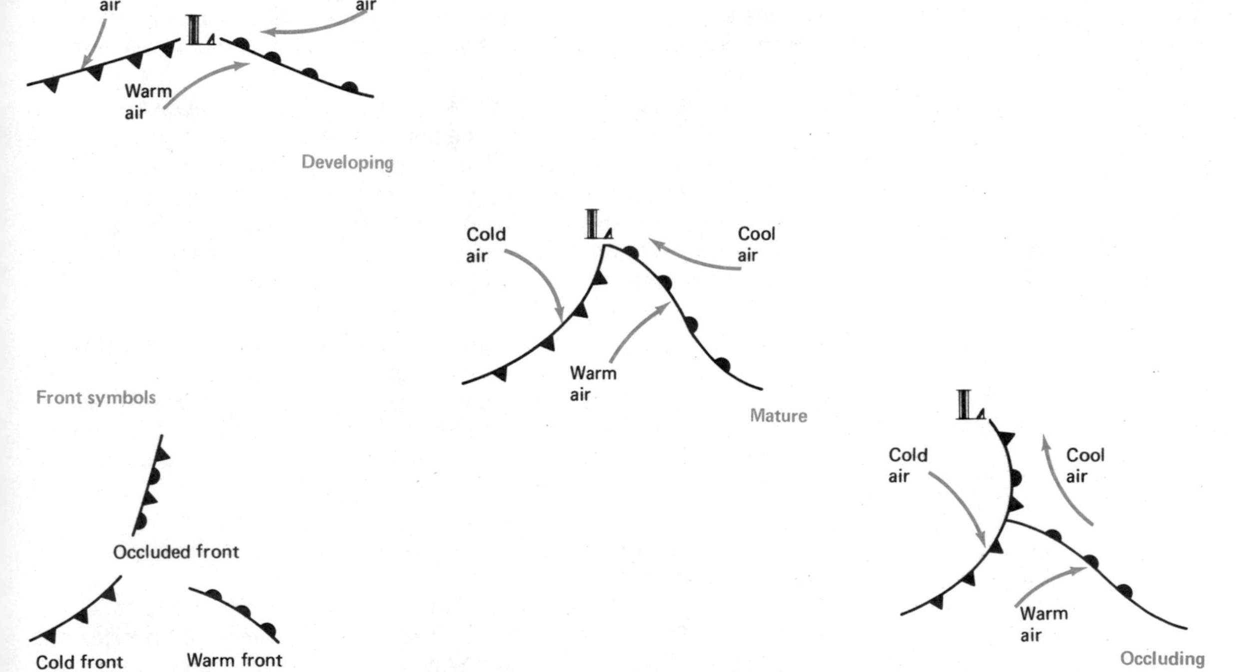

FIG. 4-8. Stages in the life history of a typical cyclonic storm over a continent in the middle latitudes. The low pressure area at the center usually moves slowly to the east as the drama unfolds. The bumps or points on a front indicate the direction the front is moving.

In summary, a front between air masses is a zone that usually has the following characteristics:

1. Fairly abrupt horizontal temperature change.
2. Comparatively low air pressure.
3. Inwardly blowing surface winds.
4. Surface air drawn upward by and into the upper-air circulation.
5. Increasing likelihood of condensation and precipitation from rising air.

SURFACE MAP FORECASTING

The forecaster trying to predict on the basis of surface weather observations is actually making rather sophisticated use of a very simple assumption: the atmospheric battle will continue as it has progressed today.

At its most simplistic level, the belief in continuity claims that tomorrow will be like today. If the wind today is southerly, tomorrow's wind will blow from the south. If the air pressure is going down, it will continue to decrease. If today is warmer and sunnier than yesterday, then tomorrow will be even more warm and sunny.

pressure tendency change of air pressure, up or down, during a period of time

It should be apparent that any atmospheric trend can be reversed if the tide of battle shifts to favor another air mass. Pressure tendency, wind direction, temperature, and cloud cover all change when a front passes. The movements of fronts are the heart of a more sophisticated concept of atmospheric continuity. The revised version of the forecaster's creed says two things: present conditions will continue unless a front arrives, and fronts will continue to move in the future as they have in the past. If a cold air mass is advancing across the plains at 40 km per hour (25 mi/hr), it will continue to move at that speed. A place 200 km ahead of the front can expect its present weather to persist for about five hours and then change rapidly.

The success of this kind of forecasting depends on reliable estimates of front speed. Suppose that a front took two hours to cover the 100 km (60 mi) between stations A and B. The front passed point B at 3 P.M., and point C is another 100 km down the road. Therefore the front should arrive at point C about 5 P.M. The forecaster in this example has a legitimate reason to worry if the front has not checked in by 6 P.M. A late front indicates that the assumption of constant motion is not valid.

To protect themselves against late or early fronts, forecasters attach still another bit of fine print to their creed. The additional paragraph specifies that the pressure along the front will stay constant even though a front may change speed or direction. The new rule enables forecasters to detect changes in frontal motion even at places that are not near the front. Suppose that the pressure at station B dropped steadily until it reached a low point of 992 mb just as the front passed. If the pressure at point C is 994 mb and decreasing at a steady rate of 1 mb every hour, a forecaster could predict that the front is about two hours away. If, however, the pressure goes down at the rate of only 0.5 mb each hour, the front is probably moving only half as fast and should not arrive until 7 P.M.

intensification or deepening decrease in the air pressure at the center of a low; deepening lows usually move more slowly but produce more violent weather than constant lows

Unfortunately, the assumption of constant air pressure along the front is also unwarranted. The continuity creed must get still another qualification, one that allows pressure to fall or rise in the future as it has in the past. If the pressure in the low has been decreasing, it will continue to decrease, the conflict between air masses will probably become more intense, and the center of the battle will move more slowly (Fig. 4-9). If a new low-pressure area appears, it will begin to draw air masses toward it and possibly create new fronts.

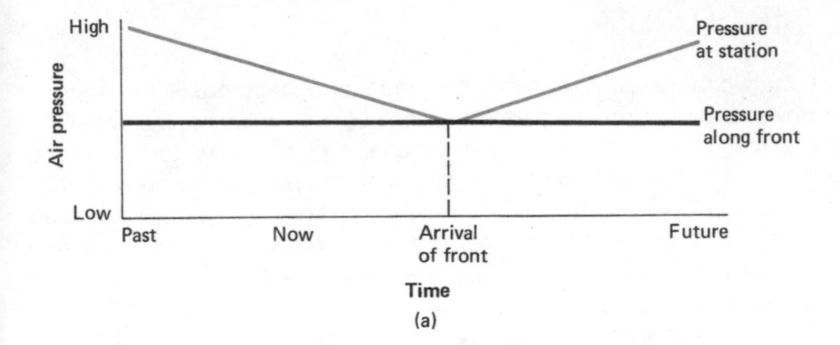

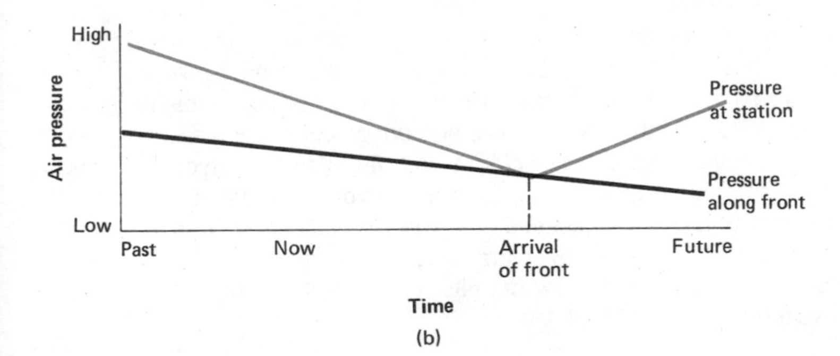

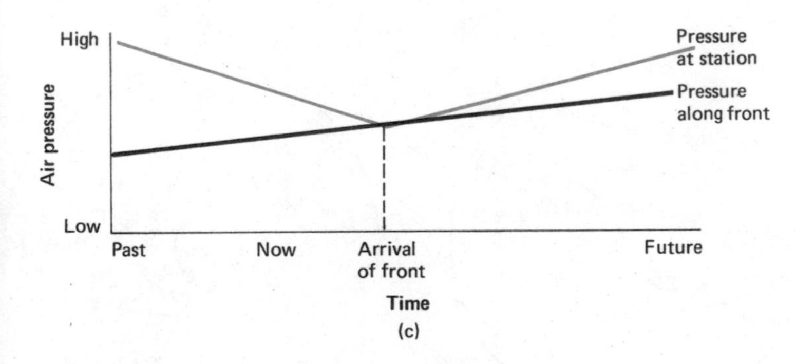

FIG. 4-9. Forecasting the time of arrival of a front by comparing the pressure changes at the station and at the front: (a) If the pressure along the front is constant, the time until the front arrives (in hours) is equal to the millibars of pressure difference between the station and the front, divided by the number of millibars of pressure decrease per hour at the station. (b) If the pressure along the front starts to decrease, the front will probably go slower and arrive later than if its pressure remained constant. (c) If the pressure along the front starts to increase, the front will probably go faster and arrive sooner than if its pressure remained constant.

Each new level of sophistication complicates the analysis of a surface weather map and multiplies the time required to make a forecast. Some kind of computer assistance becomes necessary if the forecast is to appear before the predicted events. Computer forecasting, though, is more effective if an entirely different approach is used. Surface forecasters usually stop at a rough estimate of changes in the pressure pattern. Computer analysts make their weather predictions by using measurements from the upper atmosphere and a procedure called dynamic forecasting.

Coriolis effect tendency for objects in motion on the earth's surface to be deflected from their intended paths (see Fig. 3-6)

westerly wind blowing from the west (see Fig. 3-14)

long wave snakelike meandering of the upper-air westerlies in the middle latitudes

index cycle regular progression from nearly straight to meandering and back to straight westerly flow in the upper atmosphere

Dynamic forecasters start with the idea that surface pressure patterns, wind flows, and fronts are formed and guided by movements of the upper air. Winds at high elevations in the middle latitudes originate as updrafts near the equator, veer to the right as they move toward the pole, and blow generally from west to east. The flow is very irregular, however, because a pure westerly wind is unable to carry surplus equatorial heat to the poles and therefore does not accomplish the primary objective of the wind system of the earth.

The upper-air flow continually changes its behavior because it cannot satisfy all the demands on it at the same time. Like a politician keeping a low profile when the people demand reform, a westerly wind allows heat to build up on the equator side. As the temperature difference increases, the air begins to swing north and south in wider and wider arcs, and the north-south motion carries heat toward the poles and brings cold air toward the equator. Running to the capitol with complaints and returning with favors may satisfy the constituency, but it tires the politician. When complaints subside, the back and forth movement ceases. The wind likewise tends to shift back to a simple westerly flow whenever the temperature difference is reduced. Of course, a simple westerly flow permits heat to build up again, which starts the cycle once more (Fig. 4-10).

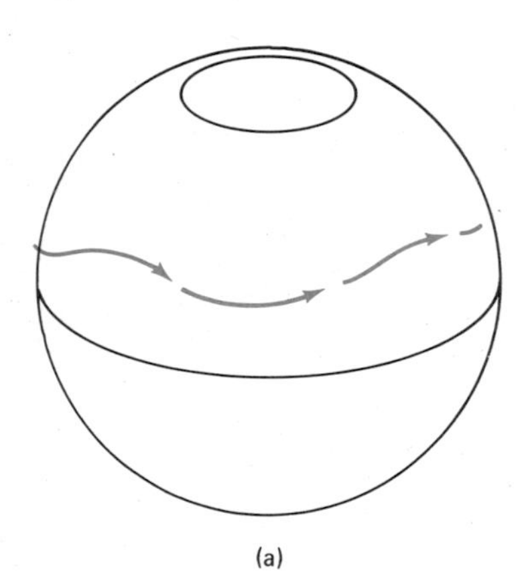

(a)

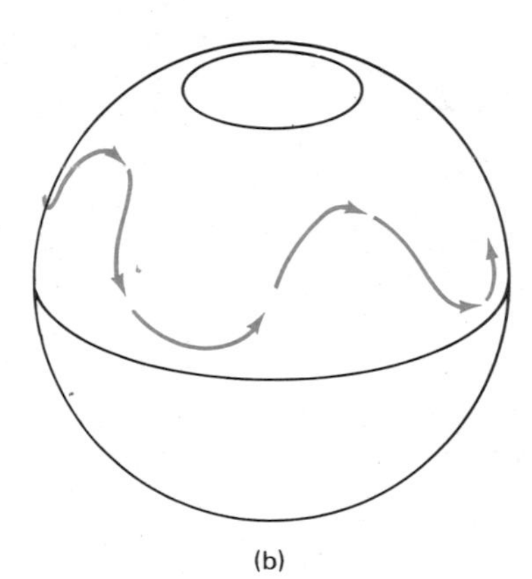

(b)

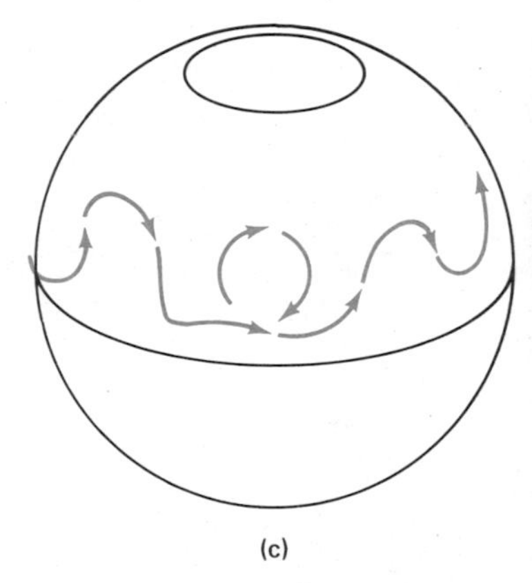

(c)

FIG. 4-10. The Index Cycle. The upper-air westerly flow alternates between a fairly direct west-to-east flow and a widely meandering stream that swings far to the north and south on its way toward the east. "Jet stream" is the common term for the fast-moving core of the westerly flow. On some days the westerly flow is complicated, with more than one jet. The cycle usually passes through three distinct stages: (a) High Zonal Index. The westerlies blow as a smooth flow from west to east. Interchange of tropical and polar air is minimal. (b) Low Zonal Index. The westerlies blow as a series of wide arcs from north to south. Considerable interchange of tropical and polar air occurs. (c) Cutoff. The westerlies take a shortcut across one of the loops. Weather on the ground changes rather abruptly near the cutoff.

The meandering wind in the upper atmosphere crowds together and spreads apart at predictable intervals along its path; and the meandering creates areas of low and high air pressure on the surface (Fig. 4-11). When air high above the United States turns to the left, it usually speeds up and spreads out. Divergence of upper air always causes a void that draws surface air upward. The result is a zone of low pressure near the ground. Air masses come in toward the low, and fronts appear.

jet stream core of the upper-air westerly, a narrow band of wind moving at very high speed

divergence spreading apart

When the upper-air westerlies in the Northern Hemisphere turn to the right, on the other hand, they slow down and crowd together. Upper-air convergence always produces sinking air and high pressure near the surface. The usual result of this temporary subsidence is clear and dry weather.

convergence coming together

Obviously, knowledge of the present and future position of winds in the upper atmosphere would aid a forecaster immensely. Fortunately, the meandering westerlies respond to temperature and pressure conditions that can be measured and mapped. Scientists of the dynamic forecasting school have developed an elaborate set of equations to predict the future state of the upper atmosphere. Their equations and computers enable them to forecast specific weather sequences a few days in advance and general patterns a month or two ahead.

numerical method analysis of weather by computer modeling of the atmosphere according to a handful of fundamental equations

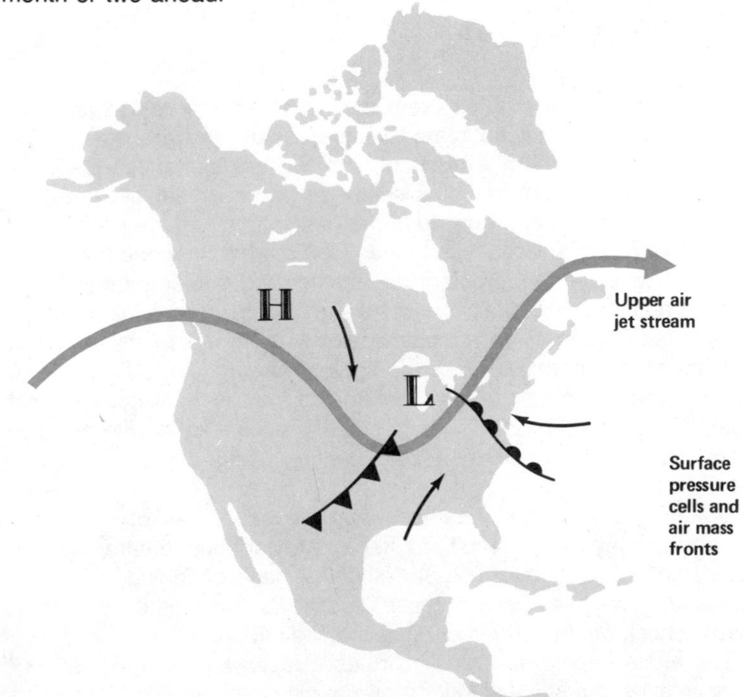

FIG. 4-11. The relationship between upper-air meanders and surface air pressure cells in the Northern Hemisphere. Leftward-turning air spreads apart, pulls surface air upward, and creates an area of low pressure near the ground. Rightward turns cause the upper air to converge, subside, and produce high pressure on the ground. When an upper-air meander is cut off, areas of low and high air pressure near the ground lose their reason for existence and simply disappear.

Dynamic forecasting is a powerful tool, but it may be near its limit of improvement , and its predictions may not get much more reliable than they already are. In the first place, random errors in the initial measurements have a cumulative effect. A local fluctuation of half a degree of temperature at the time of observation may mean an error of 20 degrees in a 10-day forecast. Second, upper-air meanders sometimes shift abruptly. It is easy to predict that a series of huge north-south loops will sooner or later change to a smoother flow by shortcutting one or more of the loops. The problem is deciding when the change will take place. The upper-air flow will continue to support the present surface weather until the shift occurs, and then the pattern of high and low pressure on the surface changes quickly and the entire tide of battle shifts with little warning.

cumulative error error that becomes worse in time

One way to compensate for the uncertainty is to "hedge bets" like a gambler. Forecasters hedge against the weather by using a procedure appropriately called a Monte Carlo simulation. They instruct their computer to make a series of forecasts and to allow each one to change slightly by inserting a few chance variations at appropriate places. Then the forecasters compare the computer printouts and develop a composite prediction.

ANALOG FORECASTING

A third method of weather prediction emerges from the experience of forecasters who have seen many weather maps and are struck by the similarities as well as the differences. No two weather maps are exactly alike, but certain general patterns seem to reappear year after year. The repetition offers a way to predict tomorrow's weather: find the map for some past day that resembled today, and then look at what happened on the next day. Until recently, the problem of storing and remembering 100,000 weather maps made it virtually impossible to give this idea a fair test. By the time the forecaster finally finds a past analog of today's map, it already could be the day after tomorrow. Modern computer indexing and retrieval systems permit forecasters to try the historical-analog approach. Moreover, analog forecasters can begin to draw on a steadily accumulating file of satellite photos and radar images which often contain useful clues of the coming weather.

weather analog previous weather map that resembles the map for today

geostationary satellite one whose orbit keeps it above the same spot on the earth.

Unfortunately, no two weather maps are exactly alike. Furthermore, similar maps may be the result of different atmospheric features moving in different directions; therefore predicting on the basis of analogous situations is not always reliable. Despite these limitations, the method serves as a valuable check on the results of other forecasting procedures.

The analog approach also illustrates a fascinating fringe area of the forecaster's life, the study of long-term climatic changes. For example, suppose that a particular upper-air pattern occurs on eight days of a typical March. Also, suppose that a search of historical maps reveals a larger than usual number of March days with this particular flow pattern in years that were exceptionally dry, such as 1933 or 1976. Even if the reasons for the relationships between March wind and July rain were unknown, the statisti-

teleconnection linkage between atmospheric conditions in widely separated places

cal correlations among these patterns would have enormous agricultural and economic value. For this reason, hundreds of scientists spend long hours searching past weather records for clues that might enable forecasters to predict the general weather far into the future. Among the more promising hints of things to come in the United States are pressure patterns near Alaska, rain in southern Mexico, wind directions and temperatures in northern Canada, snow cover and ice persistence in Wisconsin, and temperatures of the sea surface in the North Pacific (Fig. 4-12).

temperature anomalies deviations from the expected temperature for particular place or time

The use of analogs may be better for general forecasts of the distant future than for specific forecasts of the near future. Predicting if next October will be wetter or drier than normal may actually prove to be easier than foretelling the likelihood of rain next Thursday.

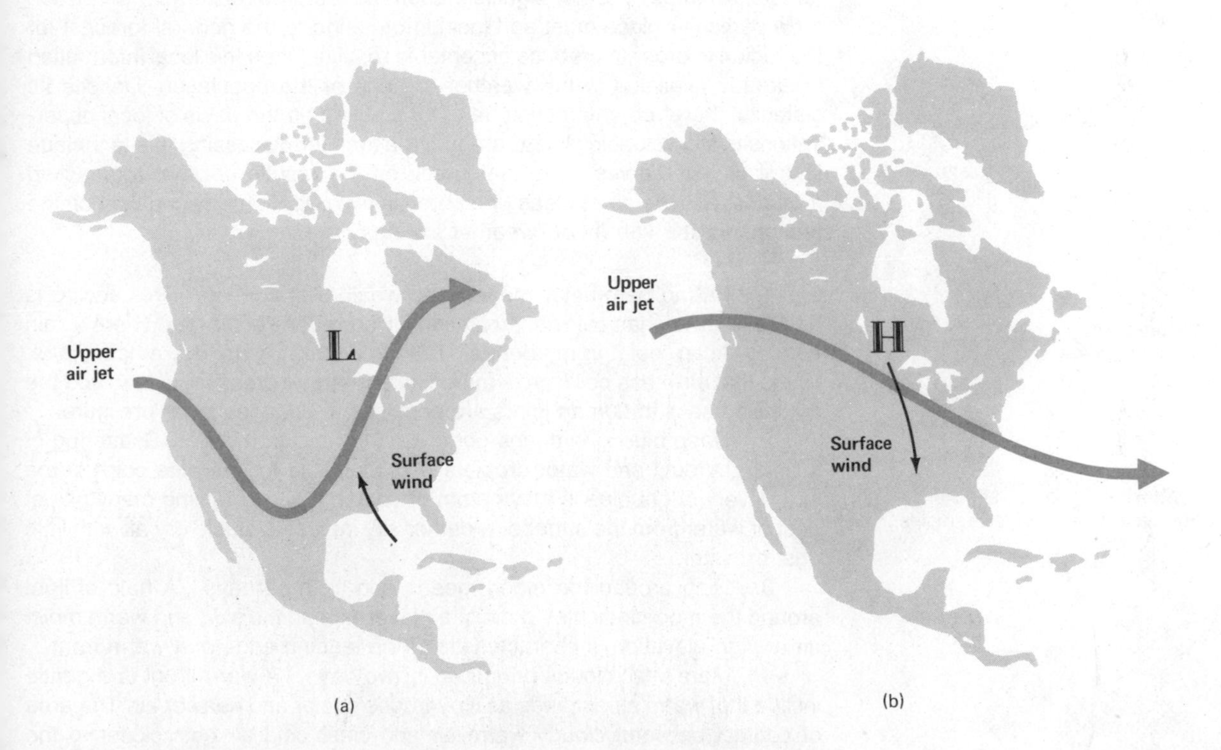

FIG. 4-12. A suggested explanation of the role of Pacific Ocean sea-surface temperatures in guiding the upper-air flow. (a) When the middle of the ocean is comparatively warm and temperatures near the California coast are low, the winds dip southward near the West Coast and then curve northward over the middle of the United States. Oregon and Washington are rainy and the East Coast is also wetter than normal. (b) When mid-ocean temperatures are low and the water off the coast is warm, the winds are likely to be north of their average position. The West Coast then tends to be warm and dry while the interior of the continent is less rainy and often cooler than normal.

COLLOQUIAL FORECASTING

The final method of predicting the weather is the oldest of all. It consists of the accumulated wisdom of generations of people who have lived in a particular place, experienced its weather, crystallized their experience as a set of "weather sayings," and handed the lore on to their children.

Anyone living in a city such as Buffalo, Chicago, Cleveland, Detroit, Duluth, Milwaukee, or Toronto can testify to the inadequacy of regional weather forecasts. The celebrated "Lake Effect"—the land/water differences in humidity, temperature, pressure, and wind—superimposes a local pattern on the general weather conditions around these cities. Complications of general patterns also occur near seacoasts, mountain ranges, urban areas, forests, swamps, or other significant surface features. A weather prediction for a particular place must add local information to the general forecast for the region in order to produce acceptable results. Often this local information is already available in the weather sayings of the population. Despite its antiquity, therefore, the technique of forecasting on the basis of local observations is still valuable. There are two reasons for discussing this technique last: it usually serves as a final check on the results of other forecasting methods, and the other ideas presented in this chapter can shed light on the reasons for the validity of "weather sayings":

heat island zone of high temperature caused by a city (see Fig. 1-14)

1. "Falling barometer means bad weather is coming." This saying is basically a mechanical modernization of an older sentence, "There's rain coming; I can feel it in my bones." Both speakers recognize, at least intuitively, that air-mass collision is more likely when air pressure is low, and the fluid-filled sacs in human joints are sensitive to changes in air pressure.

barometer instrument to measure air pressure

2. "Deep blue sky means good weather for three days." Scattering of sunlight by dust and water droplets is responsible for the blue color of the sky. A very bright blue is a symptom of unstable air containing many bits of dust or water from the surface. A darker sky indicates stable dry air with little dust or water.

scattering diffuse reflection of light from tiny dust or water particles (see p. 10)

3. "Ring around the moon means rain in three days." A halo of light around the moon indicates a moist air layer high in the sky, and warm moist air at high elevation is characteristic of the leading edge of a warm front.

4. "Mare's-tail clouds bring rain in two days." A warm front is a gentle incline that warm air ascends as it overrides cooler and heavier air. The area of contact between cloudy warm air and clear cold air gets closer to the ground as a warm front approaches and passes over a place. Cirrus clouds appear later than halo images but before rain-bearing stratus clouds.

5. "Backing wind means cold weather coming." A veering wind is one whose direction changes in a clockwise direction, blowing from the east, then from the southeast, and then the south. A veering wind shift indicates that the tide of battle is changing in favor of air from tropical areas. A backing wind shifts counterclockwise from easterly and suggests that cold polar air is winning (Fig. 4-13).

veer to turn clockwise in the Northern Hemisphere (counterclockwise in the Southern Hemisphere)

back to turn counterclockwise in the Northern Hemisphere (clockwise in the Southern Hemisphere)

FIG. 4-13. In the Northern Hemisphere a veering wind shifts in a clockwise direction and indicates that the center of low pressure has passed by on the north side of the station. A backing windshift is counterclockwise and indicates that a low is passing by to the south. The rotation of the two wind shifts is reversed in the Southern Hemisphere.

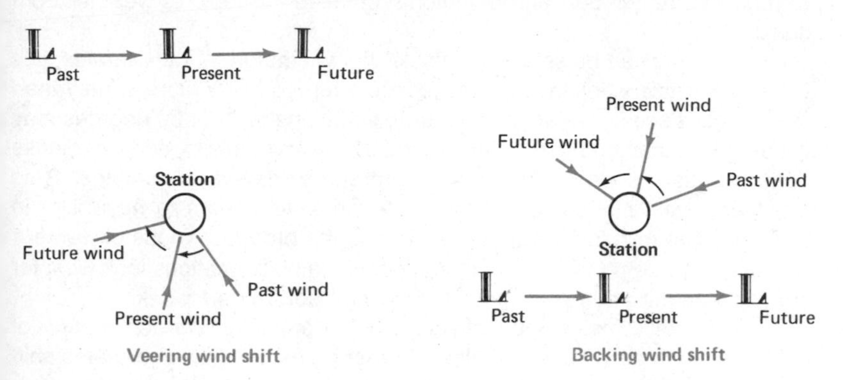

6. "Storms without warning are gone before morning." Afternoon instability and rapidly moving cold fronts are the primary causes of storms that give little advance notice of their arrival. Convectional storms and cold fronts both pass rather quickly, though they may be intense while they last.

7. "Red sky at morning, sailors take warning." Red sunrises and sunsets indicate abundant dust and moisture in the air. Dusty and humid but cloudless air can only get more unstable as the day wears on and the ground gets warmer. The companion statement, "Red sky at night, sailor's delight," is not so trustworthy because the color may foretell an approaching cold front.

8. "Green sunsets let you travel tomorrow." Pale gold and green colors in the sky at sunset are symptoms of dry air and prophets of fair weather to come.

9. "Valley winds have stopped; there's bad weather on the way." On calm nights, air loses heat by radiation on exposed hilltops and slides gently downslope into the valleys. The downhill wind is a result of weak local forces; it ceases to blow in the face of larger atmospheric forces associated with major storms.

10. "Chicago rules make Frisco fools." This gem of geographic wisdom is perhaps the most worthwhile statement about local weather patterns.

No saying on this list is universally reliable; the list is at best a tiny sampling of the more widely heard bits of North American weather lore.

model transferability applicability of a theoretical model in an environment that differs from the one in which the theory was developed

THE FRONTAL TYPES OF CLIMATE

People complain about daily weather mainly in the frontal types of climate. In the rest of the world, seasons may be distinct, but days and weeks blend into each other. Is it hot outside? So what? It has been for as long as anyone can remember, and monotony removes the topic of weather from discussion.

If anything can be said to dominate the weather in frontal climates, it is variety. A single week in a frontal climate often contains more variety than many places see in a year. Temperature variations of 20 or 30 degrees may occur in less than an hour. Shifts in wind direction, humidity, and cloudiness can be equally sudden. The common thread of day-to-day change holds together a vast number of places which otherwise seem to have little in common. The mild fogginess of Ireland and the bronze dryness of western Kansas may seem utterly distinct, but people in both locations look west for coming weather and fully expect it to change before next week.

cyclone belt zone of varying but usually low air pressure between the subtropical high and the polar region

Hadley Cell tropical circulation of air that rises near the equator and sinks about 30° N and S
(see Fig. 3-7)

jet stream zone of fast winds high above the ground

The frontal climates appear in areas of low pressure between sources of cold and warm air (Fig. 4-14). Like all other types, the frontal climates shift their positions of dominance as the sun makes its annual north-south migration. In June when the sun is north of the equator, the instability and subsidence zones are also north of their average positions. The tropical circulation is the driving force of the prevailing westerlies of the middle latitudes, so that they also shift northward. The typical summer position of the meandering upper-air flow in the Northern Hemisphere is about the latitude of Minnesota, though it may loop as far north as Alaska or as far south as Texas. Beneath the upper-air loops are their offspring, the surface cells of low and high air pressure. Southern states are controlled most of the time by tropical air masses, stable in the southwest and less so in the southeast.

inversion climate calm and cold weather associated with a cold surface
(see Fig. 2-13)

In winter the entire system shifts southward. The average position of the westerly flow is over Arkansas, though it occasionally swings north to Michigan and south to Mexico. Minnesota and the Dakotas spend most of the time under the icy control of polar air. Ohio and Iowa have occasional warm days; Tennessee and Oklahoma experience frequent warm spells. Georgia and Texas have extremely variable weather, and Florida gets a cold wave or two before the returning sun of springtime drives the weather machine northward again.

Superimposed on these seasonal shifts are two topographic variants of typical frontal climate:

specific heat energy needed to raise the temperature of one gram of a substance one Celsius degree
(see Fig. 1-5)

The maritime variant. The high specific heat of water keeps ocean temperatures fairly steady. A migratory air mass from tropical or polar areas loses its hot or cold character and gains moisture if it is forced to pass over a long stretch of ocean. The result of the extensive modification of the air is a moderate and moist mass, not very different from other oceanic air masses. A front between two oceanic air masses is a collision between marshmallows, generally devoid of the thundery, blustery excitement associated with fronts between strikingly different masses of air. A month of maritime frontal climate may bring 25 centimeters (10 in.) of rain, but the precipitation will fall

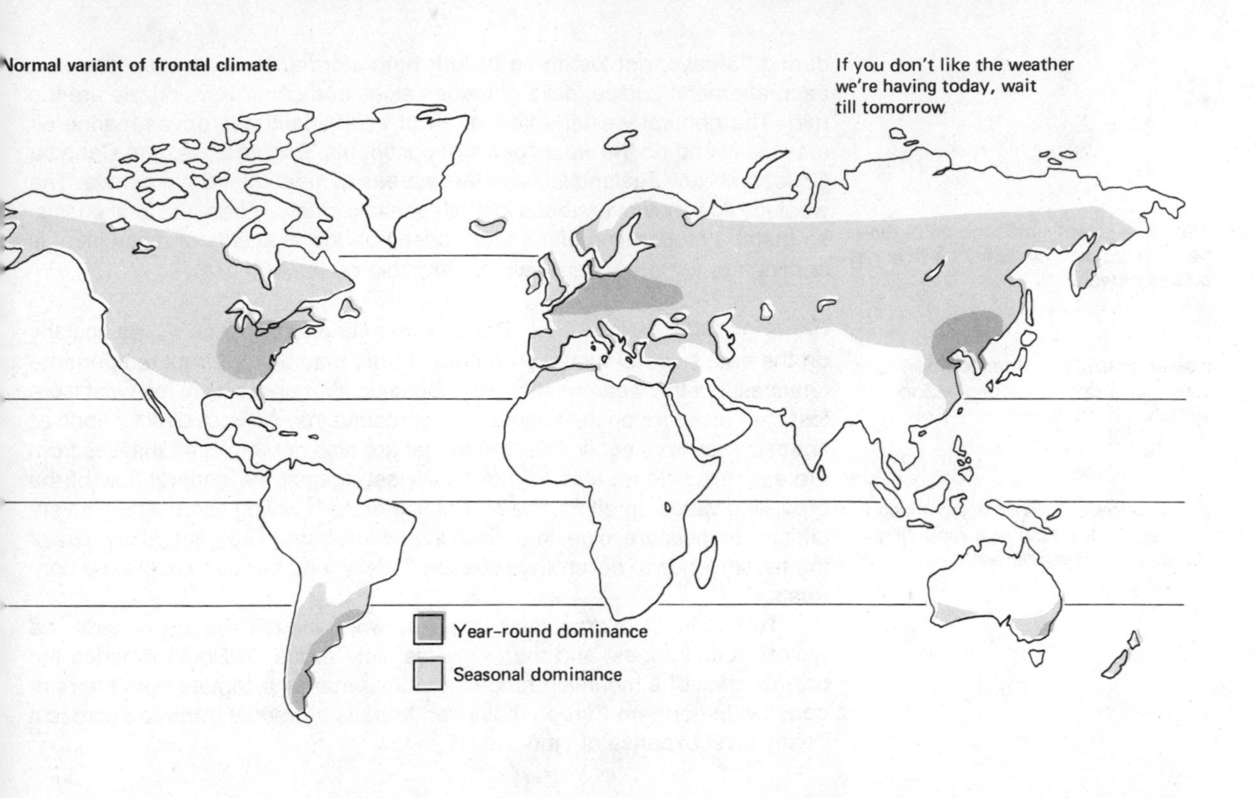

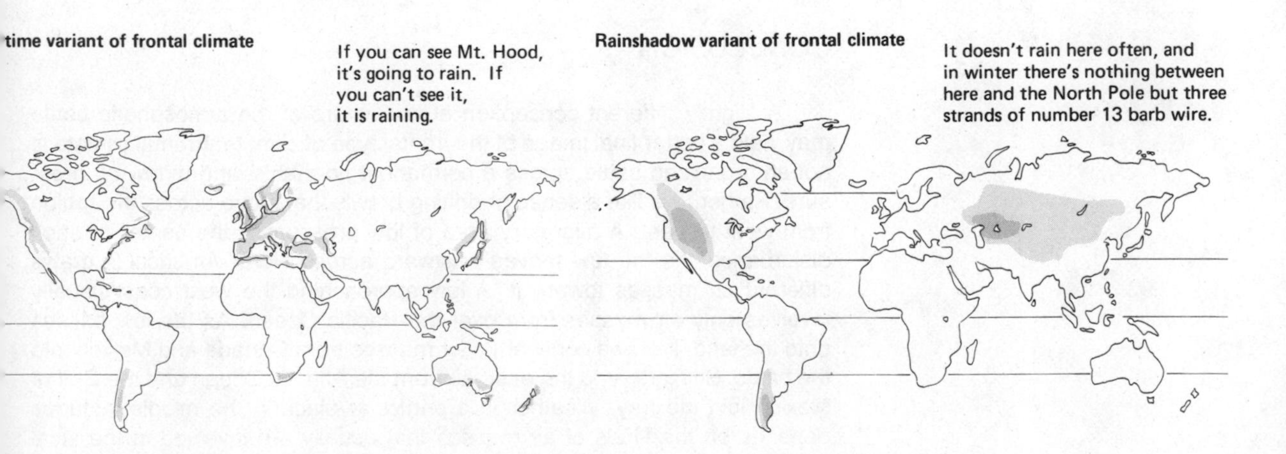

FIG. 4-14. The domain of the frontal type of climate, the changeable consequence of air-mass invasions.

during 25 days, not just three or four hard storms. Nor will a few showers each afternoon suffice; days of leaden skies and continuous drizzle are the rule. The general westerly flow of air in middle latitudes drives marine air masses inland on the west coasts of continents. England, western Canada, Chile, and New Zealand all have large areas of maritime frontal climate. The westerly flow, however, tends to push oceanic air away from the east coasts, so that the moderating effect of an ocean on the east side of a continent is confined to a narrow band, almost invisible on a world map.

maritime effect moderation of temperature caused by contact with large bodies of water

The rainshadow variant. Deep in the interiors of continents, particularly on the east sides of mountain ranges, fronts may bring abrupt temperature changes, but the weather stays dry. Oceanic air masses from the west have lost their moisture on their way over the mountains. Air from directly north or south comes from continental areas that are also dry. Moist air masses from the east are seldom able to move far west against the general flow of the prevailing winds. In effect, the mountains and prevailing winds make it very difficult for moisture to get into these inland locations. The general dryness of the terrain tends to accentuate seasonal, daily, and frontal temperature contrasts.

continentality wide temperature range and dry conditions associated with inland locations
(see p. 18)

rainshadow area of dry conditions on the leeward side of a mountain in an area of prevailing winds

Terrain helps to draw the borders between the rainshadow deserts, the typical frontal zones, and the foggy maritime areas. In South America the narrow crest of a mountain range separates maritime climate from the rainshadow. In northern Europe, however, there is a gradual transition across a nearly level expanse of land.

CONCLUSION

A slightly different conception of the nature of the atmospheric battle may yield a better final image of the frontal type of climate. Frontal climate is not an organized battle across a permanent no-man's-land of low air pressure. Rather, it is like a series of drifting brawls that move across the region from west to east. A migratory area of low pressure is the center of each disturbance. As the low moves eastward across North America, it draws different air masses toward it. A low approaching the west coast usually involves only air masses from over the Pacific Ocean. As the low passes onto the land, it draws continental air masses from Canada and Mexico into the battle. Still farther to the east, air from the Atlantic Ocean and the Gulf of Mexico join the fray. Weather in a particular place in the middle latitudes depends on the kinds of air masses that usually are involved in the local atmospheric battles.

APPENDIX TO CHAPTER 4
Common Weather Patterns

No two weather maps are exactly alike, but some basic patterns seem to recur year after year. The meandering flow of the fast westerly winds in the upper atmosphere is governed by temperature patterns, pressure patterns, and the geometry of mountains and shorelines. The arrangement of continents, ocean basins, and mountain ranges imposes a distinct regularity on the paths of the jet stream. Consequently, certain parts of the world frequently act as breeding grounds for cells of low or high air pressure. The east side of the Rocky Mountains is a particularly productive factory for the manufacture of migratory centers of low pressure.

This appendix presents seven common weather maps of North America, selected to illustrate typical weather patterns and likely consequences. Some variants of these typical patterns have regional nicknames among weather forecasters, who breathe a bit easier when they see a familiar pattern begin to develop in a time-tested manner. A particular pattern, however, does not invariably follow the sequence described in this appendix. For example, this page was written in Wisconsin on the day after the passage of an "Alberta Clipper," a familiar wintertime Alberta Low that swings south through the Dakotas before moving eastward. Alberta Clippers ordinarily do not cause an outpouring of extremely cold weather, but the temperature outside was −21 °C (−7 °F), very near the record low for the date. The writer, in his unheated cabin, is keenly aware of the probabilistic nature of the information on these maps; the reader is asked to remember that these general patterns may have many exceptions.

The Alberta Low. Cells of low air pressure frequently form just east of the Rocky Mountains and then move eastward along one of a number of paths. An Alberta Low that forms in winter and moves southeast usually brings snow and strong winds, because it pulls a tropical air mass deep into the continent and then drives a mass of polar air into the humid Gulf air. Places that ordinarily experience cold polar air are treated to a few mild days and then are hit with an often violent snowstorm. Summertime Alberta Lows usually move directly eastward, bringing rain to Canada but leaving most of the United States in undisturbed tropical air.

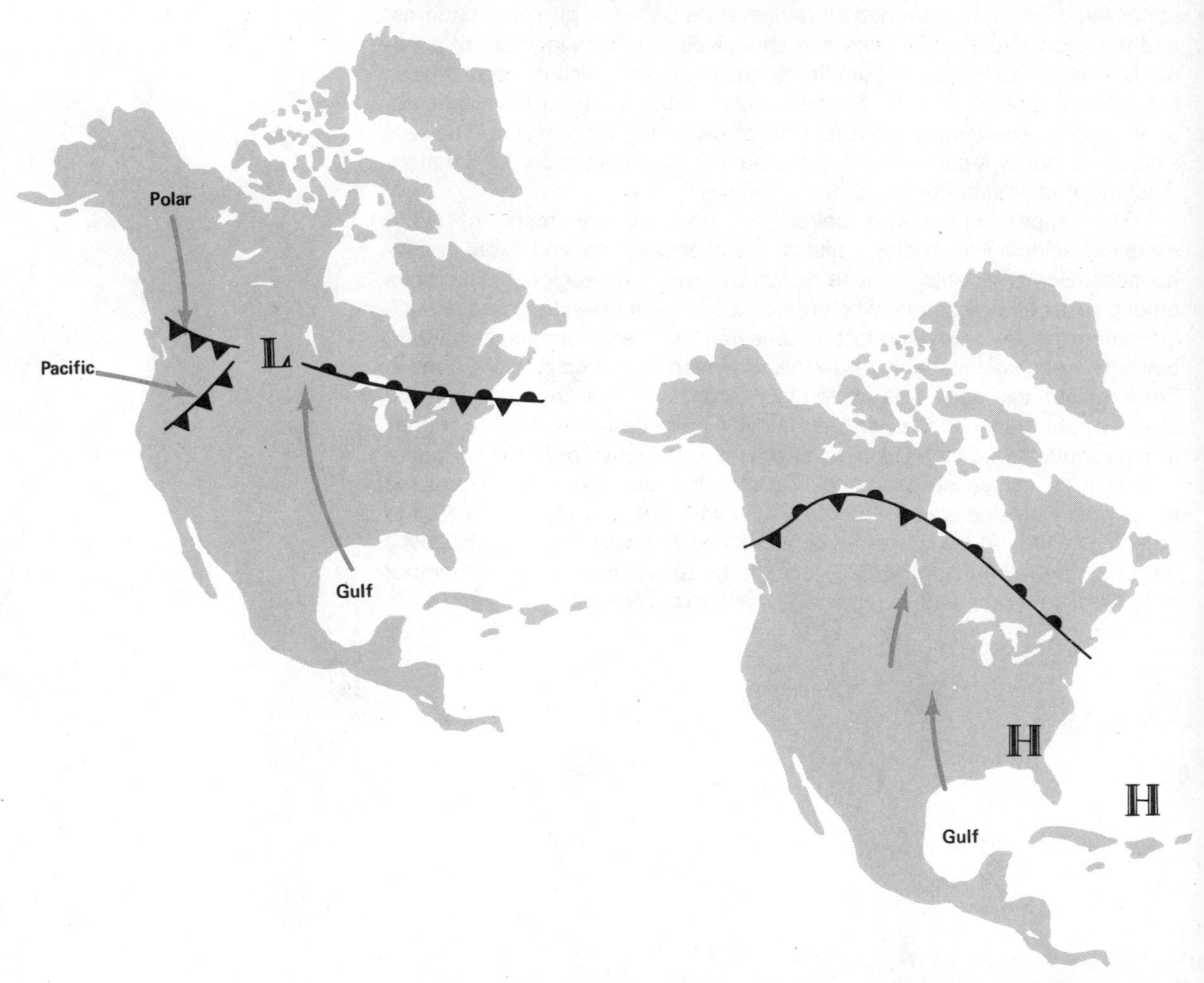

The Bermuda High. The Atlantic Ocean off the coast of Florida supports an almost permanent zone of high air pressure that occasionally expands into the southeastern United States. The encroachment is more common in summertime than in winter, because in summer the subsidence zone is north of its average position. Stifling heat and prolonged drought prevail throughout the southeast when the Bermuda High is well developed. Strong winds blow northward on the west side of the High and carry warm and humid air deep into the interior of the continent. The front between Arctic and tropical air moves far to the north and brings rain to parts of central Canada that usually are dominated by cold and stable polar air.

The Colorado Low. The area just to the east of the Rocky Mountains is a favorite breeding ground for low-pressure areas and associated cyclonic storms. The pattern shown on the map commonly appears in springtime. Strong southerly winds bring moist Gulf air northward on the east side of the low, but the cool land surface keeps the air fairly stable. The center of the low usually remains near the mountains for a day or two and then moves eastward. It picks up speed as it passes the Great Lakes, moves through the St. Lawrence Valley, and finally joins the large area of low pressure over the warm waters of the North Atlantic Drift. Figure 4A-1 could be a map of a Colorado Low as it passes Chicago.

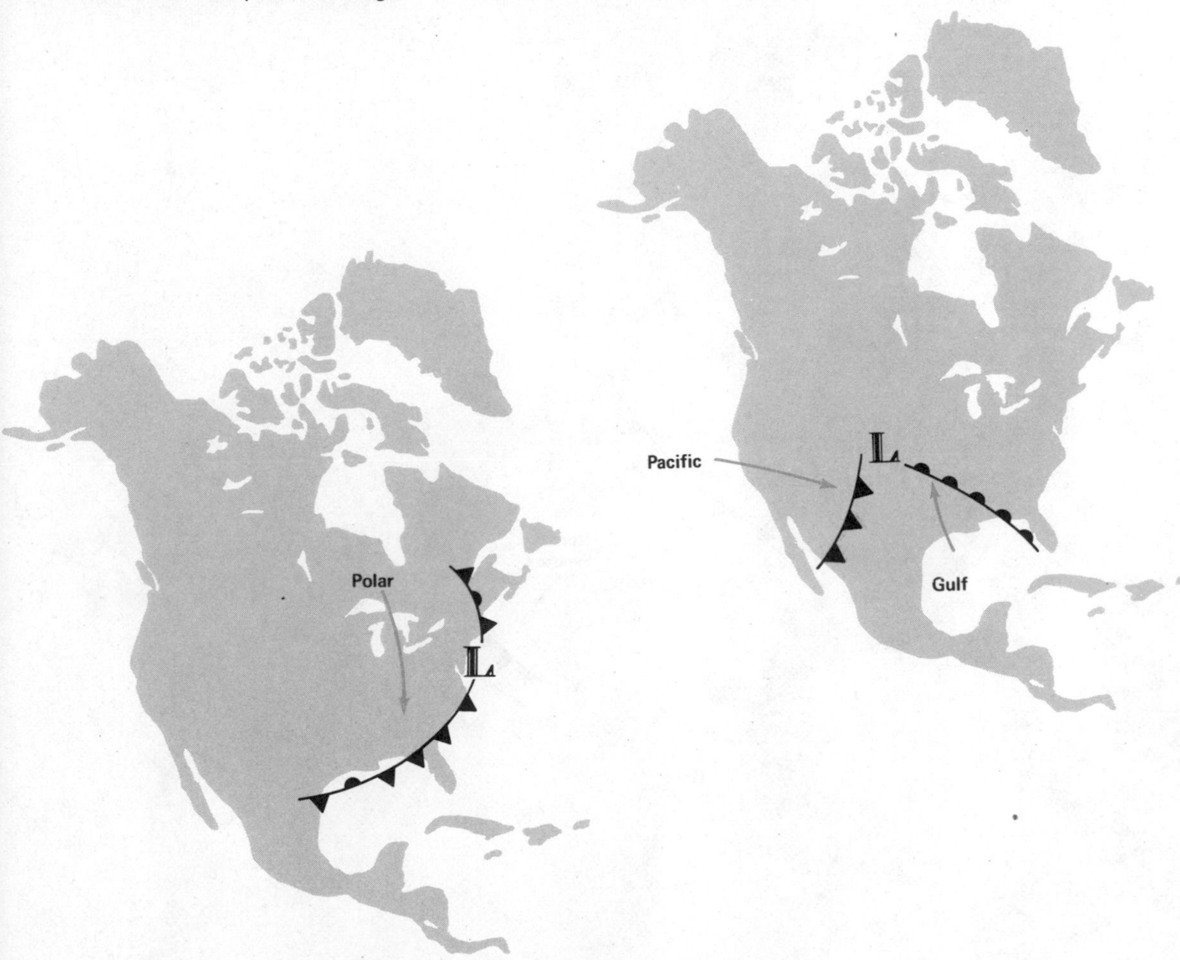

The Eastern Low. A slowly moving Eastern Low is a common fall and winter occurrence. The Low may be the remnant of a northward-moving Gulf hurricane weakened by its passage over land and transformed into a mid-latitude cyclonic storm. Strong northerly winds on the west side of the Low bring unseasonally cold weather far into the southeastern states. The interior of the country usually experiences crisp weather, with high air pressure, clear skies, calm air, and little chance of rain. An occluded front over New England brings stratus clouds and a chance of significant accumulations of snow or freezing rain, depending on the temperature of the air that is forced to rise over the oncoming polar air mass.

The Indian Summer High. "Indian summer" provides a welcome interlude of warm autumn weather before winter comes. The sun shines through blue skies and provides unseasonal warmth, while high air pressure hinders the invasion of humid air from the oceans. Air over the middle of the United States is bounded on the north and south by a pair of fronts that are stationary or at best slowly moving. The southern rampart is marked by a front between dry continental air and a humid air mass over the Gulf of Mexico. Another front crosses central Canada and separates cold polar air from the mild air over the United States. These fronts cannot advance into the United States until the air pressure in the middle begins to fall.

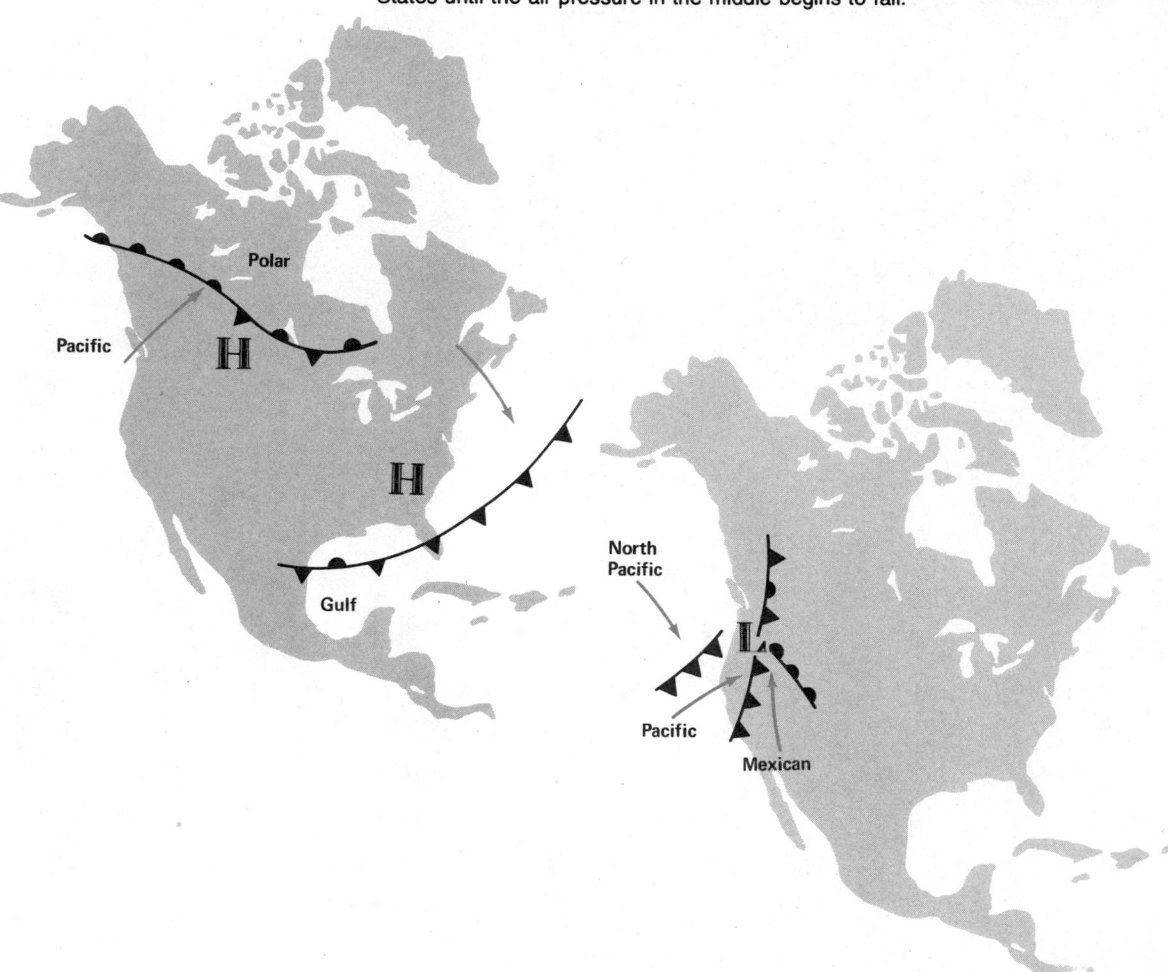

The Northwest Low. An intense center of very low pressure is a common occurrence over Oregon or Washington. The mountains act as a temporary barrier to the eastward movement of the Low. It remains near the coast and draws Pacific air onto the land. Persistent stratus clouds and intermittent rain accompany weak fronts between air masses from different parts of the Pacific Ocean. Sometimes the Northwest Low slowly dissipates; more commonly, it crosses the mountains, reorganizes, and moves eastward onto the Plains. Northwest Lows are less common in summer than in winter, because the northward shift of the subsidence zone brings high pressure and dry air into Oregon and Washington.

The Santa Aña High. The Santa Aña is a hot and extremely dry wind that originates in an area of dynamic high air pressure and subsiding air over the Great Basin. Subsiding air is dry and has little chance to pick up moisture from the deserts of Nevada and Arizona. Moreover, the Santa Aña is usually strong enough to prevent the incursion of humid air from the Pacific Ocean. The drying wind can be a serious hazard to the densely populated coast of California: dust storms, desiccated crops, excessive evaporation from reservoirs, and raging wildfires are common associates of a persistent Santa Aña.

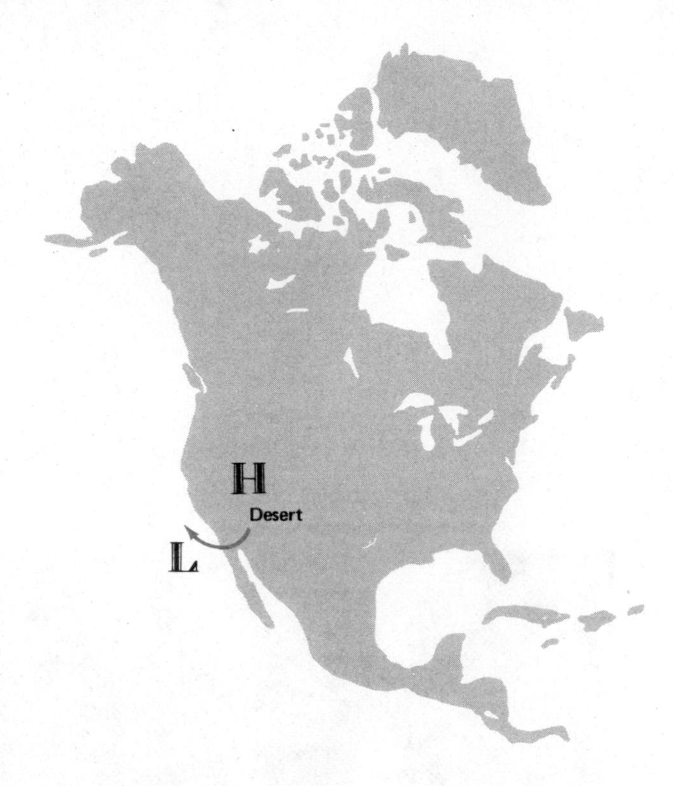

Chapter 5

Climate: Expected Patterns of Weather

It's no coincidence that the words climate and environment were long considered synonyms.

Philip Gersmehl

It is easy to justify a summary chapter on climate. The causes of atmospheric phenomena are interesting enough in themselves. Weather has a direct effect upon personal comfort. The atmosphere affects people indirectly by placing constraints on agriculture, construction, and recreation. For a student of the environment, however, the major reason for studying climate is that the weather has a profound effect on soil, vegetation, water supply, landform, and thus on many aspects of human interaction with the earth. A clear understanding of climate is a big help in making sense out of other environmental patterns. Once the need for a general understanding of climate is acknowledged, the problem of how to organize the study emerges.

climate weather over a long period of time; the conditions, both typical and abnormal, which might be expected at any given time

weather state of the atmosphere at a given time

CLASSIFICATION

A classification system consists of arbitrary rules that separate things into easily remembered groups. People develop classifications whenever things become too numerous to be remembered separately. The need for classification therefore depends on the experience of the individual. For example, many people have only a single mental image for snow. For them, a single word is sufficient. A child in Indiana needs two words, because the world of a midwestern youngster contains two varieties of snow, packable and useless. The ability of some snow to be formed into usable snowballs is the basis for a simple classification scheme. A skier recognizes a dozen different varieties of snow and groups them according to crystal shape and temperature, the two characteristics that determine the appropriate ski wax for a particular time. An Eskimo lives in still more intimate contact with snow and has half a hundred words for various kinds of frozen water. Many Eskimo snow words are classifiers, words designed to group similar types together for specific purposes.

The idea of grouping for a purpose is the key to designing a classification. The choice of criteria for the categories depends on the purpose of classification. A skier's classification of snow is not very useful for a hydrologist, whose major interest is the volume of water available in frozen form. A hydrological classification of snow therefore has categories based on variables such as depth, density, and exposure to sun and wind. Avalanche forecasters prefer to emphasize shear strength, mass, and slope angle in designing a snow classification to suit their particular purposes.

No single classification can serve all purposes. In fact, for the general student of the environment, any classification of climate may have outlived its usefulness the moment it is learned. The reader of this book needs an understanding of climate as a foundation for analyzing other environmental features. The foundation, however, can be built in a variety of ways. Some approaches may have greater appeal than others for people with particular backgrounds, interests, and psychological characteristics. Therefore this chapter presents two different general-purpose climate classifications, a means of translating from one to the other, and brief outlines of a number of special-purpose classifications that occasionally appear in newspapers, on television, or through other popular media.

A GENETIC CLASSIFICATION OF CLIMATE

Genetic classifications emphasize causes or origins; things are grouped according to how they were formed. The basic ideas of a genetic classification of climate have permeated the previous chapters on weather processes. A combination of these ideas yields a classification of climate with four basic types, four seasonal hybrids, and three topographic variants.

pure climate climate dominated by a particular atmospheric process

Each of the four basic types is the result of a single major atmospheric process:

unstable air air that tends to move vertically
(see Fig. 2-3)

The instability type. Warm and showery weather is the rule in regions near the equator. Rain arrives mainly in brief afternoon storms, the consequences of unstable air during the warmest hours of the day. Frost and drought, two big climatic threats to tender forms of life, are virtually unknown in the instability climate.

subsidence downward movement of air
(see p. 56)

The subsidence type. Intense sunshine and sinking air conspire to make summer extremely hot and dry. Winter is cooler but just as rainless as summer, because settling currents still dominate the air. The subsidence type occurs in the middles and on the west sides of continents about 30°N and 30°S.

front zone of contact between dissimilar air masses
(see Fig. 4-4)

The frontal type. Changeability is the distinguishing feature of mid-latitude weather. Sunshine is more direct in summer than in winter, so that summer usually has warmer ground, less stable air, and more precipitation than winter. All seasons, however, are characterized by frequent invasions of air masses from other regions, and each invading air mass brings a change in the weather.

albedo reflectivity of a surface
(see p. 11)

stable air air that resists vertical motion
(see Fig. 2-2)

The inversion type. A year at the pole consists of a light season and a dark one, both cold and dry. The summer sun has little effect on the frozen ground, because it shines from near the horizon through a thick atmosphere and onto a highly reflective surface. The cold surface keeps the air cold and stable, unable to hold much moisture, but still unwilling to let humid air from other regions enter and disrupt things.

complex climate climate seasonally dominated by two different atmospheric processes
(see p. 58)

Each of the four complex or hybrid climates consists of a regular seasonal alternation of two basic types:

Hadley Cell tropical circulation of air rising near the equator and sinking about 30°N and S
(see Fig. 3-7)

The instability/subsidence hybrid. Winter in this frost-free climate is a dry season, not a cold one. The seasonal shift of the Hadley Cell brings a long rainy season to places about 10° from the equator and a short one to places about 20° from the equator. The hottest weather usually occurs just before the onset of the cooling summer rains. Of course, the rainy summer comes in January for Southern Hemisphere areas with this climate.

The subsidence/frontal hybrid. Hot dry summers alternate with mild wet winters on the west coasts of continents about 35 to 40 degrees north and south. The warm season is characterized by the persistent dryness of sinking air, whereas the cool season brings humid onshore breezes and occasional collisions between polar and tropical air masses. Most of the winter precipitation comes gently, with a minimum of sound and fury.

The instability/frontal hybrid. Hot days and afternoon showers characterize the fairly monotonous summers on the east sides of continents about 30°N and 30°S. Winter is mild and changeable, with occasional invasions of polar air, but the polar air is usually travel-weary and has lost much of its icy character by the time it reaches the subtropical coasts. Frost is rare and snow seldom persists in the instability/frontal hybrid.

The inversion/frontal hybrid. Cold and dry weather ranges far into the middle latitudes during the winter, but the summer sun causes the inversion climate to shrink back into its icy refuges on the high plateaus of Greenland and Antarctica. Like other areas with a hybrid climate, this zone is one of gradual transition. Places on the side away from the equator experience 11 months of polar frigidity, broken by a short period of changeable weather. Toward the equator the hybrid gives way to nearly pure frontal climates.

Finally, each of the three topographic variants is a predictable exception to the general patterns described above:

The highland variant. Highlands are usually cooler and more humid than surrounding lowlands. Extensive mountain ranges often are precipitation doughnuts with dry centers, because the outer ramparts of the mountains extract the moisture from incoming winds. Differences in solar intensity on different slopes further complicate the climatic pattern of mountains.

The rainshadow variant. Places that lie behind mountains in regions of prevailing winds are drier than would be expected if the mountains were not there. Mountains act like custom agents, confiscating the moisture from incoming air masses before allowing them to cross the barrier. Mountain walls surround some of the driest areas on earth, such as Death Valley in California, the Dead Sea of Israel, or the Takla Makan of central Asia.

rainshadow dry area on the leeside of a mountain

The maritime variant. Coasts are milder and more humid than continental interiors. Water responds more slowly than land to energy surpluses or deficits. Winds carry some of the moisture and moderation of oceans or lakes onto nearby land. Terrain and prevailing wind direction affect the degree to which the maritime effect penetrates a continent.

maritime effect moderation of temperature by proximity to water body (see p. 17)

continentality extremes of temperature due to inland location

Figure 5-1 shows the areas influenced by the four basic types in the extreme seasons; areas of hybrid climate can be identified by comparing the two maps.

January

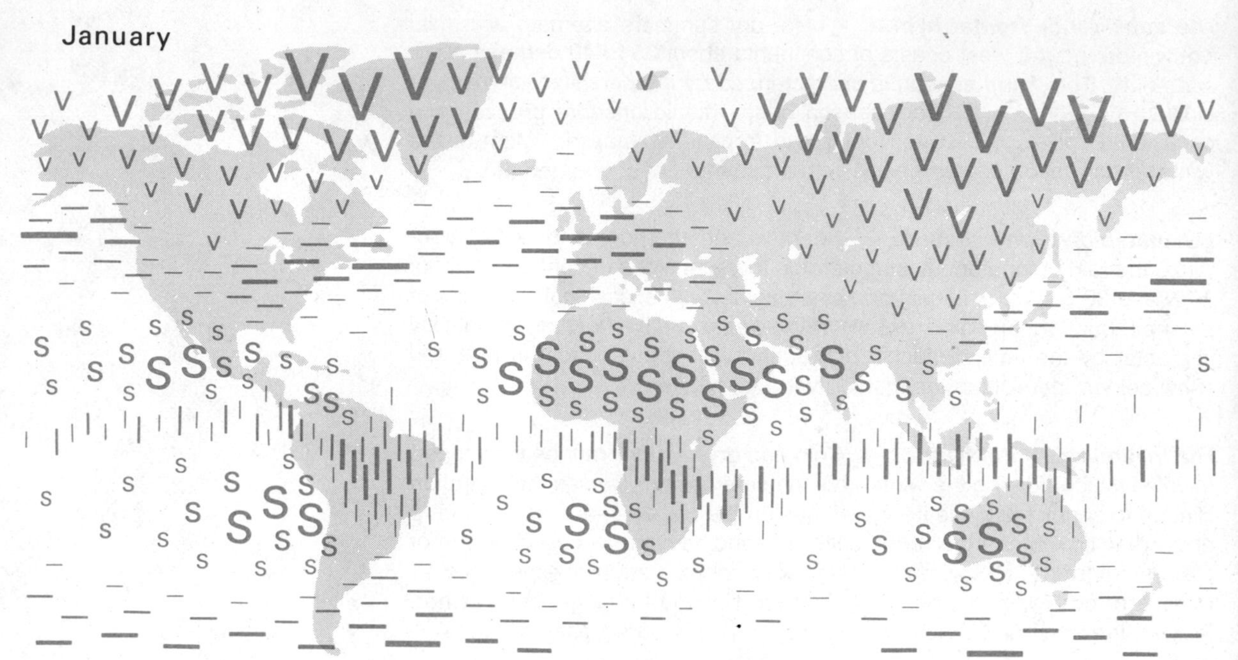

Climate Types

Symbol	Type
V V	Inversion
—	Frontal
s S	Subsidence
‖	Instability

July

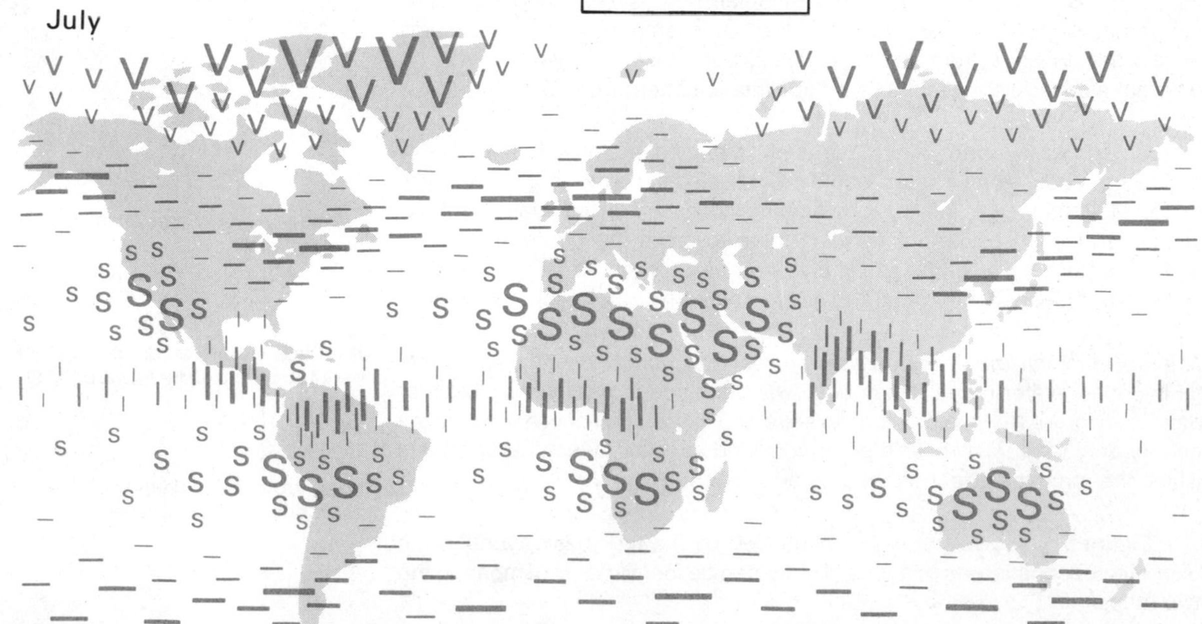

A DESCRIPTIVE CLASSIFICATION OF CLIMATE

Descriptive classifications put climates into groups on the basis of actual weather data: temperature, humidity, wind, and so on. The Köppen classification is probably the best known descriptive one; it has appeared in modified form in nearly every geography textbook and world atlas published in the twentieth century. In a way, the Köppen classification is a fortuitous relic of an extinct thought-world. In the early twentieth century there was widespread belief in the existence of discrete "natural regions," distinctive combinations of climate, soil, and vegetation that differed sharply from neighboring regions. Köppen tried to find critical values of temperature and precipitation that coincided with vegetation boundaries on the earth. His extensive correspondence, travels, and numerous map revisions testify to the difficulty of the task. Many climatologists have tried to refine his temperature and rainfall criteria to produce a closer match with vegetation patterns.

Unfortunately, maps of vegetation, soil, and climate simply do not correlate precisely. Boundaries between different types are occasionally abrupt, often gradual, but sometimes almost imperceptible. Furthermore, boundaries of soil and vegetation types may coincide perfectly in one place and yet diverge widely elsewhere. Therefore modern versions of the Köppen classification seek simply to streamline what has proved to be a useful and easily remembered set of categories. Similarities are still evident when a modern climatic map is compared with separate maps of soil and vegetation. The similarities are even more striking because modern readers know they are looking at independently derived maps of three separate characteristics of the environment.

Most versions of the Köppen classification use a letter code (e.g., Cfa) to identify each climatic type. The first letter is the most general and significant one. All climates fall into one of five major groups:

(A) Hot forest climates, which have adequate rainfall for trees and little chance of frost.

(B) Grassland and desert climates, which are too dry for forest vegetation.

(C) Warm forest climates, which often have frost but do not get cold enough for snow to persist.

(D) Snow forest climates, which get cold enough for snow to persist for more than a month each year.

(E) Tundra and ice climates, which are too cold for forest vegetation.

FIG. 5-1. Climates according to the genetic classification used in the text. (top) January, when the sun is in the Southern Hemisphere. (bottom) July, when the sun is in the Northern Hemisphere and the climatic belts are shifted northward.

The second letter in the Köppen code serves different purposes in different areas. A lower-case letter follows A, C, or D and indicates the seasonal pattern of precipitation in a forest climate:

(f) Rain in all months, with no distinct dry season.
(w) Rain mainly in summer, with a dry winter.
(s) Rain mainly in winter, with a dry summer.
(m) Intense monsoon rain, with a short dry season.

A capital letter follows B or E and shows the severity of dryness or cold:

For the dry B climates:
(S) Too dry for most trees, but wet enough for grasses.
(W) Too dry even for most grasses.

For cold E climates:
(M) Too cold for trees, but too warm for persistent snow.
(T) Too cold for trees, but warm enough for some vegetation.
(F) Too cold for most forms of life.

Additional letters of the code usually are refinements of the temperature symbols. At this level of classification, various versions of the Köppen system have different criteria and different codes.

TRANSLATIONS

The genetic and descriptive classifications are two means for organizing the same thing; each has advantages and drawbacks. For example, the genetic classification is inherently vague: the reader needs supplementary maps of temperature and precipitation in order to have a reasonably precise idea of the climate in a particular place. The descriptive classification is easy to map, because the boundaries between types are clear and precise. This precision, however, gives a misleading impression of concreteness: some readers seem unable to dispel the erroneous notion that all areas with the same map symbol have identical weather. The genetic approach emphasizes gradual changes of climate from place to place, but transitional features are difficult to map clearly. Since this text emphasizes the reasons for environmental patterns, it uses genetic ideas to help explain patterns of climate. Maps of climate in popular atlases frequently are based on the Köppen classification. Figure 5-2 is a translation dictionary that shows how a genetic and a descriptive classifier would describe the various climate types of a hypothetical continent. Both of these classifications are essentially academic, designed to impart information about climate to people who seek understanding for its own sake. Both are too general to be used in making environmental decisions; practical purposes demand specialized classifications.

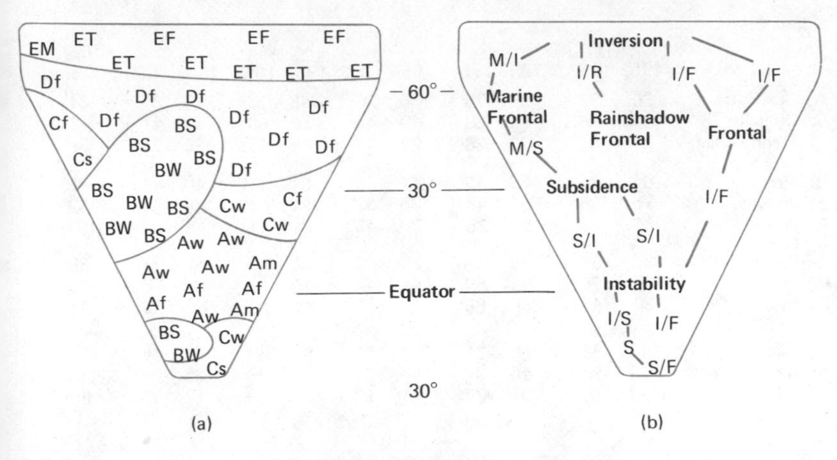

FIG. 5-2. Hypothetical continents that illustrate the typical positions of climate types as classified: (a) descriptively and (b) genetically.

A PROBABILITY CLASSIFICATION FOR FARMERS

Food production is a gigantic game of chance, even in an age of advanced agricultural technology. The weather has a particular set of strategies it can use against the farmer. In turn, food producers can choose from a number of alternative land uses and production techniques. Each crop has its own environmental preferences. Only an incurable optimist would try to grow oranges outdoors in Minnesota. Likewise, unirrigated corn in Wyoming would be economic suicide. Both environments are clearly too severe for the proposed crop, but corn in east-central Nebraska is a debatable possibility. Suppose a particular variety of corn requires about 60 cm (about 25 in.) of rain and 100 frost-free days in a row. The Nebraska summer is much longer than necessary for this corn. The average rainfall of 70 cm per year in east-central Nebraska also exceeds the requirements of the crop. In a year with average rainfall and a normal growing season, corn farming is profitable at all places east of the dotted line on Figure 5-3. Unfortunately, averages mask a key feature of frontal climates: the weather varies from day to day and from year to year. Farmers who want to maximize their chances of success in the struggle against weather can use a classification of climate based on probability.

game theory technique of analyzing human activity as if it were a game with a recognizable opponent, a set of rules, and a procedure for keeping score

growing season frost-free period with temperature high enough for plant growth

The user of a probability classification searches the records of local weather over a long period of time, preferably more than 25 years, in order to learn how frequently the weather used various strategies at its disposal. For example, 50 years of records reveals that east-central Nebraska received more than 65 cm of rain in 30 years, 55–65 cm in 12 years, and less than 55 cm in eight years (Fig. 5-4). Growing corn that needs 60 cm of water would be wise strategy about 60 percent of the time, questionable in about one-fourth of the years, and disastrous one year out of six.

statistical probability likelihood of a particular event, calculated by dividing the number of times it occurred in the past by the total number of observations

FIG. 5-3. An isoline map of average rainfall in Nebraska for the period 1951 through 1960. The averages, however, mask a great deal of variability, as the accompanying table shows.

Station	Yearly rainfall (cm) 1951	1952	1953	1954	1955	1956	1957	1958	1959	1960	10-year Average
A. Box Butte	36	33	32	25	30	28	48	33	46	31	34
B. Ogallala	61	38	43	28	40	40	56	66	51	41	46
C. Trenton Dam	61	35	33	28	23	23	48	43	45	38	38
D. Merriman	64	35	45	46	41	38	61	43	40	48	45
E. Brownlee	76	36	50	43	38	38	81	64	48	51	52
F. Broken Bow	69	32	51	36	28	43	81	61	56	53	51
G. Red Cloud	97	61	58	58	56	43	84	66	61	71	65
H. Seward	99	79	58	69	41	43	76	79	84	76	70
I. Niobrara	89	46	69	66	45	41	81	53	82	79	65
J. Tekamah	112	86	56	71	56	51	84	58	94	66	73
K. Weeping Water	109	79	48	94	46	61	94	86	96	89	80
L. Falls City	137	86	51	97	74	53	89	109	117	107	90

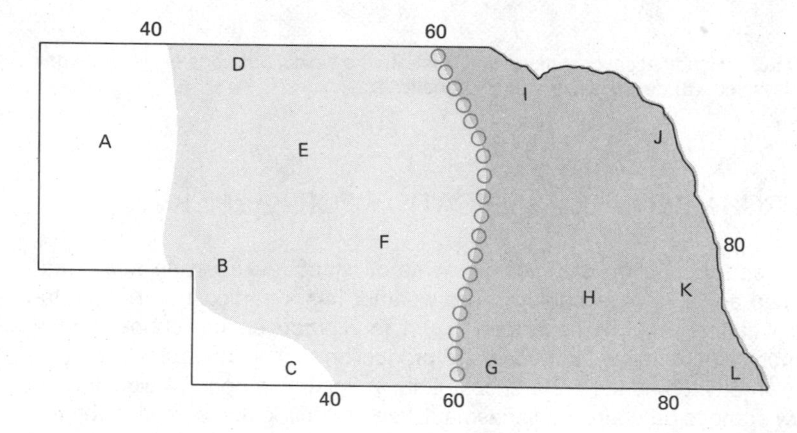

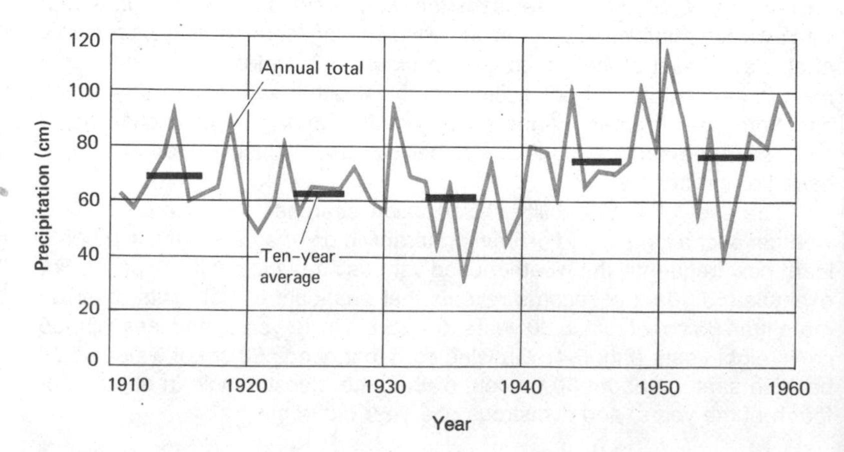

FIG. 5-4. Rainfall at Crete, Nebraska, for the period 1911 through 1960.

There are two reasons for this broad range (55–65 cm) of questionable rainfall. First, corn 'yields increase gradually with increasing rainfall, so profitability is not a yes/no proposition with a clear dividing line. Second, timing is important. Fifty centimeters of rain, arriving exactly when needed, will produce more corn than 70 cm falling mainly in a few ill-timed gully-washers interspersed with long periods of dry weather.

confidence limit expression of the degree of uncertainty about a particular measurement or prediction

Despite their limitations, probability figures are better than the averages used in most general-purpose climatic classifications. Moreover, it is easy to adjust a probability classification to adapt to new crops or market conditions. Suppose plant breeders develop a new variety of corn which can thrive with only 50 cm of rain a year. The chances of success with the new corn in east-central Nebraska are predictable with Figure 5-4. It will fail one year out of 10 and be questionable another 10 percent of the time. If corn prices rise, it may pay to take a chance on success in areas with questionable weather. High profits in good years enable farmers to absorb losses in bad years and still come out ahead.

Farmers can use probability calculations to consider other relevant conditions, such as date of planting, chance of early frost, need for supplemental irrigation, amount of sunshine, or chance of insect attack. For example, if a particularly voracious beetle appears in great numbers only after five consecutive weeks with at least 3 cm of rain in each week, then it pays to check the weather records to see how often this particular combination of events occurs. Suppose that 40 years of records contain four years with the proper conditions for a beetle infestation. The chances of an outbreak in any given year are 4 in 40, or about 10 percent. Of course, in order to determine what weather the beetle prefers, one must know the personal habits of the beetle quite well. Getting familiar with the beetle is a sizeable research undertaking, but it may pay for itself in reduced crop damage, lower insect-control costs, and less pesticide buildup in the environment. On the other hand, the farmer may evaluate the risk and the cost of insect control and choose to absorb an occasional loss or to plant another crop. In any case, the probability analysis has given the farmer a vital insight into the typical strategies of the environmental opponent in the food-producing game.

contingent probability likelihood of an event occurring, if its ooccurrence also depends on another event with its own probability

A RECURRENCE INTERVAL CLASSIFICATION FOR ENGINEERS

Forecasters of weather catastrophes frequently deal with events that are even rarer than a beetle invasion. Their techniques are similar to those of the probability analysts, but they use a slightly different vocabulary to express their findings. Consider the case of a resort town that has had three hurricanes in 87 years. The chance of a hurricane next year is 3.4 percent (3 divided by 87), and the town should expect 3.4 hurricanes in 100 years or about seven hurricanes during the next two centuries. If these hurricanes occur randomly, then any period of 30 years would probably contain one hurricane. A similar mathematical exercise tells the inhabitants of Austin, Texas, that storms with at least 20 cm (8 in.) of rain in 24 hours have a

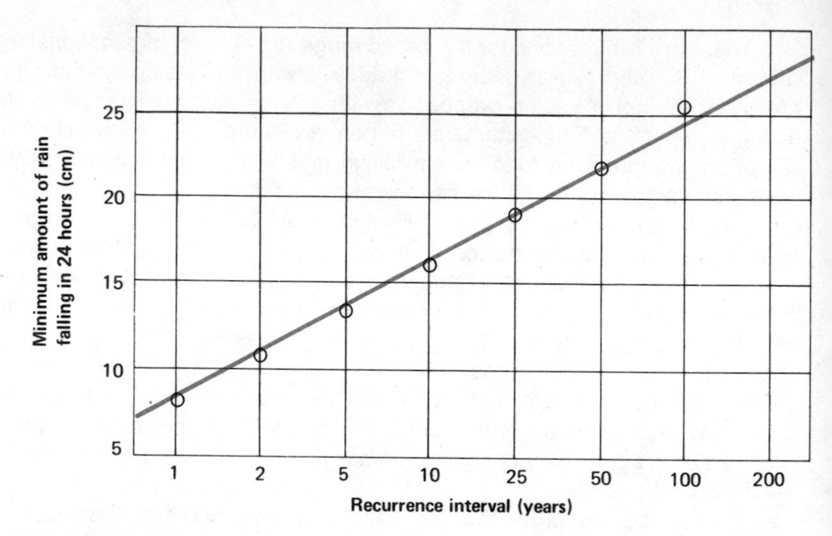

FIG. 5-5. Recurrence interval for storms of different sizes at Austin, Texas. A large storm has a long recurrence interval and is not very likely next year. On the other hand, storms bringing at least 10 cm of rain occur quite frequently.

recurrence interval period of time that probably contains one event equal to or larger than a particular size

recurrence interval of 25 years (Fig. 5-5). Knowledge of the expected recurrence interval for a particular kind of event can be very useful in designing structures to withstand environmental shocks.

An investigator must remember five things in order to apply the idea of recurrence intervals properly:

1. The size figure represents the threshold value for a particular recurrence interval. The five-year storm in Austin brings *at least* 14 cm of rain, not exactly 14 cm.
2. The given relationship between recurrence intervals and sizes of events applies only to a local area. A storm with a recurrence interval of 10 years delivers different amounts of rainfall in different places (Fig. 5-6). The hundred-year flood is not the same size on different rivers. The 20-year snowfall in Vermont is many times as deep as in Alabama.
3. The idea of a recurrence interval cannot be used to predict exact times of occurrence. The appearance of a 20-year storm in 1980 does not mean that the next storm of the same size or larger is not due until 2000. On the contrary, it could appear in 1981. In fact, the chance of it arriving in 1981 is exactly the same as its likelihood of coming in 2000: 5 percent, or 1 chance in 20.
4. The graph of recurrence intervals is based on the assumption that climate itself is not changing, but the Nebraska rainfall figures in Figure 5-4 indicate that the 1920s and 1930s as a whole were

drier than the 1940s. Calculating probabilities or recurrence intervals on the basis of the 1921–1940 record would significantly underestimate the likelihood of big storms in 1941. A long record makes an estimate of probability more reliable, but even this is no guarantee that the future will resemble the past.

5. The consequence of climate may change even if the climate remains constant. For example, paving a surface prevents rainwater from sinking into the soil and thus increases the likelihood of a severe flood. The 10-year storm may produce a flood that formerly had a recurrence interval of a century or two, and a hundred-year rain may bring a flood of unprecedented size.

Graphs of recurrence intervals are essential tools for engineers, whose work must withstand freakish blizzards as well as run-of-the-mill snowstorms. It sometimes makes economic sense to build structures just strong enough to withstand the 25- or 50-year storm, because larger stresses occur so infrequently that it is cheaper to repair the damage than to design for the storm. The decision to build for a storm of a particular recurrence interval is an engineering compromise called the storm-rating of a structure. The choice of a storm-rating (or design storm) for a particular structure usually depends on the cost of building and repair. A bridge on a major highway may be built to withstand the hundred-year flood, whereas a minor boat dock might have a design flood with a return period of only 20 years. Structures whose failure would be catastrophic (e.g., dams upstream from cities) may be designed to withstand the thousand-year storm.

storm-rating recurrence interval of the event a structure is built to withstand

design storm maximum storm a structure is designed to withstand

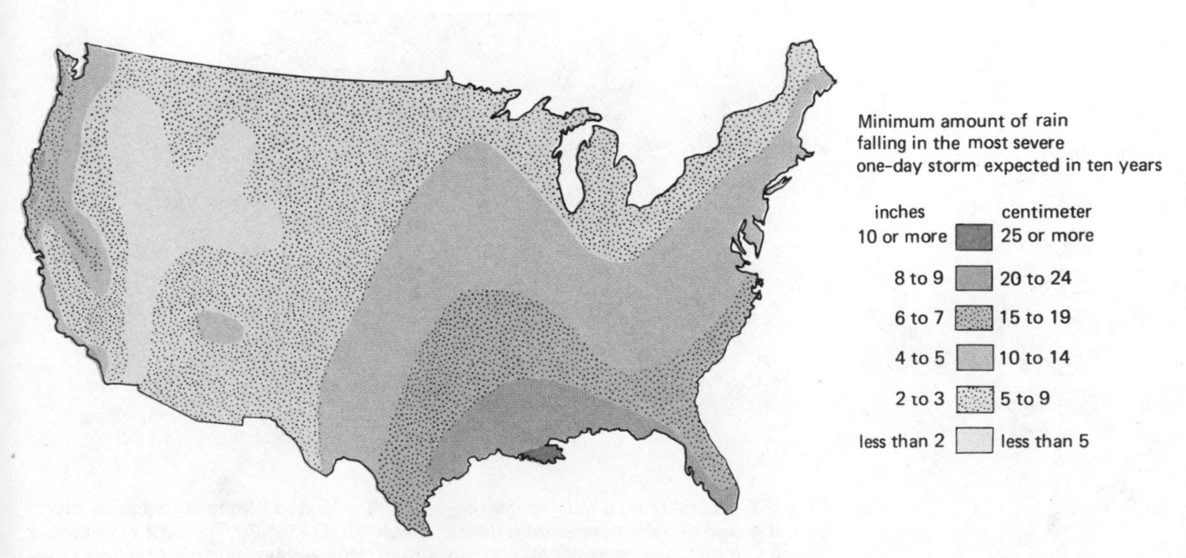

FIG. 5-6. The ten-year 24-hour rainstorm in the United States.

A DURATION SUM CLASSIFICATION FOR FURNACES

It is more expensive to heat a house in Omaha than in Boston, although both cities have 190 frost-free days and an average annual temperature of 11° C (51° F). House furnaces do not consider annual temperatures when they try to estimate the size of the winter task before them. Rather, they look at the total temperature deficit that has to be offset. Keeping a room at a comfortable temperature of 20° C (68° F) when the outdoor temperature is 0° C requires a "payment" of 20 degrees of heat per day. A day with an average outdoor temperature of 15° C (59° F) has a heat deficit of 5 degrees. A 10° day has a 10° heat deficit. A furnace sees little difference between four days of 15° C, two days at 10° C, and one day of 0° C; all add up to 20 degree-days of heat ($4 \times 5 = 2 \times 10 = 1 \times 20$).

degree-day deviation for one day of one degree from an arbitrary standard

It is possible to calculate degree-day totals using any arbitrary base temperature. A base of 20° C makes the sample calculations easy to follow, but the annual total of heating degree-days based on 18° C gives a more reliable estimate of the total cost of heating in the United States (Fig. 5-7). Omaha has a total of 3600 heating degree-days, whereas Boston has only 3100; a house in Boston needs 15 percent less energy to keep it warm.

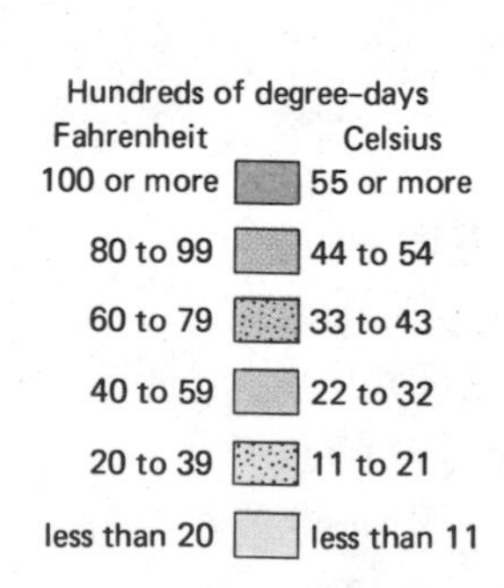

FIG. 5-7. Total annual heating degree-days in the United States. The figures represent the sum of daily temperature deficits below 18° C (65° F). The cost of heating a house in a particular area is equal to the sum of degree-days multiplied by the cost of fuel needed to offset one degree-day.

The heating degree-day total for a place, like any other measure of its climate, can vary from year to year. Most weather stations keep a running total of degree-days throughout the heating season. A total that is significantly less than normal indicates a mild winter. Homeowners will probably buy less fuel than they would in a typical year, and fuel distributors should reduce their own purchases in order to avoid having a surplus at the end of the heating season.

Analyses of probabilities and recurrence intervals can also be made by using the idea of degree-days. Heating engineers can determine how much extra capacity to build if they are told that the normal annual total of degree-days is 3100 but one year out of ten has more than 3400 degree-days. The city can then prepare emergency procedures for meeting the extra fuel demands of occasional severe winters.

The typical shape of the heating degree-day graph for an entire year adds an interesting and practical footnote to the discussion. The average minimum temperature for the coldest month in Boston is −4° C and the annual total number of heating degree-days (based on 18° C or 65° F) is 3100. Suppose that the engineers decided to lower the base temperature for the calculations by two degrees, from 18° C to 16° C. The reduction goes only 9 percent (2/22) of the way from room temperature to the minimum outside temperature of −4° C, but it reduces the number of heating degree-days for the year by nearly 19 percent. A slightly lower thermostat setting therefore can save a disproportionately large number of heating degree-days. The saving of fuel is even greater than the numbers suggest, because furnaces will not operate at all during the mild days of spring and fall. The transition seasons are times of great fuel waste when thermostats are set at or above 21° C (70° F). Doors and windows stay open on mild days while furnaces vainly try to make up a minor heat deficit. A small reduction in thermostat settings may reduce Boston's annual fuel bill by 30 percent (Fig. 5-8).

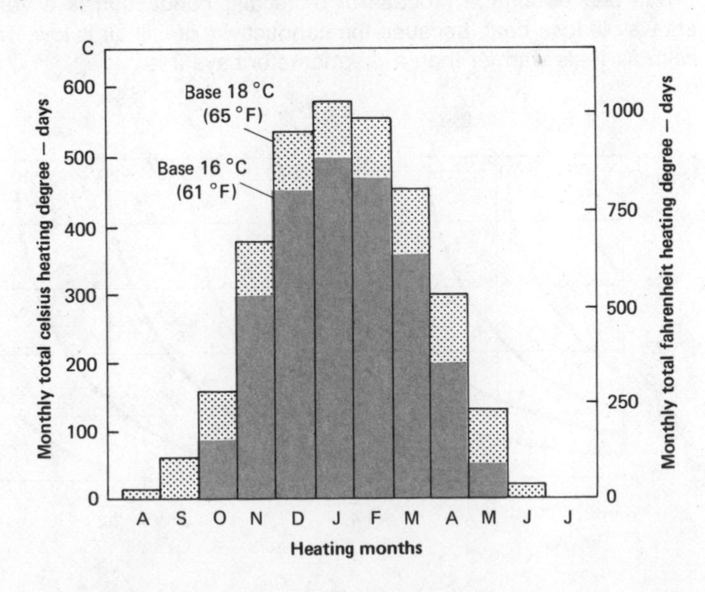

FIG. 5-8. The annual heating degree-day curve for Boston. The dotted area represents the number of heating degree-days that could be "saved" by setting the base temperature at 16° C (61° F) instead of 18° C (65° F).

The success achieved by heating engineers has persuaded other people to try the idea of degree-days. An obvious adaptation is the use of cooling degree-days as an aid in designing air-conditioning systems. Vineyard owners keep sums of growing degree-days based on 10° C and damage-days based on maximum temperatures above 40° C. Calculations of chilling degree-days help orchard owners predict the size of a peach crop, because peach trees need a certain amount of cold to induce them to form blossoms. It is also possible to stretch the idea of duration sums to include the calculation of humidity-days, precipitation-days, sunshine-days, and so on. Research along these lines so far has yielded more intriguing speculation than it has solid findings, but the possibilities are encouraging.

A PHYSIOLOGICAL CLASSIFICATION FOR HUMAN BODIES

wind chill expression of the combined effect of temperature and wind

boundary layer layer of calm and altered air near a surface

conductivity ability of a material to transfer heat from one molecule to another

One of the easiest ways to die of exposure is to consider 7° C (45° F) too warm to pose any threat to human existence. A drizzly rain and moderate breeze can turn a 7°-hour into a potential killer. Likewise, high humidity can make 35° C (95° F) more oppressive than a dry 50° C (122° F). The combination of weather factors often seems designed to make humans more miserable than they anticipated. To forewarn us, weather analysts have devised ways of describing and evaluating the combined effects of two or more atmospheric conditions.

One way of combining weather variables is the wind chill table, originally designed to aid Antarctic explorers (Fig. 5-9). The wind chill chart is based on the idea of a boundary layer, a zone of modified air near a surface. On a cold but calm day, warm human skin heats the air near it, which in turn passes heat to the next layer, and so on until the heat is carried far from the body. This bucket-brigade process of molecular conduction is a very inefficient way to lose heat, because the conductivity of still air is low. Therefore calm air feels warmer than a thermometer says it is.

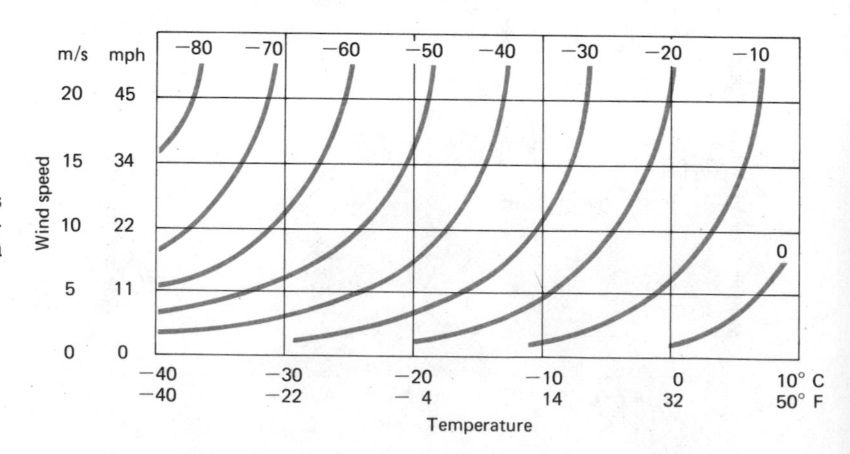

FIG. 5-9. Wind chill. The numbers on the slanting lines indicate the effective temperature (in °C) of air at a given temperature and wind speed.

A wind, however, pulls heated air away from the surface. Cold air replaces skin-warmed air and demands additional heat. The faster the air moves, the more heat it takes from the body. The resulting decrease in skin temperature may feel uncomfortable, but it has a valuable side effect: cold skin loses heat slowly because radiation and turbulent transfer of heat depend on temperature (see p. 13). Extremely low air temperature and/or very strong wind may lower skin temperature below freezing. Frozen skin is unable to produce any heat on its own. In effect, the body sacrifices its outer layers and makes them into a kind of insulation, an attached boundary layer to help keep inner tissues warm.

frostbite freezing of skin cells

Rain and snow complicate the situation in two ways. First, evaporation or melting takes heat from warm skin and thus tips the balance in favor of heat loss. Second, water can interfere with the insulating ability of hair or clothing. A coat keeps a body warm because it creates a layer of dead air spaces in the cloth. Filling these spaces with water enables clothing to conduct heat away from the body more rapidly.

Similar combinations of factors make calm and humid days in summer feel more oppressive than dry breezy days with the same temperature. Lack of wind keeps a body surrounded by a boundary layer of humid air that interferes with the air-conditioning mechanism built into human skin. Perspiration which cannot evaporate is not very useful in removing excess heat from the body. The most widely used measure of the oppressiveness of muggy days is the Temperature-Humidity Index (THI) (Fig. 5-10). Like the wind chill table, the THI tries to express a combination of conditions in terms of its effective temperature on the human body. Probabilities of particular wind chill or THI values are useful tools in evaluating the climatic comfort of a place.

relative humidity water in air, expressed as a percentage of the amount the air could hold at its temperature
(see Fig. 2-1)

latent heat transfer removal of heat when water evaporates from a surface
(see pp. 8 and 10)

THI Temperature-Humidity Index expression of the combined effect of temperature and humidity

FIG. 5-10. The Temperature Humidity Index, calculated by measuring the wet-bulb and dry-bulb temperature of the air with a psychrometer and putting the results in the formula:

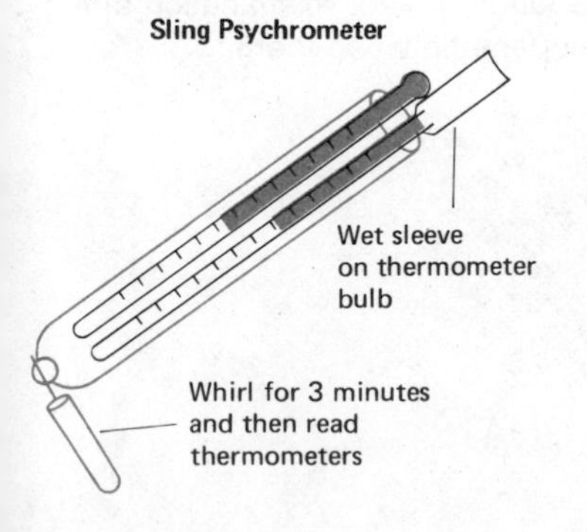

Formula

$$\text{THI} = \frac{4\,(T^{\circ F}_{\text{wetbulb}} + T^{\circ F}_{\text{drybulb}}) + 150}{10}$$

Consequence

THI = 70 A few people are uncomfortable
75 Half the people are uncomfortable
80 Most people are uncomfortable
85 Some people may suffer heat stroke

A WATER BALANCE CLASSIFICATION FOR CROPS

The Thornthwaite classification of climates rests on the same foundation as the physiologic ones, except that green plants are used as the reference organisms. The relationship between moisture supply and demand underlies the exceedingly simple logic of the Thornthwaite classification: the suitability of rainfall in a particular place should be evaluated on the basis of the amount of water a crop needs to survive there. High temperature and dry air force California cabbages to drink more than Jersey ones in order to offset their moisture losses.

water balance balance between precipitation (income) and potential evapotranspiration (debts) (see Fig. 14-2)

potential evapotranspiration (PE) evaporative power of the air; the amount of water that could evaporate from a surface and transpire from its vegetation cover if an unlimited supply of water were available

Although a water balance classification is simple in theory, it is difficult to put into practice. The amount of moisture needed by a green plant depends on a large number of conditions. Temperature is important, but its effect is modified by wind speed, relative humidity, air stability, cloudiness, day length, ground slope, and plant characteristics. A number of investigators have tried to develop equations that express the relationships among the major factors. Most of the suggested formulas are complex, and some require data from specialized measuring devices.

The ability of soil and plants to store water confuses the issue even further. Storage makes some of yesterday's rain available for use in offsetting today's needs. A deep soil with a large capacity for water allows plants to survive a long dry season, whereas even a short drought can be lethal to plants rooted in a shallow soil. The complete water balance for a place depends on soil as well as climate, although it is possible to estimate the evaporative power of the air and to use that estimate to work out the moisture account for that place (Fig. 5-11).

CONCLUSION

Specialized classifications of climate are useful tools for planning but lose much of their value when they are put on a world map. The resulting map is too complex to use because the classifications were designed for a different purpose. One of the simple descriptive or genetic classifications of climate is probably best for the mental map that is needed to help interpret other features of the environment. With that, we now turn rather abruptly to look at the solid earth, to build the foundation for a later examination of the interaction between earth and air at the surface between them.

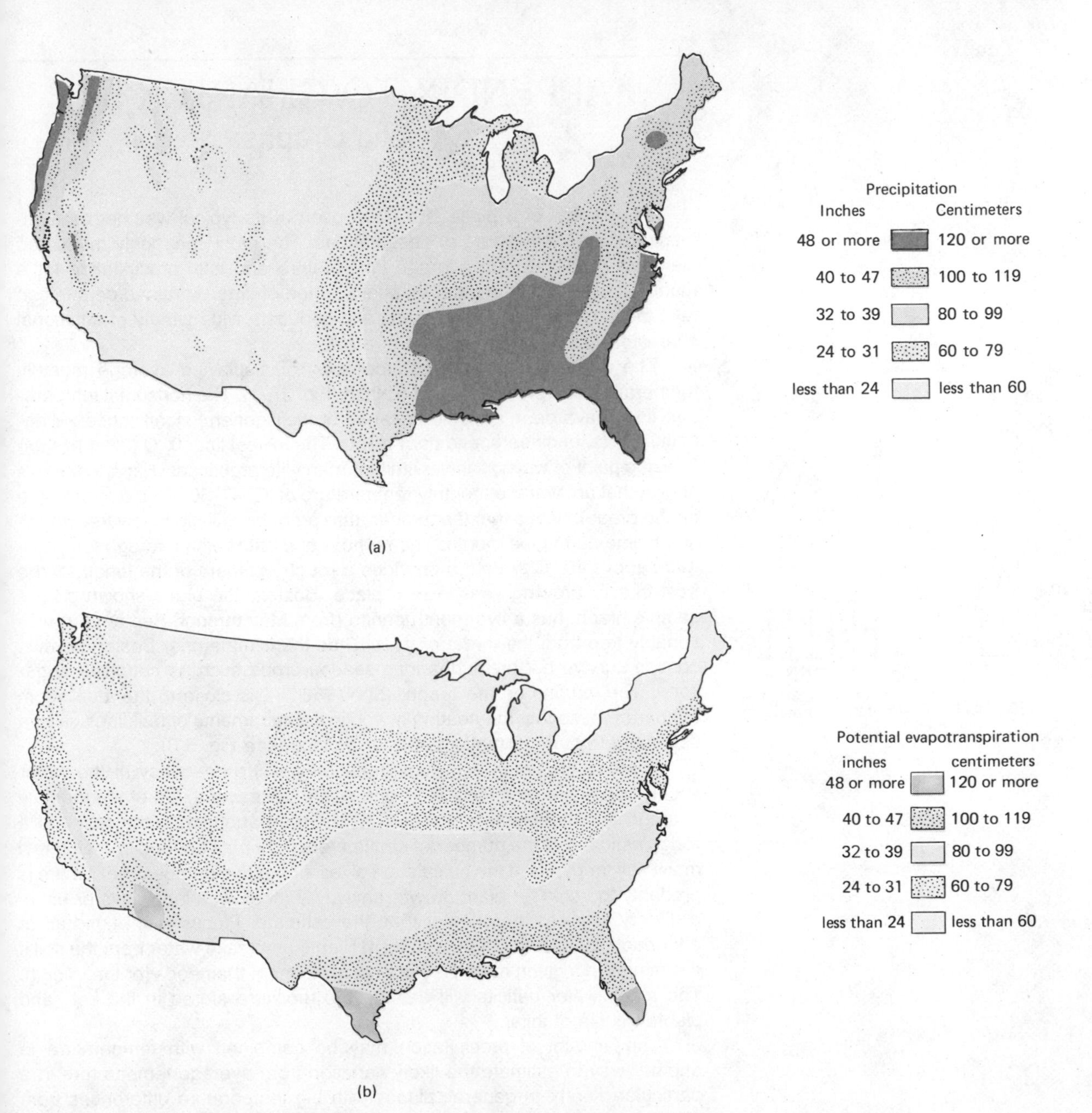

FIG. 5-11. The moisture account of the United States. (a) Annual precipitation (income). (b) Potential evapotranspiration (expenses). The annual account for a given place can be determined by comparing the two maps to find out whether the place has a moisture surplus or deficit. Monthly, daily, or even hourly data would make the moisture account more accurate and therefore more useful in making land use decisions.

APPENDIX TO CHAPTER 5
Climatic Graphs

The climate of a place is a description of its typical weather as summarized from many years of observations. The most commonly measured elements of climate are average temperature and total precipitation for a month, a year, or some other standard period of time. A trained person can take these two kinds of information and deduce a wide variety of additional characteristics about the climate of a place.

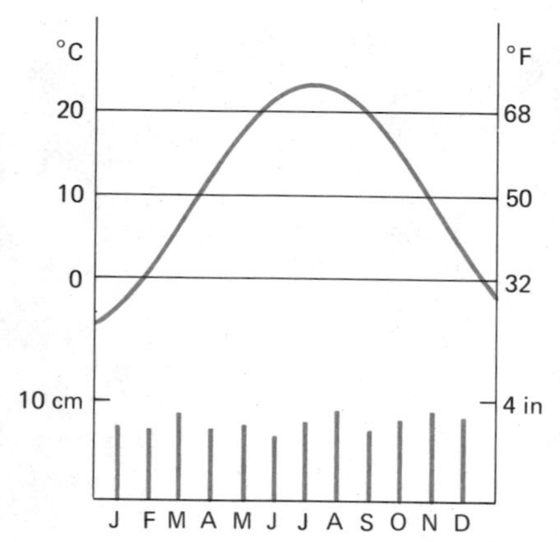

FIG. 5A-1. A standard climatic graph.

The climatic graph is a common way of displaying average monthly temperature and precipitation of a place (Fig. 5A-1). The horizontal temperature lines have been adopted because of their general significance for agriculturalists, engineers, and corn plants. The lowest line, 0° C (32° F), is the freezing point of water, a lower limit for many life processes. Experience has shown that an average monthly temperature of 10° C (50° F), the middle line on the graph) means that the temperature probably will dip below freezing at some time during the month. The number of months with average temperatures above 10° C therefore provides a rough estimate of the length of the frost-free or growing season in a place. Boston, the place shown on the sample graph, has a five-month period (from May through September) reasonably free from the threat of frost. Thus the climate near Boston permits corn to survive but precludes long-season crops such as cotton or sugar cane. The top line on the graph, 20° C (68° F), is close to the ideal room temperature, so that the heating or cooling requirements of buildings can be estimated from the climatic graph for a place (see Fig. 5-8).

The climatic graph also aids in estimating the adequacy of the water supply in a place (see Figs. 5-11 and 14-2). Take two-thirds of the monthly average temperature in degrees Celsius, then subtract 3; the result is a rough estimate of the number of centimeters of rainfall needed to maintain a moist soil for plants. If the calculation gives a negative number, the climate is probably too cold for plant growth anyway. Places with long days or especially dry air need more water than the estimate. Places with humid air or abundant clouds can get by with less. Plants must take water from the soil if the rainfall in a given month is not enough to meet the needs for that month. Too many water deficits will deplete the moisture stored in the soil, and plants will die of thirst.

The amount of precipitation may be combined with temperature in another way to estimate the likely variation from average temperature in a particular month. In general, places with big temperature differences from January to July also have fairly large day-night changes. Temperatures vary even more widely in dry months, because a dry soil heats and cools more rapidly than does a wet one, and a dry atmosphere contains few clouds and little water vapor to interfere with solar heating and radiative cooling. Places whose annual temperatures are moderated by heavy rainfall or nearness to an ocean will have relatively slight day-night changes in temperature.

A quick test. By now, you should be able to match the climatic graphs below with the appropriate places on the map. Anyone who can match them correctly without having to reread the chapters has an adequate understanding of climate for the rest of this book. The hints on the following two pages may serve as a useful summary of basic principles of climate.

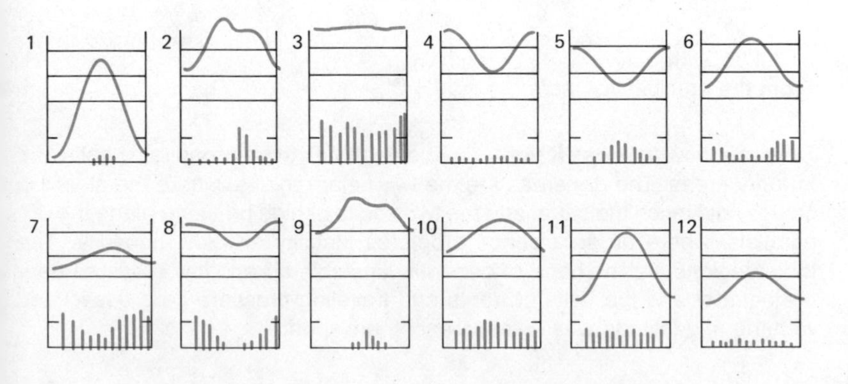

HINTS FOR LOCATING PLACES BY ANALYZING THEIR CLIMATIC GRAPHS

From the temperature curve:

In which hemisphere is it? Those months with the lowest temperatures would be the winter season of the place. To thwart the almost irresistible subconscious urge of U.S. citizens to link winter with December and January, it is customary to refer to the cool months as the low-sun season and to describe summer as the high-sun season.

How far from the equator is it? All other things being equal, a high temperature through the year indicates an equatorial location and a low annual average is typical of polar areas, because the total amount of annual incoming energy decreases with distance away from the equator. Temperature range between winter and summer, however, usually increases with distance away from the equator.

How far from the coastline is it? Places near large bodies of water have more moderate climates than inland places at the same latitude. The annual average temperature of the two places is similar, but the coastal place has less difference between winter and summer. The maritime effect decreases with distance from the coast—slowly across plains with onshore winds, more rapidly with offshore winds or rugged terrain.

How far above sea level is it? Thermometer readings usually decrease 3 to 7° C for every 1000 m of elevation (2 to 4° F per 1000 ft). A place with a narrow temperature range (usually indicating equatorial location) but a low annual average (usually indicating polar location) may be on an equatorial mountaintop.

From the precipitation bars:

How much water does it get? The earth has three broad rainy belts and four dry areas. The dry areas are the two polar regions, where the air is too cold to hold much moisture, and the two zones of high pressure along the 30° parallels, where air subsidence promotes stability and low humidity. The moist regions are the band of generally unstable air and low pressure near the equator and the belts of fronts and traveling pressure cells associated with the mid-latitude westerlies in each hemisphere.

When does it receive its precipitation? Distinct seasonality of rainfall usually is associated with the latitudinal shift of the dry subtropical areas of high air pressure. The subsidence areas move a few degrees poleward during the high-sun season and inflict dry summers on their mid-latitude neighbors. A shift toward the equator in the low-sun season brings drought conditions within a few degrees of the equator.

How far from the coastline is it? High air pressure dominates cold landmasses and hinders the onshore movement of moisture-carrying winds from the nearby oceans. Therefore inland places usually have more rainfall in summer than in winter.

From other maps:

Are there mountains near it? Mountains force winds to rise, cool, and drop moisture that they can no longer hold. Mountain ranges that lie in the path of prevailing winds create rainier than normal conditions on the windward slopes, colder than expected weather on their summits, and dry rainshadows behind them.

What kind of vegetation is there? The amount of rain needed to support a particular vegetation type depends primarily on the average temperature of the place and the seasonal pattern of rain. As a general rule, a place can support a forest if at least one month has a temperature above 10° C (50° F) and if its yearly rainfall (in cms) is greater than $2T + P/4$ where T is its average annual temperature in °C and P is the percentage of its rain which falls during the three warmest months. Half that much rain is generally adequate for grasses; deserts receive less.

Chapter 6

Plate Tectonics: The Not-So-Solid Earth

If the continents were not made of an especially light kind of rock, we would all need gills.

Robert Wallace, U.S. Geological Survey

An observer by a muddy stream after a storm might ask, "How long will it take for all the land to be washed out to the sea? Why are land and sea separate in the first place?" Though it may seem odd, part of the answer lies in the linear arrangement of high places on the earth. Even the words humans use to describe mountains are linear—Cascade *Range,* Western *Cordillera,* Carpathian *Arc,* Blue *Ridge.* There are mountains under the sea, some protruding above the water as islands, but they also are usually organized in linear patterns. Island names frequently include words like range, arc, and chain, because isolated islands, like isolated mountains, are rare.

cordillera long backbone-like chain of mountains

The elongated arrangement of high places is a significant clue to the nature of forces within the earth. However, the earth reveals more about itself when it shakes than when it stands still.

INTERNAL STRUCTURE OF THE EARTH

A geologist labors under two handicaps that do not hinder atmospheric scientists:

Time. Atmospheric processes take place in a matter of days or years, while many geologic events require millennia. Extremely slow processes force a mortal scientist to choose between two less-than-desirable alternatives: speeding up a process in the laboratory or trying to reconstruct past events on the basis of present observations.

Mass. Most rocks are hard, heavy, opaque, and buried from sight beneath tons of soil. Humans can penetrate only a few thousand meters into the interior of the earth, hardly comparable with the easy access to the atmosphere provided by satellites, aircraft, balloons, and telescopes. As a result, the interior of the earth usually must be observed indirectly.

Vibration can be a useful substitute for sight or touch. Some materials transmit vibrations more readily than others, just as some objects are opaque and others transparent to visible light. A particular material may reveal its identity by bending, reflecting, or changing the velocity of vibrations in a unique fashion.

seismograph instrument to measure earthquake vibrations

Measurements of vibration during earthquakes indicate that the rigid outer crust of the earth is only about 10 to 100 km thick, proportionately thinner than the shell of an egg (Fig. 6-1). Below the rocky outer shell is a plastic (deformable) material that is able to flow under pressure, a fact of profound importance for the surface crust. Like the circulation of air in response to temperature differences in the atmosphere, convection also apparently occurs within the earth, though on an immensely slower scale. Hot mantle material rises in some places, moves laterally while it cools, and sinks into the interior of the earth in other places. The fate of the material after it descends into the earth is less well-known than its rise, its horizontal motion, and its eventual return to the interior.

lithosphere rigid outer shell of the earth

asthenosphere semisolid part of the upper mantle beneath the rigid lithosphere

convection system set of movements caused by inequalities in temperature and density (see Fig. 3-1)

P-wave push-pull or compressional wave, as when people in a single-file line surge forward and back

S-wave side-to-side or shear wave, as when a guitar string is strummed

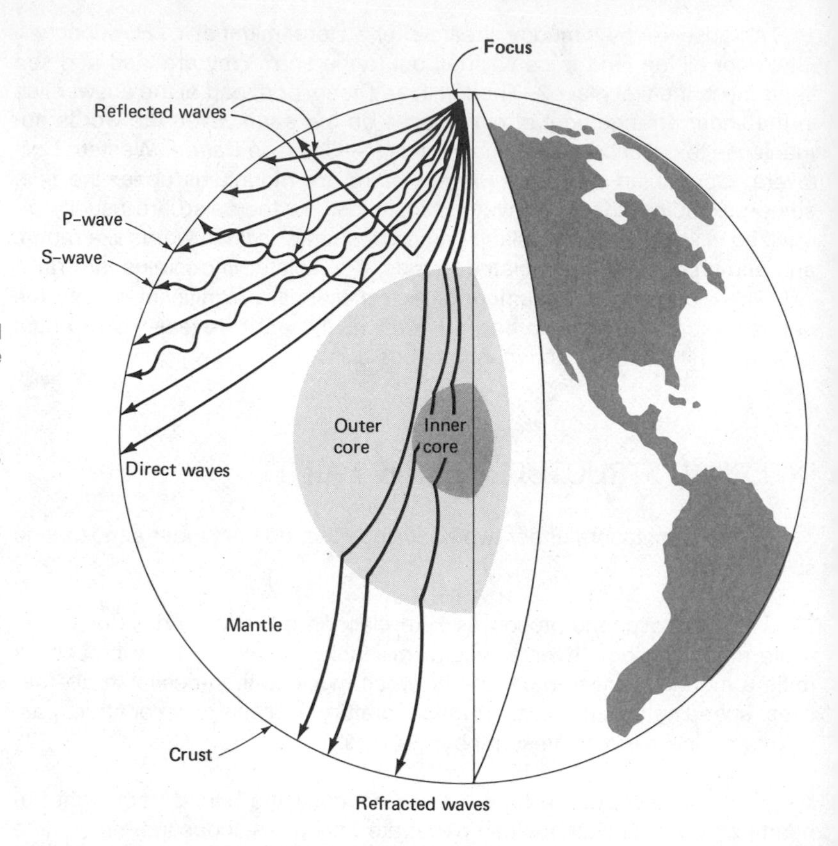

FIG. 6-1. Internal structure of the earth. Different kinds of waves from an earthquake move differently through dense, plastic, or cold materials than through substances that are light, solid, or warm. (Adapted from Gabler.)

EVIDENCE FOR CRUSTAL MOTION

School children frequently bemuse themselves by mentally piecing together the coastlines of Africa and South America and concluding that they fit fairly well. This little jigsaw exercise formerly would be discounted for its preposterous implication that entire continents must have moved. Scientists would not accept continental motion unless they had two things: proof of an initial connection of the continents and evidence for a force strong enough to tear landmasses apart. In the past few decades such proof has been rapidly accumulating.

fossil rock containing or consisting of remains or imprints of former life forms

Fossil life forms provide one way to reconstruct past events. Two continents that were formerly connected should have similar fossil records. Life forms should appear or go extinct at about the same time in the records from different parts of a connected landmass. Different continents, on the other hand, have different kinds of plants and animals, and the arrival and extinction of species would not be synchronized. The fossil records of Africa and South America show similarities in the distant past and differences in more recent times, implying past connection and present separation.

The kinds, structures, and arrangements of rocks furnish another clue to the history of continents. Geologic processes on separate continents take different courses, and maps of present-day structure on the two continents would be unrelated. Geologic patterns should mesh if landmasses were once connected. Once again, the evidence points to a connection in former times (Fig. 6-2).

stratigraphic column arrangement of rock types in a place according to their ages of formation

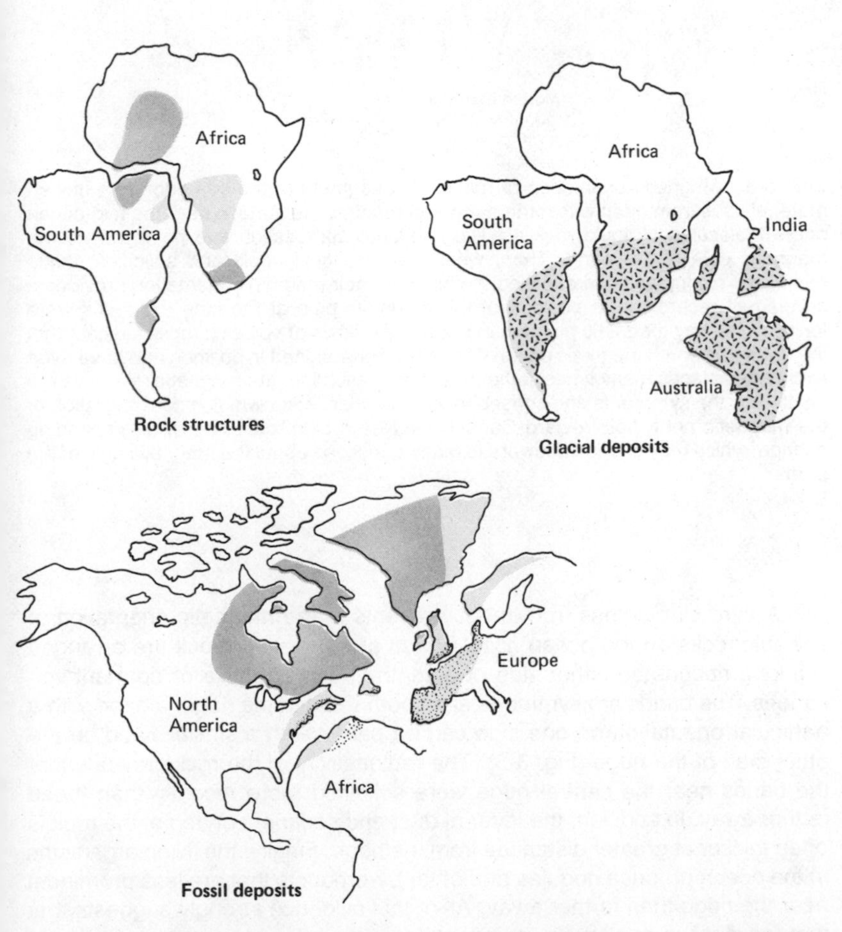

FIG. 6-2. Similarities of rock structure on opposite sides of an ocean imply that presently separate continents were formerly connected. (Adapted from both Morley and Wilson.)

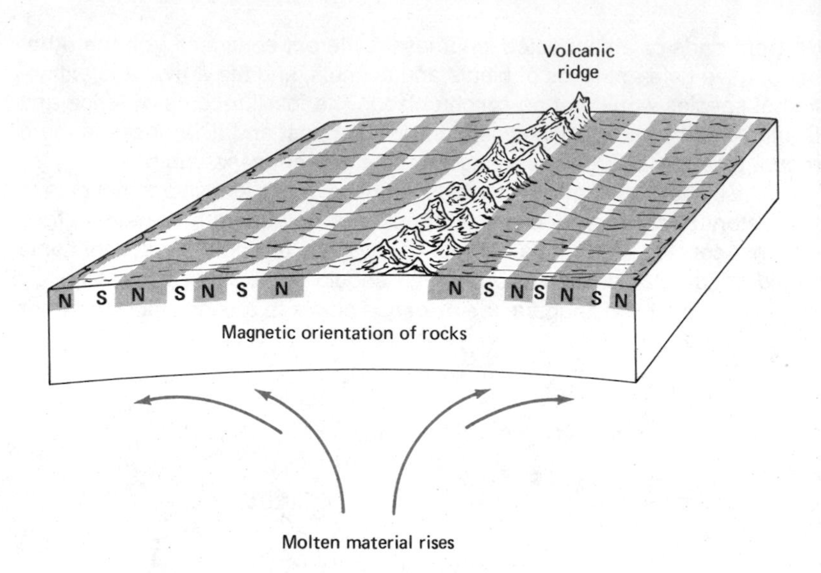

FIG. 6-3. Magnetic orientation of rocks on a segment of the sea floor. Hot molten material rises from deep within the earth and reaches the surface near the mid-ocean ridge. Molecules of liquid rock act like miniature compasses and point toward the magnetic pole of the earth. Then, when the hot liquid cools and solidifies, these microscopic compasses are locked in position. Their magnetic orientation provides a permanent record of the position of the magnetic pole at the time the rocks were formed. The tiny magnetic pointers in successive flows of volcanic rocks indicate that the north and south magnetic poles of the earth have shifted in position and have even exchanged places many times in the past. The reasons for the magnetic reversals are mysteries; the symptoms and consequences are also unknown. A recent migration of the magnetic north pole toward Canada may be a significant hint of an impending change, which may provide answers to many questions about the magnetic field of the earth.

magnetic orientation tendency to point toward the north or south pole of a magnetic field

polarity north or south magnetic orientation

A third clue comes from measurements of the magnetic orientation of volcanic rocks on the ocean floor. Bands of magnetized rock are arranged like long ribbons on either side of a central ridge of active or dormant volcanoes. The bands are symmetrical on both sides of the ridge: a band with a particular orientation on one side can be paired with a similar band on the other side of the ridge (Fig. 6-3). The radioactivity of the rocks reveals that the bands near the central ridge were solidified more recently than those farther away. In addition, the layer of dust and sediment on top of the rock is often thicker at greater distances from the ridge. Finally, the living organisms in the ocean produce nodules and other by-products that are less prominent near the ridge than farther away. All of this evidence strongly suggests that the sea floor is continually being created at and spreading outward from ridges in the ocean floor. The spreading sea floor, in turn, pushes the continents apart.

magnetic reversal reversal of north and south poles of a magnetic field

mid-ocean ridge or rift linear group of volcanic hills on the ocean floor

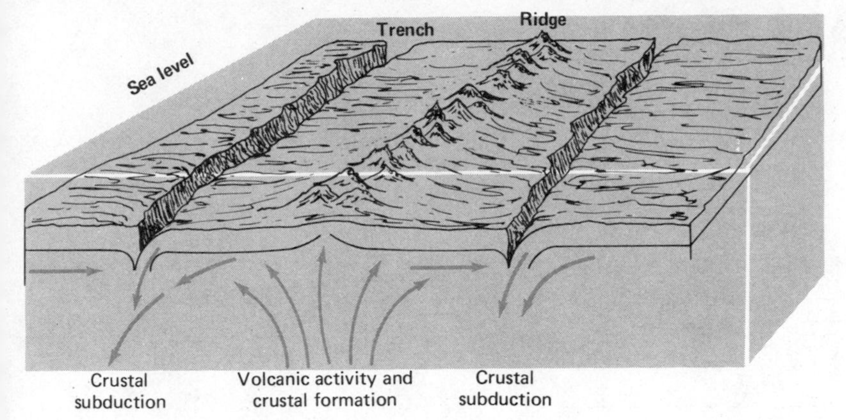

FIG. 6-4. Ocean topography and crustal movements. The vertical scale on this diagram is exaggerated to emphasize elevation differences.

OCEANIC EFFECTS OF CRUSTAL MOTION

The surface configuration of the ocean floor is a consequence of the motion of the earth's crust. The mid-ocean ridges lie beneath approximately 3 kilometers of water (about 10,000 ft), but they are the shallowest parts of the open oceans. The sea floor is closer to the water surface only near the edges of continents and islands.

continental shelf zone of shallow water near the edges of a continent

According to one theory, convection currents of semisolid material beneath the crust are responsible for the shallowness of the ocean above the volcanic ridges. Upward currents push the ocean floor up near the ridges. Most of the rising material then flows horizontally away from the ridges, and the sideways motion pushes the crust apart. Some of the rising material solidifies to become new crustal rock in the middle of the ridge. Farther away, the ocean is deeper and the bottom smoother because upward pressure is lacking. Still farther away from the ridge, the subterranean currents move downward, and the surface crust also bends down abruptly. The result is a narrow trench where the ocean bottom may be more than 10 km (6 mi) below sea level.

rift zone (ridge) place where new crust is being formed while older crust spreads apart

The downward plunge and disappearance of rigid crust near the trenches suggests an alternative to the convective-push theory of crustal motion. According to the subductive-pull theory, the sinking of solid crust near the trenches literally pulls the crust apart at the rift zones, and the rising material near the ridges is a consequence, not a cause, of the crustal separation there. The priority of interior chickens and crustal eggs is currently under vigorous debate.

subduction zone (trench) place where the crust descends into the earth's interior and melts

The consequences of crustal motion are better understood than the causes. The rocky crust is being formed at the ridges, distorted where pieces of crust slide past one another, and destroyed at the trenches. These changes do not occur silently; nearly all volcanoes and earthquakes occur near the edges of independently moving pieces of rigid crust (Fig. 6-5).

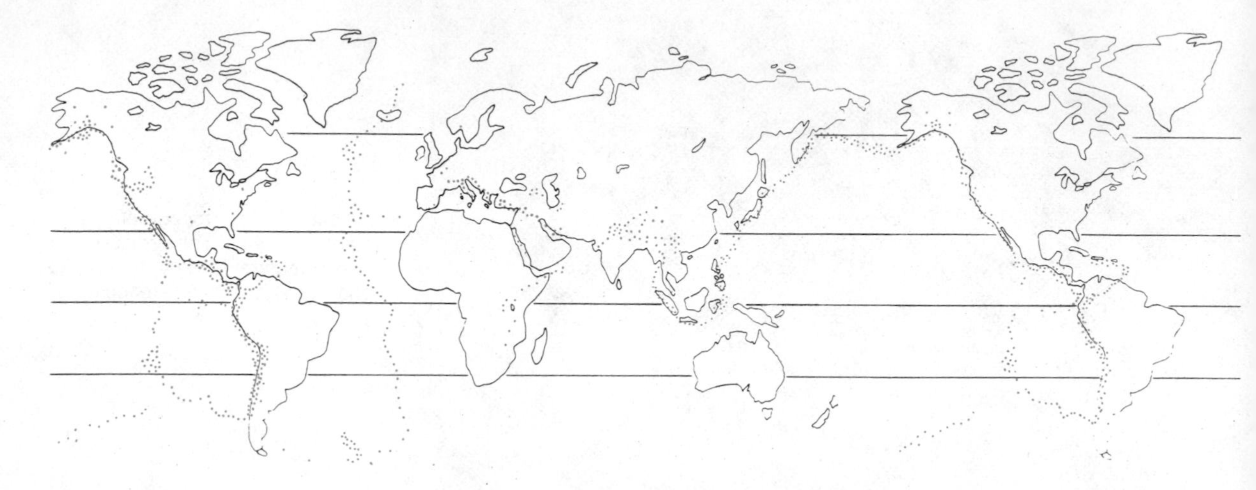

plate independent slab of rigid crustal rock

FIG. 6-5. Distribution of major earthquakes. The dots that represent earthquakes occur in clusters and lines along the margins of the separate slabs of crust, which cover the earth like the armor plates of a medieval knight. The most famous zone of earthquake activity is the "Ring of Fire" which surrounds the Pacific plate—the largest single piece of solid crust. Another zone of intense earthquakes stretches along the southern border of the Eurasian plate, from northern Italy to northern India and Indochina. A long line of milder earthquakes marks the mid-Atlantic Ridge.

volcano place where molten rock material pushes upward from the interior of the earth and erupts onto the surface

seamount underwater mountain, often with a flat top planed by wave action

The Hawaiian Islands in the middle of the Pacific plate are exceptions to the rule that volcanoes occur at plate boundaries, but they still help confirm the theory of crustal motion. For some unknown reason there seems to be a local "hot spot" in the mantle beneath the Pacific Ocean. The crust is moving from an ocean ridge southeast of Hawaii toward a trench northwest of the islands. The southeastern end of the island chain is presently over the hot spot. Periodic earthquakes and volcanic eruptions on the southeastern islands are evidence of the underground activity. The fiery formation of the islands to the northwest took place long ago, when the part of the crust they now occupy was over the hot spot. Since the time of their birth, the crust has moved away from the area where internal currents rise, and the volcanic activity on the northwestern islands has ceased. Still farther to the northwest is a chain of underwater mountains, the sunken and burned-out remnants of volcanic activity that occurred even farther in the distant past. Their watery silence foretells the ultimate fate of the present islands (Fig. 6-6). Fortunately for Hawaiian realtors, the process takes millions of years, so that the forecast of future submersion need not dissuade anyone from buying a retirement home in Honolulu.

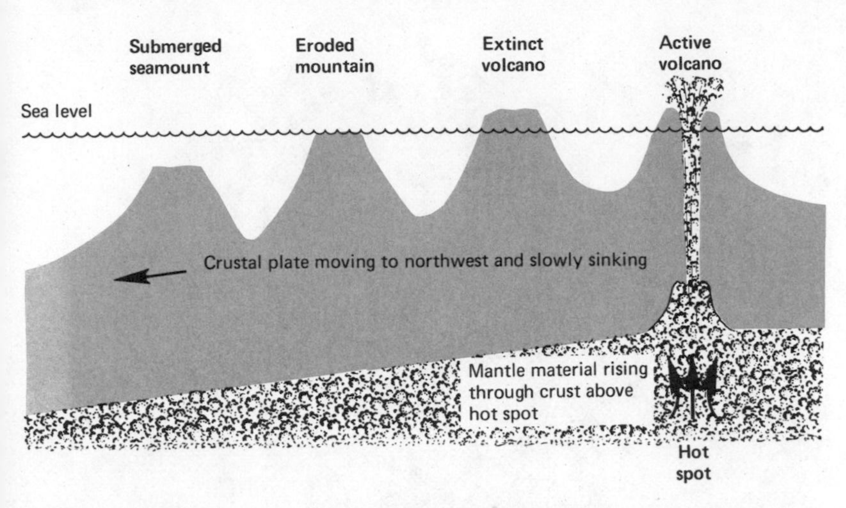

FIG. 6-6. Volcanic islands, past and present. Different islands in the Hawaiian chain represent different stages in the evolution of a typical island in the group.

CONTINENTAL EFFECTS OF CRUSTAL MOTION

The effect of crustal movement on continents is more complex than its effect on the ocean floor. Most continental rocks are significantly lighter than the rocks underlying the ocean waters. In effect, the continents are like light-rock boats floating in a heavy-rock pond (Fig. 6-7). Geologists formerly thought the continental boats were able to move slowly through the heavier crustal material around them. The phrase "continental drift" was used to describe their incredibly slow migration. Scientists now prefer the analogy of boats stuck in the ice of a frozen lake. The ocean floor consists of frozen rock material which can slide across essentially similar but unfrozen mantle material beneath. As the frozen crust moves sideways, it carries the trapped continental boats along with it.

Whenever two crustal plates collide, one slides beneath the other, sinks into the hot interior of the earth, and melts. Continental rocks, however, are too light and buoyant to continue down along with the heavy rocks below and around them. The light rocks usually remain with the part of the crust that stays on the surface.

The linear arrangements of many mountains are consequences of crustal motion. In effect, ranges of high mountains are the crumpled fenders of titanic but incredibly slow collisions between crustal plates. The extensive chains of mountains stretching from Alaska through California and Mexico to southern Chile are being pushed up as westward-moving American pieces of crust overtake the slower Pacific plate. Japanese volcanoes mark a zone of head-on plate collision on the far side of the Pacific. The complex mountain pattern in southern California is a result of conflicting forces unleashed when a small crustal plate was trapped between the giant American and Pacific plates.

FIG. 6-7. Density differences in crustal rocks. The continents stand above sea level because they are actually masses of light rock "floating" on the heavier rock that underlies the ocean basins.

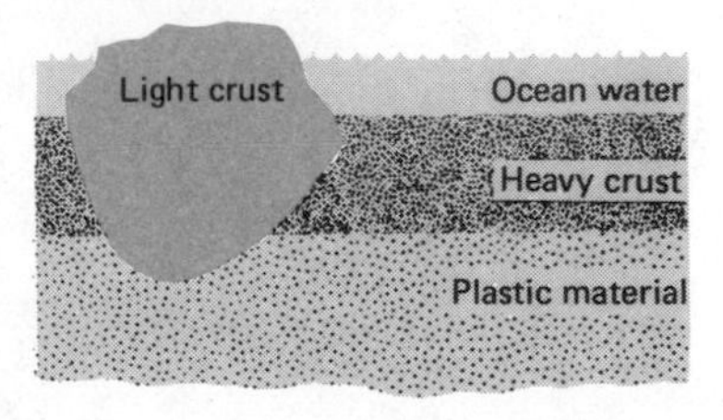

sial light rock made of silica-alumina compounds (also called silicic, felsic, granitic, or continental rock)

sima heavy rock made of silica-magnesia compounds (also called ferro-magnesian, mafic, basaltic, or oceanic rock)

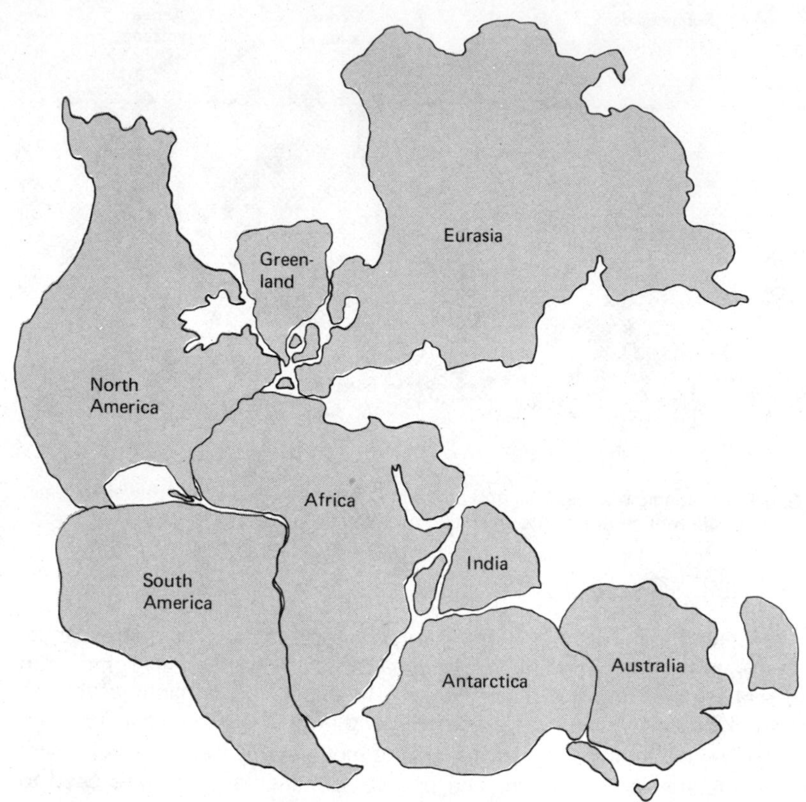

FIG. 6-8. Pangaea, the ancient supercontinent.

PANGAEA, THE ANCIENT SUPERCONTINENT

An ancient supercontinent can be reconstructed by mentally reversing the motion of the pieces of crust and running their history backwards (Fig. 6-8). If Pangaea existed, did it begin as a supercontinent, formed when the earth was shaped, or was it also a combination of smaller continents? The Himalaya Mountains of south Asia may offer a clue. Evidence from the Indian Ocean suggests that the subcontinent of India is moving northward on its own separate piece of the crust. The lofty Himalaya is being lifted as the Indian plate slowly pushes into the Asian landmass. In a similar manner the Alps and other mountains around the Mediterranean can be interpreted as evidence of a collision between Africa and Europe. The Appalachians are not near present-day zones of plate collision, but they may indicate a continental collision in the remote past. A band of volcanic rocks from Minnesota to Oklahoma appears to be the healed scar of an ancient rift, a zone of former crustal spread and volcanic action. Similar rifts can be found in Africa and the Middle East, while the Ural Mountains in Russia may be well-worn souvenirs of the ancient marriage of Europe and Asia (Fig. 6-9).

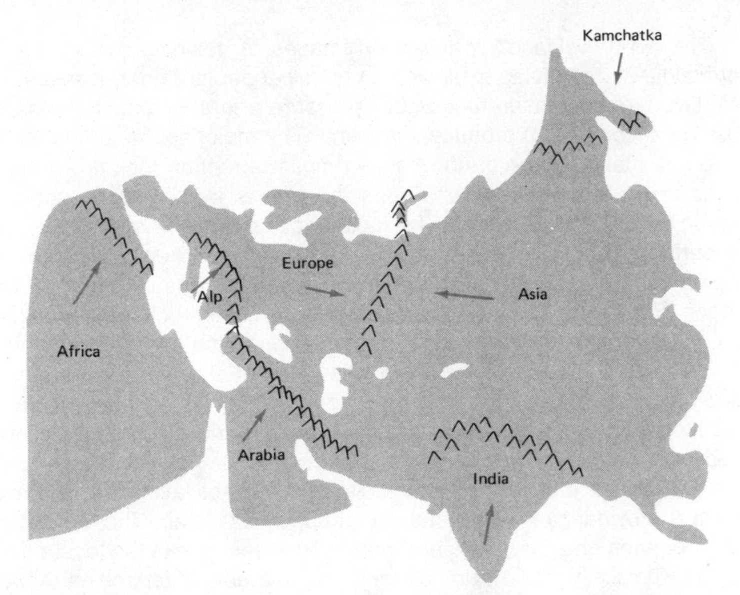

FIG. 6-9. Eurasia may represent the temporary union of several formerly separate landmasses.

Some early proponents of this grand hypothesis of crustal movement, continental collision, and mountain building offered a conveyor belt analogy of continental origin. The conveyor belt is the heavy ocean-bottom kind of crustal material that rises in one place, moves horizontally, and sinks in another place. The light continental material rides on top of the conveyors, while both conveyor and cargo are driven by forces beneath the solid crust. At times the motion of crustal plates tears continents apart. At other times the conveyors bring the light continental material together and pile it up where crustal plates collide and sink.

plate tectonics theory that the crust consists of discrete "plates" or "rafts" of rigid material that can shift in position and carry the continents with them

Pangaea, the former supercontinent, could have been such a temporary accumulation of continental rock. Its sequel, Child of Pangaea, may appear in a few hundred million years if the Americas continue to move to the west while Eurasia drifts eastward.

VOLCANIC FEATURES

The analysis of crustal activity has been very general so far. A sharper focus is needed for smaller features on the surface of the earth. At the scale of individual mountains, landforming processes can be separated into two groups. The first group consists of the internal processes that create volcanoes, cause earthquakes, and bend and lift the rocks of the earth's crust. The second group includes the external processes that break rocks apart, move loose material, and reshape the surface of the earth. Proper understanding of the external processes requires information about soils and vegetation, which are discussed in Chapters 8 through 14. Further study of the consequences of external processes is therefore postponed to Chapter 15, and this section will treat only those surface features which are primarily the handiwork of internal earth forces.

tectonic force internal force that deforms the crust

gradational force external force that shapes the land surface

lava molten rock on the earth's surface

magma molten rock beneath or in cracks inside of the earth's crust

extrusion molten rock that penetrates the solid crust and flows out onto the earth's surface

intrusion molten rock that penetrates the solid crust but does not go all the way to the earth's surface

unstable air air that tends to rise all by itself
(see Fig. 2-2)

The word "volcano" conjures up images of belching smoke, shaking mountains, and terrifying explosions, but volcanoes are not necessarily violent. Two factors, molten rock under pressure and a weakness in the rigid crust, must combine to produce a volcano. The major source of molten rock is the hot interior of the earth. A rising mass of molten rock flows through cracks in the rigid crust and widens the cracks if the rock is not strong enough to resist the pressure. The molten rock may penetrate all the way to the surface in places where rocks are weak or broken. Elsewhere, the magma merely penetrates part of the solid crust but does not make it to the surface (Fig. 6-10). The hidden movements of molten rock may be revealed later when removal of the overlying rock exposes the hardened remains.

Volcanoes associated with the heavy material under the oceans are seldom explosive. Headline-making eruptions near Iceland and Hawaii are actually gentle oozings of lava through cracks in the crust. Great billowing clouds of steam may appear when the molten rock enters the sea. Fires mark the abrupt end of forests engulfed by red-hot lava. Tall columns of smoke are evidence of the instability of superheated air. These seemingly violent phenomena, however, are merely overreactions of the local environment to a basically placid extrusion of molten material. Even the earthquakes that occur near the ridges are usually gentle shiftings of soft, warm rock. The emerging lava is the heavy kind of rock material found on ocean bottoms.

Trenches, on the other hand, often produce a more violent type of volcano. The earth explodes into activity when this kind of volcano erupts. Earthquakes are intense, and superheated gases blow cinders, ashes, and partially melted rocks out of the volcanic crater. These gases form only when continental rocks melt. For this reason, explosive volcanoes are common only where the sinking crust tries to pull continental rocks into the interior of the earth. A map showing where continental rocks are close to zones of crustal sinking serves as a good guidebook for a tour of historic volcanic disasters—Vesuvius, Tarawera, Krakatoa, Pelée, Usu, Páricutin, and others.

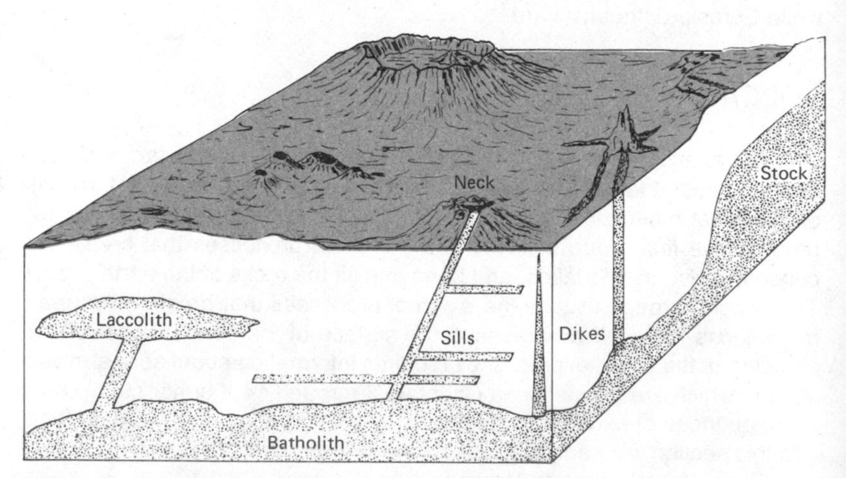

FIG. 6-10. Intrusions of molten rock into the solid crust.

FIG. 6-11. A shield volcano. The gently sloping hill is a consequence of outpourings of hot oceanic lava. (Courtesy of G. A. MacDonald, Geological Survey.)

Volcanoes cause different surface features whose shapes depend on the nature and intensity of the eruptions. The flow of lava from a shield volcano is usually a slow but relentless oozing down a gentle slope (Fig. 6-11). Mobile creatures have little to fear from such a lava flow; they usually can walk away from the hot rock material, which rarely advances faster than a few kilometers per hour. The burial or incineration of trees, soil, and buildings leaves a desolate landscape, but the destruction is not permanent. Volcanic rocks break down to become exceptionally rich soils, often the best soils in some tropical regions. Thus shield volcanoes can be a blessing to the inhabitants of a place.

shield volcano gentle extrusion of molten rock

However, volcanic material does not necessarily move at a leisurely pace. The gas-powered dust that burst out of Pelée and Vesuvius gave thousands of people barely enough time to notice the awesome spectacle of a fiery cloud approaching at interstate highway speed. Explosive volcanoes of this type may build mountains overnight, rip islands apart, and create gigantic waves that travel thousands of miles and destroy coastal cities on the far side of the ocean.

tsunami immense ocean wave created by a volcano or earthquake

Intermittently explosive volcanoes are even more awesome. Molten rock cools and hardens in the volcano during periods of quiet. Gases and steam may build up pressure beneath the plug and eventually blast it out. The resulting rain of cinders and volcanic bombs produces a characteristic steep-sided cone around the volcanic crater (Fig. 6-12).

cindercone violent extrusion of semisolid material and gas

FIG. 6-12. A cindercone—the steep-sided hill is a consequence of explosive eruptions of continental lava, gas, ashes, and volcanic bombs. (Courtesy of F. M. Byers, Geological Survey.)

One historic volcano, Krakatoa, discharged great clouds of dust, an element of potential destruction even more fearsome than hot lava, cinders, earthquakes, or ocean waves. Krakatoa erupted with megaton bomb force. Its cloud of debris rose high into the atmosphere and reached the stable layer above the tropopause. The dust in the upper atmosphere produced spectacular sunsets around the world for years; it also blocked some of the incoming solar energy and has been linked with a documented cooling of global climate. A series of such cataclysms could cause widespread famine by shortening the growing season in many important agricultural areas.

tropopause upper limit of the occasionally unstable layer of air near the ground

stratosphere stable layer of air above the tropopause (see Fig. 2-12)

OTHER TECTONIC FEATURES

Vertical and horizontal movements of the crust inevitably place it under a great deal of stress. The rigid personalities of crustal rocks do not permit them to tolerate much strain without bending or breaking. The degree of crustal deformation depends on the kinds and arrangements of the rocks, their temperature, their depth below the surface, and the intensity and duration of the stress placed on them. Different combinations of these factors can produce both bending and breaking in the same complex geologic structure.

diastrophism movement of crustal rock under stress

A visitor to the Grand Canyon sees many layers of rock that obviously originated under water. Fossil seashells on hilltops can imply that at one time sea level was thousands of meters above its present position. While this interpretation is possible, modern theories of crustal movement suggest an easier alternative: deposit the material in the ocean first and then raise the continent to its present height by crustal collision or mantle movements.

Upward movements of molten material usually push the crust up into an arch without much deformation. Intrusions of molten rock between solid rock layers can also bend overlying rocks into an arch or dome. Applying sideways pressure on the rock layers is another way to form an arch, although the rocks on the sides usually get deformed by the stress. Sideways pressure can also bend the crust downward, but sags are more commonly created by the added weight of sediment or glacial ice.

anticline or dome upward fold of rock

syncline or basin downward fold of rock

Great but steady sideways pressure causes flat layers of rock to fold or bend like a pile of rugs shoved against a wall. The fold shapes are clues to the amount of stress placed on the rocks (Fig. 6-13). Most stresses act in one direction and form long ridges and valleys. This geologic linearity has profound implications for human settlement. Often the slopes are too steep for use and only the narrow hilltops or valley bottoms are suitable for agriculture. The linear ridges restrict human movement, because travel across the geological grain is very difficult. Gaps through the ridges often assume great significance as transportation arteries; Chapter 17 contains a discussion of the effect of Appalachian geologic history on modern transportation.

truncation removal of the top layers of folded rock by erosion (see Fig. 17-1)

FIG. 6-13. Geologic folds—layered rock is deformed gradually under sideways pressure. Notice that a series of linear ridges can be formed on at least five different kinds of geologic structures with different implications for a groundwater hydrologist or mining engineer.

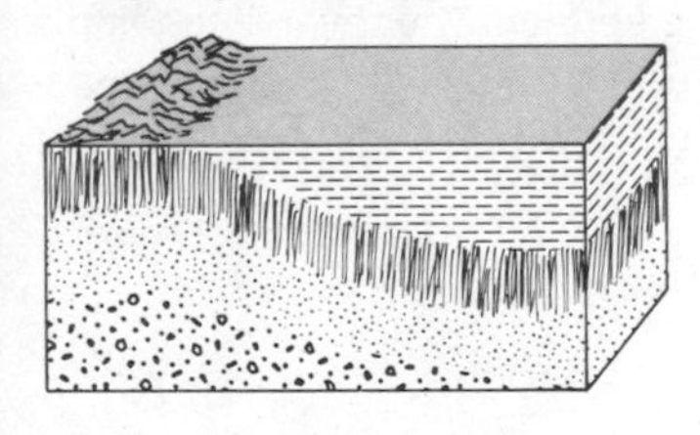
Monocline

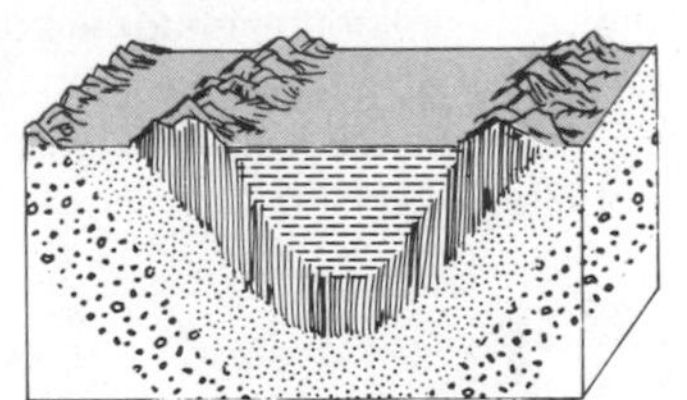
Syncline

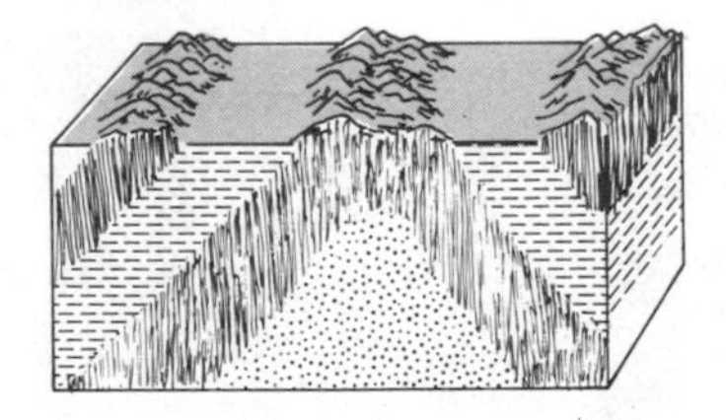
Anticline

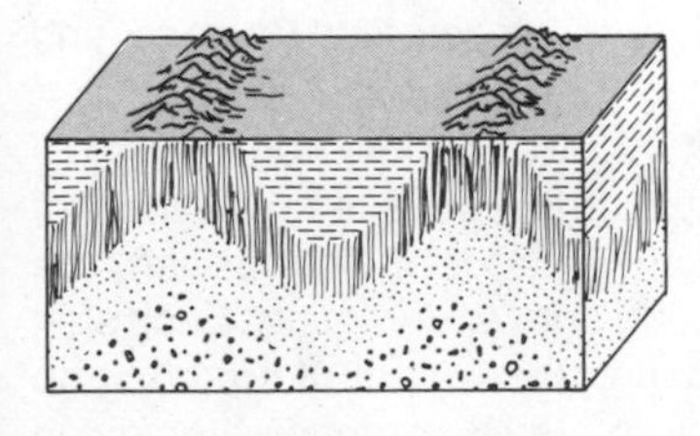
Wave train

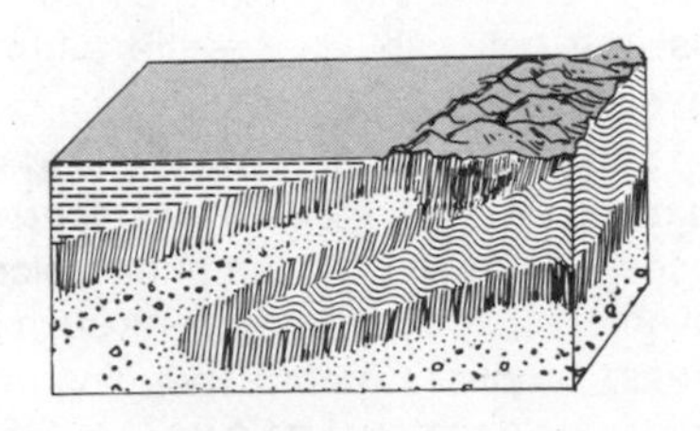
Overturn

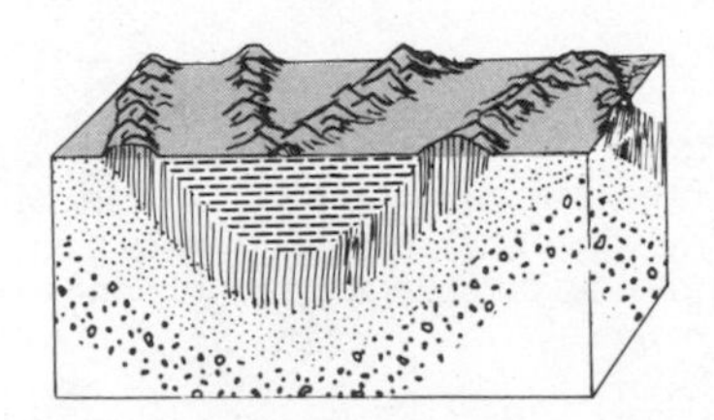
Plunging fold

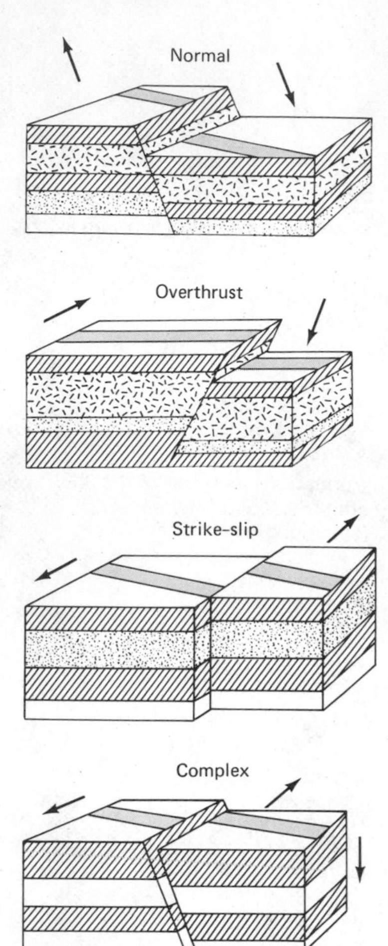

FIG. 6-14. Geologic faults—rocks broke or slipped along planes of weakness.

Rocks under extreme stress often break and slip along weak areas. Faulting, a general term for breaking or sliding of rock, can take many forms (Fig. 6-14). The key difference between folding and faulting is a matter of timing. Folding is gradual but persistent, slowly overcoming the resistance of rocks and molding them to its will. Faulting is more like a karate chop, a sudden release of forces that were marshalled over a long period of time. The visible results of faulting, like those of folding, are usually linear. Some famous faults, such as the San Andreas Fault in California, stretch for hundreds of kilometers in nearly straight lines (see frontispiece of this chapter). Vibrations of the rigid crust accompany the sudden release of tension as rocks shift along a fault. The inhabitants of the surface experience the vibrations as violent shaking of the ground. At present, earthquakes occur without obvious warning, but they may become predictable when scientists learn how to interpret the changes in rock tension that apparently always precede an earthquake.

The geographical patterns of arching, folding, and faulting are closely related to plate motions. Small sideways shifts are common when rocks located near the ocean ridges adjust to the spreading of the sea floor. Large faults occur where crustal plates slide past one another. Truly gigantic earthquakes, however, usually testify to the great contortions of rocks near the trenches, where the crust plunges into the interior of the earth. Violent explosions of superheated gases also trigger earthquakes where continental rocks are pulled down into the hot mantle. Like corks pushed down into a pool, the light continental material bobs to the surface whenever it can find a weakness in the rock above. Small wonder that the world map of catastrophic earthquakes should highlight areas in which continents lie near zones of crustal sinking.

ISOSTASY

The plasticity of the mantle is indirectly responsible for yet another class of crustal movements. The melting of the great ice sheets that once covered much of North America took a great load off the continent. Even now, thousands of years later, the continent still heaves an occasional sigh of relief as it shifts into a more comfortable position. The adjustments of the crust are noticeable as gentle earth tremors in places like Chicago and Cleveland.

isostatic adjustment vertical movement of crustal material to compensate for changes in weight on the mantle

Similar adjustments to changing weight occur at all times and in nearly all places. Material can be redistributed on the earth's surface by a variety of processes: crustal movements, volcanic activity, erosion, deposition, ice movements, and human actions such as mining or river damming. Each shift of mass produces a corresponding imbalance in the weight of the crust on the less rigid material below. The deformable material responds by moving slightly or by changing its density to compensate for the changes in pressure

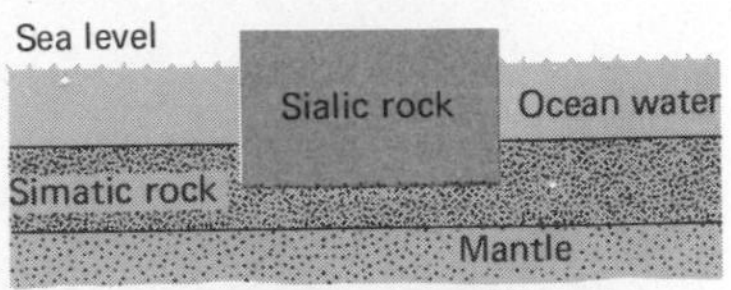

(a) Continent in position

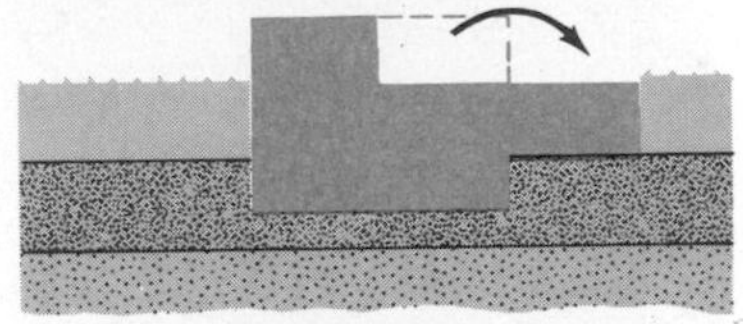
(b) Erosion and deposition

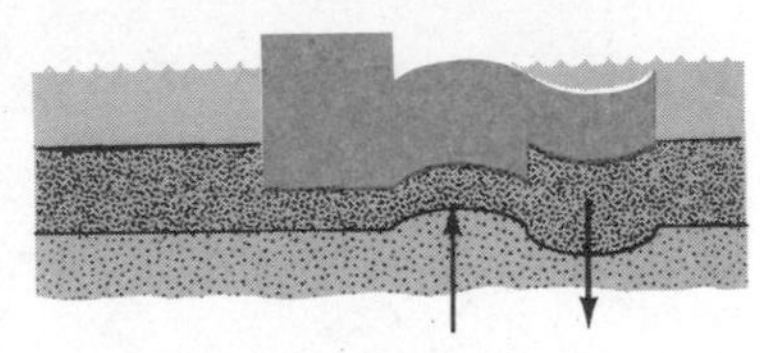
(c) Isostatic adjustment

FIG. 6-15. Isostasy. The ability of the mantle to move allows the heavy parts of the crust to sink and pushes the light parts upward.

(Fig. 6-15). The mantle movements, in turn, may disturb crustal rocks and cause rather violent earthquakes in places where even mild earth movements are rare. Unanticipated earthquakes are particularly damaging, because they strike people who thought they lived far from the areas prone to earthquakes, and therefore had no reason to expect the earth to move.

CRUSTAL MOTION AND WORLD CLIMATE

This chapter on plate tectonics has outlined a theory that explains the existence of separate continents and ocean basins and interprets the patterns of mountain ranges, earthquakes, and volcanoes. Extinct volcanoes, worn-down folded mountains, and inactive faults are symptoms of past crustal activity. The consequences of past crustal movements provide clues that help unravel some mysterious features of ancient geography. For example, fossils provide clear evidence of lush forests in places that today are barren deserts or icy wastelands. Scientists used to explain these fossil puzzles by postulating drastic changes in the pattern of climate. The complexity of the system described in Chapters 1 through 5 implies that reconstructing the weather machine in an entirely different way would be a formidable task. Nevertheless, scientists had to accept the idea of drastically different climates in the past, because the alternative of continental motion seemed so absurd.

Ironically, acceptance of the idea of crustal motion not only provides a mechanism that can move continents from one climate zone to another, but it also delivers fringe benefits in the form of three plausible theories for worldwide climatic change. Collisions of crustal plates usually involve only oceanic rocks and therefore are not too explosive. One plate simply dives beneath the other and quietly melts. If the sinking crust tries to drag continental material down, however, it lights a fuse that leads to cataclysmic volcanic eruptions. Properly timed volcanoes can produce persistent clouds that reduce solar radiation and cause worldwide cooling.

The slow building of a mountain range is less abrupt than a volcano but may have a greater effect on climate. The physical bulk of a new mountain alters the pattern of wind flow. A mountain range also accumulates precipitation on its windward side and creates a rainshadow on the leeward side. Changes in wind and moisture affect the patterns of temperature and pressure. In turn, changes in temperature and pressure may further alter the patterns of wind and rain. The system is so immense and complicated that the exact outcome of a given change may not be predictable.

epeirogeny upward movement of the crust on a continental scale

orogeny upward movement of crust to form a mountain range

In a still more subtle but perhaps even more significant manner, the slow rearrangement of continents and ocean basins changes the circulation patterns of the ocean and the atmosphere. The cumulative effect of these changes may be as profound as the melting of a polar ice cap or the creation of immense ice sheets over presently forested continents.

In these three ways the great internal forces of the earth have an impact on its climate. However, the situation is not one-sided; all of the products of tectonic activity are susceptible to modification by climate. The atmosphere begins to work on a mountain as soon as the earth raises it. Daytime heat and nighttime cold help to wedge rocks apart. Sunlight warms the surface and assists in the chemical breakdown of the rocks. Rain falls and washes soil to the sea. Winds blow sand and dust away from high places. The land-shaping work of the atmosphere is aided or hindered by other features of the earth, such as soil, vegetation, and water, which are partly under the direction of climate. It is helpful to examine these topics in greater detail before studying the surface topography of the earth itself.

APPENDIX TO CHAPTER 6

Geological Time

Suppose that a large storm occurs once every decade and removes two millimeters of soil from an Arkansas hillside. A simple calculation suggests that the weather and the Arkansas River will need only 4 million years to carry most of the state of Arkansas to the sea. The forecast rests on a basic assumption: all processes will continue to operate in the same manner as they presently work. Unfortunately, processes in the real world vary in speed, efficiency, and effect. Therefore it is necessary to have independent measures of time and process in order to unravel the geologic history of a place, and an understanding of the past is needed in order to understand the forces presently at work on the earth and how they interact with each other. This appendix consists of brief descriptions of basic geologic principles that enable scientists to reconstruct the past:

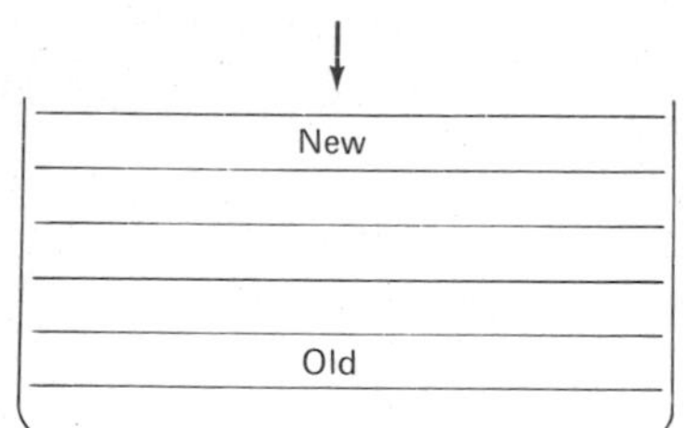

FIG. 6A-1.

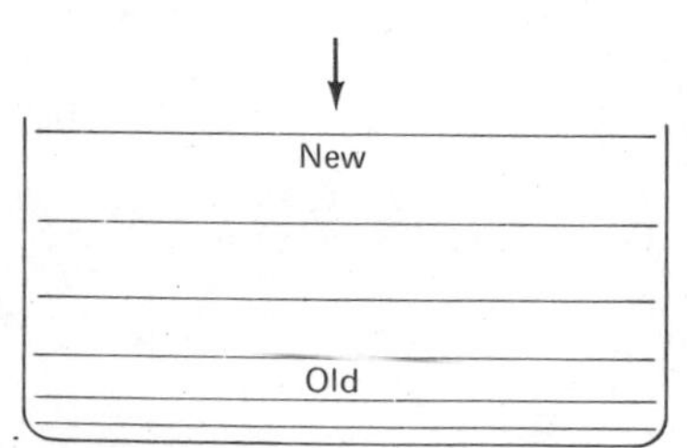

FIG. 6A-2.

Superposition. Material settles to the bottom of calm water whenever rivers and currents bring debris from surrounding land. Undisturbed sedimentary layers are accurate measures of relative time: each new deposit is laid on top of the previous ones, so that layers near the bottom are always older than those near the top (Fig. 6A-1). Ten meters of an incompressible material that accumulates at a constant rate of 1 cm per year would represent 1000 years of history. Unfortunately, additional deposits usually compress the older layers of sediment. A meter of compressed sediment near the bottom usually represents more years of deposition than a meter of sediment near the top (Fig. 6A-2).

Lamination (varving). Some deposits consist of distinct alternations of material that indicate seasonal changes in the depositional environment. Streams may carry coarse sand during flood season and tiny particles in times of low flow. Plants may die, sink to the bottom, and discolor the sediment during a particular time of the year. The resulting alternations of coarse and fine layers or dark and light bands are like tree rings: each pair of bands indicates one year of deposition (Fig. 6A-3). Unfortunately, varved sediments are rare because the annual rate of sedimentation is usually so slow that minor disturbances such as worms or currents can mix the layers and obliterate the regular seasonal alternation of deposited material. The justly famous exceptions are deep, calm, nutrient-poor lakes with reliable sediment deposition, weak currents, and little life.

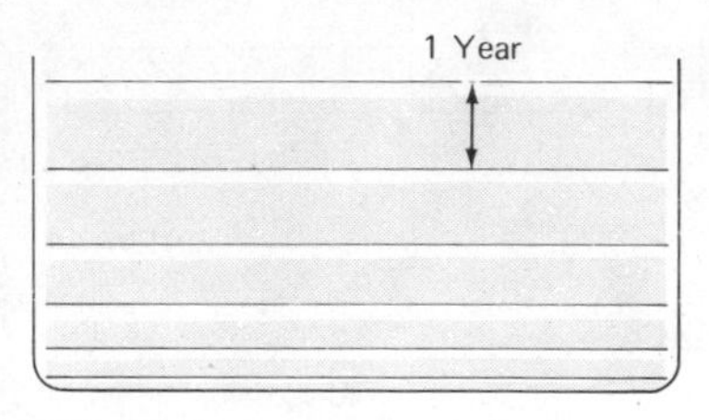

FIG. 6A-3.

Lithologic correlation. Distinct sequences of sedimentary layers or lava flows allow scientists to piece together the history and structure of large areas. A well drilled in eastern Iowa may pass through white limestone, shale, and then tan limestone. A nearby well may strike shale, tan limestone, and sandstone. A simple correlation would suggest that a layer of white limestone has been eroded away from the second area and that the first well would have entered a sandstone layer if the drillers had gone farther down (Fig. 6A-4). Lithologic correlation is easy in Indiana, where flat layers of sedimentary rock merge gradually into one another. Correlation is difficult in California where tectonic forces have rearranged the rocks into a complex mass of faulted and folded layers. Correlation becomes almost impossible when rocks are nearly homogenous, with no distinct "marker layers," or where rocks are extremely variable in origin, with no single formation that extends over any sizeable area.

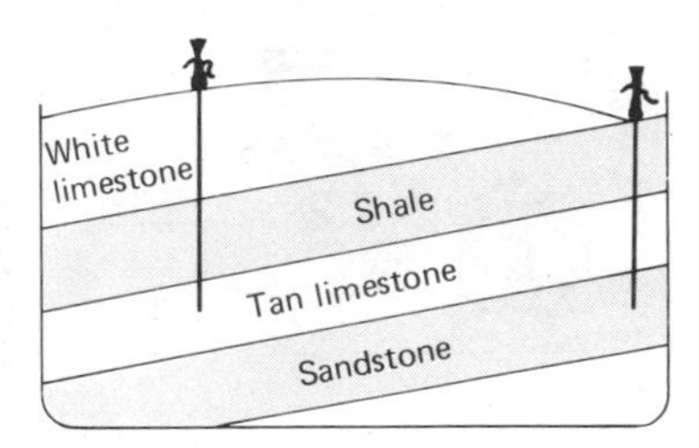

FIG. 6A-4.

Fossil correlation. Life forms sometimes leave records of their existence in the form of shells, bones, teeth, footprints, or other imprints in rock layers. Some animals and plants apparently lived for only a short period of geologic time, and their fossils appear only in rocks that formed during the proper moments of earth history (Fig. 6A-5). The remains of organisms that were widespread in space but short-lived in time have earned the title of "index fossils." The presence of a particular index fossil in a rock enables scientists to put the rock into its proper place on the Geologic Time Scale at the end of this appendix (Table 6A-1).

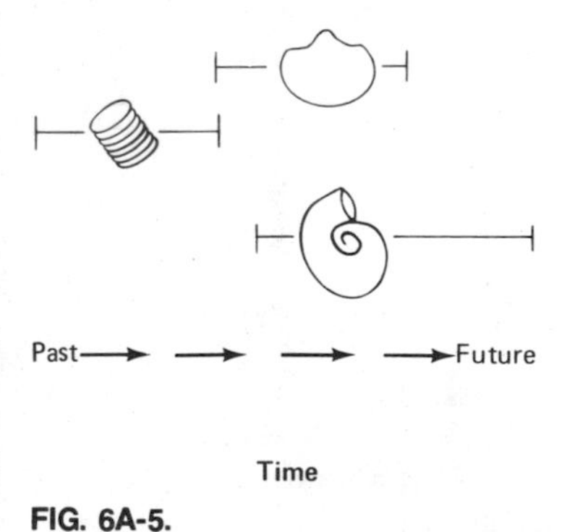

FIG. 6A-5.

Geologic unconformity. The steady accumulation of sedimentary layers is often interrupted by sudden changes in climate, diversions of rivers, lifting or lowering of the land surface, or biologic changes. One consequence may be an abrupt shift of rock types (Fig. 6A-6). Intermediate layers may be absent, and index fossils above and below the discontinuity may indicate the passage of long periods of time between the deposition of successive layers of rock. Clearly visible unconformities occur when one layer of rock is tilted, folded, faulted, or eroded before another layer is deposited on top.

Beds deposited horizontally, tilted, eroded, and buried

FIG. 6A-6.

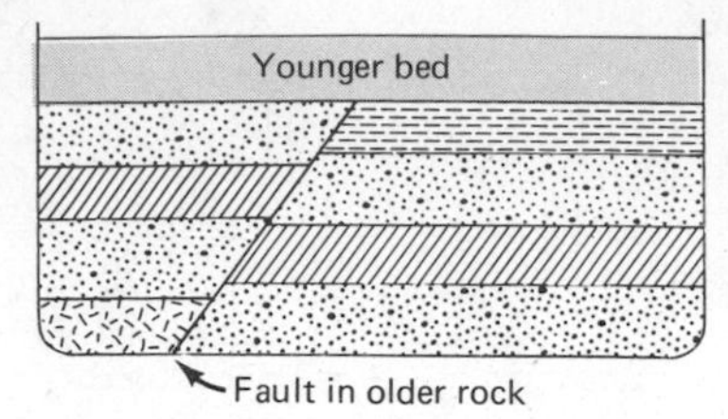

FIG. 6A-7.

Faulting. A geologic fault is younger than the material it breaks and older than any material that bridges the fault (Fig. 6A-7). The degree of displacement along a fault gives a clue to the amount of time the faulting required. Most faults involve less than a meter of rock slippage at any one time, and the episodes of slippage may be separated by decades of inactivity. A hundred meters of displacement thus may represent a few thousand years of earth history.

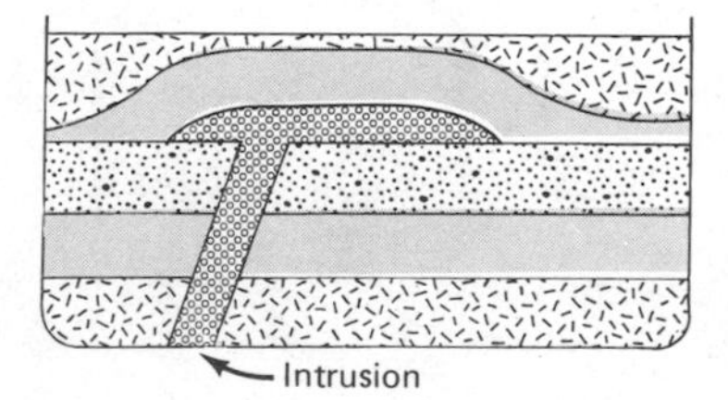

FIG. 6A-8.

Intrusion. Formerly molten rocks cannot be dated by fossils, because the intense heat of liquid rock incinerates any organisms that may have the misfortune to be in the way. Nevertheless, intrusions of molten rock provide a useful means of establishing the relative ages of rocks, since a vein of molten rock is inevitably younger than any other rock it penetrates (Fig. 6A-8). Molten rock that flows horizontally between rock layers usually leaves scars on the surrounding rock by heating and recrystallizing the adjacent material. Unaltered material on top of formerly molten rock must have been deposited after the hot rock cooled and hardened and therefore is younger than the formerly molten rock. Widespread volcanic lava flows or ash deposits are especially valuable signposts in the geologic history of many regions of the world. They can be dated by comparing the fossils in younger and older rocks, and, in turn, they help to tie the chronologies of different districts together.

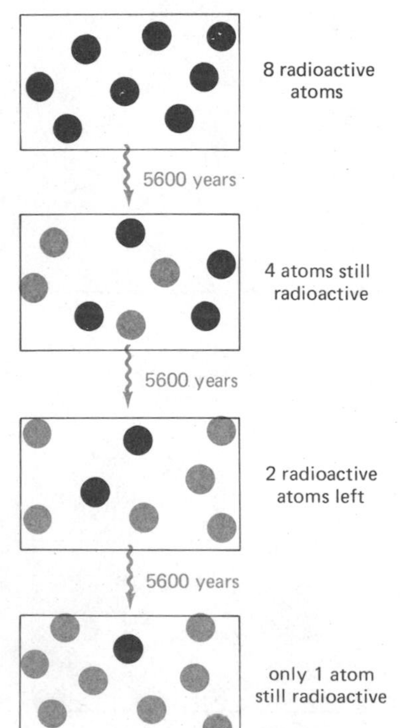

FIG. 6A-9.

Radioactive dating. Radioactivity adds a dimension of absolute time that often is missing or hopelessly jumbled in the sedimentary record. Cosmic rays high in the atmosphere or natural radiation deep within the earth will cause some elements such as carbon, potassium, or uranium to become radioactive. For example, carbon-14 is formed when cosmic rays strike carbon-12 atoms in the atmosphere. The amount of carbon-14 produced depends on the amount of carbon dioxide in the air and on the intensity of cosmic radiation. Once the initial conditions are known, the amount of carbon-14 that a plant will pick up during its life can be predicted. Burial of the organism by sediment eliminates the atmospheric supply of carbon-14; the carbon-14 already in the plant begins to decay.

A radioactive atom is a natural time bomb with a fuse of unknown length. Sooner or later it will go off, usually by emitting some energy or material and by changing into another substance. The exact time of radioactive decay of a particular atom is not predictable, but a whole mass of atoms behaves according to a statistical rule: every radioactive element has a measurable half life, defined as the amount of time it takes to reduce 50 percent of the radioactive atoms to the nonradioactive state. For example, half of the radioactive atoms of carbon will change to stable nitrogen in about 5600 years (Fig. 6A-9). Half of the remaining radiocarbon atoms will decay in another 5600 years, and so on until after 56,000 years, only 0.01 percent of the original atoms of carbon-14 will still be radioactive. The decay of such a small number of radioactive atoms cannot be detected because of the general background radiation of the universe. Therefore radiocarbon dating works only with organic sediments that are less than about 50,000 or 60,000 years old. Even in that range, the dates are uncertain because the rate of carbon-14 formation in the upper atmosphere is known to vary.

The half life of uranium-238 is about 4.5 billion years; potassium-40 decays to argon with a half life of 1.5 billion years. Dating with these long-lived elements works best with fairly old rocks, which were formed from molten material deep within the crust and were sheltered from the atmosphere by durable materials above.

A quick test: A combination of relative and absolute dating techniques can be used to decipher the history of a local rock structure. Using the clues described above, arrange the rocks in Figure 6A-10 into the proper order of formation from oldest to youngest. One rock type has already been drawn in as a clue.

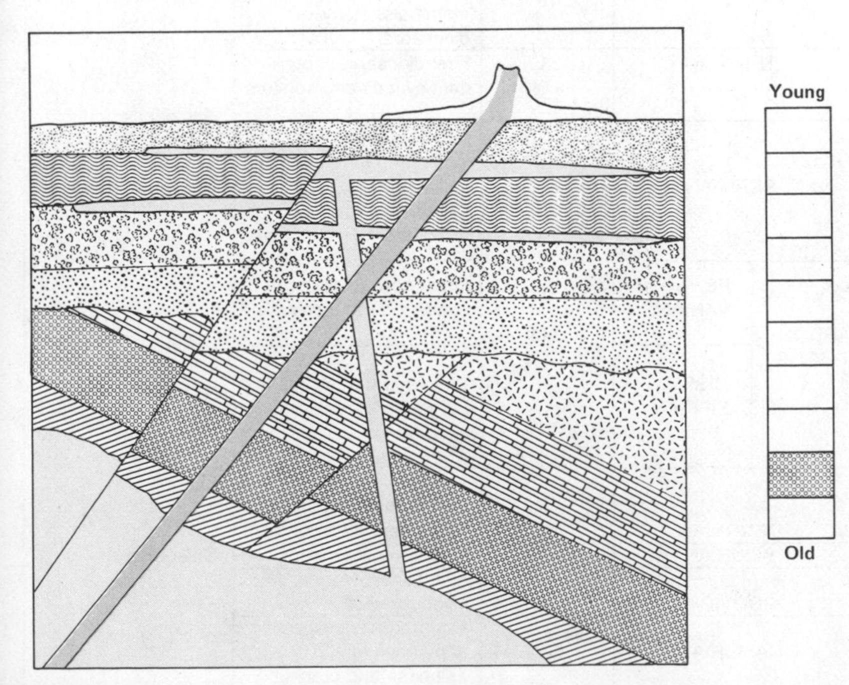

FIG. 6A-10.

TABLE 6A-1 THE GEOLOGIC ERAS

ERA	PERIOD	EPOCH	MILLIONS OF YEARS AGO	DISTINCTIVE FEATURES	MOUNTAIN BUILDING REVOLUTIONS
CENOZOIC	QUATERNARY	Recent	.01	Modern man	
		Pleistocene	2	Early man; glaciation	Alpine and Cascadian
	TERTIARY	Pliocene	13	Large carnivores	
		Miocene	25	Abundant grazing mammals	
		Oligocene	36	Large running mammals	
		Eocene	58	Modern types of mammals	
		Paleocene	63	First placental mammals	
MESOZOIC	CRETACEOUS		135	First flowering plants; climax of dinosaurs, followed by extinction	Laramide (Rocky Mtns.)
	JURASSIC		180	First birds, first true mammals. Many dinosaurs.	
	TRIASSIC		230	First dinosaurs; abundant cycads and conifers	
PALEOZOIC	PERMIAN		280	Extinction of many kinds of marine animals, including trilobites. Continental glaciation in Southern Hemisphere.	Appalachian Hercynian (Europe)
	CARBONIFEROUS	PENNSYLVANIAN	345	Great coal swamps, conifers. First reptiles	
		MISSISSIPPIAN		Sharks and amphibians. Large-scale trees and seed ferns.	
	DEVONIAN		405	First amphibians; fishes very abundant	
	SILURIAN		425	First terrestrial plants	Caledonian
	ORDOVICIAN		500	First fishes. Marine invertebrates	
	CAMBRIAN		600	First abundant record of marine life. Trilobites and brachiopods dominant	
PRE-CAMBRIAN	PROTEROZOIC				Archaean
	ARCHEOZOIC				

Chapter 7

Weathered Rocks: Surface Earth Materials

Soil is but rock on its way to the ocean.

B. C. McLean, USDA·Soil Conservation Service

Home, for most forms of terrestrial life, lies within a few meters of the border between the earth and the atmosphere. Forces of both terrestrial and solar origin help to shape the environment at this interface. This chapter deals with the different kinds of rocks, their susceptibility to destruction, the products they form when they break down, and the ultimate disposition of their remains. Rocks are the ultimate sources of gems, fuels, building supplies, and industrial raw materials. These glittering geological celebrities, however, should not blind a geographer to the important roles of less well-known rocks. The unsung geological heroes are the common rocks that build mountains, become soils, and support life.

interface boundary zone between two unlike regions

ROCKS AND MINERALS

The study of rocks is often unnecessarily dry and dusty. It is difficult to strike a balance between the need to know about the nature of surface materials and the tendency to put too much emphasis on structure, chemistry, and history. The ability to read a simple map of surface material is adequate for many purposes, such as analyzing the parent material of a soil or investigating the responses of a wild or cultivated plant. However, knowledge of surface rocks is not enough for a water-resources engineer or a mining company. Furthermore, the pattern of surface material may change as the terrain itself changes through time, and only an acquaintance with underlying structure can aid in predicting the results of changes. Finally, the study of surface materials is often a welter of detail which actually becomes easier to remember when it is organized according to the historical processes that brought it into being.

Fortunately, the present characteristics of surface materials give useful clues about their origin. Earth materials can be somewhat arbitrarily separated into minerals and rocks. Minerals are, at least in theory, chemically pure. In turn, minerals occur together in more or less predictable mixtures called rocks. Some rocks contain mostly one mineral, but a rock like granite has specks of three or four different colors, each representing a different mineral with unique chemical characteristics (Fig. 7-1).

mineral earth substance with consistent chemical composition

rock mixture of one or more minerals

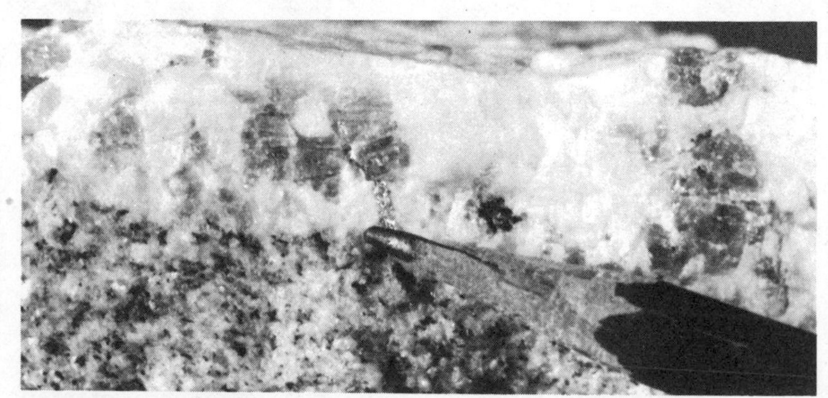

FIG. 7-1. Granite, a common rock formed from the cooling and hardening of molten rock. The salt-and-pepper appearance indicates a mixture of different minerals within the rock. The gray glassy substance is quartz, the white crystals are feldspar, and the black flecks are mica and other dark minerals. The crystals at the top of the picture cooled more slowly and therefore grew larger than those at the bottom.

The processes at work on the earth give rise to thousands of different combinations of elements and minerals, but some common combinations occur over and over again. In fact, about 20 rocks, made up of 10 minerals, account for more than 99 percent of the exposed material on the earth's surface. Each of these common rocks and minerals has its own weathering characteristics, its own potential for decomposing into fertile soil, and its own implications for the inhabitants of the earth.

FROZEN LIQUIDS—THE IGNEOUS ROCKS

All rocks on the earth's crust were molten at one time or another. Many rocks still bear the unmistakable imprint of their liquid past. The environment in which a hot liquid cools and solidifies has an effect on the characteristics of the product.

igneous rock rock formed when melted material hardens

Picture a large body of magma deep underground as it begins to cool. At a critical temperature the material changes from the moderate freedom of the liquid state to the rigidity of the solid state. Atoms that are well-matched usually arrange themselves into orderly rows (Fig. 7-2). Atoms whose electrical personalities are incompatible must be separated, or the resulting crystals will have internal weaknesses caused by hostile neighbors. The distinctive chemical substances that appear during this sorting of atoms are the minerals.

magma hot melted rock in the interior of the earth

crystal geometrically regular shape formed when atoms are arranged in a regular pattern

The temperature of a molten mass at the time of its chemical organization also has an effect on the final product. Different combinations of atoms are attracted to each other at different temperatures. For example, one of the earth's most common minerals, feldspar, consists of a combination of aluminum, silicon, and oxygen, plus a few atoms of sodium or calcium. If

viscosity "thickness" or resistance to flow of a liquid

FIG. 7.2 Crystals of halite, the natural form of sodium chloride, common table salt. The regular shape of the crystals is a reflection of the orderly internal arrangement of the atoms that make up the mineral. The rules of crystal growth dictate that if a few of the atoms of a particular mineral arrange themselves in a cube, as they do in common table salt, then a large crystal will also look like a cube.

feldspar begins to form at a high temperature, it attracts and contains more calcium than sodium. At a comparatively low temperature of formation, sodium substitutes for calcium in the feldspar. A slow and undisturbed cooling will yield a mixture of sodium feldspars, calcium feldspars, and many intermediate forms, each crystallized at a different temperature.

sialic or felsic material light earth substance composed mainly of silicates of aluminum

A magma made of magnesium, silicon, and oxygen cools in a different manner. The crystals formed at all temperatures contain the same elements, but their proportions differ according to the temperature at which they were formed. At high temperatures each silicon atom likes to be linked with two magnesium and four oxygen atoms. Rapid cooling of this hot mix produces a mineral called olivine. A lukewarm silicon atom is content with only one magnesium atom and three oxygen atoms. When cooled, this combination produces a mineral called pyroxene. The change from Mg_2SiO_4 (olivine) to $MgSiO_3$ (pyroxene) leads to the temporary unemployment of one magnesium atom and one oxygen atom. The rejected pair then seeks out a molecule of silica (SiO_2) and forms another pyroxene.

simatic or mafic material heavy earth substance composed mainly of silicates of magnesium and iron

The feldspar sequence is a progressive response to temperature, whereas the olivine-to-pyroxene change is abrupt. Both sequences can be interrupted if crystals are removed or somehow made inaccessible to the remaining molten material. For example, suppose that the temperature of a molten mass of material decreases only slightly and then the liquid is removed. Crystals that have already formed will stay behind and produce a dark rock rich in olivine and calcium feldspar, the high-temperature minerals. The rest of the liquid, when it finally crystallizes, will form a light rock that lacks olivine but contains much quartz (pure silica), mica, and sodium feldspar.

Liquid magmas often cool while they move through cracks in the solid crust. Like geological litterbugs, they leave a trail of many different rock types behind them (Fig. 7-3). Dark-colored, heavy rocks are formed at the ocean ridges, where hot mantle material moves quickly toward the surface. Dark igneous rocks also occur where hot magma rises rapidly through cracks in the deformed crust near the margins of the crustal plates. Cool magmas yield light rocks made of quartz and other latecomers to the crystallization party. Typical locations for such rocks include places where continental rocks are melted and regions where magmas move slowly upward, cooling as they rise and therefore leaving the heavy high-temperature minerals behind.

ocean ridge or rift area where crustal plates are pulled apart and molten rock rises from the interior of the earth
(see p. 118)

crustal plate piece of earth's crust moving slowly over the plastic mantle below

There is a kind of wry justice in one of the important consequences of the separation of magma into different rocks. The late-forming minerals tend to be more resistant to chemical decomposition on the earth's surface. Quartz, a mineral formed out of silica left when the olivine-to-pyroxene shift is interrupted, is the most durable common mineral on the earth's surface. Minerals that formed at high temperatures are less stable than quartz. They decompose readily but also yield soils of exceptional fertility. Human food production thus depends partly on the temperature of molten rock in the dim recesses of geologic history.

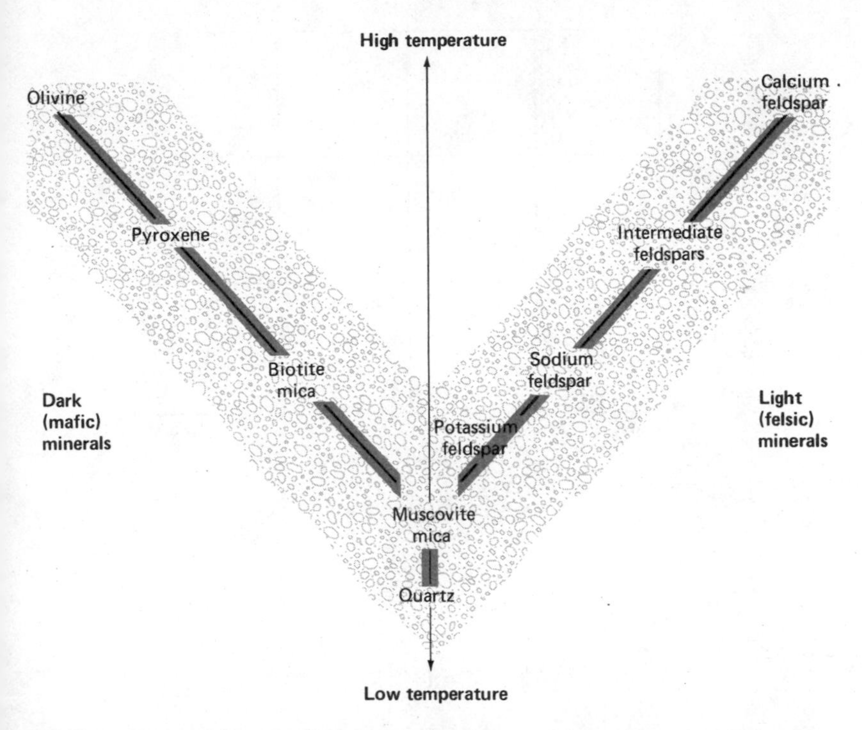

FIG. 7-3. Bowen's Reaction Series, an arrangement of the various minerals in order of their formation as hot magma cools and hardens. Minerals near the top of each sequence are formed first; quartz appears last of the common igneous minerals.

lava molten rock on the earth's surface; magma becomes lava when a volcano erupts

The rate of final cooling and crystallization of molten material adds another dimension to the classification of igneous rocks. A volcano may spew forth a molten mass of material with a comparatively high or low temperature. Each kind of lava has its own mineral composition, but regardless of its initial temperature, eruption causes the liquid to cool quickly. The resulting formation of rock is like an unannounced quiz to a student: the day of judgment arrives without warning and leaves little time for preparation. Extremely rapid cooling immobilizes the liquid atoms in a state of almost total disarray known as glass. More commonly, the outer edge of the liquid rock cools first, and the hard shell insulates the rest of the liquid and gives it a short time to get organized. The result is an igneous rock that consists of tiny crystals, too small to be seen with the unaided human eye but too large to be called totally disorganized. By contrast, individual crystals have time to become quite large when molten rock cools slowly. Slow cooling ordinarily occurs only where melted rock is covered by a thick insulating blanket of rock.

amorphous state noncrystalline arrangement of atoms

obsidian volcanic glass, usually black and shiny

plutonic rock igneous rock with large crystals formed deep underground

In summary, the size of the crystals in an igneous rock indicates how fast the rock cooled from its molten state. The chemical composition of the rock reveals the temperature of the magma at the time the minerals became organized (Fig. 7-4).

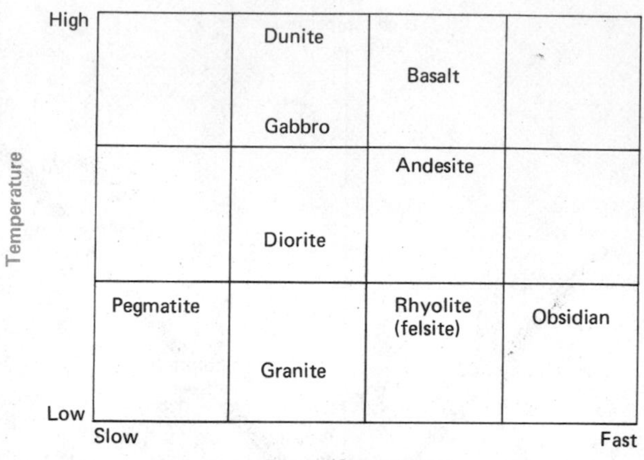

FIG. 7-4. Igneous rocks, classified on the basis of magma temperature and rate of cooling.

THE FOUR FATES OF IGNEOUS ROCKS

Four things can happen to igneous rock after it has cooled and solidified. Its destiny is imposed on it by the awesome internal forces that create mountains and arrange the continents. These forces push crustal plates around on the surface of the earth. The motion, in turn, can cause the rock to persist, remelt, change, or weather.

Frequently the motion of the crust is sideways and gentle. When the entire caravan of rocks moves as a unit, an individual piece of rock has no way of knowing it is in motion. The temperature of the rock stays nearly constant as long as the rock does not approach the surface above or the molten mantle below. The price of longevity, however, is an uneventful existence.

subduction downward movement of a crustal plate
(see p. 119)

Rock that is forced to descend gets warmer as it moves closer to the molten material below. In extreme cases the heat can melt rock. Collisions of crustal plates can force rock downward at the breakneck speed (for a geologic process) of 1 to 10 cm per year. Remelting is comparatively quick, and the individual chemical substances separate and go their own ways.

metamorphic rock rock that is altered by heat and/or pressure

A gentle increase of temperature permits individual atoms to move without disrupting the entire crystal structure. The result is a gradual rearrangement of the minerals and, sometimes, a subtle alteration of their chemical structure. The alterations produce a different kind of rock called a metamorphic rock, one that shows evidence of being changed by heat and pressure. Pressure on the rock helps to heat it and guides the recrystallization. For example, metamorphism rearranges the jumbled crystals of granite into layers which reflect the direction of pressure. The result is a banded rock

FIG. 7-5. Metamorphism of granite. Pressure usually aligns the mineral crystals in layers and changes granite (left) into gneiss (right).

called gneiss (pronounced, believe it or not, "nice," Fig. 7-5). Further heating and compression produces an even more highly metamorphosed rock called schist. Granite, gneiss, and schist represent arbitrary stages in the progressive metamorphism of a rock; the dividing lines between them are basically imaginary. When in doubt, it may be appropriate to use a transitional expression such as "granite on its way to becoming gneiss."

The fourth possible fate of igneous rock is to be lifted by internal forces and exposed by the removal of overlying rock. In terms of its entire geologic history, a rock spends only a few fleeting moments on the surface of the earth before it is weathered and removed. This instant out of its eternity, however, is singled out by mortal scientists as the most significant time in the history of a rock. Life depends on an endless exchange of nutrients among soil, water, air, plants, and animals. These nutrient cycles, in turn, rely upon rock weathering to provide a continual input of rock material. Before examining the weathering process in detail, however, it may be helpful to review a process which sometimes permits a rock to relive its "moment in the sun."

weathering physical and/or chemical decomposition of rock

biogeochemical cycle circulation of elements through living organisms and their nonliving environment (see Chapter 12)

REINCARNATION—THE SEDIMENTARY ROCKS

Earth debris enters a second rock cycle after rain and rivers have washed soil and pieces of rock off the land and carried them to the oceans. A stream rushing down a hill can move sizeable boulders, because speed gives water considerable force. A river begins to slow and lose its driving force when it reaches a level stretch of land or enters a quiet lake or ocean. A progressive loss of speed and energy leads to a neat sorting of the load of earth material. The largest and heaviest materials drop out first, followed by smaller and smaller particles as the stream weakens (Fig. 7-6).

sediment material deposited by some process

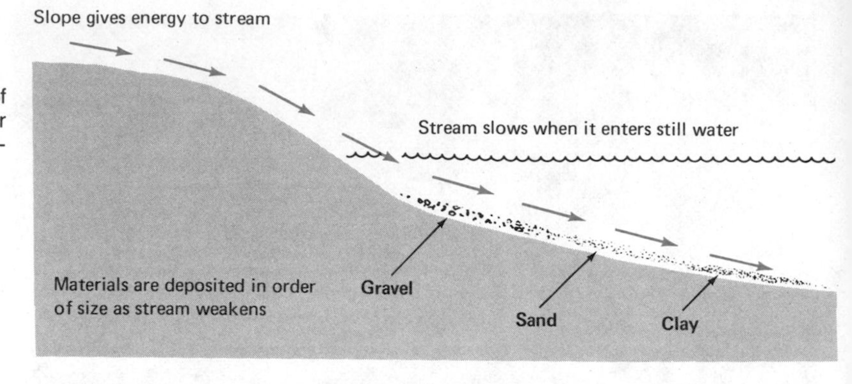

FIG. 7-6. Sequential deposition of materials as a stream enters a larger body of water and slows to an eventual halt.

clastic sedimentary rock rock made of cemented particles:

gravel—conglomerate
sand—sandstone
silt—siltstone
clay—shale

calcareous material earth substance containing calcium carbonate

Centuries of relentless removal of material from highlands and deposition in low places can produce a sizeable accumulation of debris. The sheer weight of the top material is often sufficient to squeeze the bottom layers into a compact mass that resembles rock. Moreover, buried sediments are often invaded by chemicals that cement the particles together and form an unquestionably rocklike substance (Fig. 7-7).

Living things enter the process of sedimentary rock formation in a variety of ways. The dissolved minerals in seawater are nutrients for marine life. Many organisms, from tiny diatoms to large clams and snails, are able to extract silica or calcium carbonate from the water and use it to build their cell walls or shells. When these organisms die, their hard remains may sink to the bottom and accumulate. The chemical composition of skeletons and

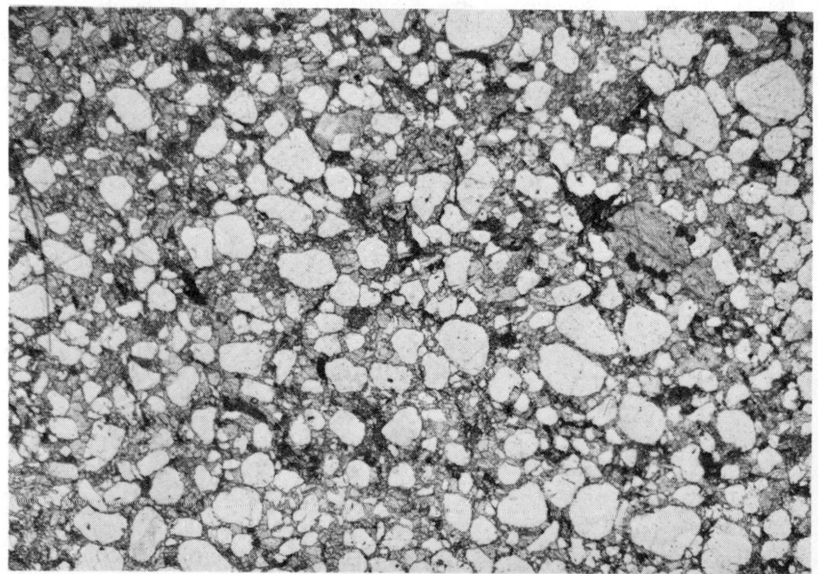

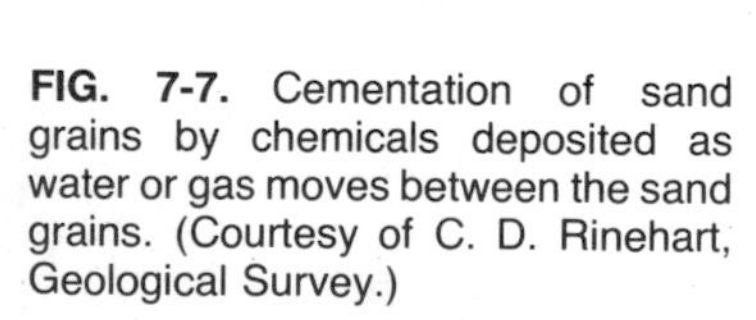

FIG. 7-7. Cementation of sand grains by chemicals deposited as water or gas moves between the sand grains. (Courtesy of C. D. Rinehart, Geological Survey.)

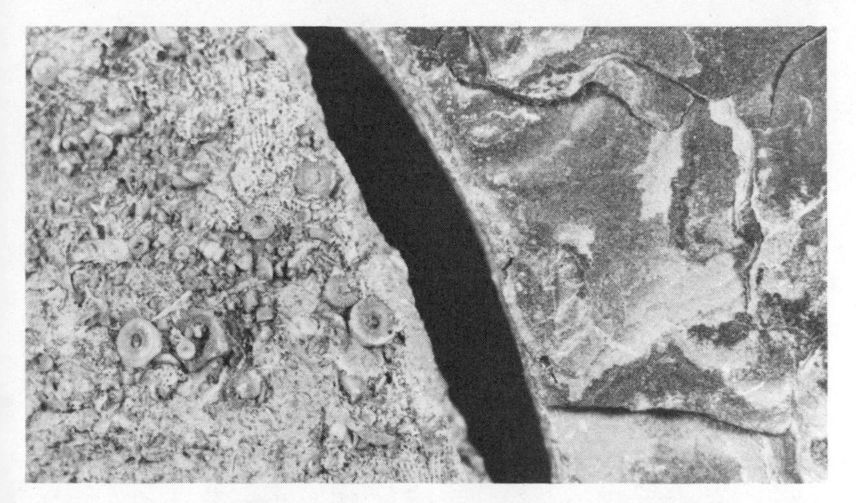

FIG. 7-8 Several kinds of limestone. The marine-shell origin of limestone is most evident in the left-hand sample.

carboniferous material earth substance containing organic matter

shells makes the deposits self-cementing under pressure (Fig. 7-8). The remains of buried plants may likewise harden into a sedimentary rock called coal. Petroleum is also of sedimentary origin, although its liquid state and mobility may mask many of the details of its formation.

Unfortunately for people seeking easy answers, sedimentary rocks are just like igneous rocks in one key respect: they do not fall into neat categories with sharp dividing lines. An arbitrary size limit of 0.05 mm is the boundary between sandstones and siltstones, but a fine sandstone and a coarse siltstone may be more alike than a coarse and fine sandstone. Mixtures of materials are more common than homogeneous substances. For example, sand deposits may harden with the remains of marine organisms to produce sandy limestone (or limey sandstone). Organic material frequently occurs with deposits of sand or mud; the mixtures eventually harden to form tar sands or oil shales. Despite their differences, however, most sedimentary rocks do show layering, particle orientation, and other evidence of their sequential deposition (Fig. 7-9).

chemical sedimentary rock rock made of precipitated chemicals:
calcium carbonate—limestone
magnesium carbonate—dolomite
silicon dioxide—chert
calcium sulfate—gypsum
sodium chloride—halite

FIG. 7-9. Horizontal coal seam. A thick layer of plant remains was buried under other sedimentary deposits that prevented the decay of the organic material. Pressure caused the plant matter to become more and more compact and rocklike. (Courtesy of C. E. Erdmann, Geological Survey.)

Sedimentary rocks, like their igneous comrades, may persist indefinitely underground, dive to a fiery destruction in the interior, slowly change under heat and pressure, or rise to the earth's surface. The first two fates usually are of little significance for inhabitants of the surface. The third alternative, metamorphism, generally produces rocks that are harder and more resistant to weathering than their sedimentary ancestors (Fig. 7-10). Consequently, metamorphic rocks such as marble and slate are popular building materials. The fourth alternative, however, is the most important one. Sedimentary rocks, like igneous or metamorphic rocks, touch life most directly during the short interval of time before they are carried off to the sea.

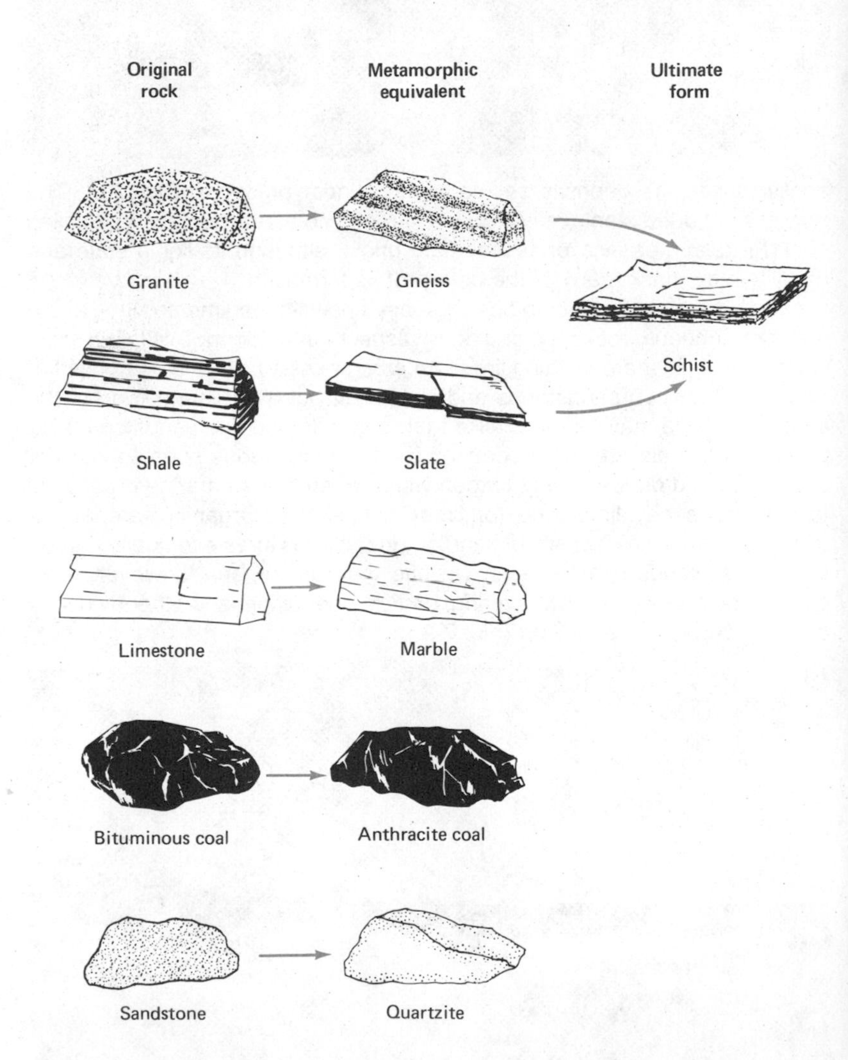

FIG. 7-10. Metamorphic rocks, arranged according to original rock and degree of alteration.

ROCK WEATHERING

Rocks must be destroyed in order to perform their greatest service to life. The breakup or weathering of rock is a necessary rite of initiation into the biological cycles and landforming processes on the earth's surface. To get an idea of how weathering operates, consider what happens to the quartz, mica, and feldspar in a freshly exposed piece of granite. Rain passes through the atmosphere and strikes the rock. Some of the water combines with carbon dioxide from the air and produces carbonic acid. This weak acid has the ability to disturb the crystalline composure of various minerals. Living organisms assist the carbonic acid by adding chemical by-products of their existence to the water. The acidic water reacts with feldspar and converts it to clay and several soluble chemicals. Water percolates through the weathered material and carries the dissolved chemicals away, but the weathered clay is a more durable combination of substances than the original feldspar.

The mica in granite is also more durable than feldspar, so that it persists for a while before it also succumbs to chemical weathering. Quartz is very resistant to the action of water or to its compounds. Soft-drink producers take advantage of the stability of quartz when they package their slightly corrosive liquids in glass bottles made of quartz sand. The unwillingness of crystalline quartz to decompose means that the quartz crystals in granite remain while the mica and feldspar crystals disintegrate (Fig. 7-11).

carbonic acid H_2CO_3, made of $H_2O + CO_2$

exudate substance given off by plant tissues

silicate clays end products of chemical weathering of many igneous minerals

quartz resistant mineral formed of silica (SiO_2)

FIG. 7-11. Weathered igneous rock. Chemical destruction of the feldspar grains causes the rock to crumble. (Courtesy of W. T. Schaller, Geological Survey.)

FIG. 7-12. Cracks in rock serve as symptoms of and avenues for mechanical weathering by frost action. Liquid water enters the cracks, freezes, and exerts outward pressure on the rock. (Courtesy of F. E. Matthes, Geological Survey.)

chemical weathering disintegration of rock by chemical alteration of constituent minerals

mechanical weathering breakdown of rock by physical separation of pieces

erosion removal of weathered material by gravity, either unaided or through its agents of water, wind, or ice
(see Chapter 16)

The destruction of a smooth surface by selective decomposition of weak material allows another form of weathering to take place. Water collects in the tiny pits that form in the granite. The pits may become deep cracks when the water freezes and expands (Fig. 7-12). The cracks tend to be self-perpetuating because they serve as avenues for water and plant roots to enter and wedge the rock still farther apart. Mechanical weathering by frost and plant roots also exposes additional surfaces to the agents of chemical weathering.

Thus granite slowly disintegrates until it is only a crumbly mass of loose material. Eventually the zone of weathered rock gets so thick that rainwater is no longer able to penetrate the loose material and reach the unweathered rock below. This might be the end of the story, if another process did not intervene. Gravity exposes additional surface by causing the weathered clay and loose quartz crystals to move downhill. Some of this downhill movement is accomplished directly, in the form of landslides or mudflows that carry loose material away and expose fresh rock. More commonly, gravity enlists the aid of other agents such as wind, rivers, or glaciers to help it transport weathered rock down from high places and toward sea level. The thickness of the layer of weathered rock thus depends on a complex balance among the forces of disintegration, the resistance of the rock, and the processes that remove loose material and carry it away.

CLIMATE AND WEATHERING

The action of acidic water, known as hydrolysis, is the most important chemical disrupter of rocks but not the only one. Because hydrolysis operates only where water is available, a precipitation map furnishes one clue to the distribution of water-weathered materials. Some other means of rock breakdown must be employed if there is no rain or other water source. A temperature map provides another clue, because the rate of chemical activity decreases as temperature goes down. A Georgia granite would weather much more slowly if it were moved to Maine.

hydrolysis weathering of rock by combination with either the OH^- or the H^+ formed when water dissociates

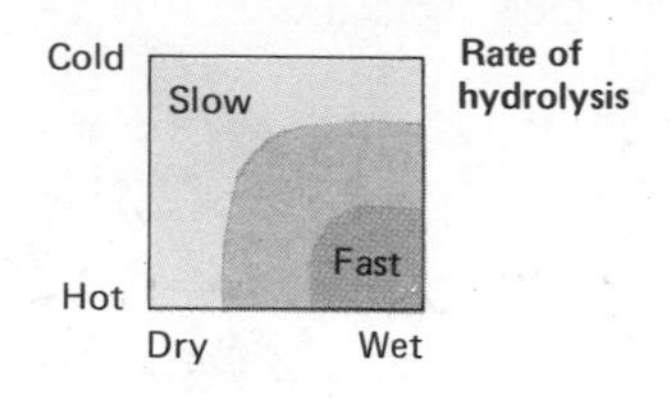

Under tropical conditions, even quartz yields to the chemical attack by water. Large particles break down to tiny ones quickly. Clays weather into simple forms, as silicon and oxygen are wrested from the original mineral and carried off in solution, leaving behind the very resistant compounds of iron and aluminum.

kaolinite the most highly weathered silicate clay

In an arid environment, granite is persistent but not eternal. Feldspar does not break down readily in the absence of water. The quartz grains serve as armor-plating against the attack of other weathering agents. Nevertheless, the borders between mineral grains are areas of weakness. Different minerals expand and contract at different rates. Tiny amounts of water seep into the cracks during occasional rains. Eventually the granite becomes a mass of loose grains of quartz, mica, and feldspar.

Other chemical processes assist water in chemical weathering. Iron in a rock is susceptible to the same kind of oxidation (rust) that seems to attack automobiles and to destroy them shortly after the last payments are made. Oxidation is usually less powerful than hydrolysis; it waits its turn in humid regions and attacks iron and other materials left over from the previous weathering of weaker crystals. Like other chemical processes, oxidation is slow and inefficient in cool places. Dry conditions, however, do not hinder oxidation as much as hydrolysis or other forms of chemical weathering. Oxidation may become the dominant chemical force in a desert, even though it too is less efficient there than in rainy places. The red color of many tropical and desert soils is a symptom of the presence of oxidized iron.

oxidation weathering of rock by combination with oxygen (a somewhat narrower definition than a chemist would offer)

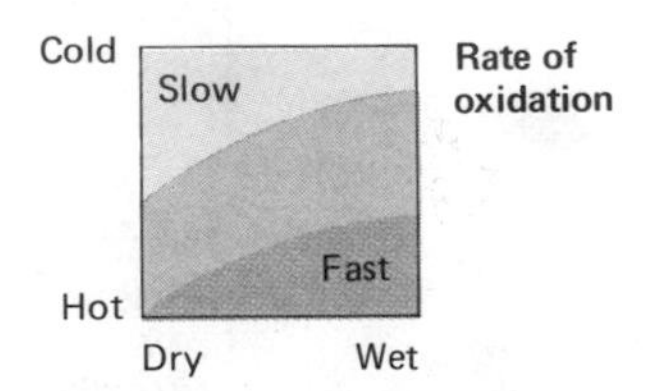

sesquioxide resistant compound of oxygen and iron or aluminum

The mechanical breakdown of rock, like chemical weathering, has a distinct geographical pattern. The areas of maximum freeze-thaw activity, however, do not coincide with the areas of rapid chemical weathering. A Brazilian granite does not experience ice wedging, because it never freezes. An Antarctic granite may never thaw, and therefore it too is immune to frost action. Freezing and thawing occur most frequently on a tropical mountaintop, where the temperature goes above and below the freezing point nearly every day of the year. The effect of frost on rocks, though, is greatest in the spring and autumn of a frontal climate, because the week-long cycles of freezing and thawing give ice time to penetrate deeply and therefore to exert great force on the rock. The symptoms of frost action are most obvious in cold regions where chemical weathering is so weak that mechanical processes rule by default. Many Arctic explorers and mountain climbers have been frustrated by seemingly endless and impassable expanses of jagged boulders formed by frost action.

permafrost permanently frozen layer of soil

frost-wedging separation of rock by water expanding as it freezes in confined places

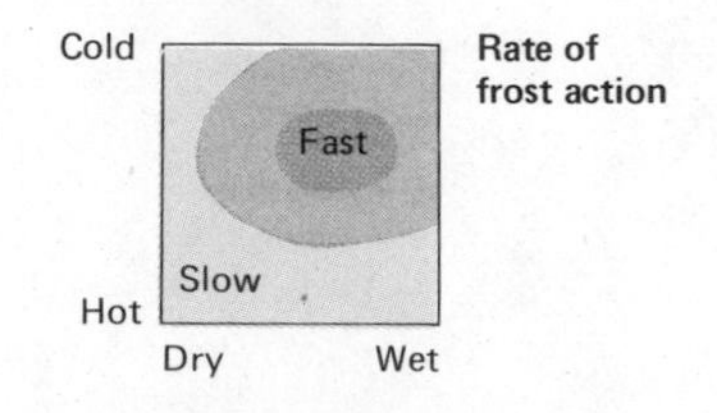

Other kinds of mechanical weathering also have distinct geographical patterns:

plant-wedging separation of rock by plant roots exerting pressure as they grow into cracks

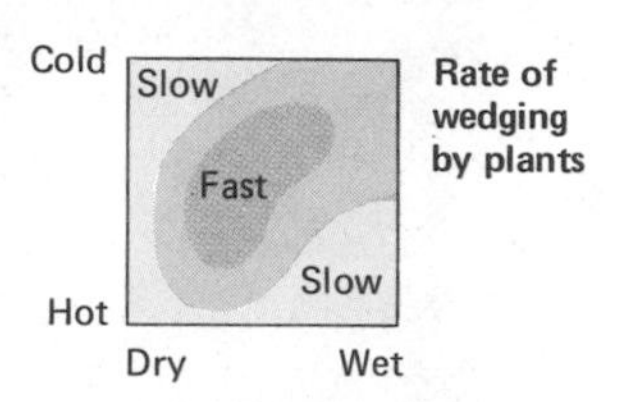

abrasion mechanical wearing due to friction with moving material

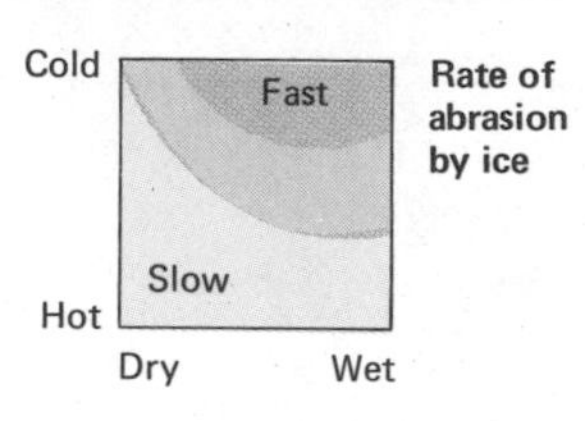

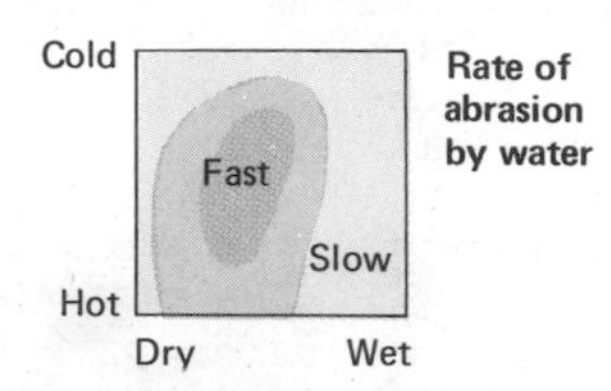

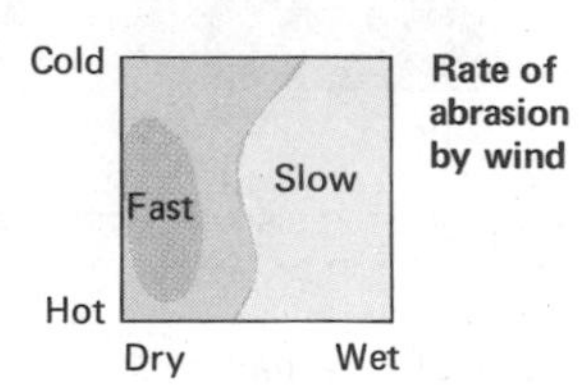

1. Plant roots are strongest where plants grow best, in warm and moist environments. The effect of plant roots on rocks, however, is most obvious in cool or dry places, where chemical activity is slow and shallow soils are the rule.
2. Grinding and scouring by moving ice requires exposed rock, a source of water, and a temperature that is not too high. Equatorial mountains are usually too warm for ice to persist, whereas polar peaks do not receive enough snowfall. Therefore the effects of ice scouring are most obvious on mid-latitude mountains.
3. Running water wears away at the sharp edges of rocks only where there is a surplus of water, a source of abrasive material, and an unprotected surface. Abrasion of rocks by running water is not noticeable where the soil is deep or where plants form a continuous cover.
4. Sand blasting by strong winds can occur only when a source of sand and a suitably exposed surface are present. These two ingredients commonly occur together only on beaches or in arid places, where there are no plants to hold the soil or to protect the rock (Fig. 7-13).

FIG. 7-13. Rock monuments in a desert area. The wind removes particles that have been loosened by chemical and mechanical weathering. (Courtesy of N. H. Darton, Geological Survey.)

Mechanical weathering is a surface phenomenon. Wind, water, and ice abrasion can work only on exposed rock. Plant roots reach only a few meters into the ground. Freezing and thawing affect only the surface meter or two, while the ground beneath stays either frozen or unfrozen. Any characteristic of a rock that helps to expose more surface to the weathering agents therefore tends to aid the breakdown of rock. For example, water can penetrate porous rock more deeply than dense rock. Sedimentary rocks are especially vulnerable to invasion, because cracks between layers often become pathways for water or plant roots.

porosity total amount of empty space between solid particles in rock or soil
(see Fig. 8-8)

joint crack or split in rock

Chemical disintegration is more rapid in slightly protected places than on exposed surfaces. A granite monument on a pedestal will last a long time, even in a rainy climate. The atmosphere is often dry, water evaporates quickly, and a polished surface provides little shelter for even the hardiest organisms. On the other hand, the soil is usually moist. Concentrations of acidic water and organisms quickly etch even the most polished surface of a buried rock.

Chemical weathering operates at different rates on different kinds of minerals. The shapes of crystals, the arrangement of minerals in a rock, and the chemical nature of minerals all help to govern the rate of decomposition. Feldspars and pyroxenes split easily and decompose readily. Flat layers of mica are planes of weakness that water can exploit, though the sheets themselves are fairly durable. Quartz crystals persist because they are both difficult to break and almost impervious to chemical attack.

cleavage plane flat surface or surfaces along which certain crystalline minerals break

The more different minerals a rock contains, the more likely a weak mineral will weather and cause the whole rock to crumble. A sandstone made of quartz sand grains cemented by silica (another form of quartz) is much more durable than one cemented by calcite and therefore vulnerable to attack by acid water. However, a homogeneous rock is not necessarily resistant. A pure limestone is exceedingly vulnerable, because the entire rock can dissolve in acidic rainwater. Consequently, limestone and its cousins simply disappear into solution wherever chemical weathering is active.

carbonic acid acid formed when carbon dioxide from the air dissolves in water

The dissolution and removal of limestone gives rise to a unique lumpy terrain known as karst. Low areas appear where acidic water dissolves the rock. The low areas are self-enlarging because water runs down the slopes and concentrates its attack on the valley bottoms. Frequently water penetrates deeply into cracks in the limestone and creates a maze of caverns and underground passageways. Progressive removal of limestone enlarges a cave until the walls are no longer able to support the roof. The ceiling of the cavern collapses, and a pit or sinkhole appears on the surface of the earth. In its most extreme form, karst consists of almost impassable expanses of steep-walled craters and potential collapses. More commonly, karst makes its presence known in the form of irregularly rolling terrain with numerous closed basins and a pronounced lack of long-lived surface streams (Fig. 7-14).

karst terrain landscape characterized by underground rivers and sinkholes on the surface

sinkhole crater formed when the roof of a cavern collapses

doline depression formed as limestone slowly dissolves away

FIG. 7-14. Karst features. On the left is a sinkhole that marks a place where limestone has been dissolved and the land has sunk. On the right is a cave entrance: the stream flows into the cave and emerges more than a kilometer away. (Right photo courtesy of W. H. Moriss, U.S. Forest Service.)

SUMMARY—THE GEOGRAPHY OF WEATHERING PRODUCTS

regolith layer of unconsolidated material on the surface of the earth

The kind of loose material produced by weathering at a particular place is the result of two things: the original rock and the weathering processes that operate in the local environment. The patterns of rock structure are more complicated than those of weathering processes. Nevertheless, a study of the spatial arrangement of surface material yields five general conclusions:

1. In high-latitude regions, mechanical weathering predominates because chemical activity is very slow. Outcrops of rock are common since dense surface materials such as granite, gneiss, and some limestones can resist frost penetration and breakdown. Moreover, abrasion by ice, water, and wind often removes weathered material from a surface. Except in places where rocks are exceptionally soft or where deposition occurs, soils in cold environments are shallow, coarse, and stony.

loam mixture of particle sizes (see Fig. 8-1)

2. In humid mid-latitude areas, mechanical and chemical processes are moderately active. Most rocks eventually succumb to the combination of weathering agents, although sandstones and igneous rocks usually persist longer than shales and limestones. Mid-latitude soils are often mixtures of particles of different sizes. Feldspars weather down to very tiny clay particles, whereas quartz usually remains as fairly large grains of sand and gravel.

3. In hot and humid areas, intense chemical weathering reduces nearly all rocks to tiny particles of iron and aluminum oxides and very simple clays. The weathered zone may exceed 200 m in depth in vulnerable rocks. Red sandstones and quartzites are the most persistent rocks; limestones and dark igneous rocks weather quickly and deeply.
4. In places with seasonally dry climates, weathering often proceeds even farther below the surface than in humid forest regions. Different processes attack different parts of the rock in different seasons. Chemical weathering is active during the wet season, whereas mechanical and abrasive processes dominate when plants are dormant in the dry season.
5. In very dry places, chemical processes that depend on water are slow and sporadic. Their inefficiency allows some chemically susceptible rocks, such as limestone or granite, to stand up as resistant ridges. Mechanical abrasion is a noticeable sculptor of exposed rock surfaces, and soils tend to be coarse and shallow.

complex or hybrid climate climate characterized by a seasonal alteration of two or more basic types (see Fig. 3-9)

The different kinds of surface material in different environments are the foundations for the biological and soil-forming processes of the next few chapters. The result of the interaction of weathered rock, atmospheric processes, and living things is soil, a key to the environmental wealth of different parts of the earth.

APPENDIX TO CHAPTER 7
Rock and Mineral Identification

Mere verbal description cannot substitute for hands-on examination of actual three-dimensional rocks. Therefore the authors suggest that this appendix be read only in the presence of a set of properly identified specimens. Even this is not an ideal situation, because a small laboratory sample does not reflect its original surroundings, and the overall setting and appearance of a rock is an important clue to its identity. Knowledge that a rock came from a lava flow or a sedimentary cliff considerably simplifies the taxonomic problem.

Petrologists and mineralogists, the scientists who study rocks and minerals, have developed a precise vocabulary which is used to identify and describe earth materials in terms of color, hardness, specific gravity, and the way a specimen breaks (Table 7A-1 at end of appendix). Many of those rules treat rocks and minerals at a level of detail that is not necessary for the general student of the environment. A geographer who can recognize about ten minerals and two dozen rocks is able to cope with 99 percent of the boulders, streambeds, roadcuts, and outcrops that are likely to be encountered in the field. Most of the complex vocabulary of the petrologist consists merely of technical terms for the intergrades between rock types (e.g., hornblende syenite is a rock that looks like granite but contains less silica and more magnesium than granite; thus it is about one-sixth of the way along a transition between granite and gabbro).

Simple aids to identification of common earth materials

Six quick tests simplify the job of identifying the common rocks and minerals:

Does it have large crystals that are tightly meshed together? It is probably a slowly cooled igneous rock. Slow cooling allows the crystals to expand and fill all of the available space.

Does it react with dilute hydrochloric acid? A strong reaction characterizes calcite, limestone, and marble. Dolomite and some shales, sandstones, and conglomerates may react weakly.

Does it lie in layers? Most layered rocks are sedimentary. Shale, sandstone, coal, and slate are usually layered. Limestone, dolomite, schist, quartzite, marble, gneiss, chert, and conglomerate are often layered but occasionally appear massive. The rapidly cooled igneous rocks—pumice, basalt, rhyolite—may have been formed by repeated lava flows and thus have a layered appearance.

Does it split easily? Many minerals have distinctive ways of breaking. Mica and many layered rocks break along one flat plane. Feldspars, dark minerals, and gypsum split along two different planes. Galena and halite break into cubes and rectangular boxes, whereas quartz and obsidian break along curved chips that resemble clamshells.

Is it transparent? Calcite, halite, and quartz crystals are usually transparent or cloudy. Mica, gypsum, dark minerals, obsidian, quartzite, marble, and schist are occasionally clear on thin edges.

How hard is it? The hardness of a rock or mineral refers to its resistance to scratching, not its ability to withstand breaking. The hardest mineral, diamond, will scratch anything. According to a commonly used scale of hardness, the diamond is given a number 10, and nine other common minerals are arbitrarily chosen to represent lesser hardness. An unknown rock that can scratch fluorite but is scratched by a knife blade has a hardness between 4 and 5.

ORDINAL SCALE OF HARDNESS

Hardness Number	Mineral	Common Material with About the same Hardness
10	Diamond	
9	Corundum	
8	Topaz	
7	Quartz—	
6	Feldspar	Glass
5	Apatite	Knife blade
4	Fluorite	Nickel
3	Calcite	Copper penny
2	Gypsum	Fingernail
1	Talc	Soft chalk

Minerals made of pure chemical compounds

Quartz—the only common mineral able to scratch glass; forms clear hexagonal crystals when it is pure; crystals may also be pale blue, violet, yellow, or pink; opaque white, gray, or brown forms are common.

Feldspar—white, gray, pink, or blue-green crystals that break along two different planes, with a distinctive satiny sheen on the flat faces.

Mica—clear to black papery sheets that are soft and easily separated.

Dark minerals (pyroxene, amphibole, hornblende, olivine)—dark green to black crystals that break along two planes; heavier than feldspars.

Calcite—clear to cloudy crystals that have a variety of shapes; all break in three planes to form lopsided boxes, and all react strongly with dilute acid.

Hematite (iron ore)—red or black iron oxide that is heavy and dull to shiny; some varieties are magnetic.

Bauxite (aluminum ore)—yellow, white, gray, or red aluminum oxide that is often mottled.

Galena (lead ore)—heavy gray crystals that break into cubes and boxes.

Pyrite—iron compound that justly deserves its name "fool's gold."

Igneous rocks solidified from molten material

Rapidly cooled lavas. Noncrystalline or with crystals so fine they cannot be distinguished by the unaided eye. These rocks are often confused with sedimentary rocks. The presence of embedded crystals and air pockets is a sure clue to an igneous origin.

Obsidian—volcanic glass; usually black or brown; breaks with chips that resemble seashells; almost transparent on thin edges.

Pumice—volcanic froth; usually pink, tan, or gray; sometimes layered; great amount of trapped air makes it very light in weight (many samples can float in water).

Felsite or rhyolite—light-colored volcanic lava; usually gray, tan, or pink; contains quartz, and therefore is harder and lighter than basalt.

Basalt—dark volcanic lava; usually dark green to black; does not contain quartz and therefore it is softer and heavier than felsite; trained geologists will recognize many intermediate rocks between felsite and basalt, depending on chemical composition.

Porphyry—basalt or felsite with occasional large crystals, usually of feldspar.

Cinder—man-made igneous rock, often found near steel mills and along railroad tracks.

Slowly cooled magmas. Crystalline, with individual crystals large enough to be seen with an unaided eye. The crystals of different minerals have sharp angles with no empty space or cement between the crystals.

Granite—interconnected crystals of quartz, mica, dark minerals, and white or pink feldspar.

Gabbro—interconnected crystals of dark minerals; usually dark green to black; heavier but softer than granite.

Pegmatite—granite with exceptionally large crystals, often more than a centimeter and occasionally as long as a meter.

Sedimentary rocks deposited and hardened into rock

Cemented sediments. Usually layered, though the layers may be too thick to appear in a laboratory specimen.

Shale—cemented clay that is usually soft and likely to break into blocks or layers; may be black, gray, brown, tan, or red; dark gray or black shale (oil shale) may contain organic matter.

Sandstone—cemented sands whose grains may be rounded or angular; nearly all colors occur, but red and tan are most common; cementing materials include soft calcite, intermediate iron oxide, or hard silica.

Conglomerate—cemented gravel or mixtures of particles that may be rounded or angular; colors are variable; hardness depends on constituent rocks and kind of cement.

Brick—man-made shale.

Mortar—man-made sandstone.

Concrete—man-made conglomerate.

Organic sediments. Burnable rocks made of the remains of plants and

Peat—fibrous organic matter; the plant parts are often visible; brown, soft, often wet.

Lignite—hardened peat; brown but firm and dry.

Bituminous Coal—chemically altered organic matter; black and fairly soft.

Chemical sediments. Sometimes layered but often massive, with few

Limestone—calcium carbonate; always reacts strongly with dilute acid; many colors but frequently white or gray; often contains visible seashells.

Dolomite—calcium carbonate with magnesium carbonate; looks like limestone but has a weaker reaction with acid.

Chert—silicon dioxide; able to scratch glass; often found in layers in limestone; usually white, but may be stained red, yellow, or gray by impurities; one variety is known as flint.

Gypsum—calcium sulfate; long clear to white crystals, sometimes stained yellow or red; can be scratched with a fingernail; usually found in deserts.

Halite—sodium chloride (rocksalt); clear crystals that break to form perfect cubes; salty taste is a dead giveaway.

Metamorphic rocks altered by heat and pressure

Gneiss—granite whose grains have been aligned by pressure; alignment may be straight or contorted; minerals may be mixed or segregated in distinct bands.

Slate—shale that is compressed and hardened by pressure; rings when struck with fingernail; breaks in thin platy layers.

Schist—slate or gneiss that is compressed still more; mica flakes clearly visible on flat layers.

Quartzite—sandstone that is cemented so firmly that rock may break through grains rather than around them; a very hard rock, able to scratch glass.

Marble—limestone that is recrystallized by pressure but still reacts with acid; shiny crystals often visible.

Anthracite coal—bituminous coal that is altered by pressure; hard, black, shiny, but still burnable.

TABLE 7A-1 ROCK IDENTIFICATION TERMS USED IN ROCK AND MINERAL KEYS

Texture—the pattern or weave of a specimen
- Granular—individual mineral crystals visible (e.g., granite).
- Aphanitic—individual mineral crystals not visible (e.g., basalt).
- Glassy—no crystal structure (e.g., obsidian).

Luster—the way a specimen reflects light
- Metallic—like polished metal (e.g., pyrite).
- Iridescent—like polished pearl.
- Vitreous—like glass (e.g., quartz).
- Greasy—between glassy and dull.
- Dull—a lack of luster (e.g., shale).

Cleavage—the way a specimen splits in flat planes
- Sheetlike—in one direction (e.g., mica).
- Prismatic—in two directions (e.g., hornblende).
- Cubic—in three directions at right angles (e.g., halite).
- Rhombohedral—in three directions not at right angles (e.g., calcite).

Fracture—the way a specimen breaks if it has no cleavage
- Conchoidal—in seashell-like chips.
- Even—smooth but not straight.
- Uneven—somewhat rough.
- Grainy—very rough.
- Hackly—jagged, almost barblike.
- Splintery—needlelike.

Streak—the color of the powdered specimen, obtained by rubbing the specimen across an unglazed ceramic tile.

Fluorescence—the color of the specimen when illuminated with ultraviolet light.

Specific gravity—the weight of the specimen, expressed as a proportion of the weight of an equivalent volume of water.

Chapter 8

Exchange Processes: Soil As A Storehouse

Only by dinte of awesome strivings can we make this barren grounde repay oure toile.

Mauver, USDA Soil Conservation Service

Most of the land and the sea floor is covered by an irregular blanket of broken and altered rock, a legacy of the weathering processes described in Chapter 7. The layer of loose material is usually shallow or absent in cold or dry regions and where wind or water have carried weathered rock away. The surface layer is thick in warm and humid climates and where loose material has been deposited. Life on the land depends on the top few meters of this thin layer of weathered rock, but the productive potential of soil varies widely from place to place. In an age when famine again menaces parts of the world, the reasons for differences in soil productivity are of more than mere academic concern.

SOIL TEXTURE AND STORAGE CAPACITY

Textural terms—silt, sandy clay loam, loamy sand, gravelly clay, and so on—are part of the basic vocabulary of soil scientists (Fig. 8-1). The texture of the material near the surface is the usual "surname" of an individual soil, such as the Cecil clay of the Carolina hills or the Fayette silt loam of the Corn Belt. An emphasis on texture is understandable, because the texture of a soil is usually the most significant factor in estimating its agricultural potential or engineering characteristics.

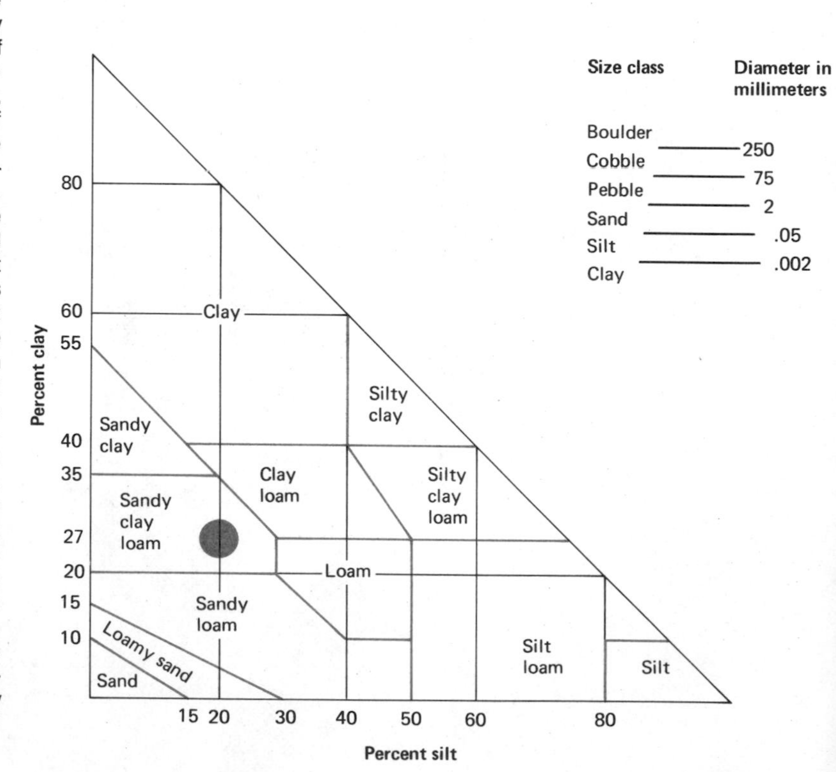

FIG. 8-1. A modified soil textural triangle. Texture refers to the sizes of the pieces of weathered rock in a soil. It is extremely rare for a soil to be composed entirely of particles of only one size. To describe a mixture of various particle sizes, scientists take out all the pebbles, large rocks, leaves, roots, and other remains of plants and animals. Then they mix the soil with water and observe the particles as they sink through the water. The heavy sand particles settle to the bottom first, followed by the silt and then the clay. The percentages of silt and clay in a soil are the basis for its textural name: a soil with 27 percent clay and 20 percent silt (and therefore 53 percent of its screened weight in sand-sized particles) would be called a sandy clay loam (the dot on the graph). A soil with many organic remains or large pieces of rocks gets an appropriate adjective in front of its textural name, for example, peaty sand or cobbly clay loam. Loam is an ancient word associated with productive soils. Soils with loamy textures (mixtures of different particle sizes) generally have better moisture characteristics and nutrient storage ability than soils composed predominantly of coarse sand or fine clay particles. (Adapted from standard USDA textural triangle.)

A plant root does not consume soil particles whole. Rather, it seems to select its food from a host of tiny bits of material floating between or adhering to the soil particles. The geometry of a handful of soil is like the maze of small coatrooms and hallways in an old theater or sports arena. Solid particles form walls around tiny empty spaces, and electrical charges on the surfaces of the soil particles act like coat hooks on the walls (Fig. 8-2). Plant nutrients and other tiny substances are "hung" on the electrical "hooks." Surplus nutrients float around in the empty spaces between soil particles. Like unhung coats on the floor, they can be intentionally removed by something that wants them or can be accidentally swept away in the general confusion.

crystal lattice regular geometric arrangement of atoms in most solid substances; the atoms are held in place by electrical forces

Most plant nutrients have a positive charge and are attracted to the negative charges on the surfaces of soil particles. The total amount of storable plant food can therefore be estimated by measuring the number of negative electrical charges on the surfaces of the particles in a given mass of soil.

FIG. 8-2. A microscopic view of soil clay. A typical clay particle looks like a torn piece of paper. Weathering has reduced the clay to its present size and disrupted the crystal structure along the edges of the particles. Each sheet of clay has a few positive and negative electrical charges on its flat faces and a concentration of negative charges around the torn edges, where mated atoms were separated and electrical bonds are unsatisfied. A grain of sand, if magnified as much as this photo, would be about as big as a bedroom but would actually have fewer electrical storage sites than the clay shown here. (Courtesy of W. C. Lynn, Soil Conservation Service.)

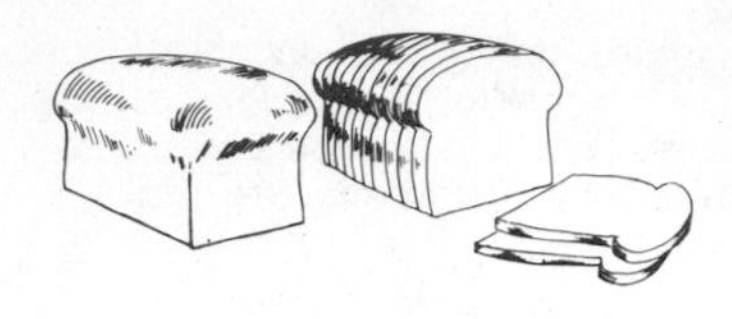

FIG. 8-3. Specific surface, a fundamental concept that is sometimes difficult to visualize because it seems to contradict an intuitive understanding that many people have. A useful analogy is an uncut loaf of bread: it has a volume of one loaf and a surface area equal to the sum of four sides and two ends. After the loaf goes through the slicer, its total volume is still one loaf, but the surface area has increased to four sides, two ends, and fifty or so inner surfaces. Subdividing a large soil particle, likewise, increases the surface area while the solid volume remains the same. Most natural soils have about the same mass of solid material in each cubic centimeter of soil, but the amount of surface area, and thus the potential fertility, varies widely from one soil to another.

specific surface amount of surface area per unit of volume

Soils with coarse textures generally have less capacity to hold nutrients than fine-textured soils, because coarse particles have comparatively little surface area (Fig. 8-3). A handful of sand may have several square meters of surface area, about the size of a Ping-Pong table. An equal volume of clay might have as much surface as a football field.

EXCHANGE PROCESSES

ion single atom or group of atoms with an electrical charge

The kinds of materials that are stored on soil particles can vary from place to place or from time to time. Calcium usually occupies most of the storage space on the surfaces of soil particles that weathered from a calcium-rich rock. Adding potassium fertilizer to a calcium-rich soil causes potassium to change places with some of the calcium on the soil particles (Fig. 8-4). The displaced calcium is then free to move elsewhere in the soil, perhaps to replace other substances on other soil particles. Alternatively,

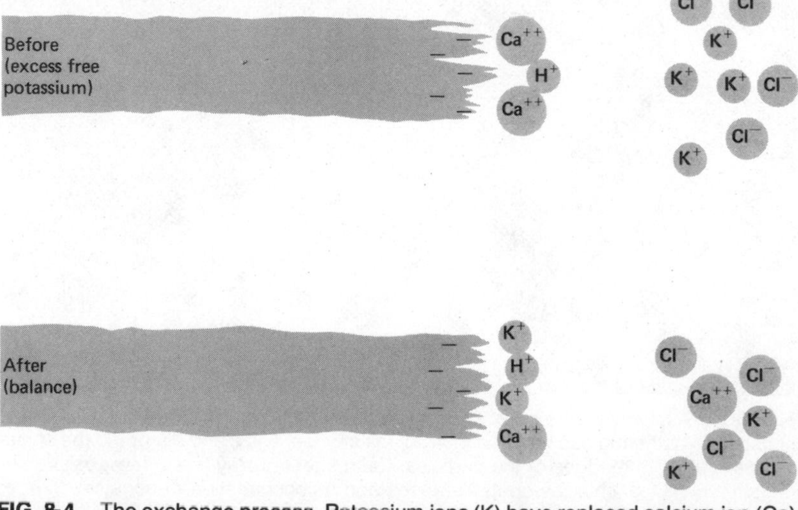

FIG. 8-4. The exchange process. Potassium ions (K) have replaced calcium ion (Ca) on the soil particle. The difference in number of plus signs by the chemical symbols for potassium and calcium reflects innate differences in the electrical activity of the two substances.

either calcium or potassium may be taken into a plant root or carried away by water moving between the soil particles.

leaching removal of nutrients by water percolating through a soil

Unattached potassium will continue to replace calcium on the soil particles until the attached and free proportions of both substances come into balance. When that occurs, the total amount of potassium or calcium attached to the soil particles remains fairly constant, although individual replacements are occurring at all times. If a plant root removes some of the free potassium from a soil, the balance between free and attached potassium is upset, and potassium attached in the storage areas will be replaced by free calcium until the balance between free and attached potassium and calcium is restored.

percolation movement of a liquid through a porous solid

porosity percentage of total volume not occupied by solid material (see Fig. 8-8)

Thus, the exchange process in soil accomplishes two things: it provides a means for the temporary storage of plant foods and other electrically charged particles, and it maintains a balance between the proportions of loose and attached ions of various substances. The continuous exchanging of substances on the negatively charged surface of soil particles is so important that soil scientists use the term exchange capacity to evaluate the ability of a soil to store electrically charged particles. Exchange capacities of soils range from less than 1 milliequivalent (a measure of electrochemical activity) per 100 grams of coarse sand to more than 100 milliequivalents per 100 grams of fine clay.

exchange capacity total number of exchange sites (surface electrical charges) in a mass of soil

LEACHING

Water percolates downward through the empty spaces in a soil after a heavy rain. A few of the water molecules will break apart (Fig. 8-5). The hydrogen part of the former water molecule has a positive electrical charge, so that it can replace plant nutrients on the negatively charged exchange sites. The exchange process tries to bring the proportions of hydrogen and other substances into balance.

However, each new rain brings a fresh supply of hydrogen. Like a millionaire at an auction, the rain can keep outbidding the other substances for the exchange sites on the soil particles. Detached nutrients usually dissolve in the remaining water and are carried away. In very rainy areas, leaching by rainwater can remove most of the exchangeable plant nutrients and leave a soil quite unproductive. Leaching is especially severe in hot climates, because high temperatures cause water to dissociate more rapidly and the surplus hydrogen can remove nutrients faster than in a cool soil.

Some large plants, as if recognizing the threat to their future food supply, send down deep roots to bring some of the leached nutrients back up into their stems and leaves. The nutrients are released near the top of the soil when these plants lose their leaves or die and their remains decay. In this way a given bit of plant food may be brought back to the upper layers of the soil many times. Removing the plants interrupts the recovery of leached nutrients. For this reason, cleared fields in hot and rainy places are prone to a rapid reduction in fertility.

FIG. 8-5. Dissociation of a water molecule. At higher temperatures a larger proportion of the water molecules are dissociated.

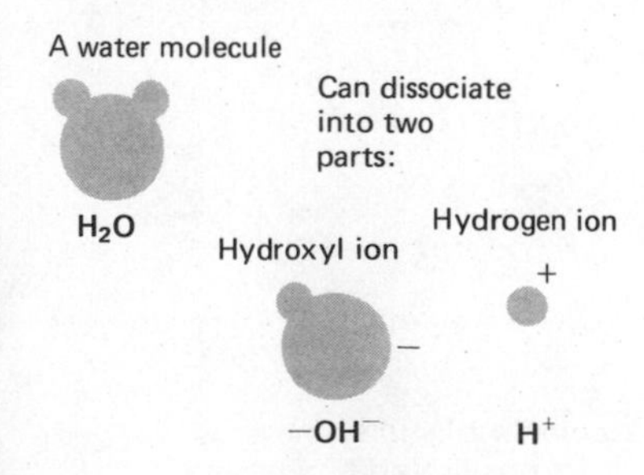

pH AND BASE SATURATION

pH hydrogen ion concentration or exchange acidity:

1 = very acid
7 = neutral
14 = very basic or alkaline

liming adding lime (calcium carbonate) to a soil in order to neutralize its acidity (raise its pH)

base saturation percentage of total exchange capacity occupied by bases (elements such as calcium, magnesium, sodium, or potassium)

The chemical antagonism of hydrogen ions and plant nutrients suggests two quick but useful measures of the fertility of a soil. One common method is to determine the amount of hydrogen that is not locked up in water molecules. The result is usually expressed as a pH number from 1 to 14. A soil in which the amounts of free hydrogen and other ions are in balance has a pH of 7. A pH above 7 indicates an unleached soil with little hydrogen but a surplus of plant nutrients and other substances, whereas a pH of less than 7 is a symptom of a leached soil with a surplus of hydrogen ions. The fact that pH also denotes acidity makes it doubly useful, because plants differ in their abilities to tolerate soil acidity. Most crops do best with a soil pH around 6, although some (such as potatoes or blueberries) prefer a lower pH, and others (such as alfalfa or sugar beets) would rather have a pH around 7.

The base saturation of a soil is the percentage of its total exchange capacity that is occupied by bases—positively charged substances other than hydrogen or aluminum. Base saturation is like a fuel gauge that measures the liquid in a fuel tank but cannot guarantee that all of the liquid is usable fuel. A base saturation of 100 percent may indicate all calcium, all sodium, all potassium, or some combination of these three and a dozen other elements.

Exchange capacity and base saturation must be combined to yield a complete measure of soil fertility. A large tank that is half full may actually contain as much fuel as a completely filled but small tank. One hundred grams of soil with an exchange capacity of 9 milliequivalents and a base saturation of 50 percent holds as many nutrients as a soil with an exchange capacity of 5 and a base saturation of 90 percent. The two soils may seem quite similar for a plant. A farmer would prefer the first soil, however, because it is capable of responding favorably to a fertilizer treatment, whereas the second is near the limit of its nutrient storage ability. A soil with a low exchange capacity in a hot rainy climate is nearly a hopeless case; it is able to hold a plant upright, but the plant must be fed by continuous applications of fertilizer. Even though the exchange sites may be easily filled with fertilizer minerals, the total amount of stored plant foods will never be great because the exchange capacity is low. To compound the problem, such a soil will probably have an insufficient ability to store moisture.

SOIL MOISTURE TENSION

polar molecule molecule with distinct positive and negative ends

A single water molecule is like a miniature magnet with distinct positive and negative ends (Fig. 8-6). The electrical configuration of a water molecule governs its behavior in soil. The positive end is attracted to the negatively charged exchange sites on a soil particle. The connection leaves the negative end of the water molecule sticking out into the empty space between particles, where it can attract another positively charged substance or the positive end of another water molecule. A typical soil particle in a humid

FIG. 8-6. Water molecules. If the chemical formula for water were really just H_2O, it could not exist as a liquid on earth. Most substances whose isolated molecules are as light as a single water molecule are ordained to be gases. The difference between a molecule of water and a molecule of nitrogen or oxygen or carbon dioxide is subtle but profound. A water molecule is electrically unbalanced because the two positively charged hydrogens are not located on exactly opposite sides of the oxygen. In a glass of liquid water the positive end of one water molecule is attracted to the negative end of another, and that one to another and another, so that the individual molecules of water are restrained by their neighbors from flying off into the air too easily.

The electrical restraint among molecules is related to temperature. The electrical bonds between partners are fairly rigid and long-lasting when water is cold enough to be frozen. Ice crystals usually have three or six sides because the angle between the two hydrogens is close to one-third of a full circle.

The bonds between molecules in the liquid form of water are short-lived and continually shifting as the molecules move about. As water is heated, the molecular activity increases, the bonding force becomes less capable of restraining the motion of individual molecules, and the water flows more readily.

Individual molecules of water vapor are so far apart and moving so rapidly that the attractive force between molecules is almost inconsequential. Evaporating water is a good air conditioner because it takes so much energy to overcome the electrical attraction between molecules of liquid water. The release of this energy when water vapor condenses is what gives thunderstorms and hurricanes their awesome power.

The electrical imbalance in water molecules also enables water to exert an electrical pull on substances with either a positive or a negative charge. Thus water is able to dissolve more materials than most other liquids. Its chemical versatility makes water an efficient aid in rock weathering (and an invaluable transportation agent in living things).

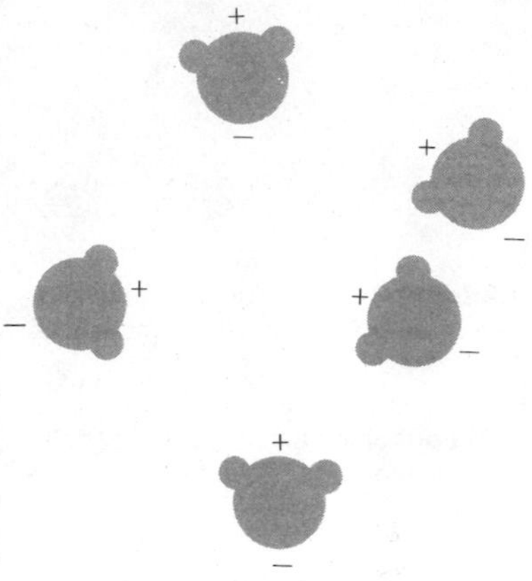

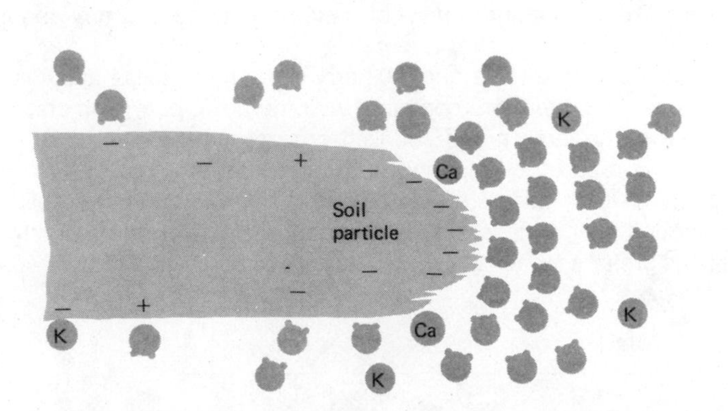

FIG. 8-7. Layers of water around an idealized soil particle. A real soil particle would be proportionally much larger than shown here and would have many more layers of water molecules. Other electrically charged particles usually are scattered among the water molecules in a typical soil.

environment may have many layers of water molecules around it, all aligned with their negative ends pointing outward (Fig. 8-7).

Like spectators in the front row at a carnival show, the innermost water molecules are held most tightly to the soil particle. The attractive force weakens with each additional layer of water around the soil particle. It is easy to remove water from a very wet soil. As the soil dries, however, the remaining moisture is held more and more tightly, and it becomes progressively more difficult to take water out of the soil. The force required to pull water away from soil particles is called soil moisture tension, usually expressed in bars.

soil moisture tension force holding water to a surface; the wetter a surface is, the lower the force on the outermost layers of water

bar measure of pressure or tension; a bar equals 1000 millibars

Water molecules are quite sensitive to differences in tension. Again like carnival spectators, they usually leave areas where their attention is not held very tightly and move toward places where the attraction is more powerful. The migration of water toward areas of greater moisture tension is the reason that sap rises in trees, towels draw water upwards, and water and food pass through the cell membranes of plants and animals. A given substance, however, has only a limited amount of attractive force which it can use to pull water.

A typical plant, for example, can exert only about 15 bars of moisture tension. If water is held in a soil with more tension, it is for all practical purposes unavailable. The plant wilts and dies of thirst even though the soil does contain some water.

wilting point amount of water held in a soil after plants have removed all they can

field capacity amount of water held in a soil after gravity has removed all it can

The pull of gravity on soil moisture is only about one-third of a bar, much weaker than the force that plants can exert. Nevertheless, gravity is important for the health of many plants. A soaking rain fills the pores near the surface of the soil. Water in the middle of large pore spaces is held with very little tension, so that gravity can pull it downward in the soil. The removal of water allows air to penetrate the soil and to keep plant roots and other soil organisms from suffocating.

The texture of a soil governs its ability to store water (Fig. 8-8). A clay has great electrical pull. Most of its tiny pores are still filled with water after gravity has removed all it can. The presence of too much water in a clay soil may cause plant roots to suffer from a lack of oxygen. A coarse sandy soil, on the other hand, can hold very little water in its large pores against the pull of gravity. From the point of view of a plant, a sandy soil may seem like a desert only a short time after a rain.

Fortunately for farmers and mulberry trees, most soils have a variety of particle sizes and therefore contain a wide range of pore diameters. Water drains out of the large pores soon after a rain, and the removal of water allows air to reach the plant roots. Smaller pores hold water longer and provide moisture for plants during rainless times. The need for both air and water around their roots is the reason that most plants under most climatic conditions prefer a loamy soil, with a mixture of particles and pore spaces of different sizes.

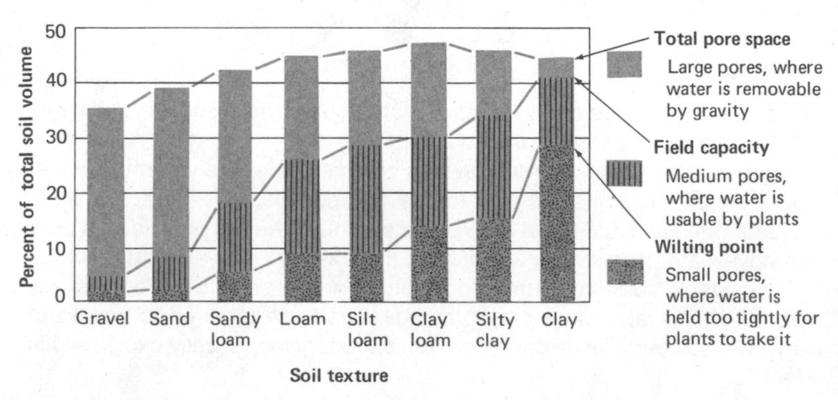

FIG. 8-8. Pore space, water available to plants, and water removable by gravity, all expressed as percentages of the total volume of a soil.

THE WATER TABLE

Gravity continually pulls surplus water downward, and in most soils the pore spaces are completely filled with water at some distance below the surface. The top edge of this saturated zone is called the water table. Like a bank savings account, the level of the water table in a particular soil depends on the balance between the rates of deposit and withdrawal (Fig. 8-9).

water table upper edge of a completely saturated zone in the soil

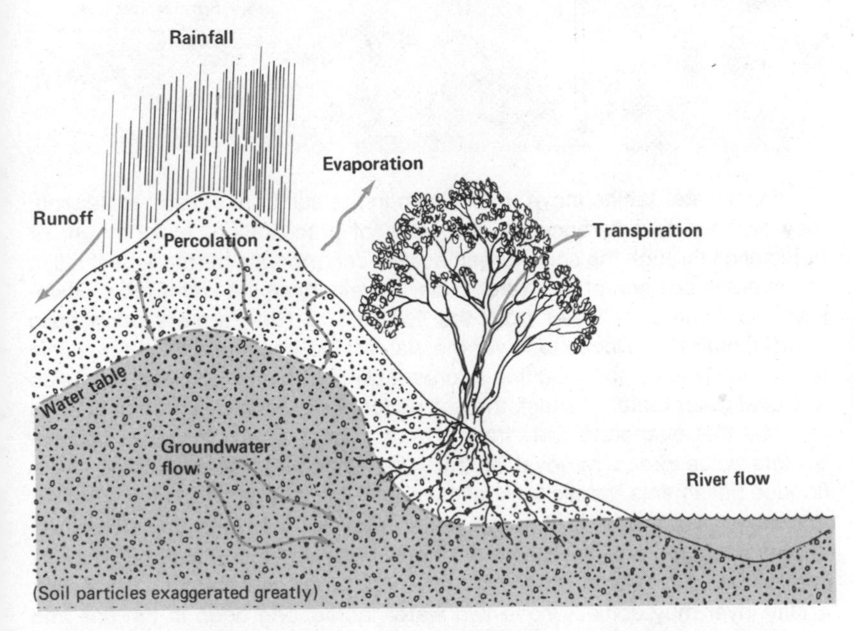

FIG. 8-9. Processes that affect the water table. Rainwater falls on the ground surface; plants remove water from the soil; soil moisture percolates downward to the water table; water in the saturated zone of the soil moves downward and sideways toward river valleys; and the rivers carry water back to the ocean where it evaporates and becomes rainwater again. The water table responds to the balance among a variety of additive and subtractive forces.

Factors That Raise the Water Table by Aiding Inflow	Factors That Raise the Water Table by Interfering with Removal
Rainy weather	Fine soil texture
Coarse soil texture	Cold weather
Flat Land	Flat land
Downhill location	Vegetation removal
Good soil structure	Frozen subsoil
Irrigation	Poor soil structure
Factors That Lower the Water Table by Interfering with Inflow	**Factors That Lower the Water Table by Aiding Removal**
Dry weather	Coarse soil texture
Steep slope	Hot weather
Fine soil texture	Uphill location
Frozen soil	Vegetation growth
Pavement	Artificial drainage
Poor soil structure	Good soil structure

FIG. 8-10. Depressed water table around a well. In this example a new well removed water and lowered the water table enough to cause an old well and a nearby stream to go dry.

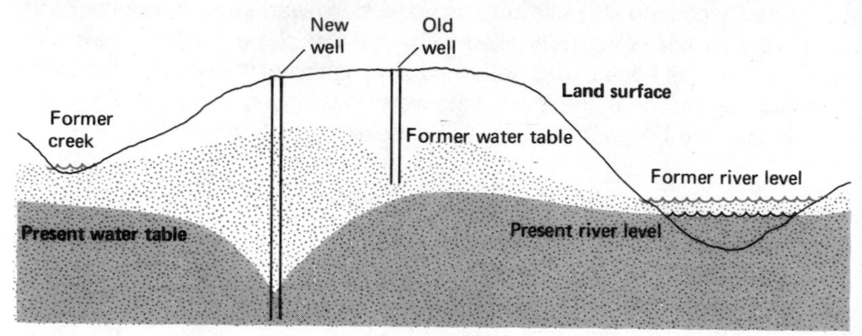

Most water tables move up or down in the soil from season to season; they respond to differences in the rates of precipitation, evaporation, or percolation through the soil. Human activity can produce permanent shifts in the average position of the water table. A well draws the water table downward as it removes water from the soil (Fig. 8-10). Farmers often build artificial drains or ditches to lower the water table and thus make it possible to farm in swampy soils. Cutting a forest down, on the other hand, can raise the local water table, because the soil no longer loses the hundreds of liters of water that evaporate daily from a mature tree. To the dismay of homeowners, widespread removal of trees in residential areas often leads to flooded basements, septic tank failures, or the appearance of new streams.

permeability ability of a material to allow water to flow through it

perched water table saturated zone on top of an impermeable layer in the soil

Water cannot move through dense rock or soils with tiny pores. An impermeable material such as rock or clay beneath the surface may make it impossible for gravity to drain even the large pores of a thin soil. A soil with a clay layer may actually have two water tables, one deep in the soil and another perched on top of the clay layer. Perched water also may seep out where a permeable rock layer is exposed on the side of a hill (Fig. 8-11). Like

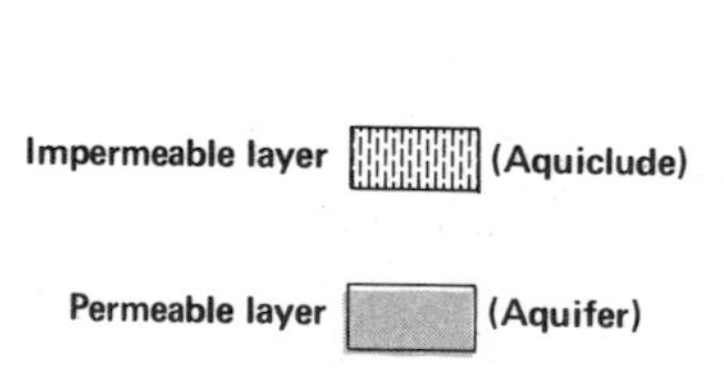

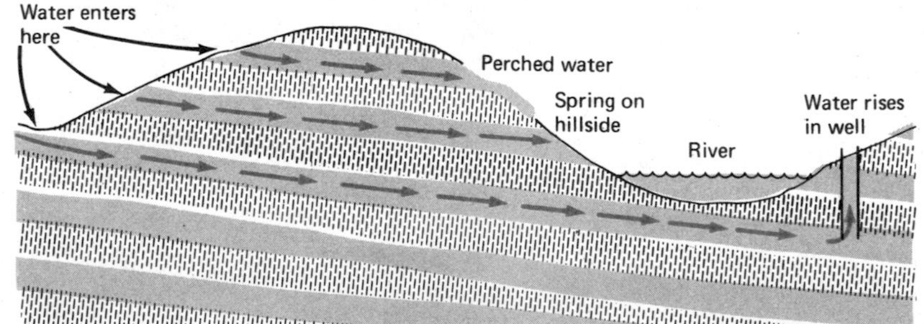

FIG. 8-11. Results of a combination of permeable and impermeable layers of rock or soil. The spring occurs on the hillside because water perched above an impermeable layer cannot flow down through it and therefore seeps sideways onto the surface. A flowing or artesian well occurs when water enters a confined layer at a high elevation and cannot escape. Downhill pressure of water seeping through the aquifer forces water up into the well.

ground water tables, perched water tables often move up and down, and a seepage or spring may respond to the variations in soil moisture by disappearing in dry weather and reappearing after a storm.

SOIL DRAINAGE

A soil is likely to be poorly drained if it is very fine in texture, with small pore spaces and strong moisture tension, or if it is near or below the water table for extended periods of time. A poorly drained soil is unfavorable for the growth of many plants; therefore it is useful to be able to recognize the symptoms of poor drainage before investing in a garden, lawn, or farm with a soil that will doom green plants to a miserable existence under the constant threat of suffocation.

The color of iron compounds in a soil layer is a handy indicator of average moisture conditions. A soil that is red (or black or brown with red spots) is usually well-drained. Air can enter the soil pores easily, and oxygen combines with iron in the soil to form rusty red iron oxides. Yellow colors indicate intermediate drainage with occasional moist periods. Pale gray or blue-gray colors are symptoms of poor drainage; the iron in a soil cannot rust when water occupies the pores and air is excluded. Mottles (mixtures of different colors) often occur where a soil is partly aerated and partly saturated, such as near the water table or in other places where pore sizes are unequal and water remains in the smaller pores most of the time.

oxidized iron yellow or red compounds of iron and oxygen
(see p. 147)

reduced iron blue or gray compounds of iron without oxygen

Soil color is not a perfect guide to moisture conditions. The bizarre reds, whites, yellows, greens, and purples in a desert soil are hereditary rather than environmental. The color of a soil with little moisture or life is usually a consequence of parent material rather than the moisture conditions of the soil.

The black color of decaying organic matter can mask the colors of iron compounds and parent material. The odor of a black soil provides a clue to its normal moisture characteristics. Black soils that are well drained and not too dry for life have an "earthy" smell, a distinctive and easily recognized by-product of some oxygen-breathing organisms in the soil. Black soils that are usually saturated with water have a weak oily or sour smell, difficult to describe but easily distinguished from the earthy aroma of unsaturated soil. Very dry soils that inherited a black color from their dark parent material often have a faint chalklike odor.

actinomycetes aerobic (oxygen-requiring) microorganisms that impart an "earthy" smell to well-drained soils
(see p. 187)

anaerobic organisms living things that can live in the absence of free oxygen
(see p. 180)

The physical problems of poor drainage or inadequate moisture-holding capacity are common soil ailments, but they often go unrecognized in the face of the contemporary American faith in a chemical cure for all soil ills. Impeded drainage is difficult to remedy if a soil has fine texture and inherently high moisture tension. The best solution appears to be surrender: either replace the soil entirely or concede the battle and select plants that can tolerate the water. The outlook is not so bleak if a high water table or clay layer is the reason for poor drainage of the topsoil. Artificial drains and ditches can carry excess water away and give plants a chance to grow.

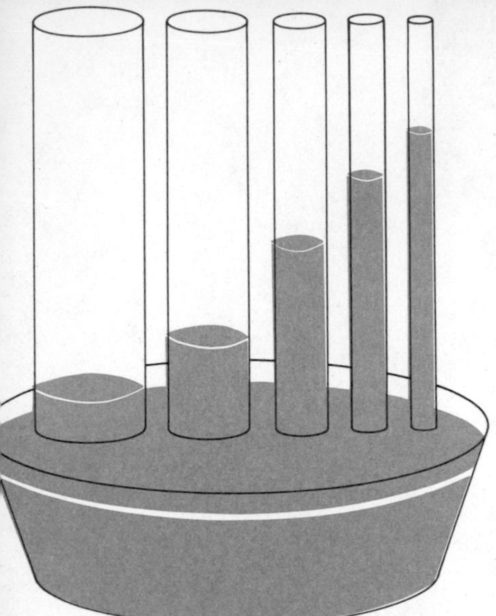

FIG. 8-12. Capillary action. Electrical attraction draws water upward into small pore spaces. It rises higher in thin tubes because moisture tension is more effective in small spaces than in large ones.

CAPILLARITY AND DESERT SOILS

High school teachers argue persuasively that pride was responsible for the downfall of Ozymandias, the legendary Egyptian monarch whose monument to his mighty works now lies toppled and broken on the barren desert sands. Pyramids and other impressive ruins are persistent reminders that Babylon, Industan, Thebes, Persia, and other ancient desert civilizations have also perished. Scholars have proposed many reasons for the disappearance of these cultures, but modern scientists have looked at the earth beneath the ruins and concluded that one of the key forces in the collapse of desert civilizations was the electricity that stores water and nutrients in soil.

Water usually moves downward in a wet soil because gravity exerts a constant force on the weakly held water molecules. The pull of gravity is not strong enough to move the small amount of water in a dry soil; the water migrates toward areas of greater moisture tension regardless of the direction of the pull. A fine-textured soil has more electrical pull and can draw water more effectively than a coarse soil (Fig. 8-12).

The movement of soil water toward areas of greater moisture tension is often advantageous. Plants increase the moisture tension near their roots when they take water from the soil. Additional water is then drawn toward the roots by electrical attraction. This automatic delivery system provides a valuable rescue service for plants during dry spells.

Nutrients dissolved in soil water often move with the water toward areas of greater tension. Plants pull nutrients toward their roots when they take water from the soil. Gravity combines with excess rainfall in humid regions to carry dissolved substances downward in the soil. In desert regions, on the other hand, material is often leached upward, because the rapid evaporation of water into the dry air increases the moisture tension near the surface. The dissolved salts cannot evaporate with the water; they remain as a salty crust on the surface of the soil, a prospect that pleases chemical companies and people trying to set new speed records on the Bonneville salt flats (Fig. 8-13).

FIG. 8-13. A desert salt flat. The surface layer contains crystals of salt that were left behind as saline water rose to the soil surface and evaporated into the dry air. (Courtesy of C. G. Hunt, Geological Survey).

Irrigation farmers, however, rightly fear the tendency for salt to accumulate near the top of the soil in dry areas. One of their major concerns is to provide adequate drainage for their fields, so that they can pump excess water onto the fields occasionally to wash some of the toxic salts away from the roots of sensitive crops. Even this remedy is not a panacea, because it adds salt to rivers and reduces the value of the water for farmers or cities downstream. Many thousands of acres of farmland have been abandoned because of salt accumulation in Arizona, Idaho, California, and other irrigated areas of the United States. A regional accumulation of salt also may have been the underlying reason for the decline of some ancient desert civilizations in Mesopotamia and the Indus Valley of present-day Pakistan.

salinization accumulation of salts near the surface of a desert soil (see Fig. 10-11)

flushing washing of salts from a soil by adding excess water

SOIL STRUCTURE

Salt accumulation also causes problems by disrupting the structure of a soil. Particles in a natural soil are seldom completely independent of one another; they usually cluster in groups of varying size and cohesiveness, like cliques in a high school homeroom. Soil aggregates are held together by a variety of means, including the same electrical forces that are responsible for nutrient storage and the behavior of water in the soil. A well-aggregated soil is usually a good soil, easy to work, and high in productivity.

Two soil particles cannot come together to form a structural aggregate until something overcomes the natural tendency of negatively charged objects to repel each other. An ion of calcium has two positive charges. In a simplified way, Figure 8-14 shows how calcium is able to attract several soil particles simultaneously. In effect, calcium can neutralize the electrical antagonism of soil particles. Aluminum, with a triple charge, is even more effective than calcium in bringing soil particles together. Water treatment plants often add aluminum compounds to muddy water to attract the negatively charged soil particles and clump them together so that they will settle to the bottom of the tank. Potassium, sodium, and ammonium are electrically weak, with only one charge. They actually tend to separate soil particles rather than bring them together.

ped group of soil particles held together by a combination of electrical, chemical, and biological forces

flocculation clumping of particles by electrical attraction to an intermediate substance

deflocculation separation of particles by electrical antagonism

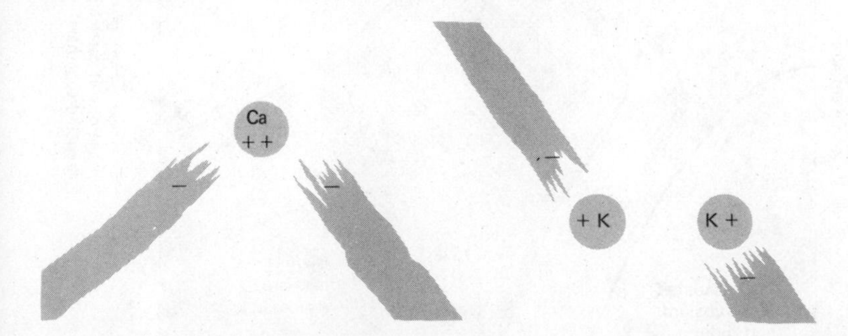

FIG. 8-14. A simplified model of flocculation. An ion of calcium attracts two clay particles and draws them together. Potassium, on the other hand, satisfies the electrical charge of soil particles without bringing them together.

mycelia threadlike bodies of fungi that hold soil particles together

Other forces assist in holding a soil aggregate together once the electrical antagonism of soil particles has been overcome. Certain organic and inorganic compounds act like cements of varying effectiveness. Soil microorganisms and plant roots wrap around soil particles and help bind them in place. Continued chemical decomposition of clumped soil particles may alter their electrical antagonism.

The structural development of a soil depends partly on climate (Fig. 8-15). Soil particles are held together primarily by calcium and magnesium in areas where rainfall and evaporation are about balanced. Aggregates are numerous but weak enough so that plant roots can enter them. Desert soils often contain salts with large quantities of sodium and potassium, which hinder the development of soil structure. Soils are also poorly structured in cool and rainy places, because calcium and magnesium are leached out of the soil. In hot and wet environments, weathering and leaching work so rapidly that only the very resistant oxides of iron and aluminum remain in the soil. These electrically powerful substances are often bound into extremely tight aggregates. A tropical soil may consist of more than 90 percent clay, yet it still feels and acts like a coarse gravelly sand, with large pore spaces and poor nutrient and water storage ability.

The parent material of a soil can modify the general relationship between climate and the development of soil structure. Coarse materials are hard to hold together in any environment. Like a bull on a slender rope, a sand grain or pebble has too much mass for the weak forces of aggregation.

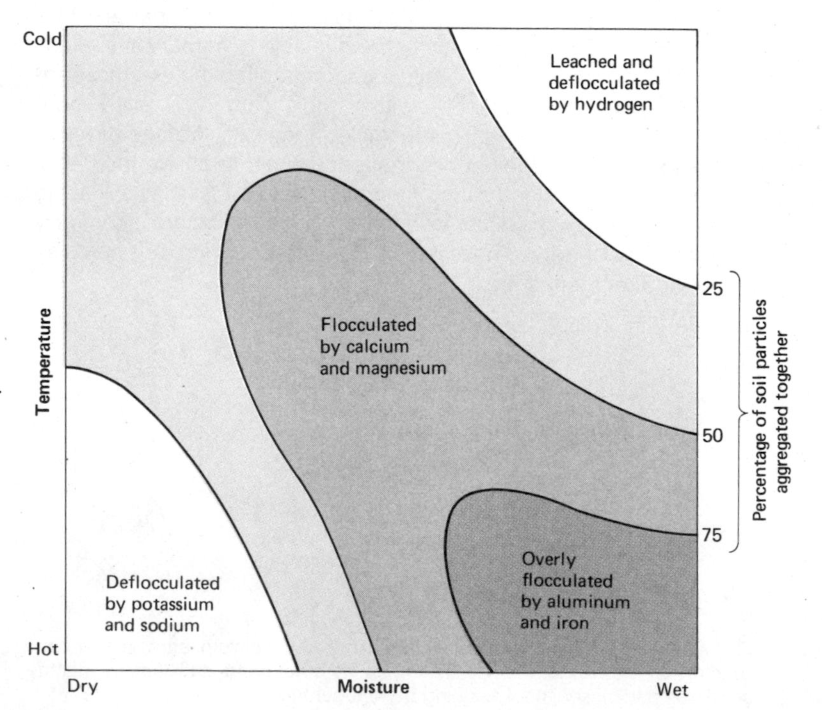

FIG. 8-15. Soil aggregation under various climatic conditions.

Chemical composition is also significant: under identical weather conditions, a sodium-rich rock will yield a soil with weaker structure than one containing aluminum or calcium.

Soil drainage plays a self-perpetuating role in structural development. Waterlogging weakens the electrical, chemical, and biological forces that bind soil aggregates together. At the same time, water cannot flow through a structureless soil as easily as through a well-aggregated one. Structureless soils stay wet, and wetness impedes the development of structure. Like a microphone held in front of its own loudspeaker, poor structure and poor drainage reinforce each other, creating a vexing problem.

gley layer poorly drained layer with massive structure

In summary, the structure of a natural soil depends on parent material, climatic conditions, land use, and pressure on the particles (Fig. 8-16). Tilth is the farmer's term for the structure of the top layer of the soil. A soil in good tilth is easy to work, drains well, stores plant foods efficiently, absorbs rainfall readily, and resists erosion. Percolating water goes around clumps of soil particles rather than flowing through them, although moisture tension will pull water into the clumps if they are dry. Nutrients inside soil aggregates are sheltered from the leaching effects of water. The storage of nutrients inside clumps of soil particles is beneficial as long as plant roots are able to penetrate the aggregates and remove the imprisoned nutrients.

tilth colloquial term for soil structure (especially its plowability)

FIG. 8-16. Common soil structural types. A complete soil structure description includes cohesiveness and size in addition to type (e.g., "strong fine granular" or "weak-to-moderate coarse subangular blocky").

a. Loose or single-grain—a structureless soil in which the particles show little tendency to stick to each other at all; common in coarse-textured materials and dry environments; very permeable.

b. Crumblike—light, rounded aggregates with considerable pore space between and within the peds; most common in the surface soils of grassland regions and in well-managed cultivated soils.

c. Granular—dense, rounded aggregates with little empty space inside the peds; typical of surface soils in tropical regions and some mid-latitude agricultural areas.

d. Blocky—compact and tightly fitted together, with less pore space than any of the above; common in subsoils in humid regions; subangular blocky peds have slightly rounded corners; angular blocky peds are sharp-edged because they are more tightly compressed.

e. Columnar and prismatic—large peds with considerable vertical extent, common in dry regions where little moisture percolates downward and few roots reach the lower layers of the soil; columnar peds have rounded tops; prismatic peds are more angular.

f. Platy—thin flat peds with considerable horizontal extent; common in areas where soil processes produce distinct layered accumulations of clay or minerals in a soil, such as the calcium layers of the Great Plains or the iron layers in New England; platy peds tend to resist upward or downward flow of water.

g. Massive—a structureless soil in which the particles all seem to stick to each other; most common in poorly drained soils where roots are few and water movement is slow; massive layers are often nearly impermeable.

Poor farming practices, however, can break down the structure of a soil. Plows and cultivating tools disrupt aggregates, particularly when soil is too wet or dry. Irrigating with brackish water adds salts that cause soil particles to separate (remember Ozymandias?). Excessive use of potassium, sodium, or ammonium fertilizers has the same effect. A plume of dust behind a moving tractor is a symptom of weak soil structure.

plinthite mixture of simple clays and oxides of iron and aluminum; it stays soft inside a soil, but when exposed it often hardens irreversibly

On the other hand, the particles in many tropical soils are already too tightly clumped together for their own good. Exposing that kind of soil to the direct sun may hasten the chemical processes and cause irreversible hardening of the aggregates, a fatal agricultural disease.

Farming does not always damage soil structure. In fact, manuring (putting animal wastes on the soil), liming (adding calcium and magnesium carbonates), and mulching (returning crop residues to the soil) may be even more helpful as structural aids than as fertility additions. Modern farmers are becoming more aware that structure is as important as fertility for the long-term health of a cornfield.

CONCLUSION

Soil is a storehouse for plant foods and water because the surfaces of most soil particles are electrically charged. Since electrical forces are responsible for many soil characteristics, it is not surprising that a measurement of electrical conductivity is one of the most useful indicators of overall soil quality.

electrical conductivity ability of a material to transfer electrons

A soil must contain water and dissolved salts in order to conduct electricity. Plants also require moisture and nutrient salts, and therefore the conductivity of a soil is a reasonable measure of its value to a plant. Moreover, brief descriptions of four places on the map of soil conductivity can summarize many of the major points of this chapter (Fig. 8-17).

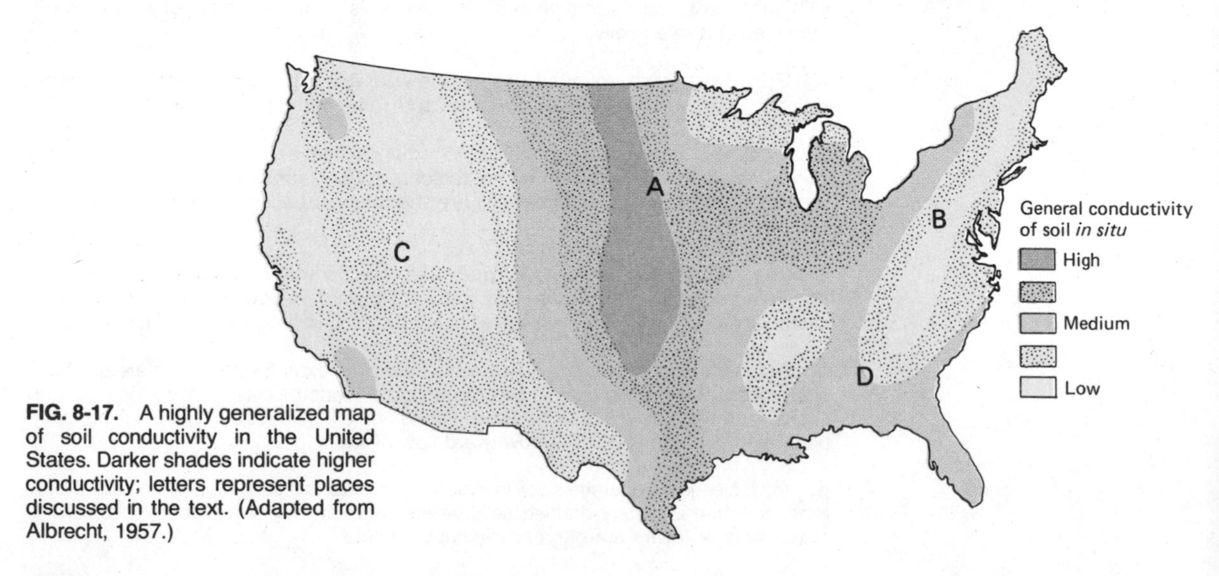

FIG. 8-17. A highly generalized map of soil conductivity in the United States. Darker shades indicate higher conductivity; letters represent places discussed in the text. (Adapted from Albrecht, 1957.)

Place A on the map is in the heart of the Corn Belt. Rainfall and evaporation are about equal on this former grassland. Leaching is not significant, and the soil contains ample supplies of calcium and other plant foods. These substances and the grass roots bind the soil particles into rounded, porous aggregates that store nutrients efficiently and drain well. A high conductivity reading reflects the availability of nutrients and moisture in this productive and easy-to-work soil.

Place B, an Appalachian hillside, has a thin soil over sandstone bedrock. Coarse particles are not able to attract and hold nutrients or water very efficiently. Place B gets more rain than place A, and leaching is rapid when water is abundant, pores are large, and exchange capacity is low. The resulting nutrient poverty keeps the conductivity reading of this soil low.

The conductivity of soil at place C is limited by lack of water. Rain falls fewer than half a dozen times in a typical year. Soil moisture rises to replace water that evaporates from the surface of the soil during the extended dry periods. Dissolved substances cannot evaporate with the water and are left behind near the surface of the soil. The result is a potentially poisonous excess of plant nutrients and other salts in the surface layer of soil.

Place D was chosen to illustrate the fact that landforms can produce local exceptions to the general climatic pattern of the map. Place D lies at the foot of a hill and is almost always saturated with water seeping down from the higher ground around it. The soil is gray and structureless, although it is formed from the same parent material as the red soils on the uplands. Excess water prevents the rusting of iron in the soil and interferes with the development of soil structure. In turn, the massive structure impedes the movement of water and thus prolongs the saturation of the soil. The excess water in the root zone is a real hazard for plants, because roots can suffocate when air is kept out of the soil. Most soils on the hills near place D are leached, but the valley soils receive water and nutrients from the uplands, and therefore the conductivity of soil is higher in the valleys than on the surrounding hills.

The first of these four soils was influenced mainly by its vegetation cover, the second by its parent material, the third by climatic conditions, and the last by its topographic setting. These four factors act through time to define the significant characteristics of any soil—its texture, structure, color, drainage, and fertility. The environment of a soil is therefore responsible for its value as a storehouse.

APPENDIX TO CHAPTER 8

Plant Nutrients And Deficiency Symptoms

More than 90 percent of the dry weight of a typical plant consists of the three structural elements—carbon, hydrogen, oxygen—which are obtained from water and carbon dioxide. The remaining 10 percent is made of nutrient elements taken from the soil.

Chemical Element	Grams in a 1000-kg Plant	Major Functions in Plant Nutrition	Deficiency Symptoms in Common Plants	Soils Likely to Be Deficient	Interactions and Other Characteristics
Macronutrients					
Calcium	2500	Cell growth, circulation	General stunting	Sandy, acid, leached	Makes soil less acid, hinders boron uptake
Magnesium	500	Chlorophyll formation	Purple or white stripes on leaves	Sandy, acid, leached	Aids phosphorus uptake, behaves like calcium
Nitrogen	3000	Photosynthesis, absorption	Yellow leaves	Inorganic, hot, wet	Taken from air by bacteria, hinders sulfur use
Phosphorus	500	Cell growth, transpiration	Purple leaves, stunting	Acid or alkali, inorganic	Combines with some metals, hinders zinc uptake
Potassium	2500	Disease resistance, fruit setting	Stunting, yellow or rusty leaf edges	Alkali, sandy, cold, wet	Clay form unavailable, hinders magnesium uptake

Micronutrients					
Boron	5	Circulation, absorption	White end leaves	Dry, sandy, alkali	Released from decay of organic matter
Chlorine	30	Disease resistance	Crinkling of leaves	Wet	Very rarely deficient in natural soils
Copper	2	Proteinsynthesis, root growth	Curled tips of leaves	Alkali, sandy, organic	Hinders molybdenum, iron, and phosphorus uptake
Iron	20	Chlorophyll formation	White new leaves	Alkali, wet	Difficult to correct deficiency
Manganese	15	Oxygen balance	Mottled leaves	Alkali, organic	Interferes with iron uptake
Molybdenum	1	Protein synthesis	Twisted leaves	Acid, sandy	Aids nitrogen fixation, hinders copper uptake
Sulfur	200	Vitamin and protein synthesis	Yellow leaves	Sandy, leached	Aids nitrogen uptake, hinders calcium uptake
Zinc	100	Enzymes and auxins synthesis	Yellow spots, white leaves	Alkali, eroded	Combines with soil elements

Animals also need cobalt, fluorine, iodine, selenium, sodium, and vanadium, which commonly occur in plants but have no well-understood role in plant nutrition. Plants also take up significant quantities of silicon and aluminum, but these two elements appear to play no role in either plant or animal nutrition. Many chemical elements, including some listed on this table, are toxic if too abundant. This table is a broad generalization; individual plants vary widely in their requirements, tolerance limits, and symptoms of deficiency or toxicity. Finally, this table represents a summary of a rapidly changing research field, and therefore it may need revision by the time it appears in print.

Chapter 9

Decomposers: Natural Sanitary Engineers

I don't go to national conventions no more; those midwestern farmers think their good soil is their own doin'.

E. W. Cole, USDA · Soil Conservation Service

The initial contact with a tropical soil is a striking experience for a tourist from the Corn Belt of the midwestern United States. An ordinary roadcut stands out like a bleeding gash, and a plowed field seems like a giant lifeless burn, shimmering in the blazing sun. The local inhabitants often verify the tourist's first impressions regarding the poverty of a red soil. Many southern soils have an almost legendary infertility. By contrast, the home ground of the tourist is deep and dark and crumbly, a soil that literally exudes richness.

It is ironic, then, that the black topsoil of Minnesota or North Dakota should be the consequence of life impaired, while the red fields of Alabama are the result of unhindered living activity. But ask any one-celled microorganism, and it will say it prefers a warm and moist environment, safe from the threats of frost, drought, or flood. The opinions of the microbes are significant, because the activity of living things in the soil is a major influence on the color and fertility of the ground. Among the factors that affect their activity are temperature, moisture, soil texture, and the adequacy of their diet.

microbe one-celled (noncellular) organism

TEMPERATURE

Up to a point, the warmer a soil is, the better the soil organisms like it. Large plants become active when their internal temperatures rise safely above the freezing point of water and their circulatory systems can begin to function. The rate of plant activity increases as the weather gets warmer. Green leaves produce the most plant material at temperatures around 30° C (86° F). Temperatures above 30° C often reduce plant growth, because too much heat lowers the efficiency of the various processes of obtaining nutrients and converting sunlight to plant material. Plants expend too much energy in hot environments just keeping themselves alive; little energy remains to add to the permanent structures of the plants.

photosynthesis sun-aided production of green plant material (see Table 13-1)

net primary productivity surplus of photosynthetic production above what the plant needs to stay alive

Whenever a plant or animal dies, it enters the sepulchral realm of the decomposer microorganisms, the multitudes of minute creatures that make their living by eating the remains of dead plants and animals (Fig. 9-1). The

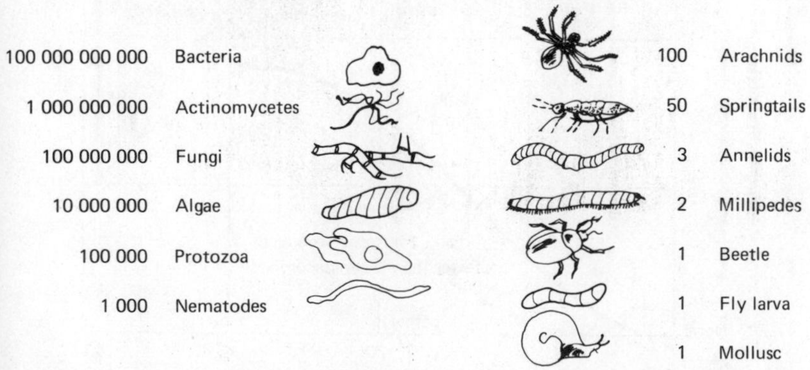

100 000 000 000	Bacteria	100	Arachnids
1 000 000 000	Actinomycetes	50	Springtails
100 000 000	Fungi	3	Annelids
10 000 000	Algae	2	Millipedes
100 000	Protozoa	1	Beetle
1 000	Nematodes	1	Fly larva
		1	Mollusc

FIG. 9-1. Typical populations of organisms in a handful of topsoil. The measure of volume is intentionally vague because soil populations vary widely from place to place.

movements of large soil animals help to break plant and animal fragments apart and to mix them with soil. Earthworms, soil insects, and other medium-sized animals can take in pieces of organic debris and digest them internally. Tiny bacteria and fungi secrete digestive chemicals into their surroundings. In this manner, complex organic materials are decomposed into simple substances that can be absorbed by the microorganisms or taken up by the roots of higher plants.

extracellular enzyme digestive chemical secreted into the environment

Like the growth of higher plants, the activity of decomposer organisms depends on temperature. Microbes decompose organic material very slowly when the average temperature is below 10° C (50° F). Occasional frost is a major threat to microbes in cool soils, because freezing can kill many decomposer organisms or force them into dormancy. As temperature rises, however, the speed of decomposition increases rapidly. Many bacteria do not reach their maximum efficiency until temperatures go well above 35° C (95° F).

There is an important balance point near 25° C (77° F) (Fig. 9-2). At temperatures below 25° C, organic matter is produced more rapidly than it can be decomposed, and some partially decayed plant and animal material remains in the soil. The partially decomposed organic matter is the humus that gives a rich black color to soils in cool regions.

humus partially decomposed organic remains; complex carbon compounds give humus an intense black color

The forces of decay get the upper hand in places where the average temperature is higher than 25° C. The great efficiency of decomposers at high temperatures allows them to break down organic matter almost as soon as it is produced. Soils in warm and moist regions rarely contain much organic matter, because plants cannot keep ahead of decomposers. Soil particles that are not coated by intensely black humus usually have a gray, tan, yellow, or red color, depending on their chemical composition.

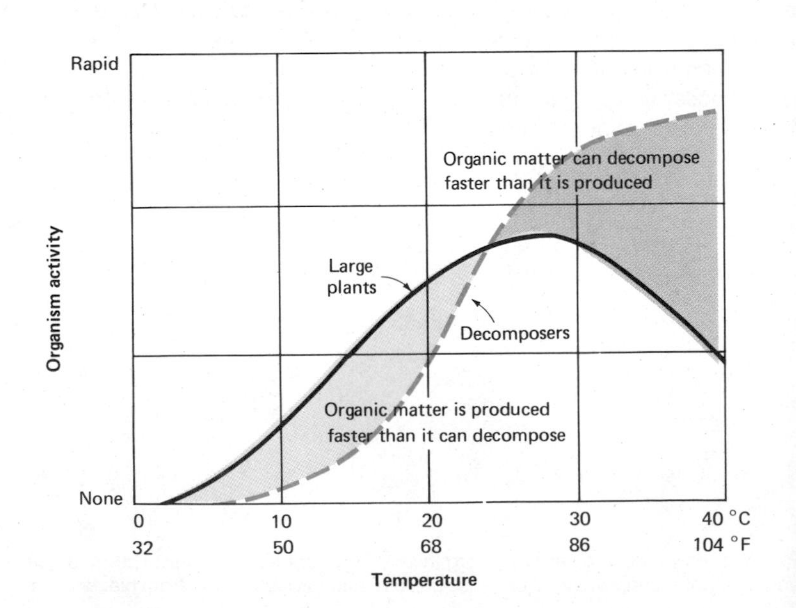

FIG. 9-2. The rate of activity of producers and decomposers of organic material.

Clearing a mid-latitude forest reduces the amount of humus in the soil by exposing the ground to the direct rays of the sun, raising the soil temperature, and thus hastening the rate of microbial activity. At the same time, the removal of trees halts the annual fall of leaves to the ground. The soil organisms are deprived of their food supply, and the soil becomes lighter in color as the remaining humus is consumed and the microorganisms begin to cannibalize each other. In former times, tan or light red or ashy gray fields were often abandoned, with good reason, as "worn-out."

MOISTURE

The simple relationship between temperature and soil humus rests on the assumption that all other conditions for life in the soil are acceptable to the decomposer organisms. However, many bacteria and fungi cannot tolerate a dry soil. Unprotected wooden fence posts soon rot in humid lands but may last for centuries in the dry soil of a desert. Decay-causing organisms are totally frustrated by the lack of water in the arid soil.

The decomposition of organic material in a soil thus depends on a combination of heat and moisture availability (Fig. 9-3). The humid southeastern states have ample warmth and abundant rainfall. Microorganisms in this favorable region can consume organic matter more rapidly than the higher plants are able to produce it. As a result, Alabama and Georgia have light-colored red and yellow soils. Soils become darker toward the north because low temperatures slow the decay of organic matter in the soil. Soils toward the west are also dark because the lack of soil moisture hinders

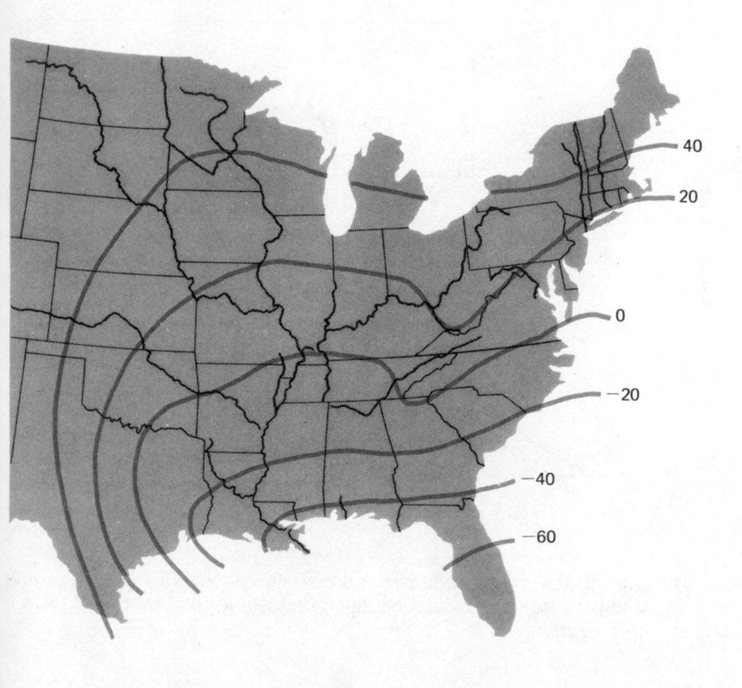

FIG. 9-3. Comparison of rates of production and decomposition of organic matter. Negative numbers indicate that potential decomposition is more rapid than production. (Adapted from Meentemeyer, 1974.)

decomposition. In extremely cold or dry areas, however, plants produce very little organic matter and soils remain light-colored by default.

Local drainage and microclimatic changes can greatly modify this general relationship between climate and the persistence of organic matter in the soil. Soils on south-facing slopes in most parts of the United States receive more energy from the sun and therefore are warmer and lighter in color than those on slopes that face to the north. The wind can deprive some soils of organic matter by hindering plant growth on exposed sites or by carrying leaves and twigs away from where they were produced. A high water table in a soil tips the balance in favor of the accumulation of humus, because the organisms that require free oxygen are suffocated if the pore spaces in a soil are completely filled with water. Some specialized bacteria are able to survive without free oxygen, but they cannot decompose organic matter as rapidly as the more efficient aerobic microorganisms (Fig. 9-4). Therefore swamps and bogs usually have black soils, rich in partly decayed organic matter.

aerobic condition presence of oxygen

anaerobic or anoxic condition absence of oxygen

muck soil with a large amount of partly decomposed organic matter

The fertile appearance of mucky soils in wet areas has tempted many people to drain swamps for agricultural purposes. Allowing air to enter formerly saturated soil, however, often hastens the decay of the organic fraction of a soil. Some bog soils may contain as much as 50 or 80 percent organic matter, and the accelerated decomposition of organic matter can lead to a noticeable shrinking of the soil and general lowering of the land surface. The lowering, in turn, may necessitate digging deeper ditches to drain a still lower layer of soil, which leads to more decay of organic matter, and so the cycle goes. Some former swamps in areas of old settlements in Ohio and Illinois are presently being drained by their third generation of

peat soil with a large amount of undecomposed organic matter

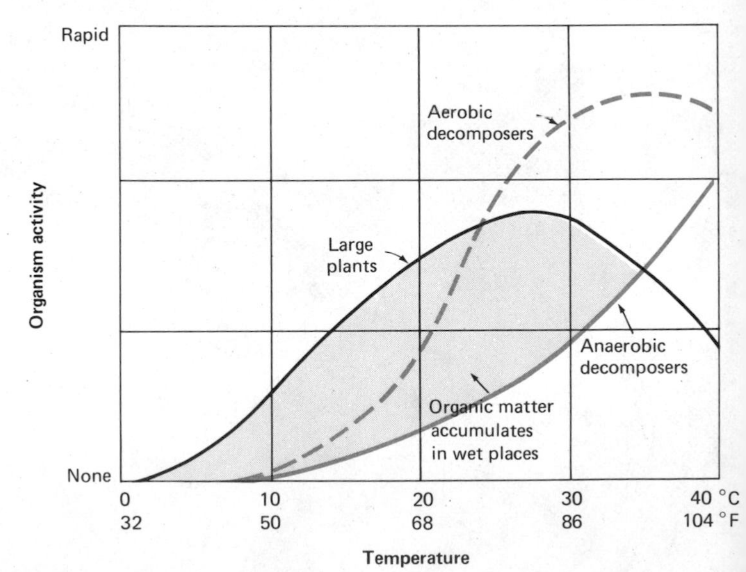

FIG. 9-4. The rate of activity of producers, aerobic decomposers (ones that must have free oxygen to breathe), and anaerobic decomposers (ones that can survive in places that lack free oxygen.)

FIG. 9-5. Installation of underground drainage tile. (Courtesy of E. E. Rowland and N. W. Weller, Soil Conservation Service.)

ditches and tile sewers (Fig. 9-5). However, crop yields on those former swamps are exceptionally high. The productivity of the land often justifies the extra expense of periodic rebuilding of drainage structures.

A similar problem of land subsidence following artificial drainage affects many cities that are located on river floodplains or on flat lands near ocean or lake shores. The most striking example in the United States is the city of New Orleans, which is built almost entirely on a former swamp between the Mississippi River and Lake Pontchartrain. The land was created by the Mississippi River carrying soil from the interior of the continent and depositing the sediment when it entered the Gulf of Mexico. Even in flood, the river could not rise more than a few inches above sea level, and therefore any land built by the river could not be anything but low and level.

delta swampy plain created where a river enters a larger body of water (see Fig. 16-28)

Nevertheless, when it became apparent that the natural advantages of the location as a trade center could offset the drawbacks of its swampy terrain, people began building dikes and drainage systems to dry the soil and protect the growing city from the effects of river floods or hurricane rains. The resulting aeration and gradual shrinkage of the soil has led to an extensive catalog of damages: displaced foundations, broken pavement, clogged sewers, and ruptured gas and water mains. The true extent of the subsidence is apparent on a map of land elevation: parts of the city now lie two meters below sea level and as much as eight meters (26 feet) below the level of the river. The simple ditches and pipes of a century ago have been replaced by a battery of enormous pumps in an increasingly costly effort to remove excess water. Of course, additional drainage only leads to further shrinkage as the soil adjusts to a drier environment than the one in which it was formed.

FIG. 9-6. A flooded rice field. Saturating the soil makes the environment less favorable for life: it aids the crop by inhibiting weeds, insects, diseases, and the decay of soil organic matter. (Courtesy of H. Bryan, Soil Conservation Service.)

paddy flooded field used for rice growing

In some tropical areas, agricultural practices have just the opposite goals. Farmers shade or deliberately flood their fields in order to retard the decomposition of organic matter and thus maintain some humus in the soil (Fig. 9-6). Tree plantations and wet-rice farming are two of the most successful ways of using tropical soils. Both systems of farming manage the land to keep the soil as cool and moist as possible. Thus the farmers inhibit competing organisms and delay the decay of organic material.

SOIL TEXTURE

North American soils include a variety of exceptions to the rule that cool or dry conditions produce humus accumulations and dark colors. Yellow and red soils can be found in Wisconsin, while Alabama has a sizeable belt of dark soils. A close examination of the atypical soils reveals a pattern that is based on one of the key ideas of Chapter 8. In general, soils that are darker than expected are ones that have high nutrient-storage capacity. The dark soil in the aptly named Black Belt of Alabama is derived from a soft limestone. The parent material is rich in calcium and yields a soil with fine texture and good structure. These traits, in turn, give the Black Belt soil a much greater exchange capacity than nearby soils formed on sandstone or granite.

exchange capacity quantitative measure of the ability of a soil to store nutrients
(see p. 161)

A high exchange capacity has two different effects on the balance between production and decomposition of organic matter. Plants grow rapidly in a fertile soil, and a soil with fine texture has many storage sites and usually is more fertile than one with a low exchange capacity. At the same time, a fine-textured soil holds more water than a coarse soil. Wet soils are often cool because evaporation of water takes heat from the soil, and the lack of warmth makes decomposer organisms uncomfortable. A high exchange capacity in a soil thus stimulates big plants, hinders decomposers, and tips the balance in favor of the accumulation of humus in the soil (Fig. 9-7).

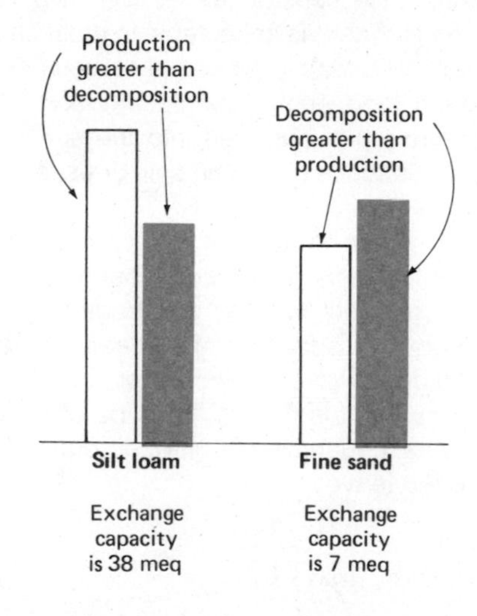

FIG. 9-7. The relationship between organic matter production and decomposition in two soils with different textures. The productivity of the finer soil allows humus to accumulate, because it stimulates plants and hinders decomposers.

The other extreme, low exchange capacity, is illustrated by a shallow, coarse, or rocky soil. Plant productivity is hindered by the limited ability of the soil to store nutrients and water. Decomposer organisms, however, usually do not require nutrients from the soil, so they consume organic remains fairly rapidly. A coarse soil therefore contains less humus than would be expected on the basis of climate alone. The poverty of such a soil is accentuated if some of the organic production is harvested and removed from the field. Soils with low exchange capacity therefore commonly "wear out" faster than soils with high storage capacity.

THE C/N RATIO

A reasonably balanced diet for soil organisms is another necessity for efficient decomposition of dead plants and animals. Few living things like to live in a place that is extremely acid or alkaline or contains a poisonous substance. The rate of decay of organic material is adversely affected by

pH quantitative measure of soil acidity; 3 is acid, 7 neutral, and 10 alkaline
(see p. 162)

C/N ratio amount of carbon divided by the quantity of nitrogen in a source of food

mineralization release of inorganic nitrogen compounds as organic matter decays

immobilization use of soil nitrogen by decomposers as they consume organic matter

cruel and unusual chemicals in the soil. Pine needles, for example, break down slowly because they contain resins and organic acids that many soil organisms consider inedible. A soft ripe apple, on the other hand, is relished by a variety of decomposers and therefore decays rapidly. Fortunately, different soil organisms have different definitions of edibility, so that nearly every natural substance has at least one potential customer. Some artifical materials, however, may persist almost indefinitely in the soil, and their presence may inhibit the decomposers.

The carbon-nitrogen (C/N) ratio of the food supply is one of the most useful indicators of dietary balance for a decomposer. Starches and sugars and other substances that contain carbon are the sources of food energy for consumer organisms. Nitrogen compounds are necessary components of protein and other complex life substances. A C/N ratio of about 15:1 (15 times as much carbon as nitrogen) is acceptable to most bacteria and fungi.

Organic material with less than 15 grams of carbon per gram of nitrogen does not provide enough food energy for decomposers to use all of the nitrogen. The excess nitrogen is released into the soil, where it becomes available to higher plants, leaches out of the soil, or escapes into the atmosphere.

Organic matter that has a surplus of carbon makes it necessary for microorganisms to take nitrogen from the soil. Their use of soil nitrogen can deprive higher plants of needed supplies of nitrogen. Yellowed leaves on garden vegetables, an obvious symptom of nitrogen starvation, may be the inevitable result of using dried leaves, sawdust, or shredded newspapers as a garden mulch. These materials have C/N ratios of 80 or 100 to 1. Soil organisms need to take vast quantities of nitrogen out of the soil in order to consume sawdust or dried leaves.

ORIGINS OF ORGANIC MATTER

deciduous tree one that has no leaves during a cold or dry season

The types of plants and animals that inhabit a region have considerable influence on the distribution of organic matter in the local soil. Forest trees lose a significant portion of their new growth in the annual leaf fall. Many trees in the middle latitudes drop their leaves almost simultaneously, on cue from the short days of autumn. The leaf fall of various trees in a humid equatorial forest is not synchronized by a distinct cold or dry season. The constitutions of most needleleaf trees provide for staggered terms of office for their leaves. Each needle operates on its own calendar and persists for a few years before dropping off; thus the tree gives the impression that its foliage is permanent.

soil profile vertical cross section of soil

soil horizon distinct layer in a soil profile

Regardless of the species, the stem and root system of a tree is far more long-lived than the leaves. As a result, most of the dead or discarded organic material in a forested area is found on the surface of the ground. Soil organisms understandably prefer to be near their food and therefore are most numerous in the top layers of a forest soil. Soils in wooded areas characteristically have a black layer of humus and organic litter lying on top of the ground, with a sharp boundary between the organic and inorganic

FIG. 9-8. A typical forest soil profile. The soil has the characteristically sharp boundary between the dark organic layer on the surface and the lighter leached mineral horizons below. (Courtesy of W. H. Lyford, Soil Conservation Service.)

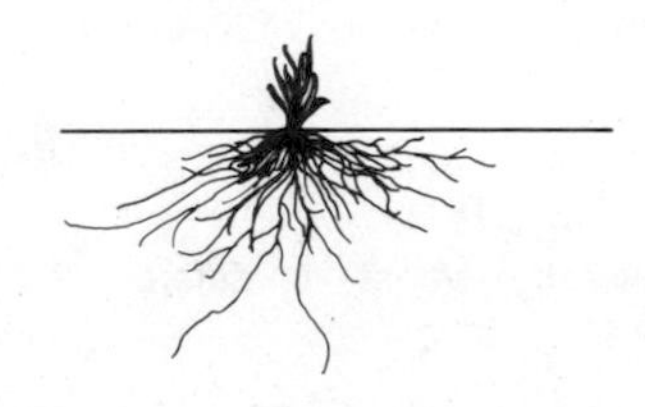

FIG. 9-9. A comparison of the aboveground and underground portions of typical plants.

layers of the soil (Fig. 9-8). The thickness of the organic layer depends on the rate of decomposition: tropical soils have very thin organic layers, whereas the organic horizon in a Canadian or Siberian forest soil may be half a meter thick.

Soils that developed under grass, on the other hand, have deep black surface layers that consist of mixtures of mineral and organic material. A typical grassland occupies a drier environment than a forest, and an average grass plant needs an extensive root system in order to gather water for a comparatively small amount of leaf surface above the ground (Fig. 9-9). The roots of grass are finely spread through the soil. Many grasses live only one year, yet a single grass plant may manufacture several kilometers of roots during its short life. When the plant dies, its extensive root system begins to decay. Death of roots is common even among the grasses that live through the winter. The widespread tangle of dead roots encourages an even distribution of soil organisms and produces a greater uniformity of soil characteristics than is found in a forest soil. Instead of the sharply separated horizons of a forest soil profile, a grassland soil has an almost imperceptible decrease in organic matter from the top down (see the chapter opening).

annual plant that lives for only one growing season

perennial plant that survives many years

BENEFITS OF ORGANIC MATTER IN A SOIL

humus partially decomposed organic material

Humus is a valuable addition to a mineral soil for a variety of reasons:

1. Decaying organic matter is a source of plant nutrients. After all, it was once a plant that had to assemble all of the mineral substances it needed for life. Humus is not a complete fertilizer, however, unless the entire plant is returned to the soil. The fertility of the soil will decrease if grain is harvested from the field and only the stalks are put back into the earth.

2. Humus has an abundance of electrical charges for nutrient or moisture storage. Adding just one gram of humus to a kilogram of sandy soil can multiply its storage capacity by a hundred times. A mixture of organic and chemical fertilizers is usually more effective than the same quantity of inorganic chemical nutrients, because the organic portion can help the soil to store the chemicals until they are needed.

buffer chemical that resists changes in acidity

3. Organic matter is a chemical buffer. It helps to smooth out variations in acidity and other chemical characteristics. Nutrients are easier to extract from a well-buffered soil, and a soil containing humus is not likely to change its chemical nature too quickly for plants or animals to adapt.

structural aggregate (ped) clump of soil particles
(see Fig. 8-16)

mycelia threadlike structures of fungus plants

4. Decomposing organic matter helps to bind soil particles into structural aggregates (Fig. 9-10). Plant decay products form fibrous masses with considerable electrical activity. Humus also serves as a food source for fungi, whose threadlike bodies trap soil particles and hold them together.

infiltration entrance of water into a soil
(see Fig. 14-6)

aeration entrance of air into a soil

5. Decaying roots serve as infiltration and aeration channels into a soil. A growing root forces its way between soil particles and wedges them apart. When the root dies and begins to decompose, it leaves an empty space which often persists and allows water and air to penetrate deep into the soil.

6. Organic matter keeps soil loose, so that seeds can germinate and roots can penetrate the soil easily. Soils without organic matter often become firm and difficult to penetrate. A loose soil also warms up rapidly in spring, because a loose soil does not freeze as deeply as a compact one.

diversity variety of different living things

biological control use of other living things to control pests

7. Organic matter stimulates the activity of soil organisms. Organisms have a physical effect on soil, because the burrowing of earthworms and insects mixes and aerates the soil. They also have a biotic effect, since a large assemblage of soil organisms include many potential parasites and predators. Diversity of soil life is usually a blessing. The multitudes of living things in a rich soil are often so busy hiding from or dining on each other that they are less likely to specialize in munching on crops. Pests become pests when their populations are no longer controlled by their natural enemies.

8. Organic matter helps to protect soil from erosion. Decaying leaves on the surface of a soil absorb much of the impact of falling raindrops. Rainwater infiltrates rapidly into a loose soil, and infiltration reduces the amount of water running off the field into nearby streams. Roots and fungi hold the soil particles together, making it difficult for water to carry them downhill.

FIG. 9-10. Soil structure and organic matter. The dark soil on the right is rich in humus, which helps to form good granular structure. (Courtesy of J. McConnell, Soil Conservation Service.)

9. Organic soils purify water as it percolates through. Disease organisms are often killed by other living things in a rich soil. Many modern antibiotic drugs (e.g., penicillin and streptomycin) come from decomposer microbes that live in well-drained soils rich in organic matter.

actinomycete microscopic organism that grows in colonies in well-drained soils rich in organic matter

10. Soil organic matter is a source of food for microorganisms that convert nitrogen and other nutrient elements into usable forms. Nutrient conversion is so critical that it deserves treatment in a separate section below.

These 10 reasons all suggest that efforts to increase the organic fraction of a mineral soil would be worthwhile. Manuring, plowing crop residues under, mulching, and crop rotation are common practices that raise the organic content of a soil. Chemical fertilizers often lead to an increase in soil organic matter by stimulating plant production.

Like any good thing, however, it is possible to have too much humus in a soil. Benefits decrease in significance when the content of organic matter goes beyond about 5 to 10 percent of the weight of a soil. Large quantities of organic matter bring disadvantages: acidity, nutrient tie-up, poor internal drainage, and unfavorable structure. The ultimate in organic soils is a black swamp that rarely drains, supports only a few forms of life, and often allows disease organisms to survive indefinitely.

soil drainage ability of a soil to get rid of excess water
(see p. 167)

amino acids nitrogen compounds, the building blocks of protein

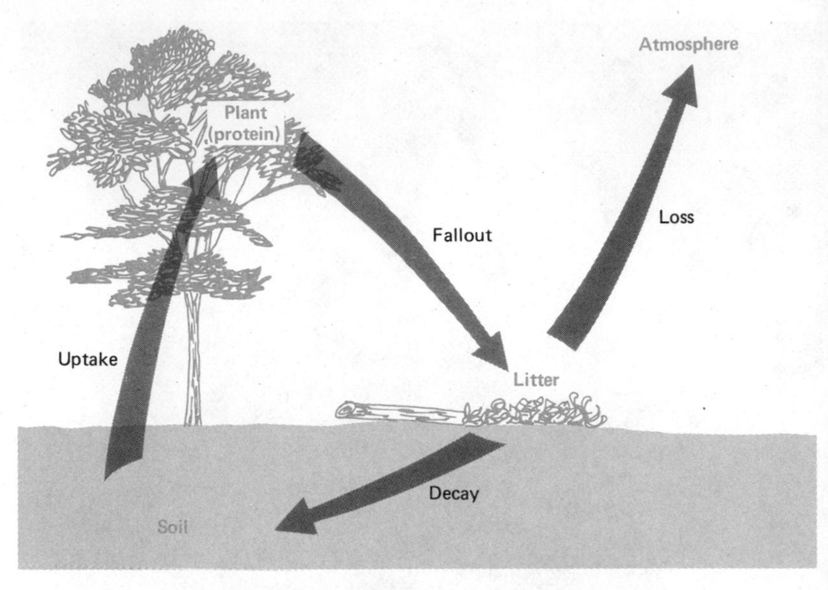

FIG. 9-11. An incomplete nitrogen cycle. Nitrogen continually escapes into the air as litter decays.

NITROGEN CONVERSION

Nitrogen gas is one of the most chemically inactive substances on earth. At the same time, all living things require nitrogen. Animals cannot use the nitrogen they breathe; they must obtain their supply from plant proteins. Most green plants are also unable to use the inert atmospheric gas. They absorb nitrogen compounds from the soil to meet their needs.

Every time a plant or animal decays, some of its nitrogen changes to inert gaseous forms and escapes into the atmosphere. The result is a partial cycle that if continued, would lead to the eventual depletion of all soil nitrogen (Fig. 9-11). Loss of nitrogen, of course, would mean the eventual starvation of many plants and animals. Most forms of life owe their continued existence to the recovery of nitrogen by the activity of two kinds of soil bacteria, the nitrogen fixers and the nitrifiers.

Nitrogen-fixing bacteria are able to use nitrogen gas directly from the air. One group, *Azotobacter,* is independent and lives alone in the surface soil. Under ideal conditions, these free-living bacteria can fix (convert) as much as five grams of atmospheric nitrogen per square meter of ground every year (about 40 lbs of nitrogen per acre of soil). *Rhizobium,* the other major group of nitrogen fixers, prefers a more cloistered life. It can survive in soil but prefers to live in swellings on the roots of clover, alfalfa, soybeans, black locust trees, and other members of the legume family of plants (Fig. 9-12).

FIG. 9-12. *Rhizobium* nodules on a clover plant. The white swellings on the plant roots harbor bacteria that take atmospheric nitrogen and convert it into forms that are usable by plants. (Courtesy of Culver, Soil Conservation Service.)

Thus protected from the outside world, the microorganisms absorb gaseous nitrogen, convert it to forms that are usable by larger plants, and trade surplus nitrogen for shelter and food from the host plant. A field of clover or alfalfa can harbor enough *Rhizobium* nodules to fix 10 to 20 g of nitrogen per square meter per year (80 to 160 lbs/acre). The activity of the bacteria makes alfalfa and clover rich in protein for livestock and gives soybeans a promising future as a human food supplement. At the same time, soybeans and alfalfa and clover aid the crops that later occupy the same field. The slow decomposition of *Rhizobium* nodules on dead legume roots releases nitrogen compounds into the soil and makes them available for other plants.

nitrogen fixation conversion of nitrogen gas to plant-usable forms

The nitrifying bacteria, *Nitrosomonas* and *Nitrobacter,* play key roles in the decomposition of organic matter (Fig. 9-13). Both are free-living organisms. Together they take ammonium (NH_4^+), which is produced when organic material decays, and convert it to nitrate (NO_3^-) compounds. The ammonium form of nitrogen has a positive electrical charge and therefore is attracted to soil particles. The nitrate form, with a negative charge, can move around freely. Plants do not have to expend much effort to get it away from the soil particles, but the nitrate form can also be leached from the soil more easily than the ammonium form.

nitrification conversion of ammonium nitrogen to the nitrate form

Like other aspects of the decay of organic matter, nitrification proceeds rapidly in hot, moist areas and is hindered in cold, acid, or wet soils. Soils in cold regions frequently seem deficient in nitrogen because the bacteria do not convert enough ammonium to the more usable nitrate form. Humid tropical soils often are low in nitrogen because it is converted too rapidly to the nitrate form, where it cannot be stored on the soil particles and therefore is carried away by the abundant rainwater percolating downward through the soil.

leaching removal of nutrients by percolating water
(see p. 161)

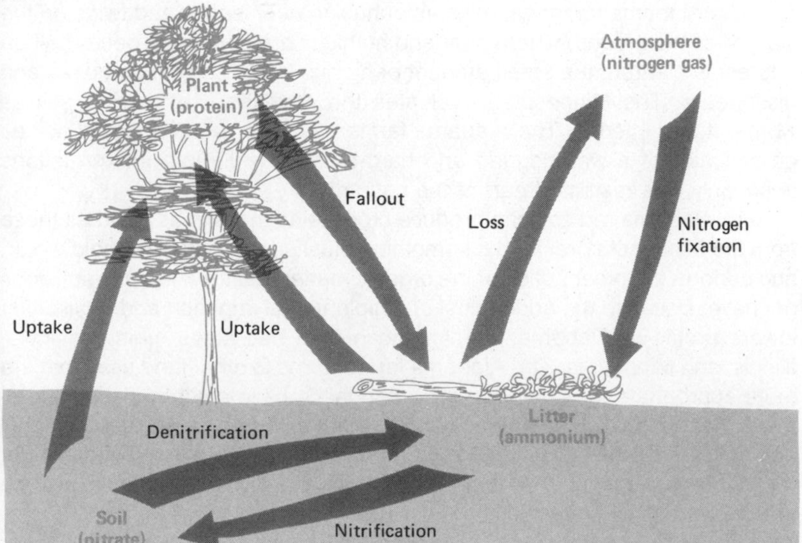

FIG. 9-13. A complete nitrogen cycle. Bacteria make atmospheric nitrogen available to plants by converting nitrogen gas to ammonium and nitrate forms.

On a more minute scale, the colorful springtime mosaic of yellow and green in a mid-latitude cornfield is really a sinister display of unequal nitrification. Corn on hillsides and ridges is green and healthy, because the well-drained soil is warm and moist and loose. Unhindered by cold or lack of oxygen, soil bacteria are actively supplying nitrate to the corn. The cool and wet soils of valley bottoms and clay areas are low in nitrogen because the chilled or suffocated microorganisms cannot convert ammonium to nitrate rapidly enough for the corn plants. As if to compound the difficulty, some anaerobic bacteria even steal oxygen from nitrate and release the nitrogen into the air. Corn plants in cold or wet sites therefore have the sickly yellow pallor that is the most obvious symptom of nitrogen starvation.

anaerobic bacteria bacteria able to exist in environments that lack free oxygen

denitrification conversion of nitrate nitrogen to the ammonium form

Soil microorganisms also play a role in the transformation of other nutrients, such as sulfur or phosphorus. Moreover, the various cycles are intermeshed; a surplus or deficit of one nutrient can alter the rate of transformation of another one (see Appendix 8). The decomposition of organic remains is thus a more complex and more sensitive process than most people realize.

CONCLUSION

This chapter ends where it began. The Minnesota prairie soil was formed on fine-textured material with a high exchange capacity and a grass cover in a cool and moderately dry climate. Organic remains are distributed throughout the upper layers of the soil, where they decompose slowly. Partly decayed humus gives a rich black color and a number of agricultural advantages to the soil.

The Alabama forest deposits an annual load of leaves and twigs on the surface of the ground, where heat and humidity promote rapid decay. When a forest is cleared, the small amount of humus in the soil decomposes and disappears. The Minnesotan evaluates the resulting light-colored soil as worn-out and infertile. The Alabama farmer rightfully regards a dark soil as an indicator of a waterlogged and hard-to-farm field. Both interpretations apply only to a restricted part of the country.

An Alabama red soil can produce crop yields that are as good as those from the Minnesota prairie, but something must supply nutrients, hold water, and perform the other duties of the organic matter that the Alabama soil does not have. Because the added cost of supplemental irrigation and fertilization lowers profits in Alabama, the American corn belt stays mainly in Iowa, Illinois, and Minnesota. The Alabama farmer turns to other land uses that are more appropriate than trying to grow annual crops in a subtropical environment. Permanent pastures and groves of nut and pine trees keep soil cool and moist and protect it from erosion. In turn, cool and wet conditions hinder the soil organisms, so that their decomposition activity is in better balance with the productive capabilities of the land.

Chapter 10

Soil: Reprocessed Dirt

Is soil a mystical living thing or a convenient way to hold plants upright while we feed them?

Alan E. Amen, USDA·Soil Conservation Service

soil earth material after it has been shaped by its environment

Soil is not just dirt. Dirt is an inert mass of weathered rock material; soil is the material after it has been shaped and altered by the environment around it. The soil in a particular place owes its characteristics to its parent material, its topographic position, the climate of the area, the plants and animals around it, and the amount of time it was exposed to these factors. Each of these five soil shapers has a different role to play in different places, and the variety of interactions gives the earth many thousands of different kinds of soils.

Soils can be classified in many ways, depending on the characteristics that are considered important by the classifier. The categories in a soil classification, like those in classifications of rocks or climates, are based on somewhat arbitrary divisions in the continuously varying features of the soil. To maintain a focus for the discussion of soils, this chapter uses a single 40-acre patch of land in east-central Iowa as a home base. A narrow focus is necessary in order to establish some principles of soil microgeography. The choice of Iowa is partly rational—Iowa is one of the major farming states—and partly personal—one of the authors owns a farm there and knows it intimately. Notetakers should remember that the purpose of this chapter is to teach about soil geography, not about Iowa; the appendix includes information about other soils found in different parts of the country.

SE$\frac{1}{4}$ of the SE$\frac{1}{4}$ of section 9
Township 81N, Range 10W
Fifth Principal Meridian

Thus reads the legal description of 40 hilly acres near the Iowa River in Iowa County, Iowa (Fig. 10-1). About 10 acres lie on a hilltop, mostly flat enough to be safe for plowing. Crops can also be planted on a few acres of level bottomland near a creek. Oak trees occupy 10 steep acres, while cattle roam over the remaining land.

Records kept by well drillers indicate a smooth and fairly level bedrock platform about 50 m (170 ft) below the ridgetop and 20 m beneath the valley bottom. None of the surface soil material is derived from the deeply buried rock, although the shape of the bedrock surface does affect the movement of underground water.

till unsorted earth material deposited by a glacier (see Fig. 16-18)

Above the rocky basement is a thick layer of loose material that was brought into Iowa County when a glacier invaded the area thousands of years ago. The mile-thick sheet of ice slowed down, stopped, melted away, and left a jumbled accumulation of clay and gravel, the fragmented remains of glacially scoured northern landscapes. In time, trees and grass covered the area and helped to build a deep reddish-brown soil during an interval of warm climate.

loess silty material deposited by wind (see Fig. 16-24)

The climate turned cool again, and the ice made another advance out of Canada, but this time it did not reach all the way to Iowa County. The recent glacier, however, did have an indirect effect on the soils of the lower 40 acres. Winds from the glacier blew across debris left by melting ice, picked up immense quantities of silt, carried it away from the glacier, and spread a thick blanket of wind-blown material over the land. This silty material, called loess, covered the reddish-brown soil to a depth of many meters.

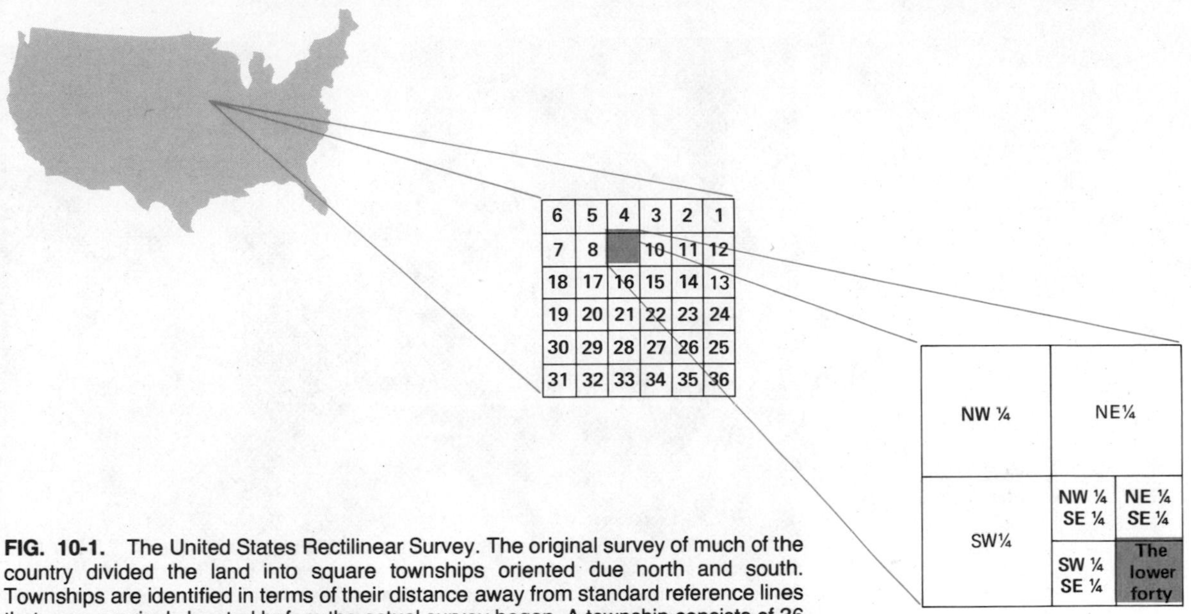

FIG. 10-1. The United States Rectilinear Survey. The original survey of much of the country divided the land into square townships oriented due north and south. Townships are identified in terms of their distance away from standard reference lines that were precisely located before the actual survey began. A township consists of 36 sections, each one mile square. The 640 acres in each section are divided into fourths, and each 160-acre quarter section is cut once more into fourths. The resulting unit of 40 acres is the basic survey unit. (The legal division of land into acres is an exceptionally difficult thing to change into even hectares—the metric unit of area—although many farmer's and surveyor's units are surprisingly compatible with the metric system. For example, a fence rod is 16½ feet or almost exactly 5 meters in length. The mixing of acres and hectares in the text reflects a likely blend of units that will emerge in time.

More recently, creeks and rivers have made valleys in the loess. Some of these valleys are deep enough to uncover the ancient soil on the glacial material beneath the silt. A path from the hilltop to the creek valley on the lower forty crosses four distinct kinds of earth material: some wind-blown loess, an ancient red soil, a thick layer of glacial till, and some recent deposits in the valley bottom (Fig. 10-2). These four parent materials, together with differences in topography and natural vegetation, have produced many different types of soil in Iowa County.

paleosol former soil that was covered by some deposit of material

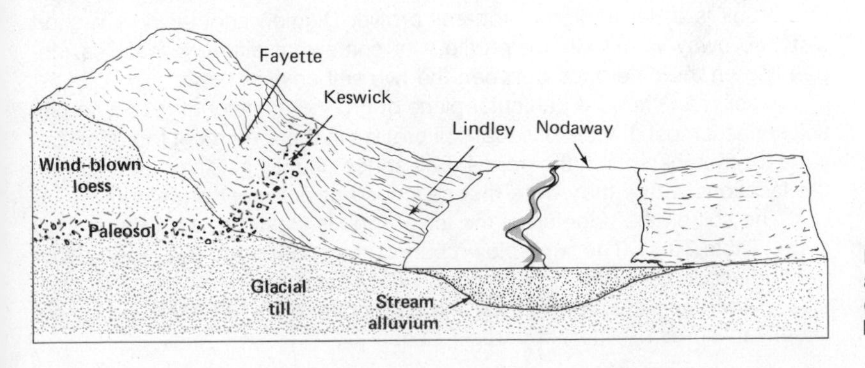

FIG. 10-2. An east-west profile across the lower forty. The four parent materials produce four distinct kinds of soil.

FIG. 10-3. A soil surveyor making a profile description of a grassland soil. The book in his hand is a key for making precise color descriptions. (Courtesy of R. L. Meeker, Soil Conservation Service.)

SOIL SERIES AND SEQUENCES

soil series basic soil classification unit

soil profile description of the layers of a soil

parent material the dirt from which a soil was made

The soil series is the basic unit in U.S. soil classifications. To determine what series a soil belongs to, a soil surveyor digs a hole and records the characteristics of soil material taken from each level of the hole (Fig. 10-3). The result is a description of the soil profile. Digging another hole a short distance away would yield a profile with somewhat different features, depending on the difference between the two soil environments.

A soil map shows a particular piece of ground as belonging to a certain soil series if most of the individual soil profiles in the area meet the qualifications for membership in the series. Soil series usually get their names from the locations where they were first described or are most prominent.

The soil on the ridgetop of the lower forty is classified as a Fayette silt loam (Table 10-1). The wind-blown parent material of this soil is the dominant influence on its texture. Fayette soil particles are overwhelmingly silt-sized, whereas soils from glacial material or recently weathered rock are usually mixtures of different particle sizes.

TABLE 10-1 TECHNICAL DESCRIPTION OF FAYETTE SILT LOAM[a]

Profile of Fayette silt loam, 1240 feet south and 1050 feet west of the NW. corner of sec. 25, T. 81 N., R. 12 W., on a 3 percent east-facing slope, in an oak-hickory type of forest.

A0—0.5 inch to 0, very dark brown (10YR 2/2) partly disintegrated, hardwood leaves and twigs; abrupt, smooth boundary.

A1—0 to 4 inches, very dark gray (10YR 3/1) friable silt loam: weak, thin, platy structure and fine granular structure; few brown to dark-brown (10YR 4/3) worm casts; very strongly acid; clear, smooth boundary.

A2—4 to 8 inches, dark grayish-brown (10YR 4/2) silt loam; 10 percent is brown to dark brown (10YR 4/3), gray (10YR 6/1) when dry; friable; weak, thick, platy structure breaks to weak, fine, subangular blocky structure; silt grains on ped surfaces; very strongly acid; clear, smooth boundary.

B1—8 to 14 inches, dark grayish-brown (10YR 4/2) and brown to dark-brown (10YR 4/3) light silty clay loam; brown to dark-brown (10YR 4/3) interiors; friable; strong, fine, subangular blocky structure light brownish-gray (10YR 6/2, dry) silt grains on ped surfaces; very strongly acid; gradual, smooth boundary.

B21—14 to 24 inches, brown to dark-brown (10YR 4/3) light to medium silty clay loam; kneads to dark yellowish brown (10YR 4/4); friable to firm; strong, fine, medium, subangular blocky structure; few, thin, discontinuous, clay films; light brownish-gray (10YR, 6/2, dry) silt grains on peds; strongly acid; gradual, smooth boundary.

B22—24 to 35 inches, brown to dark-brown (10YR 4/3) light silty clay loam; kneads to dark yellowish brown (10YR 4/4); friable to firm; few black (10YR 2/1) oxide stains on peds; strong, medium and coarse, subangular blocky structure; light-gray (10YR 7/1, dry) silt grains on peds; common, thin, discontinuous, clay films; strongly acid; gradual, smooth boundary.

B3—35 to 46 inches, yellowish-brown (10YR 5/4) light silty clay loam; friable; few fine concretions of black (10YR 2/1) oxide; weak, coarse, subangular blocky structure; light-gray (10YR 7/2, dry) silt grains on peds; few thin clay films on some peds; medium acid; gradual, smooth boundary.

C—46 to 58 inches, yellowish-brown (10YR 5/4) silt loam; friable; massive with some vertical cleavage; light-gray (10YR 7/1, dry) silt grains on vertical faces; slightly acid.

[a]This table is reproduced from the Iowa County Soil Survey, 1965, in order to show the degree of detail in a standard soil description and not to give students something to memorize for a test.

Forests occupied most of the lower forty before the land was cleared for farming. For this reason the Fayette soil is relatively light in color and low in organic matter. Farther to the west the natural vegetation was grass, and the soil has a dark color, more organic matter, and granular structure. These legacies of the former grassland are sufficiently distinctive so that the soil is classified as a member of the Tama series.

tilth farmer's term for soil structure

sequence group of soils similar in all respects except for a single soil-forming variable:

biosequence—vegetation
toposequence—topography
climosequence—climate
lithosequence—parent material
chronosequence—time

The soil in the transition zone between forest and grassland has traits of color and structure that are between the extremes of the Fayette and Tama soils. The surface layer of the transition soil is thinner and lighter in color than in the Tama series, but the soil has more organic matter than the Fayette series. The differences are so great that a separate kind of soil, the Downs series, is recognized. These three soil series form a biosequence—a group of soils formed by the same parent material, climate, and topographic setting but different vegetation.

A toposequence is a similar logical arrangement that consists of several soils with similar backgrounds except for their positions on the slopes. For example, the Fayette soil usually occupies land which is slightly hilly. Erosion of soil from a steep slope may make the surface layer of dark soil somewhat thinner than the 36 in. given in the standard description of a Fayette soil. In Iowa County the loess is thick enough so that the loss of some soil is not very critical. In western Wisconsin, however, Fayette soils are only a few meters thick over bedrock. Erosion changes a Fayette soil sufficiently to warrant putting it into another series, the Dubuque.

oxidation combination with oxygen, as when gray iron becomes rust-red iron oxide

soil drainage ability of a soil to clear itself of excess water and allow air to penetrate to plant roots
(see p. 167)

Soils on level land often are waterlogged when surplus rainwater is unable to flow away. The soil is gray rather than brown because iron in a waterlogged soil cannot rust. Soil organisms have difficulty breathing and cannot consume organic material very rapidly. The poorly drained version of the Fayette soil on flat uplands belongs in the Stronghurst series (Fig. 10-4).

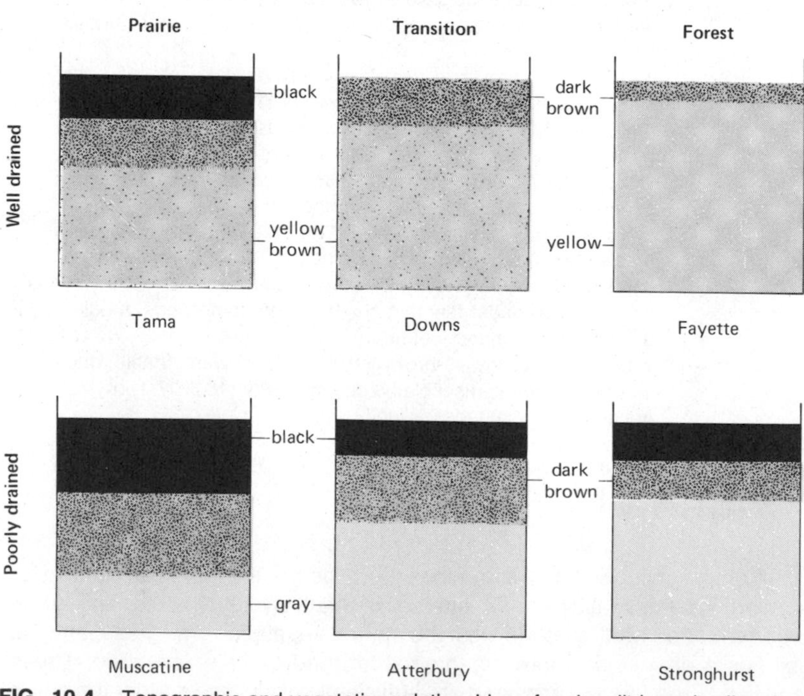

FIG. 10-4. Topographic and vegetative relationships of major silt-based soil series in Iowa County. The diagrams are generalized side views of the top 2 m of soil.

The processes of separating soils into different series could be extended through geological and climatic sequences to cover the entire country. But the purpose of this outline is not to describe all of the thousands of different soil series. Rather, this brief excursion into the Iowa fields is intended to illustrate the logic that guides soil scientists as they put soil individuals together into their most basic groups. The tour also serves as a brief review of basic soil-forming processes, and thus it underscores the idea that the characteristics of the soil in a particular place reflect the influence of the total environment.

SOIL SERIES AND CROP PRODUCTIVITY

The recognition of different soil series is not merely an academic exercise. Each soil series has its own characteristics which can be correlated fairly well with engineering capabilities and agricultural potential. For example, the combination of good management and a typical Fayette soil can produce nearly 9 tons of corn per hectare of land (140 bushels per acre). For those who grew up thinking that food comes from urban supermarket shelves, an acre is about the size of a football field and the U.S. average corn yield is about 100 bu/a. The prairie-formed Tama series, with its somewhat better structure, will reward its tiller with an additional ton of corn per hectare. The costs of production on both soils are essentially the same, so that a Tama soil will give the owner-farmer about 30 percent more money for every hour of invested labor. By contrast, there is a significant profit penalty in farming the poorly drained Stronghurst series (Fig. 10-5).

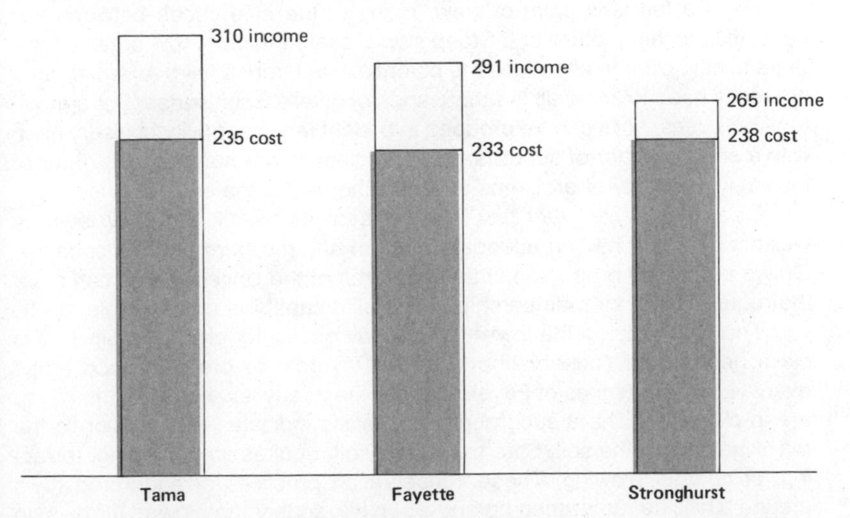

FIG. 10-5. Expected dollar income and costs of production for corn on three Iowa County soil series in 1976. A minor difference in yield can become a significant change in profitability (white area) for the farmer.

leaching removal of nutrients from a soil by percolating water (see p. 161)

paleosol buried former soil

plastic soil sticky and able to deform under pressure

These differences, however, are mere trifles when compared with the cost of owning land on the lower slopes of the deep valleys, where streams have removed the blanket of wind-blown loess and exposed the old glacial deposits. The soil in the valley of the lower forty is mainly in the Lindley series—an acid, infertile, sticky, cloddy agricultural curse. Under good management, Lindley soils yield only about 5 tons of corn per hectare, which is probably not enough to pay the costs of production. About 7 hectares of pasture on Lindley soil tries vainly to feed as many cattle as are kept on 3 hectares of Fayette land nearby.

Even the Lindley is productive when compared with the Keswick soil, a reddish clay loam on the leached top layer of glacial material. The Keswick series is literally the exhumed corpse of an ancient soil, formed on glacial deposits during an interval between glaciers and then buried under loess. The ancient soil was a product of a warm climate, which accounts for its red color and clay texture. Fortunately, there are only a few small patches of Keswick soil on the lower forty, because it is incredibly difficult to manage. Rain makes it sticky enough to trap tractors. Then, when it dries out, the soil becomes hard and cloddy, a threat to sharpened plows. As a result, farmers usually leave areas of Keswick soil in grass when they plow the nearby Fayette soils for crops.

In summary, the differences in parent material, topography, and natural vegetation have created a wide variety of different soils within a quite small area. These differences must be recognized by anyone trying to make wise use of the land in Iowa County.

LAND CAPABILITY CLASSES

land capability class group of soils with similar management problems

stripcropping alternation of row crops and grass crops, or of crops and fallow land, in narrow strips at right angles to the slope or wind

From a farmer's point of view, many of the differences between soil series cancel each other out. Steep slope, coarse texture, low organic matter, and cold climate all reduce the potential yield from a field. A farmer does not really need to know all the thousands of different soil series. For agricultural purposes, soils can be grouped into eight land capability classes, each with a small number of subdivisions. The classes are numbered in order of increasing severity of problems affecting the use of the soil.

Class I land has very few limitations on its use. It is nearly level, is well-drained, and has an adequate nutrient and moisture storage capacity. Only a very small proportion of the land area of the United States can meet the requirements for membership in this elite capability class (Table 10-2).

The best soils on the lower forty barely qualify for class II. Soils in this class need some conservation practices in order to produce good crops every year. A few acres of Fayette soil on the nearly level parts of the hilltop are in class II-e. The e-subgroup of this class indicates that erosion is the major hazard on the soil. Safe use of II-e soil requires stripcropping, terracing, or contour plowing. These conservation practices arrange crop rows around a hillside rather than up and down hill, so that the plowed furrows do not serve as channels for miniature streams trying to carry some of the hillside down into the valley after every rain (Fig. 10-6).

TABLE 10-2 LAND CAPABILITY CLASSES IN THE UNITED STATES

Capability Class	Percent of Land	Percent Used for Cropland	Characteristics of Capability Class
			Can be used for crops:
I	3	76	Few limitations on use
II	20	66	Limitations for crops are minor
III	21	49	Limitations for crops are costly
IV	12	29	Limitations for crops are severe
			Should not be used for crops:
V	3	4	No limitations on pasture or forestry
VI	19	6	Some limitations on pasture or forestry
VII	20	2	Severe limitations on pasture or forestry
VIII	2	0	Little economic use except for recreation or wildlife habitat

FIG. 10-6. Contour stripcropping of sloping land. This practice reduces erosion by keeping rainwater on the land until it has time to infiltrate into the soil. (Courtesy of W. H. Lathrop, Soil Conservation Service.)

fallow land that is left idle for a period of time

row crops crops such as corn, soybeans, or cotton that grow in widely spaced rows

infiltration entrance of water into a soil
(see Fig. 14-6)

Most of the Fayette soil on the lower forty belongs in capability class III-e. Even with erosion control techniques, it is not safe to grow row crops, such as corn or soybeans, every year on class III-e land. The row crops should alternate with oats, hay, or other grass crops that protect the soil from the impact of raindrops and help to rebuild the structure of the topsoil.

Unfortunately, the undesirable consequences of mismanagement may not become apparent for decades. At the same time, row crops are far more profitable than most other uses of Iowa County land. For these reasons, many farmers in Iowa County have unwittingly damaged their soil by growing row crops more frequently than would be ideal.

An eroded Fayette soil has a thinner surface layer with less organic matter, lower infiltration capacity, and poorer structure than an uneroded one. These changes levy a significant penalty in productivity, but the loss is largely unnoticed by the users of the land. Their crop yields, aided by fertilizers and hybrid seeds, have risen consistently over the past decades. Nevertheless, an eroded Fayette soil produces 10 to 20 percent less corn per hectare than an uneroded one could (Fig. 10-7). Even more important, from an economic point of view, row crops cannot be grown as frequently on eroded soils. The soils need more time to rest and rebuild their structures between row crop years.

Eroded Fayette soils on steep slopes go in capability class IV-e, suitable for pasture with at most an occasional year of row cropping. The most favorable Lindley soils are also in class IV-e; it is a well-deserved insult to the

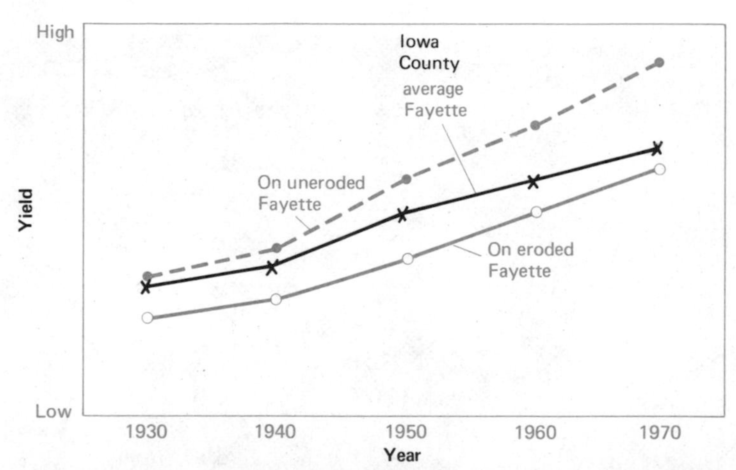

FIG. 10-7. Crop yields on Fayette soils have increased during the past four decades, but poor management has reduced the potential yield of the soil. It is impossible to maintain a soil in a completely natural state and still get a harvest from it. For that reason, conservation-as-preservation is not feasible on cropland. Nevertheless, management practices can prevent or correct some soil erosion, organic matter depletion, structure breakdown, subsoil compaction, and fertility loss on cropland. Management decisions are made by balancing the costs and benefits of alternative uses of the land.

ancestry of the Lindley soil that the best soil on glacial till is in the same capability class as the poorest soil on wind-blown loess.

Most of the Lindley series is in class VI-e or VII-e. The former is suitable for controlled grazing, while the latter is questionable even as pastureland. The erosion hazard on class VII-e land is so severe that overgrazing can lead to deep gullies, rapid loss of soil, and flooding and siltation problems downstream. Forestry and wildlife habitat are the preferred uses of class VII land, while class VIII land has such severe limitations that agriculture, grazing, and forestry are essentially impossible.

A few acres of Nodaway series soil lie near the creek in the flat valley bottom of the lower forty. The Nodaway is a deep black soil made out of material deposited when the creek floods and then slowly drains away. Texturally, the Nodaway series is a mixture of particle sizes from the three upland soils. Its black color is caused by organic material that decomposes very slowly in the frequently flooded soil. Agriculturally, the Nodaway soil belongs in land capability class II-w. The w indicates that wetness is the principal hazard on these soils, which can be as productive as the best upland soil if measures are taken to alleviate the excess water in the creek valley. Recommended practices include putting drainage tiles into the soil or digging ditches to divert floodwaters during rainstorms.

alluvium material deposited by a flooding river
(see Figs. 16-8 and 16-18)

colluvium material deposited at the foot of a hillslope

tiling putting pipes into a soil to remove excess water
(see Fig. 9-5)

Overuse of wetland soils, however, also threatens long-term repercussions that can be as damaging as the erosion losses on abused hillside soils. During the early 1970s economic pressures and Federal policies encouraged increases in the production of farm commodities. Farmers throughout Iowa County responded by ditching and tiling valley soils and by clearing additional pasture lands on the hillsides. These investments in cropland expansion are not likely to be reversed in the near future without some kind of financial assistance.

Wooded or brushy hillsides and unregulated meandering creeks, however, perform some useful services as obstacles to rapid removal of water. Clearing the slopes and straightening the creek will affect the local water table by decreasing the amount of water that sinks into the ground. The changes will also have an impact on other areas. When water from hundreds of similar creeks rushes to the major rivers, the floods that may occur downstream should not be considered as entirely "natural" catastrophes.

Ditching and tiling also have a local effect on a wetland soil. Removal of water allows air to enter and hasten the decay of soil organic material. Drainage pipes which were installed nearly two meters underground in 1950 were less than one meter below the surface of some Iowa fields in 1970. Part of the shrinkage of the soil is due to the loss of organic matter. In addition, the use of heavy agricultural machinery compresses the soil. As a result of compaction and loss of organic matter, the potential yield of the soils has declined.

aeration entrance of oxygen into a soil
(see p. 180)

These stories are not intended to suggest that hillsides and bottomlands should not be farmed. Classifying land on the basis of capability is simply a way of informing a farmer (and the public) that most farmland has limitations that influence its management and long-term productivity. Some of these limits may be overcome, but only at a cost. Some may be ignored, but only for a short time. Eventually, the innate capability of land must be recognized.

Ignoring the limits or refusing to pay the costs will lead to an inevitable decline in future yield or an increase in future costs, but unwise management is often profitable for the immediate future. Farming practices that conserve soil usually involve a financial sacrifice in the short run, and it is unfair to expect farmers to accept a drastic reduction in income without some kind of compensation. Telling farmers that conservation will help everyone in the long run does not make their personal sacrifice much easier to bear, and allowing neighbors to get rich by mismanaging the soil is adding insult to injury.

ENGINEERING PROPERTIES OF SOILS

septic drainage field a system of pipes to dispose of household wastes by spreading them through a large area of soil

Many of the fine distinctions of soil fertility and production hazards described above are insignificant when the soil is viewed as a construction medium. In designing a roadbed, dam, septic system, or house foundation, engineers consider a different list of soil features:

frost-heaving soil movement due to alternate freezing and thawing

permafrost ground that is permanently frozen

Frost penetration. The maximum depth of soil that freezes in the winter. Freezing poses a serious threat to sewer and water pipes that are not buried deeply enough. Furthermore, alternate freezing and thawing pushes soil particles around and can crack pavements, dislocate foundations, and tilt fence posts and telephone poles (Fig. 10-8). Soils in cold climates may be permanently frozen except for a thin surface layer that thaws during the warmest months of the year. A frozen soil is a firm foundation, but buildings and roads in the Arctic may inadvertently alter the energy budget of the ground beneath them and thus commit suicide by thawing their own foundations.

FIG. 10-8. Pavement cracked by frost action. Water seeped into the soil beneath the paved surface, froze, and had nowhere to expand but up through the pavement. (Courtesy of R. L. Larsen, Soil Conservation Service.)

FIG. 10-9. Pipes corroded by an acid or salty soil. (Courtesy of H. Bryan, Soil Conservation Service.)

Corrosivity. The ability of chemicals in the soil to dissolve or rot pipes, sewers, posts, or foundations. Untreated wood posts have a short lifespan in soils with active microorganisms. Acid soils dissolve some kinds of stone and concrete. Wet and salty soils are corrosive environments for metal (Fig. 10-9).

Permeability. The amount of water that a soil will permit to flow through it in a given period of time. The Lindley clay loam is nearly impermeable; its tiny pore spaces impede seepage, so that it is an excellent pond site but a miserable place for a septic tank. The permeable Fayette silt makes good septic drainage fields but its ponds usually leak.

aquifer material that transmits water
(see Fig. 8-11)

Bearing strength. The amount of weight that a soil will support without allowing it to sink more than a specified amount. Special foundations minimize settling by spreading the weight of the building over a large area of soil, but they are also costly, so that a builder prefers a foundation design that is just large enough to suit a particular soil. Remembering the famous miscalculation of Pisa, engineers usually include a sizeable safety factor in their specifications for critical foundations.

bearing strengths in kg/cm² (the bearing strength in tons/ft² is numerically about the same):

hard rock	100
soft rock	20
gravel	8
sand	4
silt	2
clay	1

footing support for a building

Shrink-swell potential. The change of volume that occurs as a soil is alternately wetted and dried. Coarse soils are usually good building sites

kaolinite clay clay with a rigid structure

montmorillonite clay clay that expands when wet; montmorillonite is more fertile than kaolinite but decomposes in hot climates

mass movement gravity-aided downhill movement of earth material (see Fig. 15-4)

under all moisture conditions. The Fayette silt has an intermediate particle size and a moderate shrink-swell potential. The Nodaway soil in the valley contains more clay and is prone to expand and contract more than the Fayette soil as the weather changes.

Slope stability. The ability to rest on a non-level surface without slipping downhill. Roadbanks or terrace walls are hard to maintain on Nodaway soils, because low strength and high shrink-swell potential make landslides or mudflows likely even on gentle slopes (Fig. 10-10). By contrast, the coarse Fayette soil can stay put even on fairly steep slopes.

Nearly all of the engineering characteristics of a soil are closely related to the texture of the parent material and the various soil layers. For this reason, soil texture forms the basis for the two most commonly used engineers' classifications. The American Association of State Highway Officials (AASHO) classifies soils into seven major groups, from A-1 for stable coarse soils to A-7 for weak and unstable clays. Other engineers use the Unified classification, which recognizes three major groups: coarse soils, fine soils, and organic soils, each of which is subdivided into narrower categories. Both of these classifications treat the soil as an impermanent material that can be moved, compressed, mixed, or changed into whatever assemblage of earth material happens to be needed for a particular purpose.

FIG. 10-10. Landslide on weak slope. When the forces holding the soil together are too weak to counter the pull of gravity, the soil slips down the slope. (Courtesy of R. L. Newbury, Soil Conservation Service.)

THE SEVENTH APPROXIMATION

The capability, Unified, and AASHO classifications are descriptive ways of grouping soils. Their fine distinctions between soil series are important for farmers or engineers for different reasons, but the classifications are too complicated for a geographer interested in the soil pattern of an entire continent. The criteria used to separate groups of soils on a national map must be simple, yet they should imply as many additional characteristics about the soil as possible. Most soil scientists prefer a genetic classification, because knowledge of the processes that helped to shape a particular soil enables a trained person to deduce a considerable amount of additional information about the soil. The modern U.S. Department of Agriculture classification (popularly known as the Seventh Approximation because someone counted that many formal revisions before it was adopted) is based on precise definitions of unique soil layers that apparently develop under particular environmental conditions (Fig. 10-11).

genetic classification classification that emphasizes causes (see p. 93)

diagnostic horizon soil layer with significance for classification

zonal soil soil with layers shaped by the climate in which it was formed

levels in the U.S. Department of Agriculture soil classification: order, suborder, great group, subgroup, family, series

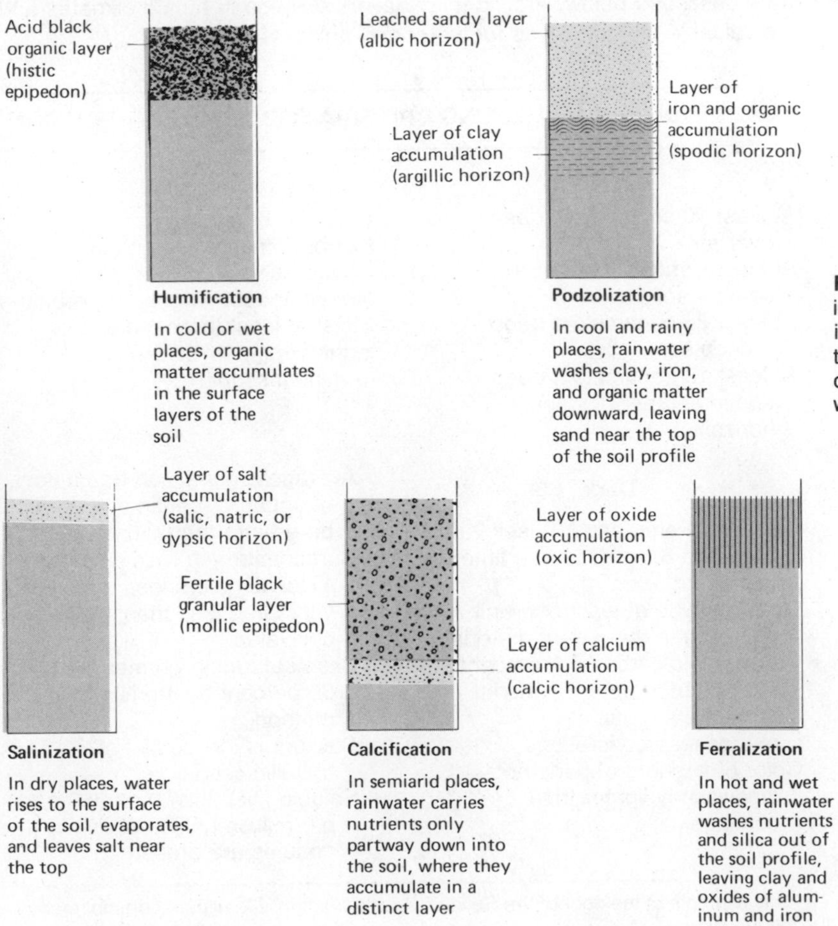

Acid black organic layer (histic epipedon)

Humification

In cold or wet places, organic matter accumulates in the surface layers of the soil

Leached sandy layer (albic horizon)

Layer of iron and organic accumulation (spodic horizon)

Layer of clay accumulation (argillic horizon)

Podzolization

In cool and rainy places, rainwater washes clay, iron, and organic matter downward, leaving sand near the top of the soil profile

Layer of salt accumulation (salic, natric, or gypsic horizon)

Salinization

In dry places, water rises to the surface of the soil, evaporates, and leaves salt near the top

Fertile black granular layer (mollic epipedon)

Layer of calcium accumulation (calcic horizon)

Calcification

In semiarid places, rainwater carries nutrients only partway down into the soil, where they accumulate in a distinct layer

Layer of oxide accumulation (oxic horizon)

Ferralization

In hot and wet places, rainwater washes nutrients and silica out of the soil profile, leaving clay and oxides of aluminum and iron

FIG. 10-11. Soil-forming processes in different environments. The words in parentheses are names for the particular soil features that are used to diagnose the soil-forming process at work.

The logic of the Seventh Approximation is evident in the full name of the Tama soil: Typic Argiudoll. The last three letters of its group name give away the most important secret: the soil is a mo*ll*isol. To belong to the exclusive Order of Mollisols, a soil must possess a mollic layer, a thick, soft, dark, fertile surface layer that meets a strict set of criteria (Table 10-3). The membership rules, though admittedly arbitrary, are rigid. They were formulated because a mollic layer, when defined in this way, is a common trademark of grass vegetation. The soil traits persist long after the removal of the natural plant cover and they affect the yield of crops and the firmness of foundations.

base saturation percent of exchange capacity occupied by nutrient bases
(see p. 162)

leaching removal of bases by percolating water
(see p. 161)

The second last syllable, *ud,* indicates that Tama is in the suborder of udolls, the h*u*mi*d* mo*ll*isols. Udolls are fertile but slightly leached of nutrients by percolating water. Leaching occurs in well-drained soils under climates with a surplus of rainfall and a moderate to high temperature, useful information for corn growers.

The "argi" part of the name puts Tama in the group of argiudolls, udolls that have a layer of clay accumulation that is defined as rigidly as the mollic layer described below. *Argi*llic (clay) layers develop in humid climates and are valuable storage areas for water and nutrients.

TABLE 10-3 THE LIMITING CRITERIA FOR A MOLLIC EPIPEDON[a]

Thick	Soft
At least 10 centimeters deep over rock	Granular or crumblike structure
At least $\frac{1}{3}$ of the depth of a shallow soil	Not both massive and hard when dry
At least 25 centimeters deep in a deep soil	Not composed largely of allophane
At least 18 centimeters deep above another diagnostic horizon	Moist at least three months of the year
	n-value less than 0.7
Dark	**Fertile**
Munsell chroma and Munsell value less than 3.5 unless free lime is present	Average organic carbon content of upper 17 centimeters must be greater than 0.6 percent
Munsell value at least one unit darker than the parent material or organic carbon 0.6 percent more than in parent material	Carbon/nitrogen ratio of upper 17 centimeters less than 17 if virgin and less than 13 if cultivated
Color not the result of manganese concretions	Base saturation greater than 50 percent by the NH_4OAc method
Color of interiors of peds not significantly lighter than coatings	Calcium is the dominant metallic cation
	Soluble P_2O_5 less than 250 parts per million unless phosphate nodules are present

[a] This example of the rigor of the Seventh Approximation is for illustration, not memorization.

This all seems very precise and very orderly. In real life, however, the problems of diagnosis are complicated by the fact that a particular soil layer does not appear abruptly on one side of an imaginary fence across a field. Soils often contain partially developed layers that do not quite meet the criteria for a particular diagnostic horizon. For example, the forested Fayette soil on the lower forty is a Typic Hapludalf, a typical haplous (simple) udic (humid) alfisol. The Order of Alfisols includes many mid-latitude forest soils, all containing an argillic horizon but no mollic layer (see Fig. 10-11 and the appendix to this chapter).

The Downs series at the boundary between forest and grassland is also a hapludalf, but it narrowly misses qualifying for membership in the mollisols. Its surface layer is a few centimeters too thin and the next layer down is half a unit too light in color to meet the criteria for a mollic layer. To indicate how close it came to being a mollisol, the classifiers call the Downs series a Mollic Hapludalf—a hapludalf that exhibits some mollisol characteristics but does not possess a strictly defined mollic layer.

biosequence set of soils having similar features except for plant cover (see p. 196)

intergrade subgroup that is transitional between great groups

The other soils on the lower forty get their official Seventh Approximation names by a similar procedure. The Lindley, with the same layering as the Fayette, is also a Typic Hapludalf. The ancient Keswick series is a slimy relative of the Lindley, an Aquic ("wet") Hapludalf. The flood-prone Nodaway soil on the valley bottom does not stay undisturbed long enough for soil-forming processes to produce any distinctive diagnostic horizons. It belongs to the subgroup of Typic Udifluvents, typical udic (humid) fluv (fluvial or river-formed) entisols (rec*ent* soils).

azonal soil old name for soil that has no diagnostic horizons

All told, the Seventh Approximation arranges more than 11,000 soil series of the United States into 10 orders, with a few dozen suborders, about a hundred groups, and several hundred subgroups. The strength of the classification is that it can describe soils very precisely with a unique vocabulary of fewer than a hundred syllables. The obvious weakness is its new and tongue-twisting vocabulary. A subtle weakness is that the Seventh Approximation tries to be a comprehensive classification, even though no one set of categories could be detailed enough to satisfy all the particular requirements of farmers and engineers and soil scientists.

Fortunately, the logic of the Seventh Approximation permits the user to work at any level of the classification; geographers usually stop at the order or suborder level. The appendix to this chapter contains maps of the 10 soil orders and brief descriptions of a few important suborders. The maps and descriptions reinforce the basic idea that a soil gets its characteristics from the environment in which it was formed.

APPENDIX TO CHAPTER 10
Soil Orders

The characteristics of a soil reflect the environment in which the soil developed. Exposure to a given environment causes inert weathered rock to undergo a variety of changes collectively called soil formation. Many of these changes may be described as balances between opposing forces:

1. Weathering produces loose material; erosion carries it away.
2. Weathering releases soluble nutrients; plants and percolating water remove them.
3. Leaching carries soluble materials downward; capillary action brings them upward.
4. Plants produce organic matter; decomposer organisms consume it.
5. Calcium, aluminum, and organic matter bring soil particles together into structural aggregates; hydrogen, potassium, and sodium separate them.
6. Percolating water carries clays downward; stationary water stops clay movement and produces massive soil structure.
7. Warm water dissolves silica and leaves iron and aluminum; cool water dissolves the metals and leaves silica.

Each of these balances is tipped one way or another by particular environmental conditions. For example, a hot and wet climate favors weathering, organic matter decay, leaching, clay movement, and silica removal. The resulting soil is rich in iron, poor in nutrients, deficient in organic matter, and aggregated by aluminum oxides. By contrast, a cool grassland produces a soil that lacks free aluminum oxides, is well-aggregated, and contains abundant organic matter and nutrients.

The various processes of soil formation tend to occur in combinations that leave recognizable imprints in the soil. The resulting characteristic layers, called diagnostic horizons, are the basis for the U.S. Department of Agriculture soil classification. It is more convenient to discuss the various diagnostic horizons in the contexts of the soils that they help to categorize. For each of the soil orders, this appendix has a brief description, a map of common occurrences, and some data for two representative soil profiles. The statistical information includes percentages of silt and clay (to permit textural classification), percentages of organic matter and base saturation, measurements of exchange capacity in milliequivalents (to indicate nutrient-holding ability), verbal descriptions of color (to help evaluate drainage), and pictorial portrayals of soil structure and layering (to give an impression of appearance and workability).

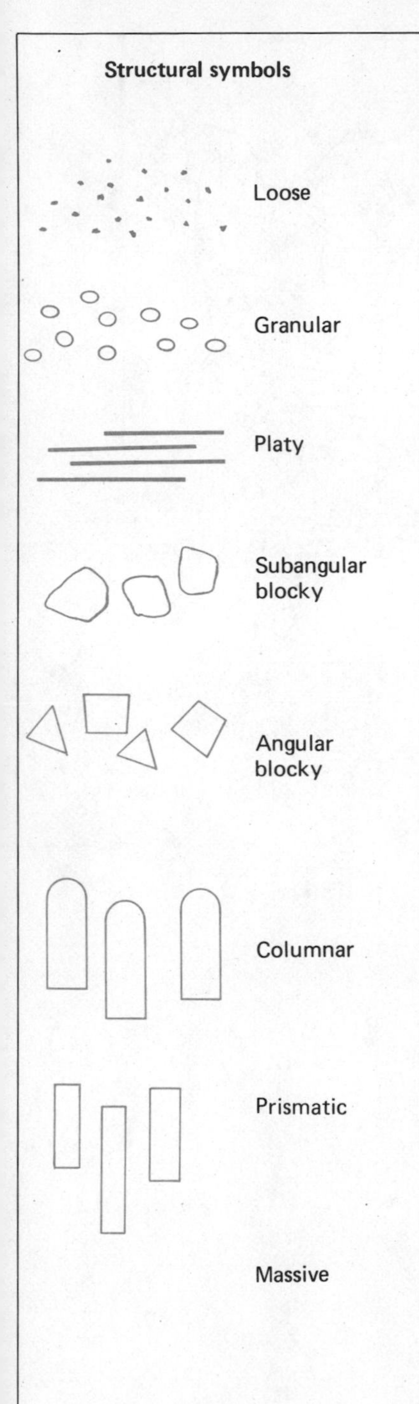

At the present time, the maps in this appendix cover only the conterminous 48 states. Soil surveying is time consuming and costly, and only about one-fifth of the United States and one-twentieth of the world have been accurately surveyed. We have left space for the addition of world maps for each soil order in later printings of this book. The maps are gradually being compiled as United Nations soil maps of various regions become available.

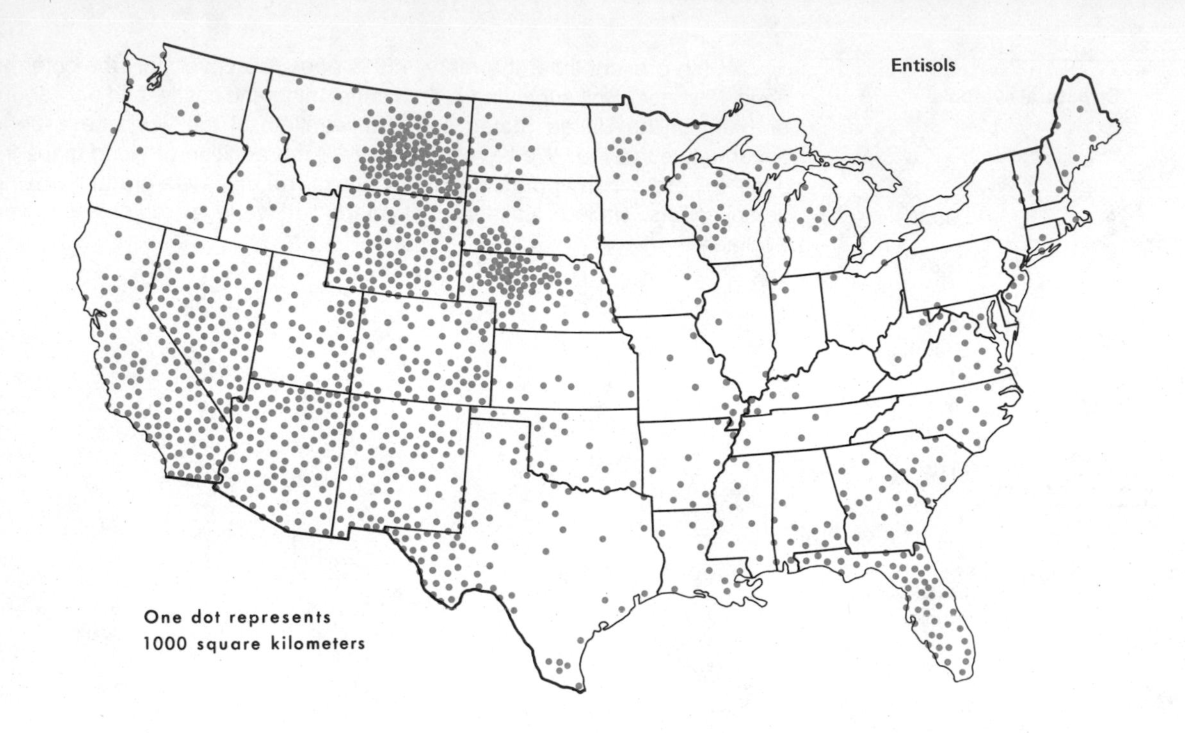
Entisols
One dot represents
1000 square kilometers

Examples of soils in the order of entisols

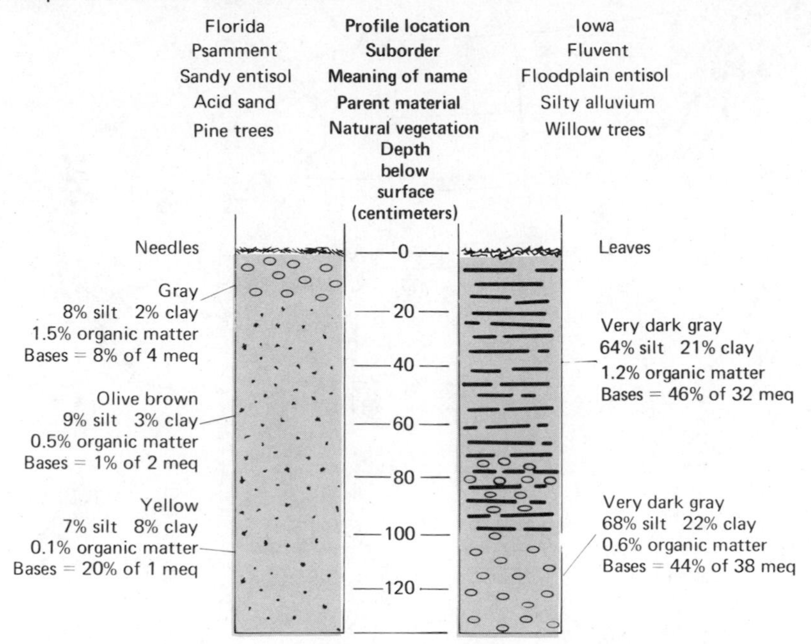

THE ORDER OF ENTISOLS

Entisols are rec*ent* soils, too young to show the modifying effects of their surroundings; therefore they have no diagnostic layers that would qualify them for membership in any other soil order. The extreme entisol is a pile of material dumped on a construction site by a bulldozer. Shifting sand, recent volcanoes, river floodplains, landslides, very cold places, and windy deserts are other likely sites for entisols. Recent soils are scattered throughout the country because they are not unique products of a particular climate. Concentrations of entisols appear in the Flatwoods of Florida, the former lake bottoms of Wisconsin, the river floodplains of the Midwest and Northeast, the sandhills of Nebraska, and the deserts of the West. The productivity of entisols varies from very low, in the case of some arid sands, to very high, as on some floodplains.

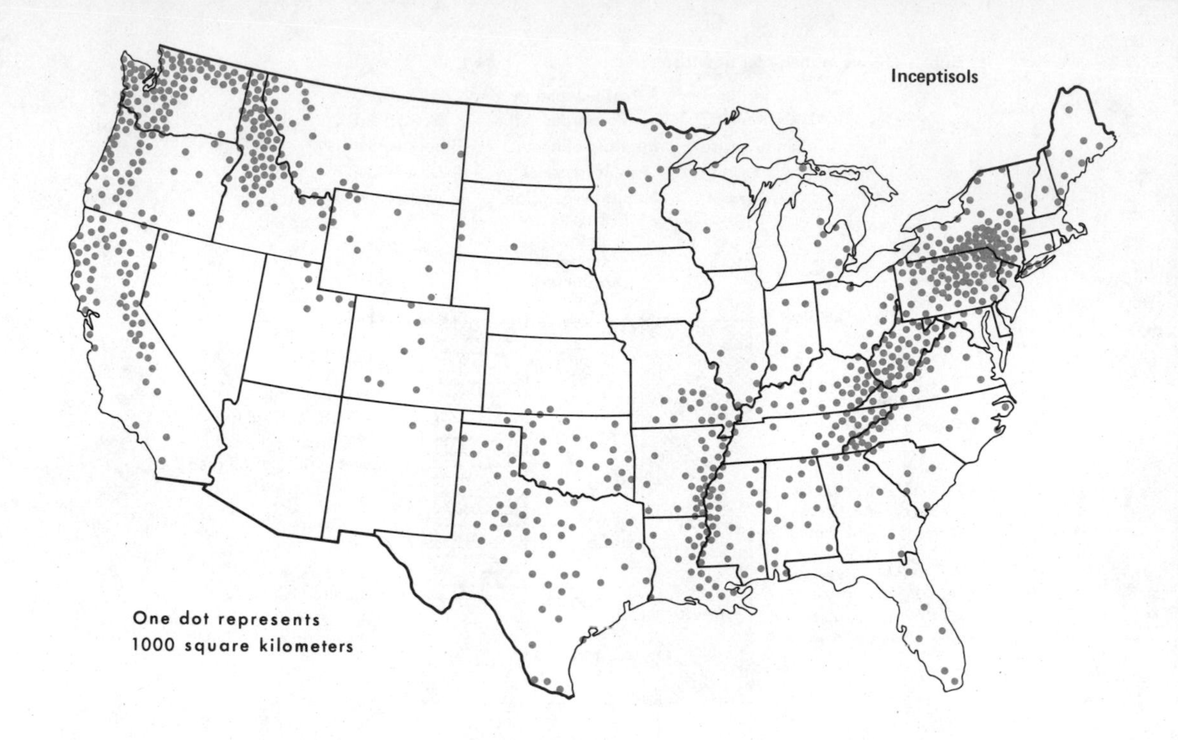
Inceptisols
One dot represents
1000 square kilometers

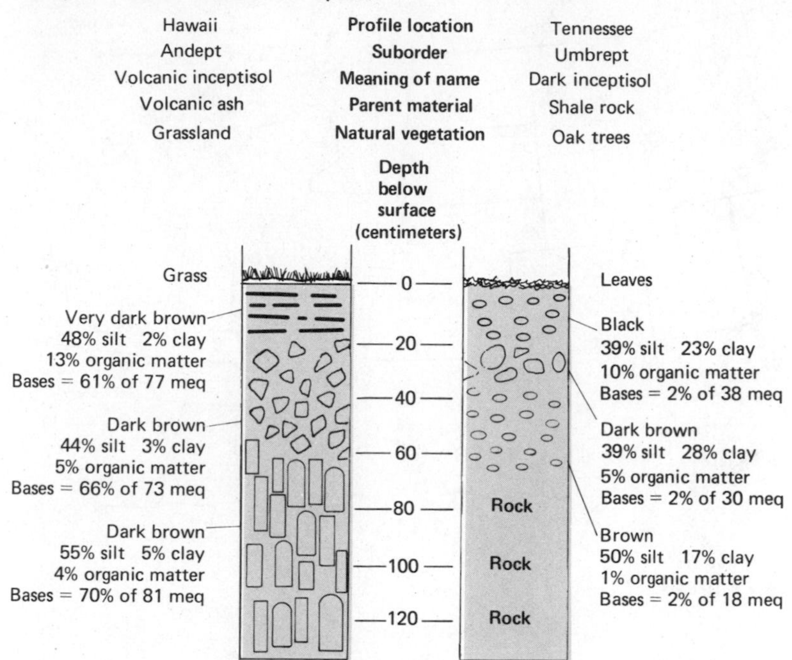

THE ORDER OF INCEPTISOLS

Inceptisols get their name from the word in*cept*ion. Climate and vegetation are just beginning to leave distinctive imprints on an inceptisol, but the features are not yet pronounced sufficiently to qualify as diagnostic horizons. At the same time, it is apparent that the soil is not a featureless mass, so that it is not fair to call it an entisol. The classic inceptisols come from mountain regions, where erosion removes soil before the profile can become mature. Other typical inceptisol sites are older floodplains along big rivers, stable sand dunes or beaches, and stony places on very resistant bedrock. Like the infant entisols, the adolescent inceptisols include a wide variety of dissimilar soils whose only common characteristic is their lack of maturity. Inceptisols on volcanic ash are fantastically fertile, while those on rocky hillsides may be economically worthless.

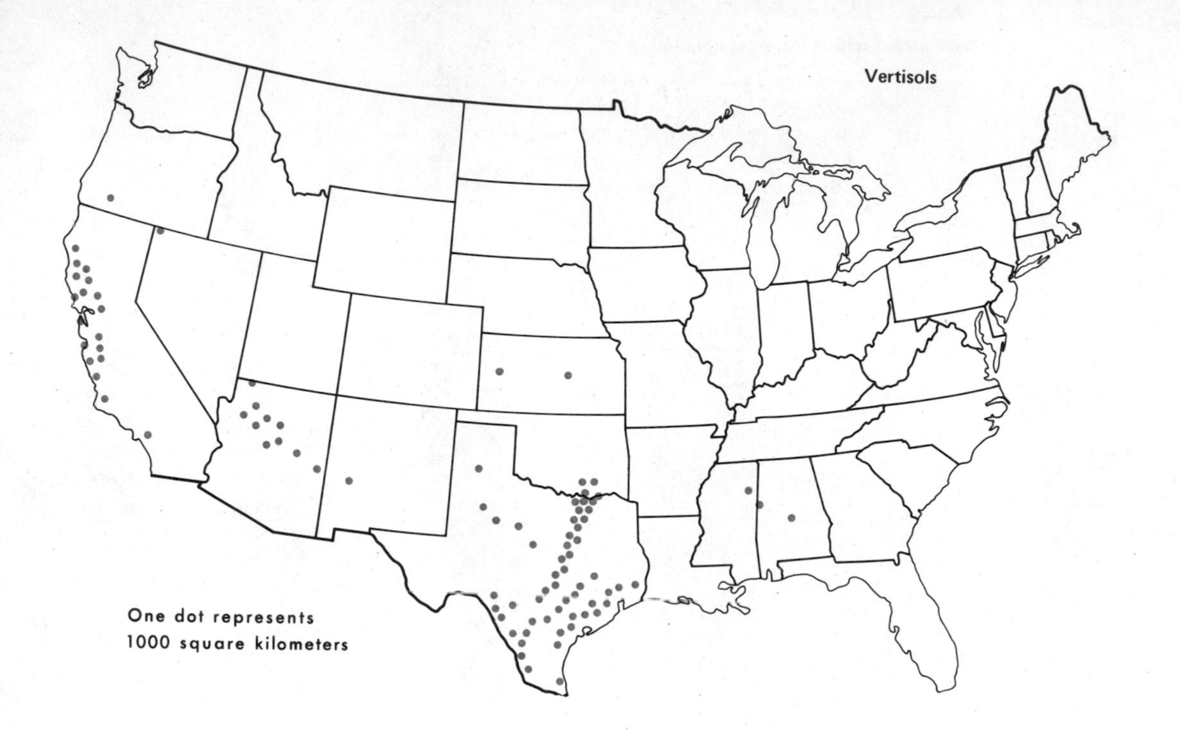
Vertisols
One dot represents
1000 square kilometers

Examples of soils in the order of vertisols

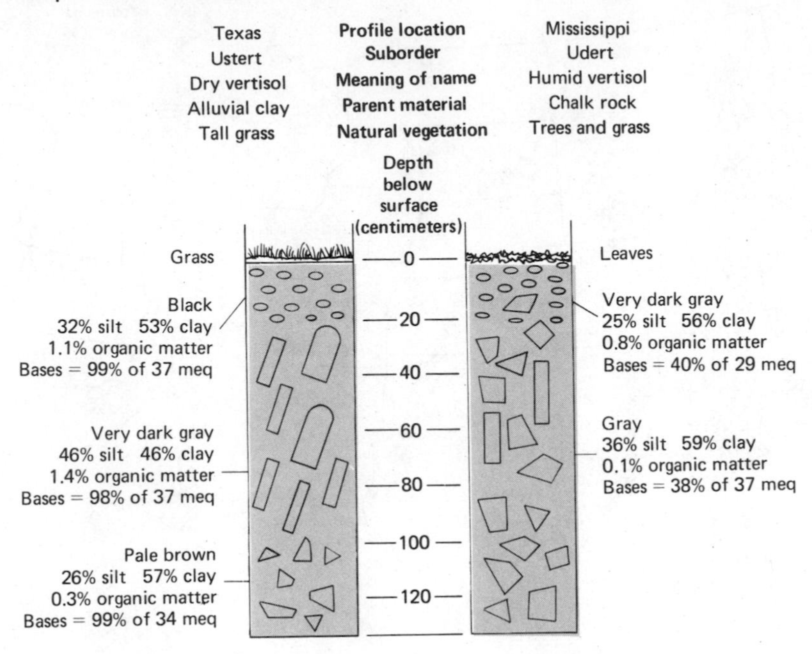

THE ORDER OF VERTISOLS

Vertisols get their name from the fact that they are self-inverting. They contain large quantities of clay that expands when moistened and shrinks when dry. The repeated swelling and shrinking churns the soil and obliterates any other diagnostic layers before they have a chance to develop fully. Vertisols occur only on clay exposed to a seasonally dry climate. Soil cracks do not get wide enough in a wet environment, and continuously dry weather would eliminate the cracks by filling them with dust. Vertisols occur in the United States only in limestone areas of the South and river plains in the West. Other countries, such as India or Egypt, have sizeable areas of cracking soils. Vertisols are fertile because expanding clays have a high exchange capacity, but their physical characteristics make them difficult to farm.

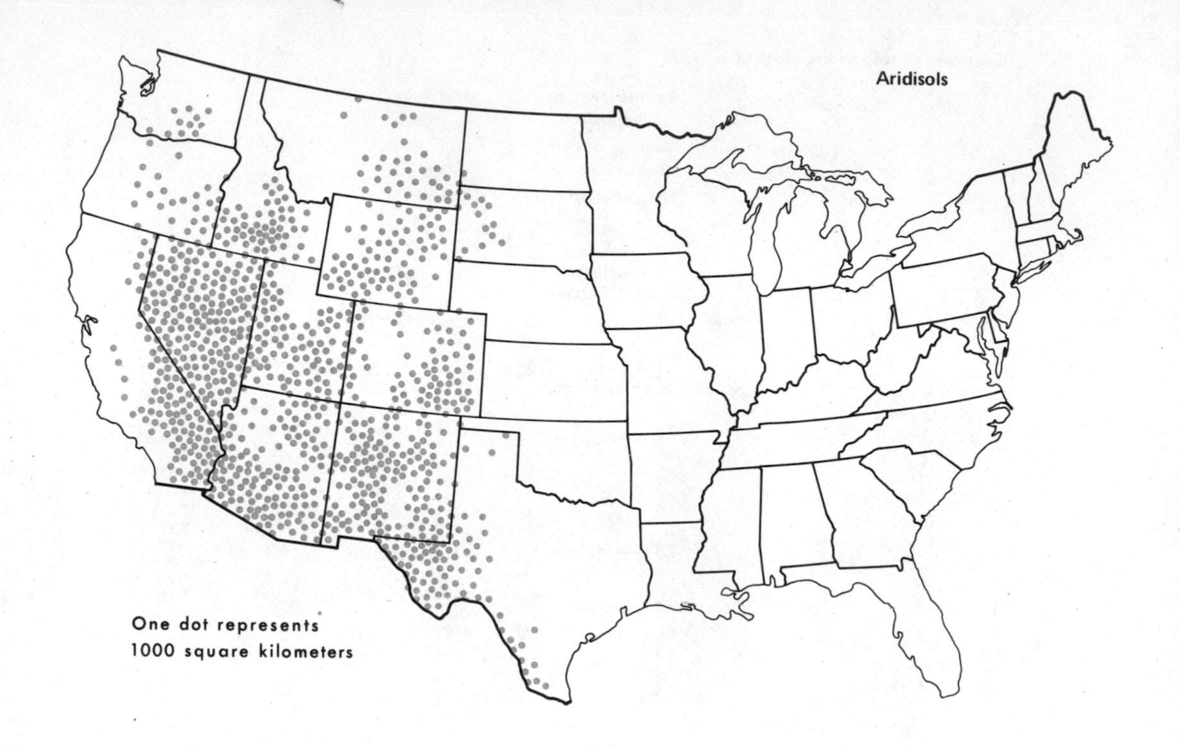
Aridisols
One dot represents
1000 square kilometers

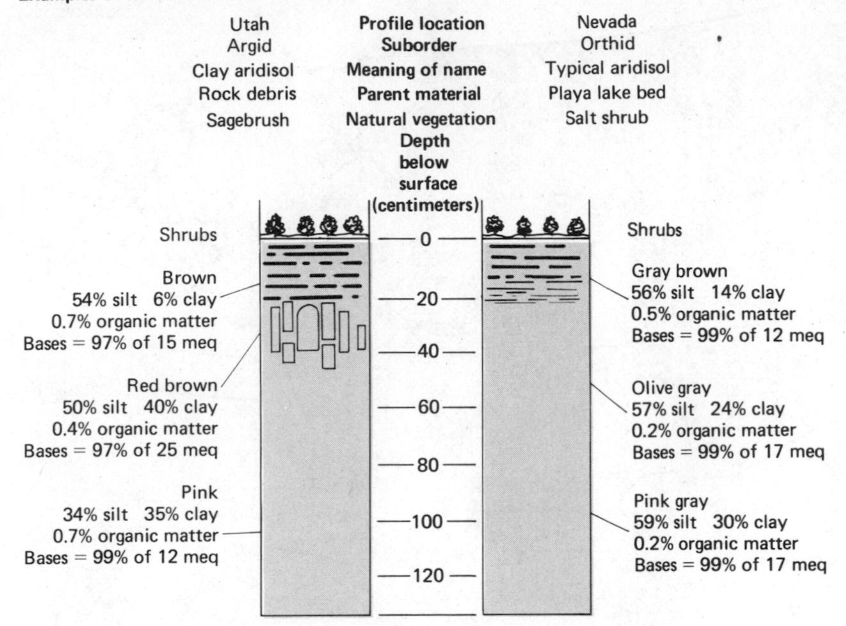

THE ORDER OF ARIDISOLS

Aridisols get their name from the *arid*ity of the regions they occupy. Their environment does not supply enough water to remove soluble minerals from the soil. The result may be one of a variety of diagnostic horizons—layers where particular chemicals have accumulated: calcic (calcium carbonate), gypsic (calcium sulfate), natric (clay and sodium salts) or salic (mixture of salts). The chemical characteristics of a particular aridisol depend on its parent material and the kind of forces at work on it. Ironically, some aridisols may be swampy at times. Many desert rivers flow out of nearby mountains when the snow melts in the spring. Then, as summer wears on, the temporary lakes dry up and leave salt flats in the valleys. Aridisols usually have no agricultural value unless they are irrigated. Even then, the threat of salt accumulation often poses a serious problem for the users of the land.

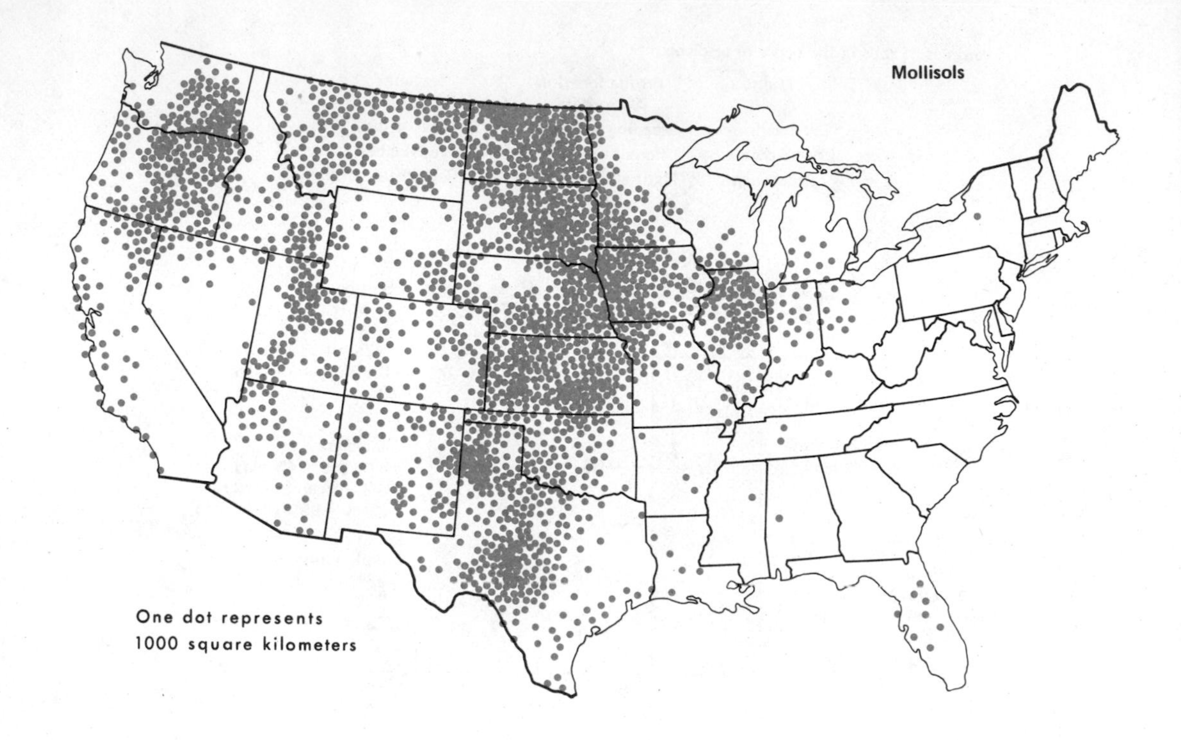
Mollisols
One dot represents
1000 square kilometers

Examples of soils in the order of mollisols

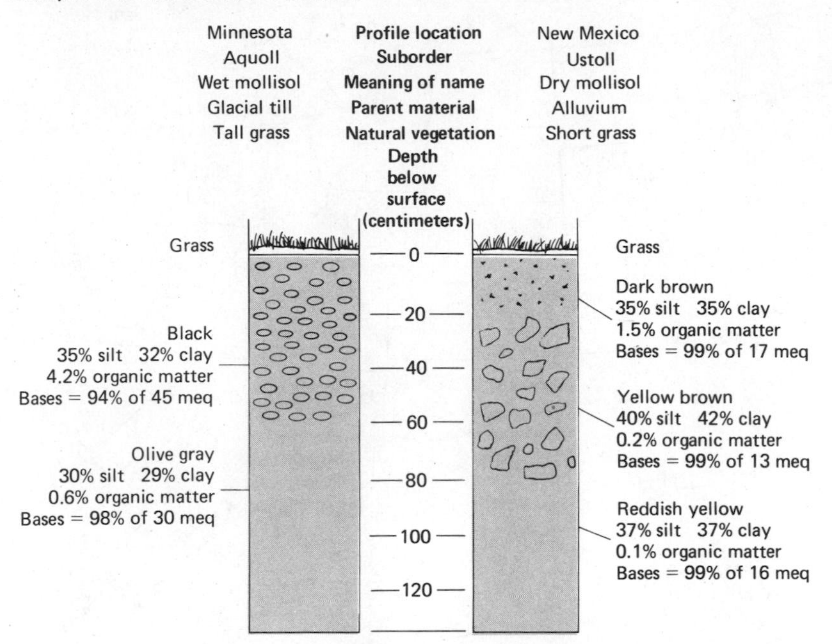

THE ORDER OF MOLLISOLS

Mollisols get their name from a Latin word meaning soft. The surface layer of a mollisol is soft, dark, fertile, and deep. However, some soils have a mollic surface layer and yet are excluded from the order of Mollisols. The exclusions include soils near volcanoes, soils with oxic or spodic layers, or soils with acid layers beneath the surface. The typical mollisol is formed under closely spaced grasses. If the climate is too dry for a dense stand of grass, the soil is likely to be an aridisol—too hard, light, or salty to be a mollisol. Rainier climates usually support trees whose leaves form a mat on the surface rather than a thick mixture of organic material and mineral soil. Earthworms are an important aspect in softening and mixing a mollisol; they add one more reason that the order of Mollisols includes some of the most productive soils in the world.

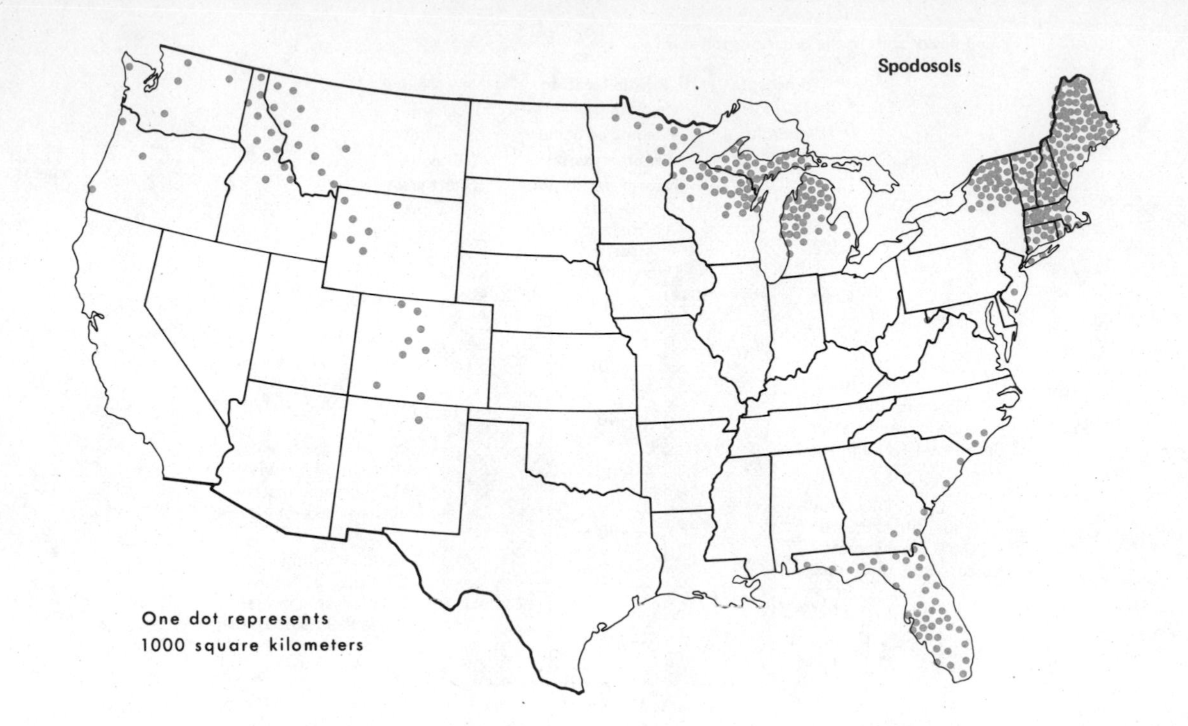
Spodosols
One dot represents
1000 square kilometers

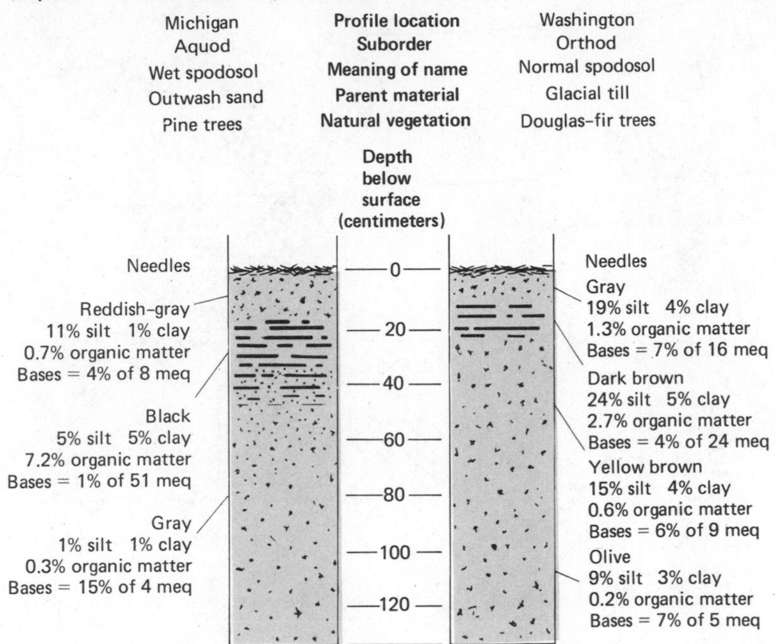

THE ORDER OF SPODOSOLS

Spodosols usually have four very distinct layers: a surface layer of partly decayed organic matter, a light upper layer of coarse leached material, a dark lower layer where organic material and aluminum from the leached layer accumulate, and finally, the unaltered parent material. The subsurface layer of organic matter and aluminum (the *spod*ic horizon) is the essential diagnostic feature. Ordinarily, spodic horizons form only on sand in rainy areas, often but not always under needleleaf trees. In the United States the proper combination of conditions occurs mainly in northern forests, high mountains, and salt marshes along some seacoasts. Soils with spodic horizons are usually very acid, low in nutrients, and poor for farming, except for acid-tolerant crops, such as potatoes, blueberries, or cranberries. Irrigation and fertilization have made some spodosols very productive, but the cost is very high.

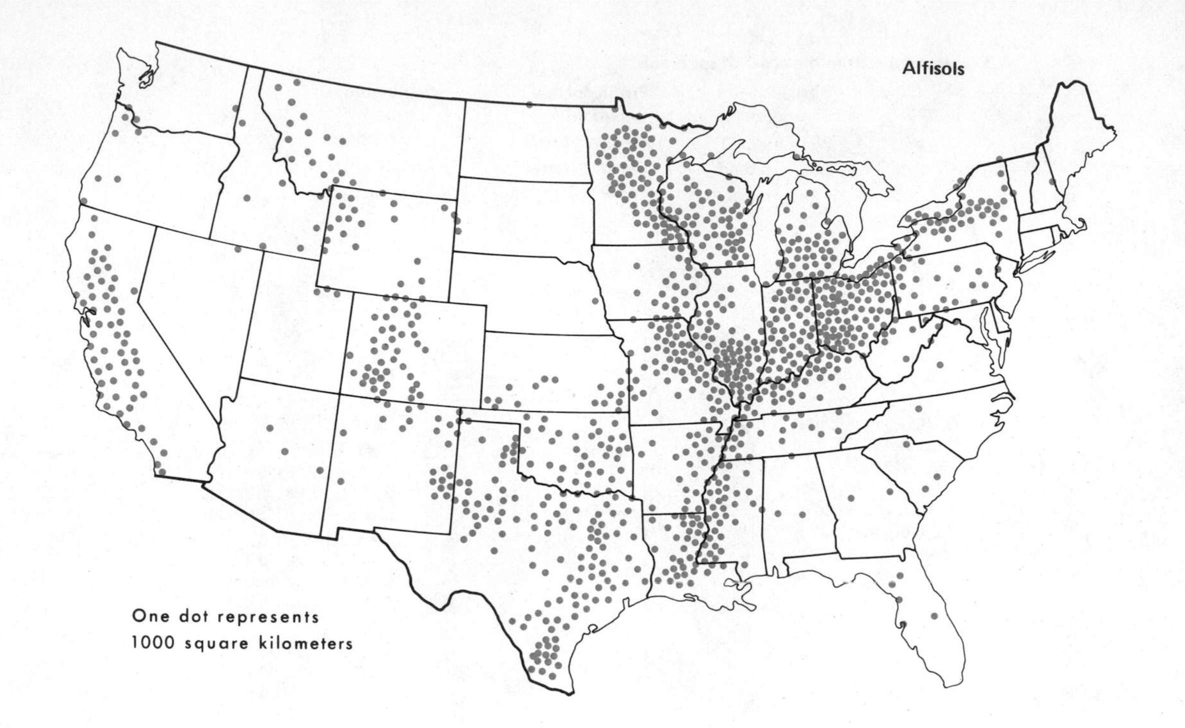
Alfisols
One dot represents
1000 square kilometers

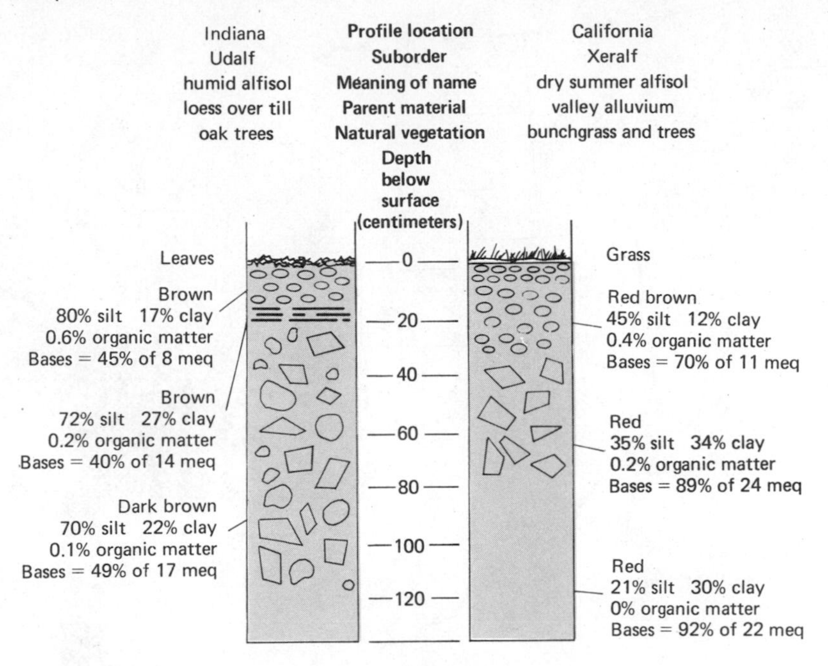

THE ORDER OF ALFISOLS

Alfisols get their name from the chemical symbols *al* (aluminum) and *fe* (iron). They are transitional soils, with more organic matter than ultisols but less than spodosols or mollisols. Alfisols occupy regions with enough water percolating through the soil to carry small clay particles out of the surface layer. A subsurface layer of clay accumulation (the argillic horizon) is the necessary diagnostic feature. The rest of the definition of an alfisol is by exclusion. The soil cannot be so wet, dry, or cold that organic matter accumulates in a histic, mollic, or spodic layer. The soil also cannot be too highly leached or weathered, so it does not occur in climates that are hot and wet. Alfisols are therefore found mainly in mid-latitude forests and some grass-and-tree savannas. Most alfisols are fertile and productive; on the average, only mollisols have more agricultural value.

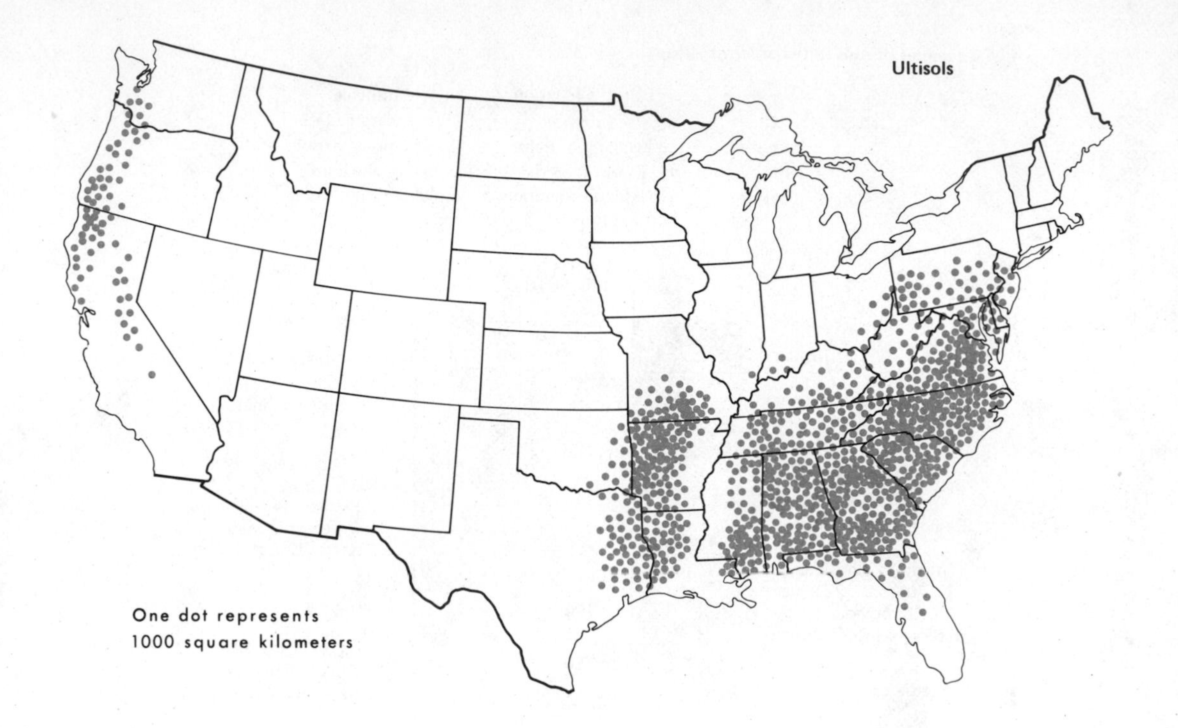
Ultisols
One dot represents
1000 square kilometers

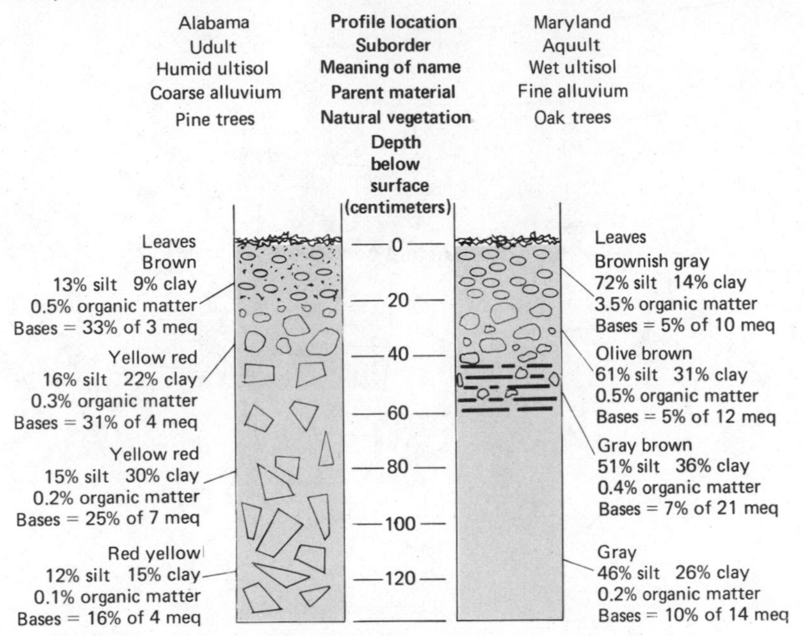

THE ORDER OF ULTISOLS

The ultisols get their name from the word *ult*imate. They represent the ultimate stage of weathering and soil formation in the continental United States. Ultisols form in environments with a long, frostless season and abundant rainfall. Chemical activity has broken most of the parent material down into small particles with simple and durable chemical structure. Percolating water has removed many of the soluble nutrients and carried a significant amount of clay downward in the soil. The diagnostic features of the ultisols are low base saturation and a subsurface layer of clay accumulation. Soils in swamps or cold places usually have too many nutrients or too much organic matter to be considered ultisols. Sandy soils cannot develop the necessary horizon of clay accumulation. The long growing season is favorable for agriculture, but farming the ultisols is usually expensive, because they are prone to leaching and erosion.

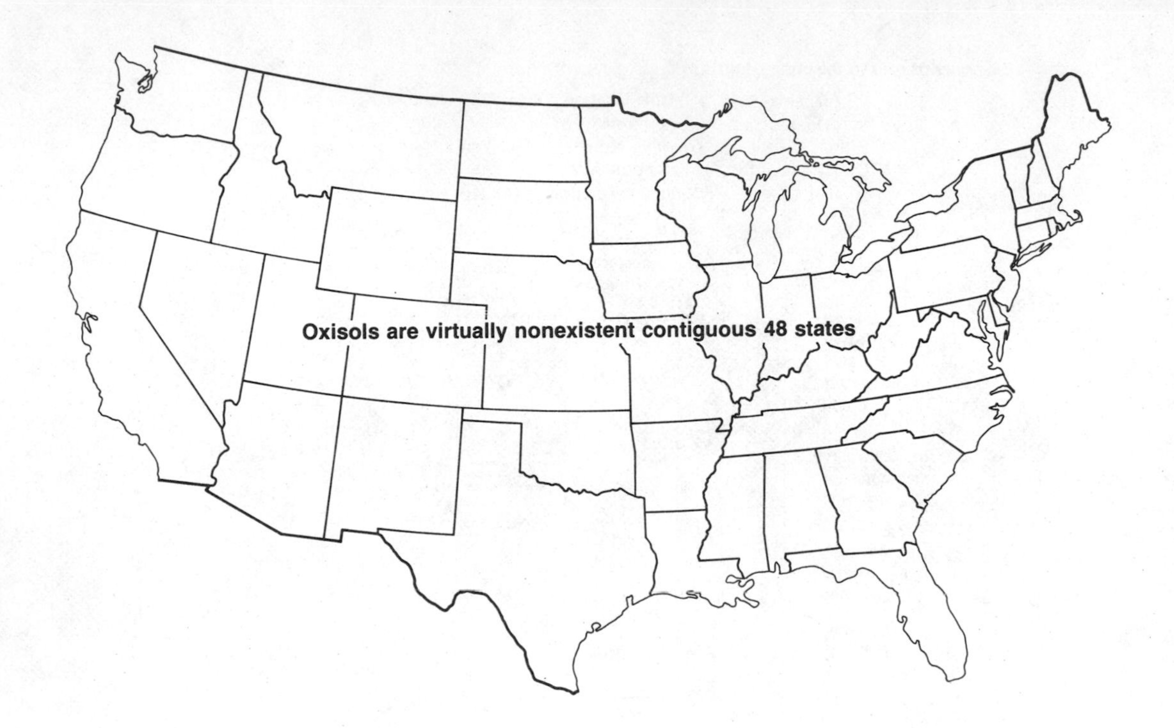
Oxisols are virtually nonexistent contiguous 48 states

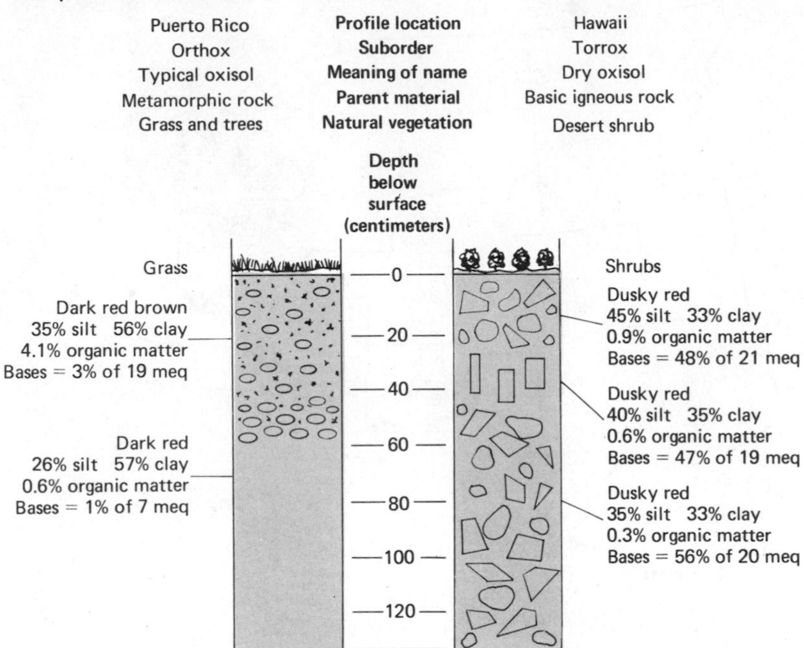

THE ORDER OF OXISOLS

Oxisols get their name from the *ox*ic horizon, a diagnostic accumulation of iron and aluminum *ox*ides. Oxisols are very rare in the continental United States but common in Hawaii. They form in climates that are usually frostless throughout the year and rainy for at least part of the year. The resistant oxides are left in the upper layers of the soil after heat and moisture have destroyed or removed most of the salt, complex clay, organic material, and plant nutrients. In turn, oxides often bind the soil particles into tight clumps. Oxisols in seasonally dry climates are prone to harden into rocklike masses when exposed to the sun or subjected to alternate wetting and drying. An oxisol in a continuously wet climate is usually soft but leached of most of its plant nutrients. The alternate hazards of leaching and hardening make oxisols very difficult to farm.

Histosols
One dot represents
1000 square kilometers

Examples of soils in the order of histosols

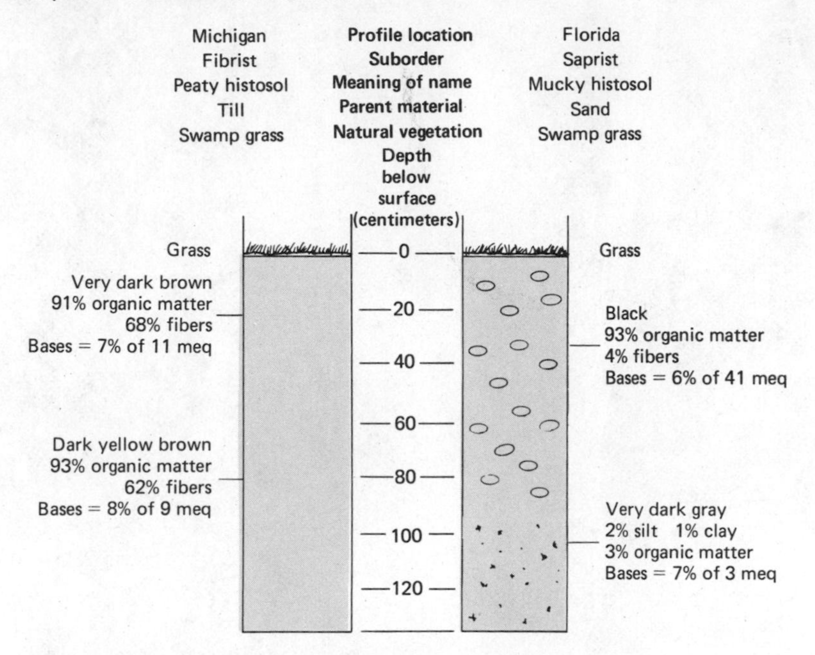

THE ORDER OF HISTOSOLS

Histosols get their name from a Greek word for tissue. Their identification badge is a layer of organic remains (the *hist*ic layer). The dead plant and animal material may be partly decayed but still recognizable (peat), or it may be decomposed into a featureless black mass (muck). Peat usually occurs in permanent swamps; muck is more widely distributed than peat, though it too is poorly drained. Organic matter accumulates because decay microorganisms cannot work well in wet soils. Artificial drainage can make muck very productive, but the high cost can be offset only by very valuable crops, such as vegetables for urban markets. Muck usually shrinks when drained, since the organic material decays when air enters the soil. Peat is coarser, more acid, and less fertile than muck. It has little agricultural value except for a few specialized crops, such as cranberries.

CLASS VII LAND

CLASS VIII LAND

CLASS VII LAND

CLASS VI LAND

CLASS IV LAND

CLASS II LAND

CLASS V LAN

CLASS I LAND

CLASS III LAND

LAND CAPABILITY CLASSES

SUITABLE FOR CULTIVATION		NO CULTIVATION-PASTURE, HAY, WOODLAND AND WILDLIFE	
I	REQUIRES GOOD SOIL MANAGEMENT PRACTICES ONLY	V	NO RESTRICTIONS IN USE
II	MODERATE CONSERVATION PRACTICES NECESSARY	VI	MODERATE RESTRICTIONS IN USE
III	INTENSIVE CONSERVATION PRACTICES NECESSARY	VII	SEVERE RESTRICTIONS IN USE
IV	PERENNIAL VEGETATION - INFREQUENT CULTIVATION	VIII	BEST SUITED FOR WILDLIFE AND RECREATION

SCS Photo/Leon Sisle

Chapter 11

Tolerance and Succession: Life on Land

Porridges can be too hot, too cold, or just right; fortunately, different diners have different concepts of just right.

U.S. Forest Service

Give "Parent Nature" enough time, and it will usually determine which are the best available plants and animals for a particular place. Furthermore, it will arrange its affairs so that the populations of living things are just large enough to use the incoming solar energy in the most efficient manner. Geographic differences in energy income, water availability, or soil quality produce differences in the kinds and numbers of living things in different environments. The related concepts of tolerance and succession are useful ways of interpreting the responses that living things make to environmental differences.

TOLERANCE THEORY

An orange tree and a college student simply do not respond to the same things in their environment. Each is affected to a different extent by different characteristics of their surroundings. Because students take their responses generally for granted, it may be better to look at the idea of tolerance from the point of view of the orange tree. Its response to some obvious environmental influence, such as temperature, can be summarized as follows.

optimum ideal range of some environmental condition

tolerance limit environmental condition that is barely tolerable; if exceeded, the result is permanent harm or even death

primary production excess of photosynthesis beyond what the plant needs to stay alive

Law of the Minimum theory that growth is restricted by the one critical element that is nearest to intolerable

First of all, an orange tree prefers the temperature to stay near 28° C (83° F) and becomes uncomfortable if the temperature goes below 22° C. A temperature of 13° C (55° F) significantly impairs its ability to produce leaves. At that temperature, the production of plant material by photosynthesis is too slow to keep up with the day-to-day repair needs of the plant. A hard freeze will probably kill the tree; 0° C is a valid lower limit for its general temperature tolerance.

In the other direction from the ideal 28° C, temperatures above 33° C begin to bother the tree. It diverts more and more of the incoming sunlight into repair and maintenance as the temperature increases beyond the optimum range. The long-term business of growing and making oranges gets less attention than the short-term problem of survival. Prolonged exposure to temperatures of 40° C (104° F) or so would probably be lethal. In general, the temperature preferences of an organism can be displayed on a tolerance graph, such as Figure 11-1.

Afraid of giving a simplistic view of its ability to live in a given environment, the orange tree points out that it does not really matter what the temperature is if the soil is too dry. The tree will be so hampered by the lack of water that subtle differences of temperature will be of little concern. In fact, if the surroundings of an orange tree were arbitrarily separated into a hundred environmental characteristics, its growth would be restricted by the one factor that is farthest away from what it considers ideal (Fig. 11-2). Too much or too little of just one necessity can make an otherwise perfect place downright miserable.

Even this venerable old way of acknowledging the effects of many different characteristics of the environment is not completely acceptable. Different parts of the environment interact, and one influence modifies the effect of another. For example, an orange tree that gets plenty to drink can tolerate

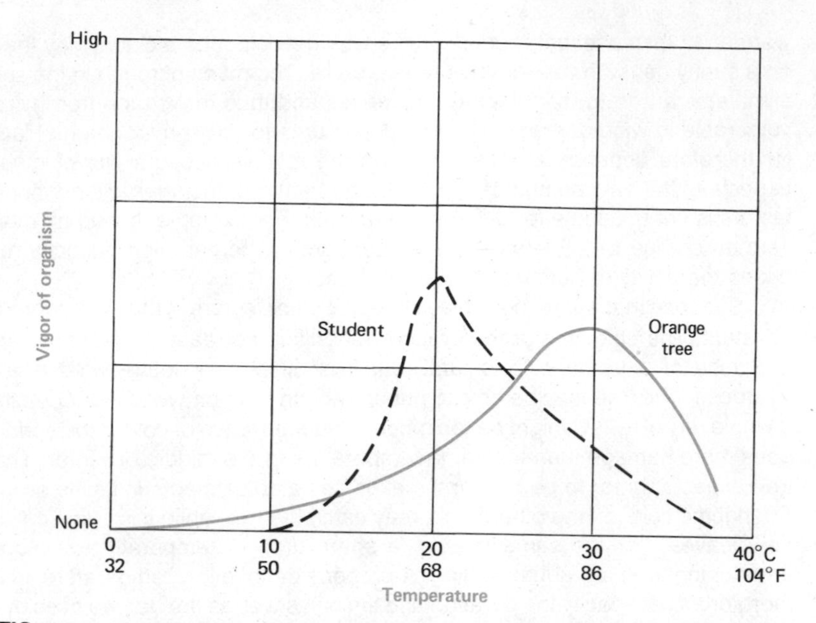

FIG. 11-1. Temperature tolerance curves of an orange tree and a student. Each organism has an optimum temperature, at which its vigor is maximum, and lethal limits, which cannot be exceeded for a long time without causing damage or death to the organism.

FIG. 11-2. Leibig's Law of the Minimum. The solid black line and left scale show the vigor of the organism. The other lines and right scale indicate the favorability of different factors at various places. The vigor of the organism cannot go any higher than the tolerability of the single factor that is farthest from the ideal at each place, even though other factors are favorable.

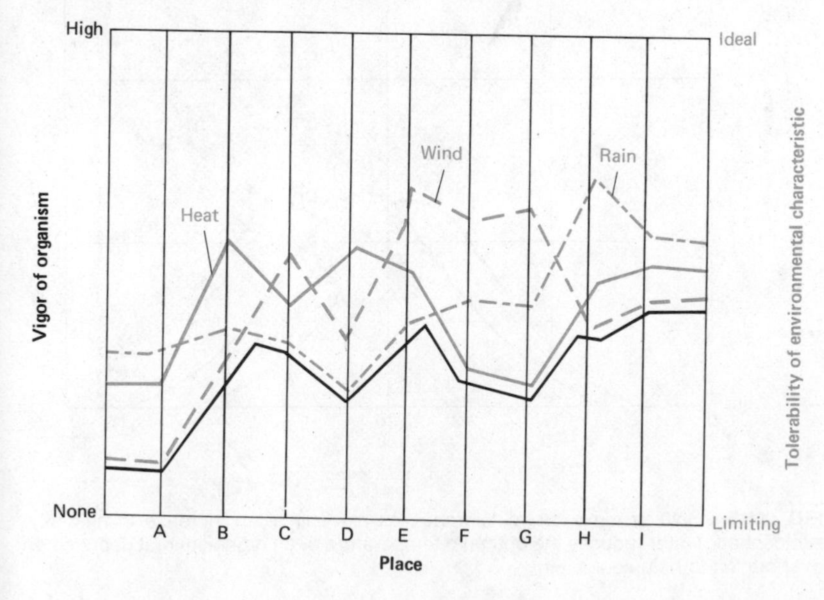

synergism combined effect of two or more factors, different from the sum of their individual effects

duration sum deviation from some arbitrary reference multiplied by the length of time it must be endured; a heating degree-day is one kind of duration sum
(see p. 104)

warmer air than one that is dried out. Leaves get colder on a cloudy day than on a sunny day with the same air temperature. Too much nitrogen in the soil stimulates a tree to manufacture big leaves and thus makes the tree more vulnerable to wind damage. The tolerance curve for one environmental factor therefore depends at least partly on the relative acceptability of other aspects of the environment (Fig. 11-3). The nature of the interaction among factors is not the same for different organisms. For example, humid air may help an orange tree tolerate high temperatures, whereas high humidity reduces the ability of humans to withstand heat.

Suppose that it were possible to keep all other factors in the environment constant. The cold limit of the orange tree still is not easy to describe. An overdose of extreme cold is fatal, but it is difficult to define what is an overdose. The tree may be able to put up with an hour or two of −6° C, while a whole day of −2° C might be too much. A sudden wave of cold in the spring does more damage than the same temperature in the middle of winter. The tree expects winter to be cold and makes some adjustments in its life style. Springtime cold, on the other hand, may catch the tree while it is trying on its new leaves. For the same reason, a sharp drop in temperature is more threatening than a gradual cooling. A properly defined low temperature limit therefore must specify the duration and timing as well as the degree of cold.

The age of trees is also important in determining their ability to tolerate cold. Old trees can withstand temperatures that would mortify a tender young seedling. Most living things are exceptionally sensitive to environmental stresses during their reproductive periods. Landscape architects

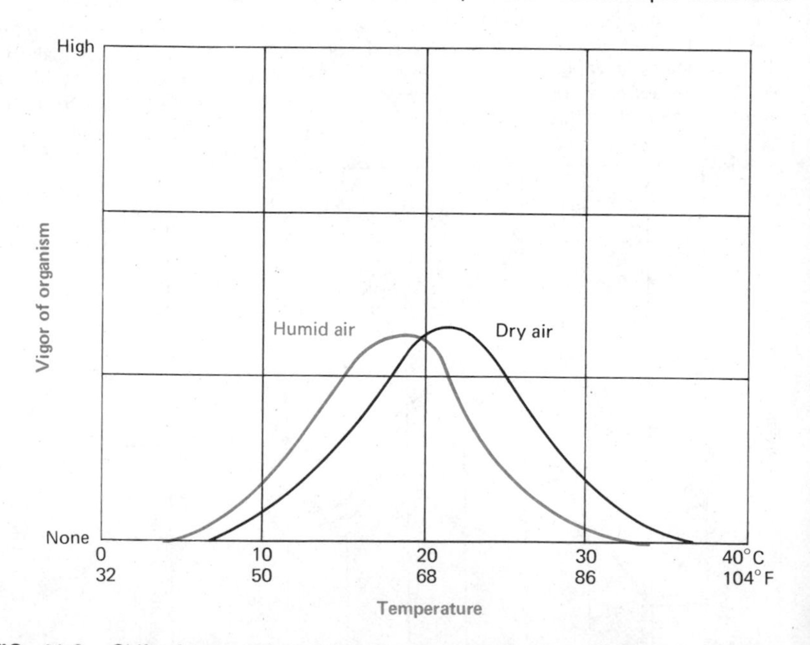

FIG. 11-3. Shift of temperature tolerance as a result of an increase in humidity. Atmospheric water reduces the optimum temperature of an organism that depends on evaporation to help cool itself.

sometimes take advantage of this fact by deliberately growing some ornamental plants far away from their homelands, in places where conditions are too severe for them to flower and produce seed. These plants may remain permanently green and youthful, unable to reproduce and therefore unlikely to die of old age.

The reproductive characteristics of a living thing combine with the variability of climate to influence the tolerance of the organism in another way. Orange trees must get rather old before they can begin to produce offspring. A generally tolerable environment becomes unacceptable for a long-lived species if conditions go beyond its limit every five or ten years. Therefore natural colonies of orange trees are unknown in places that have even an occasional severe frost.

recurrence interval period of time that probably contains one event equal to or larger than a particular size
(see Fig. 5-5)

The personal history of the individual orange tree must also be considered. A tree that has experienced near freezing conditions is able to withstand weather that may be intolerably cold for trees that have led more sheltered lives. Environmental stresses, if they are not too severe, often provoke reactions which enable a living thing to protect itself against future stresses. For example, a cold spell may kill some of the outer tissues of a tree. The result is a thickened layer of bark that helps insulate the plant against occasional low temperatures. High winds may break slender twigs and thereby make the tree more compact than before and better able to withstand the wind (Fig. 11-4).

growth form shape of an organism as influenced by its environment

FIG. 11-4. Wind-stressed trees. This species normally grows tall and thin, but the trees adopt a low-branching habit on exposed sites in order to withstand the wind. (Courtesy of L. Ellison, U.S. Forest Service.)

Finally, the pedigree of the orange tree is also significant. If the local conditions frequently dip just below freezing for a short time, some of the sensitive individuals of the previous generation would have succumbed to the cold. The resistant individuals would reproduce more effectively than the sensitive ones and pass their slight advantage on to their offspring. Trees in each generation become slightly more cold tolerant than their ancestors were. The continuous genetic adjustment enables the population as a whole to keep up with slow changes in the environmental conditions.

natural selection individuals that are better adapted to an environment are also able to produce more offspring than poorly adapted ones; if the advantage is hereditary, the next generation will include a larger proportion of better adapted individuals

In summary, there is no simple answer to a question about the temperature preference of an organism such as the orange tree. Temperature tolerance depends on the humidity of the air, the quality of the soil, the age of the tree, the kind of childhood it had, the sturdiness of its parents, the season of the year, the amount of time the tree must tolerate the cold, and so on.

Each living thing has its own set of ideal conditions. Furthermore, no two species agree on the relative importance of various environmental factors. The orange tree views frost as a major threat to its survival; therefore frost-free conditions are high on its priority list when it is looking for a place to live. A pine tree regards frost as a nuisance that may interfere with production for a short period of time. A peach tree actually requires a certain number of hours of below-freezing temperatures to enable it to produce blossoms and fruit.

Different organisms may have the same ideal temperature but radically different abilities to tolerate temperature variations. A few hardy microscopic creatures can survive temperatures that vary as much as 50 Celsius degrees from normal. Some deepwater fish, on the other hand, are extremely sensitive, unable to tolerate a temperature change of more than a single degree.

eurythermic able to tolerate a wide range of temperature conditions

stenothermic able to tolerate only a narrow range of temperature conditions; the prefixes steno- and eury- can apply to many environmental factors (e.g., stenohaline, unable to tolerate much variation in water salinity)

COMPLEX ENVIRONMENTS AND ORGANISM RESPONSES

Living things would distribute themselves according to world temperature patterns if temperature were the only critical feature of the environment. Each species would be most abundant where conditions were nearest to ideal for that species. However, its population would be low where average temperature or temperature range, duration or frequency of change were less tolerable than in the ideal place. The species would be absent wherever some aspect of temperature went beyond its tolerance limits.

habitat sum of environmental conditions an organism can tolerate

Mountain vegetation offers a good illustration of the correlation of temperature and life. Climbers routinely carry coats when starting up a mountain, even though the temperature at the base camp may be comfortably warm, because they will be moving into colder areas the higher they climb. If occasionally they stop and note the vegetation on the way up, a pattern of living things begins to emerge.

A path in the Great Smoky Mountains of North Carolina and Tennessee starts in a forest of lofty, broad-leaved trees. Farther up the slopes, the trees get shorter. Southern species disappear, and some northern needleleaf

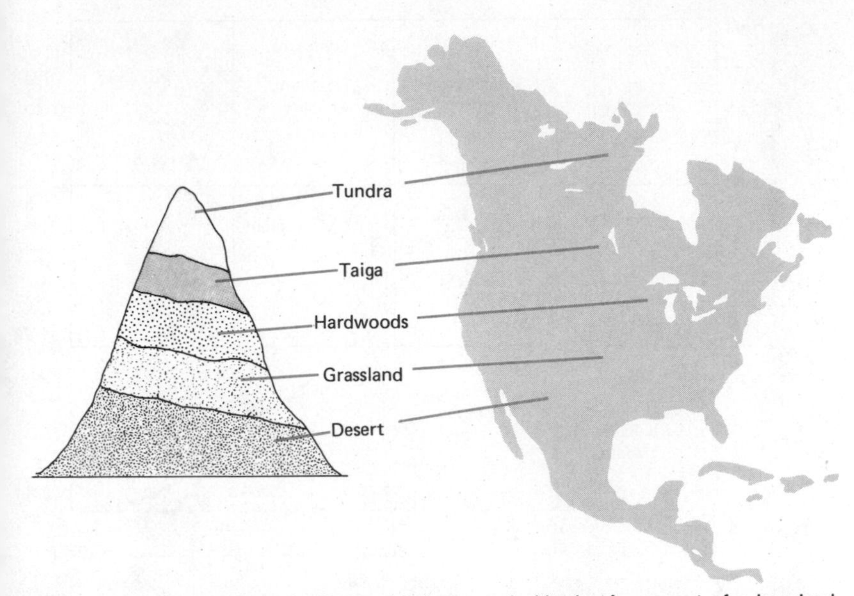

FIG. 11-5. The ecological effects of latitude and altitude. An ascent of a hundred meters up a mountainside is the **approximate** ecological equivalent of a trip of a degree of latitude toward the pole. The word "approximate" is emphasized because such factors as daylength, growing season, sun angle, and moisture availability are different, even though average air temperature may decrease in a similar manner on both trips.

trees appear. Finally, near the summit of the mountain, the forest looks like the woodlands of eastern Canada. The cold-sensitive southern species simply cannot tolerate the low temperatures so far above sea level.

Even Canadian trees cannot survive on the lofty summits of the Rocky Mountains. Hardy low plants remind hikers that conditions on a Colorado mountaintop may be as harsh as those near the shore of the Arctic Ocean (Fig. 11-5).

altitude zonation arrangement of vegetation types in distinct bands on a mountain, like a layer cake of plant associations

The interpretation of species' responses becomes considerably more complicated when other environmental variables are also considered. Two organisms with similar temperature tolerances but different responses to wind, sunlight, or available soil calcium would occupy different places. Interactions between aspects of the environment would modify the distributions of species still further.

Even if the responses to environmental factors could be measured perfectly, the theory of tolerance would still not explain the distributions of living things. Plants and animals simply do not respond only to the inanimate environment. For example, consider two plant species with the same optimum temperature. If one is taller but the other can tolerate a lower temperature, the tall plant would occupy the terrain with the ideal temperature for both. The short plant, shaded out of places with its preferred temperature, would have to be content with colder but still tolerable regions. In effect, each living thing responds to conditions in the inanimate environment as those conditions are modified by neighboring plants and animals (Fig. 11-6).

competitive exclusion if two species compete for the same resource, the more efficient species will eventually eliminate the less efficient one

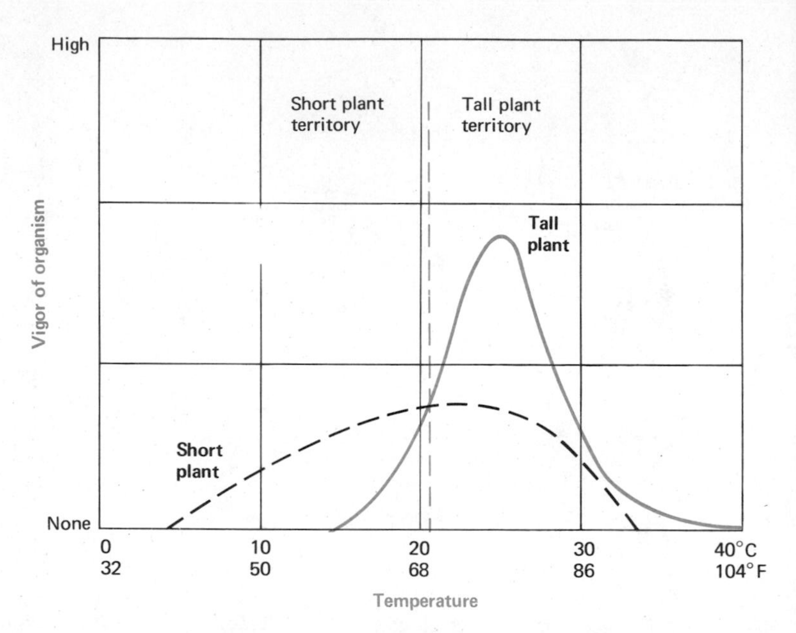

FIG. 11-6. Displacement of an organism from its optimum environment by a tall competitor. The short plant, shaded out of places with its preferred temperature, is found only in places that are less desirable but still tolerable.

SUCCESSION AS ENVIRONMENTAL MODIFICATION

outcrop exposure of bare bedrock

table air air that resists vertical motions
(see fig. 2-2).

microclimate energy and moisture conditions in a small area

lichen combination of algae and fungi, both noncellular organisms

Consider the extreme severity of the environment on bare rock in eastern North America. By day, unobstructed rays of the sun do little but heat the surface, which can get hot enough to fry eggs. By night, the surface radiates heat away unhindered, and stable air in minor depressions on the rock can get much colder than air at shoulder height. Rain and sleet beat directly on the rock, hammering it with forces that may exceed a ton per square centimeter. Winds sandblast the unprotected surface on dusty days. In winter, the snow blows away, exposing bare rock to the elements. Rain cannot soak into solid rock; in fact, the water flows off or evaporates soon after a rain, and the rock surface resembles a desert much of the time. In summary, the bare rock seems to be an intolerable environment, unable to support life. Yet despite its harshness, bare rock does get covered with living things in time.

A process called biotic succession is the key to the puzzle of living things colonizing what seem to be uninhabitable places. Succession can be defined as a progressive modification of a local environment by its inhabitants. On bare rock, conditions are beyond the tolerance limits of nearly every form of life. There is little food or shelter to persuade animals to stay on the rock. An occasional plant seed falls on the rock, only to wither in the heat or to be picked up by a passing animal. The first permanent inhabitant of an eastern North American rock is usually a lichen, one of the most tolerant forms of life on the face of the earth (Fig. 11-7).

FIG. 11-7. Lichens on bare rock. These hardy associations of tiny plants secrete an acid which etches the rock allowing the light-colored plants to grip the rock surface.

Even for these rugged associations of tiny plants, conditions on bare rock are not really ideal. The efficiency of the plants is limited by extremes of temperature and moisture, but the very existence of lichens on the rock has an effect on the local environment. Their bodies change the energy budget of the surface and make days cooler and nights warmer than before. Rainwater penetrates the space in and between living plant cells, where it remains available longer than it could on bare rock. The plants secrete organic acids that help to break the rock apart into loose sand and clay. Finally, as individual plants die, their decaying bodies combine with weathered rock to form a thin soil that acts as a nutrient and water storehouse.

In time, conditions on the lichen-covered rock become less and less severe. Temperature varies less widely, moisture conditions become more moderate, the soil provides more nutrients, and toeholds on the rock are safer and more numerous than before the first lichens came. The lichens respond to the improvement of their surroundings by growing more rapidly.

Masses of lichens can modify the local environment even more than a few hardy pioneers could. Eventually, conditions on the rock become tolerable for a more complex form of life, such as a moss, which was unable to survive on the bare rock. Without apparent recognition of its debt to the lichens, the moss begins to grow over them. Moss-covered lichens die because they cannot get enough sunlight to survive. In effect, the presence of moss puts conditions on the rock out of the tolerance range of the lichens. The interaction of moss and lichen must rank as one of the great ironies of nature: lichens work for years to make an inhospitable place better for lichens, only to bring conditions within the tolerance range of a competing form of life that comes in and eventually eliminates the lichens.

pioneer species initial inhabitant of an unoccupied site

But the days of the moss are likewise numbered. Its existence further moderates the temperature and moisture conditions on the rock. The death of moss contributes additional depth and richness to the soil. Grasses and simple flowers are standing in line, waiting for conditions on the rock to become acceptable. The tall plants shade out the moss, change the environment still more, and thus pave the way for woody shrubs and pine trees. The first trees grow rapidly and shade out the grass. Soon, they offer enough shelter so that oaks and other shade-tolerant trees can grow beneath them. Ultimately, the giant trees will shade the sun-loving pines and eliminate them from the forest.

sere sequence of plants and animals that occupy a site one after another

And so it goes, until a type of vegetation appears that can reproduce under itself and is adapted so well to the general environment that no species has a competitive advantage over it. The king-of-the-mountain kind of life is not the same everywhere. In dry regions, for example, succession may stop with grasses or small shrubs because there is not enough water to meet the requirements of tall trees. In extremely cold places, succession may not progress beyond the hardy lichens. The local environmental conditions are intolerable for larger forms of life even after the lichens have altered their surroundings as much as they can.

climax ultimate stage of succession

primary succession succession that begins with fresh rock, sand, new lake, or other previously, unoccupied site

secondary succession succession that begins with an abandoned field, burned area, or other previously occupied site

Succession may also begin with different species, depending on the initial conditions (Fig. 11-8). Lichens cannot accomplish much on the constantly shifting sand of a coastal dune. Succession on a dune usually starts with hardy grasses that spread their roots underground. The root network binds the sand together, and the death of some of the grass adds organic matter to the soil. Shrubs and trees could not tolerate the shifting sand, but they readily come in and take over once grasses have stabilized the dune. Succession then proceeds toward the ultimate kind of life in the local area.

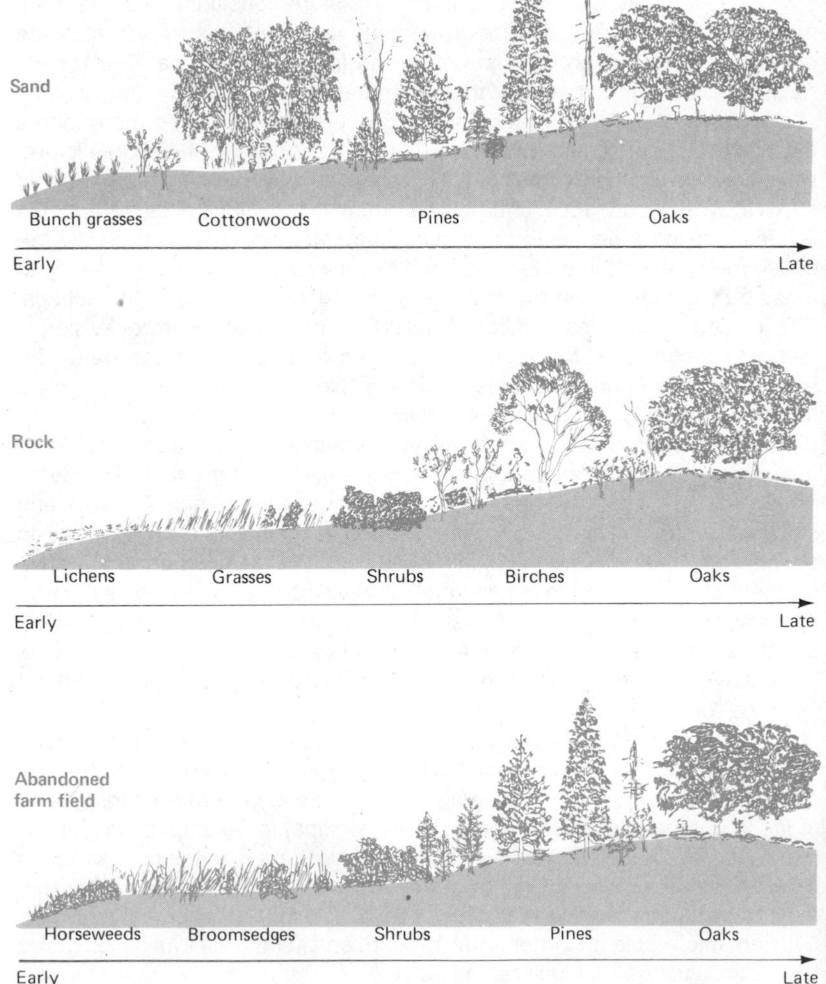

FIG. 11-8. Three kinds of succession under the same climatic condition. Organisms in each stage of succession modify the environment to permit others to enter. The total impact of the living things increases as succession proceeds. Thus the final stage can be identical despite differences in the initial conditions. Different climatic environments, however, will cause succession to proceed in different ways, and chance variations or differences in seed availability can yield different patterns of succession, even in the same general environment.

SUCCESSION AS ENERGY OPTIMIZATION

A sizeable amount of energy is wasted when lichens are the sole inhabitants of a surface in most parts of the world. Available energy is more than sufficient for their needs; the primary concern of lichens is survival in the face of harsh conditions. From an energy point of view, pioneers cannot afford to worry about individual growth and energy efficiency. Rather, they emphasize population growth as if they were trying to outnumber the potential causes of death.

***r*-selection** favoring of organisms that reproduce abundantly; most pioneer species are *r*-selected

As succession progresses, more and more kinds of life enter a local area, partly because conditions are more favorable than before and partly because the number of available places to live is increasing. The bare rock offered only sunny, hard places for prospective occupants. A blade of grass creates several microenvironments that are habitable for specialized plants and animals. One kind of tiny plant or insect may prefer the shady areas on the rock under the grass. Another may be adapted to the windy and sunny places on the top of the grass, a third to the windy and shady places on the undersides of the leaves, and so on. One insect may specialize in eating only shade-dwelling plants, while another learns to like windy places and the inhabitants there. In this manner, life becomes more diverse as succession proceeds.

niche ecological role of an organism—its food, habitat, and organisms that prey on it

diversity number of different kinds of living things that are important in a local area

Diversity of living things has three general consequences of far-reaching significance:

1. A diverse ecosystem uses energy efficiently. Various plants and animals specialize in order to avoid competing directly with each other for the same things. Competition leads to weakness, injury, or death, whereas specialization allows energy to be used in more productive ways. Shade-tolerant plants live on the minute amounts of sunlight that filter through the tree leaves. Wind-tolerant plants make better than average use of sunshine in a windy place, because they do not have to spend much time repairing wind damage. Their wind-resistant bodies, however, put them at a disadvantage in quiet nooks where they yield to flimsier but taller and faster-growing varieties. Ideally, each bit of space in a diverse area supports a form of life that is most suited to the local conditions and thus able to concentrate on using the available energy with the least waste.

competitive exclusion tendency for competition to eliminate at least one of the competing organisms from an area

2. A diverse ecosystem is able to withstand stress. A sudden cold wave can be a catastrophe in an area occupied by a single kind of organism that cannot tolerate cold. Diversity is a form of insurance against excessive casualties due to weather fluctuations, disease, predators, or other stresses. Different species have different tolerances, so that a single environmental stress rarely affects more than a small proportion of the whole ecosystem. Moreover, a complex ecosystem affords many different kinds of shelter from environmental extremes. Finally, a heat wave that stresses one organism may actually help another one that finds the average temperature a bit low.

3. A diverse ecosystem has a great mass of living things. Specialized plants and animals devote much energy to building permanent structures that enable them to use energy efficiently. For example, a woody trunk helps

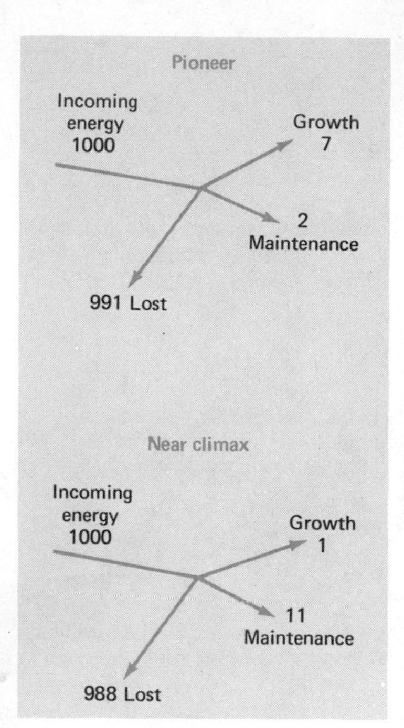

FIG. 11-9. Energy budgets of pioneer and climax vegetation. Later stages of succession use more energy for maintenance and less for growth than earlier stages do. Neither stage, however, uses more than about 1% of incoming energy (see p. 22).

to save energy because it holds leaves high in the sky and does not have to be rebuilt every year. An insect-attracting flower conserves the energy needed to manufacture an immense amount of pollen and scatter it in the wind. Burrows, fur coats, cactus barrels, and expendable leaves are all more energy-efficient than just allowing the whole organism to die in short unfavorable seasons. Eventually, however, the sheer mass of living tissue becomes so large that much of the incoming energy is used simply to maintain the structure of organisms. Growth is just barely sufficient to keep up with the occasional but inevitable deaths of parts of this large quantity of living matter.

The structural complexity of late stages of succession suggests another way of recognizing the final stage: at climax, the biomass is so great that maintenance and replacement consume all of the usable energy or materials (Fig. 11-9). Energy and materials left over for growth would be proof that succession has not reached its ultimate equilibrium. There can be no surplus for growth when the king-of-the-mountain finally arrives.

P/R ratio production of a community divided by its respiration (consumption); pioneers have high P/R ratios, whereas climax species have P/R ratios near unity

POPULATION REGULATION

There is a considerable amount of abstract truth in the concept of succession as a steady pilgrimage toward some "Promised Land" of maximum diversity, efficiency, and stability. The actual progress of succession, however, is rarely smooth and does not necessarily lead toward a predictable goal. Biomass and diversity fluctuate widely as populations of organisms enter, flourish, and die out along the way. One species arrives, proliferates for a while, and then is beaten back or eaten, reducing it to a moderate population (or even to extinction). Another species sneaks in, bides its time until conditions improve or a competitor or other natural enemy disappears, and then increases in numbers and assumes a dominant role. Still another finds an abundant source of food, multiplies rapidly, and literally eats its way to a low final population by consuming too much of its food supply. In effect, inanimate environmental conditions put a ceiling on the total biomass that can be supported in a given area, but relationships among living things are responsible for fixing the populations of each species.

Self-control is one of the most significant of these relationships. Members of the same species have similar needs and tolerance curves and therefore inevitably compete with each other if they become too numerous for the available energy, water, nutrients, and space. Competition within a species leads to stunted individuals, and stunting weakens the species and makes extinction more likely. Most higher forms of life seem to avoid crowding by developing some form of internal population control (Fig. 11-10).

FIG. 11-10. Apparent population control mechanisms.

THE GEOGRAPHY OF LIFE

The principles of tolerance, succession, and population regulation can now join the familiar patterns of sun energy, climate, soil, and terrain in an effort to explain the distribution of living things on earth. The presence of a certain organism in a particular place means four things:

1. It can survive there. No condition of the local environment is beyond the tolerance limits of the organism.
2. It can put up with its neighbors. There is no competitor that is too efficient and no predator that is too effective.
3. It was able to get there. An island may lack rabbits, a lake may have no fish, or a mountain may not support trees, but that does not necessarily mean that the organisms could not survive there. They simply may not have been able to cross the unfavorable areas between their homes and the potential site.
4. It can get enough energy to support its population.

primary producer (autotroph) green plant

herbivore plant-eating animal (see p. 24)

carnivore animal-eating animal

This final requirement is the most subtle. It is not very significant for plants, but it is often important for herbivores and quite critical for carnivores. The sunlight that strikes a hundred square meters of land in Ohio is adequate for one oak tree and a carpet of grass. Acorns from the tree can support seven squirrels, just enough to maintain the squirrel clan indefinitely if no catastrophes enter the picture. A mid-sized horde of insects and worms can also survive on the tree and grass. In turn, a family of small insect-eating birds is an ecological possibility: 200 kg of annual growth of tree and grass will keep 20 kg of insects alive and thus support about 2 kg of birds. However, a population of 20 to 50 squirrels and small birds is needed to support one fox or hawk. The total production of green plant material on the hundred square meters of land is not sufficient to feed even one top carnivore, much less keep a whole population alive (Fig. 11-11). Unless there are more oak trees or other primary producers and therefore more herbivores within easy commuting distance for the foxes or hawks, the carnivores probably will not appear on the little plot of land.

The lack of predators removes a vital check on the population of the squirrels. Throughout their long family history, squirrels have been hunted by other animals, and have responded to the pressure by learning to climb, run, hide, and breed more rapidly than they would need to in the absence of predators. It is not easy to forget these painfully learned traits of squirrelhood. Therefore the animals continue to dart about nervously, hide in the branches of the tree, and produce large families. In turn, their large families consume more than their annual share of acorns, leaving less for their children or for the next generation of oak trees. The long-term health of the green plants in an area often depends on the number and kind of carnivorous inhabitants. Carnivores impose a vital check on the populations of herbivores and thus keep plant material from being eaten faster than it can be replaced.

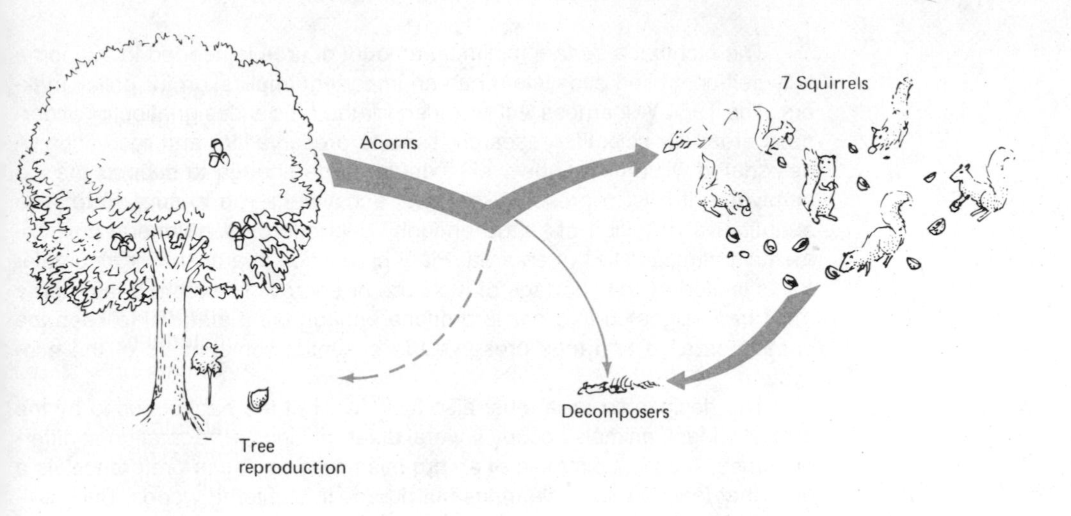

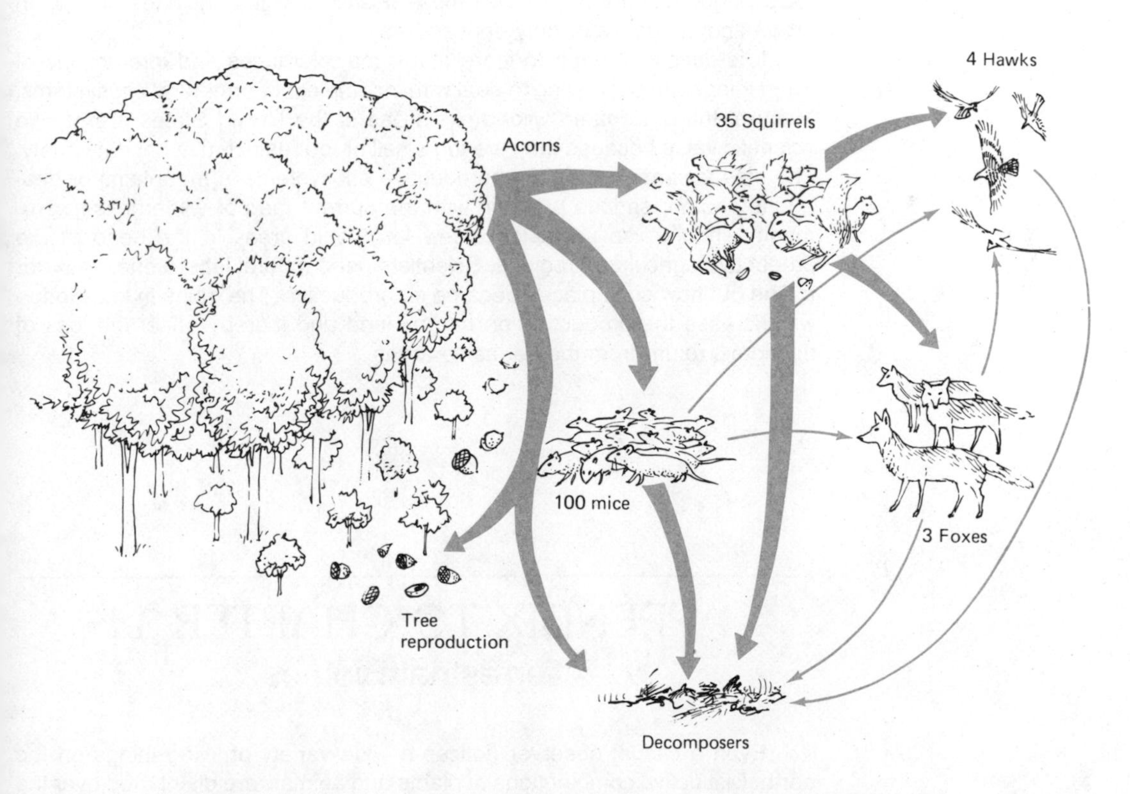

FIG. 11-11. Energy flow and population sizes of two forest food chains: (top) an isolated tree and (bottom) a sizeable forest.

CARNIVORES, ENERGY, AND WILDERNESS

The fact that a certain minimum amount of area is needed to support a fully self-controlled ecosystem has an important implication for policy makers. The 1964 Wilderness Act provided for the official designation of wilderness areas for scientific research, species preservation, and recreation. A designated wilderness, however, must be large enough to support the top carnivores if it is to preserve the local ecosystem in a natural state. The quantitative definition of "large enough" depends on the general environmental characteristics of an area. Plant production in a desert or mountaintop is limited by the shortage of moisture or energy. Therefore a large area must be designated in order to produce enough plant material to keep the carnivores alive and thus preserve the essential components of the ecosystem.

edge organism organism that lives on both sides of the boundary between two or more distinct habitats

The designated area must also include all of the habitats used by the animals. Many animals occupy several different kinds of vegetation at different times. A typical example of an organism with an "edge preference" is a deer that feeds in open clearings but breeds in sheltered woods. Such animals will be hindered by a wilderness designation that includes only open areas above treeline or only dense forests.

It is difficult enough to learn about the tolerances and interactions of living things without having to deal with incomplete or unnatural ecosystems. Many of the designated wilderness areas of the United States are of little scientific value because they are too small or too restricted in habitat variety. The bias toward scenic but unproductive ecosystems in mountains or deserts is another serious problem with the current map of wildernesses. Humanity should also set aside a few large wild areas in the heart of the productive agricultural regions. Scientists need natural laboratories in order to find out how such places became so productive. The gains in knowledge will increase the production on nearby land and thereby offset the loss of economic return from the set-aside areas.

APPENDIX TO CHAPTER 11
Terrestrial Biomes

Even a casual observer notices a wide variety of living things on the earth. Distinctive combinations of plants and animals are distributed over the continents in a manner which makes geographic sense. Organisms respond to their surroundings, which means that the regularities of climate and soil influence the distributions of living things. This appendix contains descriptions of eight major terrestrial biomes (communities of plants and animals). The analysis of life in each biome emphasizes four features:

1. Physiognomy, the general shapes of living things.
2. Productivity, the amount of new living material produced each year.
3. Diversity, the number and importance of different kinds of organisms.
4. Economic usability, the capacity to produce crops, forest products, or wildlife.

The eight descriptions are like snapshots of particular places at particular times. As such, they fail to communicate three key characteristics of the pattern of life on the earth surface:

1. The geography of life is one of gradual transitions that are sometimes so smooth that any selection of "types" is arbitrary. A Pennsylvania forest is strikingly different from a Kansas grassland, yet the "oakwoods" of Illinois have more grassland than forest species, and an eastern Iowa grassland shares more flowering plants with an Indiana forest than with a western Kansas grassland. For this reason the map of a particular biome will be misleading if the reader assumes that all parts of the biome have exactly the same plant and animal inhabitants.

2. Dissimilar plant and animal communities often occur within a small area because of local differences of solar radiation on different slopes, of moisture content at different elevations, of soil nutrients on different geologic structures, or of seed availability on different exposures. Readers of the world map of a biome should remember that living communities belonging to the biome may occupy favorable sites outside its boundary, while unfavorable locations within the mapped confines of a biome may support different kinds of life.

3. The life in a place does not always represent the theoretically "best" combination of plants and animals for the local environment. The conditions on an island, mountaintop, or valley may be appropriate for a particular group of living things, but the organisms may not be there because they were unable to cross the unfavorable surrounding area.

SELVA ("jungle," equatorial forest, tropical rainforest, broad-leaf evergreen forest)

The equatorial rainforest is one of the most complex assemblages of life on the face of the earth. An area the size of a city block may support several hundred different kinds of trees, arranged in three or more layers of treetops. Climbing vines and parasitic plants also share the territory above the ground. Undergrowth is sparse except near rivers or other openings in the forest, because the dense tree canopy blocks most of the incoming sunlight, and the ground surface does not get enough light for most green plants. In the dark and windless innards of the selva, things stay moist and the air smells of mold and decay.

The selva occupies a region without climatic seasonality; it lacks the grand annual rhythms that characterize other forests. Most of the trees have broad leaves, but the leaf fall is not synchronized by a dry or cold season. As a result equatorial forests usually appear green and lush all year long. Likewise, flowering and seed production occur at many different times of the year. Storage of food and building of shelter are not common habits among the animals.

Animals are scarce near the ground, because the green-plant mass of the selva is concentrated far above the surface of the earth. The scarcity of large ground animals is more than offset by multitudes of tree-living creatures: birds, insects, tree snakes, and monkeys. Large populations of these arboreal animals can exist because the forest is exceedingly productive. In turn, this productivity is based on abundant energy and relative freedom from stress. Frost and drought are unknown in the instability-dominated climate. The sun rises high in the sky each day, but the abundance of moisture for evaporation and cloud formation guarantees that daytime temperatures never go as high as those in North Dakota in the summer.

Indeed, the major stresses on an organism in the selva are those inflicted by other plants and animals. Fierce competition for light compels most plants to grow tall quickly or die. Rapid microbial activity and leaching deprive the soil of organic matter and nutrients. Predators and parasites are numerous because there is no season to force them into dormancy. A great diversity of life arises out of the complex interactions between environment and inhabitant, plant and animal, victim and disease. In turn, this diversity is a form of protection against rapid change. The population of a particular species may fluctuate wildly, but the selva as a whole remains almost unchanged because any one species makes up such a small proportion of the entire ecosystem. Attempts to impose single-crop farming systems in the selva have usually failed because the crops succumb to the stresses of disease, insects, leachable soil, weeds, or predators. Multiple crops, tree farming, or shifting cultivation (burn, farm awhile, and abandon) are the preferred farming practices.

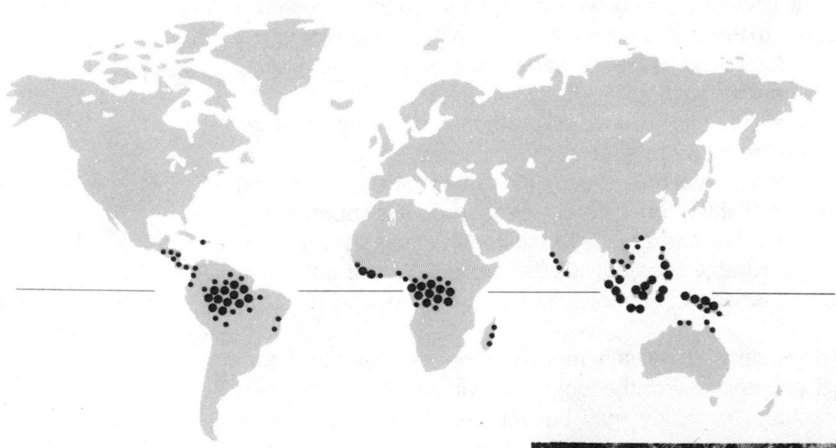

Large dots mark areas of widespread dominance; smaller dots indicate occasional occurrence, transitional mixtures, or scattered stands.

Most selvas have very many different species of plants and animals; annual productivity usually is between 800 and 4000 grams per square meter.

SAVANNA (tropical grassland, thorn forest, scrub woodland, seasonal forest)

Immense herds of large animals thunder across the plain as they seek new forage before the dry season approaches. The retreat of the instability clouds and the arrival of subsiding air mark the "winter" in the tropical savanna. Tropical grasslands lie close to the equator, so that they never experience frost. Nevertheless, the dry season interrupts growth and forces plants into temporary dormancy. The longer the dry season, the larger the number of equatorial tree species that find conditions in the savanna intolerable. The remaining trees usually scatter themselves across the landscape, leaving wide spaces for grasses and other ground vegetation. Still farther into the subsidence zone, trees become rarer and grasses shorter. Productivity and diversity decrease as the drought period lengthens.

Grassy expanses and a long dry season are an ideal combination for fires, another essential feature of the savanna. The fires help grazing animals by destroying plants that are thorny, woody, old, unpalatable, or otherwise uneaten. The burning of ungrazed plants allows all species to start out the next season on an even basis. In the absence of fire, animals would soon eat all the delicious plants and leave only the less acceptable ones to reproduce.

The savanna is much less productive than the selva, but the plant material is fresh every year and near the ground, within reach of large herbivores. Despite a common movie misconception, the savanna, not the rainforest, is the domain of the zebra, impala, wildebeest, rhinoceros, giraffe, elephant, and other large plant eaters. The availability of large prey makes life tolerable for large carnivores, such as lions, tigers, and hyenas. Many of the large animals exist in large groups with elaborate social organizations. Herbivores seem to think there is safety in numbers, whereas carnivores find that hunting in packs offers more success.

For obvious reasons, many savannas are used as grazing lands for herds of domesticated animals. Some West African towns have served as regional cattle markets for thousands of years. Large herds represent food, wealth, and prestige for people in dry savannas, where crop farming is risky and unprofitable. Nevertheless, some ecologists argue that harvesting the diverse natural game may be even more productive than raising a few species of domesticated animals. The wild animals have fewer problems with disease or predators and make better use of the range by eating a wider variety of plants than domestic cattle do. Both cattle ranching and game farming, however, must exist within the productivity limits of the savanna. The long dry season makes overgrazing a serious threat to continued use.

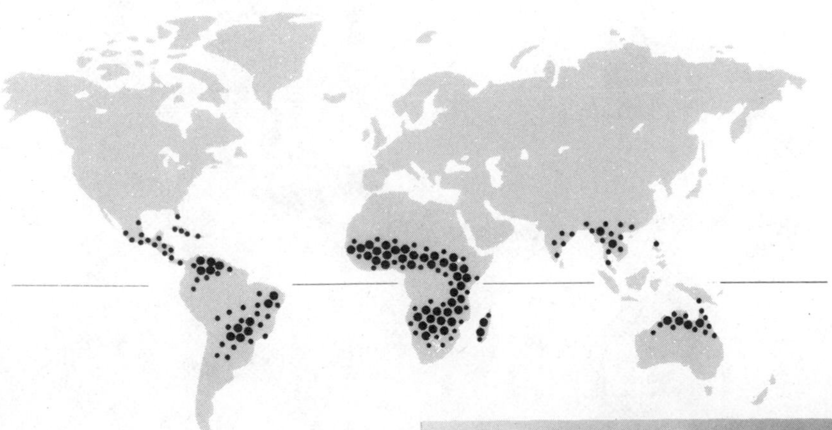

Large dots mark areas of widespread dominance; smaller dots indicate occasional occurrence, transitional mixtures, or scattered stands.

Most savannas have many different species of plants and animals; annual productivity usually is between 200 and 3000 grams per square meter.

DESERT (succulent desert, shrub or scrub desert, cactus forest)

Scarcity of water is the overriding fact of life in a desert. The stifling heat that characterizes many mental images of deserts is in part a consequence of dryness. Energy that cannot be reflected or used for evaporation must heat the environment. Dusty winds and stark landforms occur because rainfall is too scanty to support a continuous vegetation cover.

Unlike the equally stark tundra, however, the desert imposes only one great stress on life. Organisms can use a variety of strategies to combat a single environmental adversary. Some plants cope with dryness by means of structures that minimize the loss of water. Examples of such structural adaptations include thick leaves, green stems, spiny leaves, waxy coatings, and leaves that fold up during the hot parts of the day. Other plants try to avoid drought by completing their entire life cycle soon after the soil is moistened by one of the infrequent rains or floods. The key to the success of this strategy is a seed cover that dissolves only when the soil becomes wet enough to support a full cycle of life. Another way to tolerate dry climates is typified by shrubs that are bare of leaves except during a few weeks after a rain. A fourth strategy of desert living is water storage, exemplified by the barrel-shaped cactus, which is the symbol of desertness for many people. (Such a structure, however, is doomed to failure in places where temperatures go far below the freezing point.) A fifth means of coping with dry conditions is to give off chemical poisons that eliminate potential competitors. This form of territoriality lets each plant get enough water. Finally, there are plants that live on dry sites but use extremely long roots to obtain water from deep in the soil, perhaps beneath a nearby stream valley.

Animals use similar techniques of desert survival: loss prevention (lizards and insects), rapid life cycles (insects and soil animals), avoidance (nocturnal rodents and snakes), storage (camels), and territoriality (birds and coyotes). In addition, the mobility of animals allows seasonal or daily migration to get water. The occasional waterholes in a desert have a significance far out of proportion to the area they occupy. In fact, plants may find living near waterholes intolerable simply because of the many animals that pass by every day.

In summary, there are many ways to cope with dryness. Hence, there are many kinds of deserts: cactus "forests," thorn shrublands, shrub deserts, and seemingly barren places that burst into bloom after a rain. The drier a place is, the less life it can support and the fewer alternatives it can offer its inhabitants. The moist margins of a desert can support moderately successful grain farming or livestock raising, whereas the very dry core areas are able to sustain only sparse populations of nomadic herders, except in occasional oases of intensive crop and tree farming or on the artificially watered soils of irrigated areas.

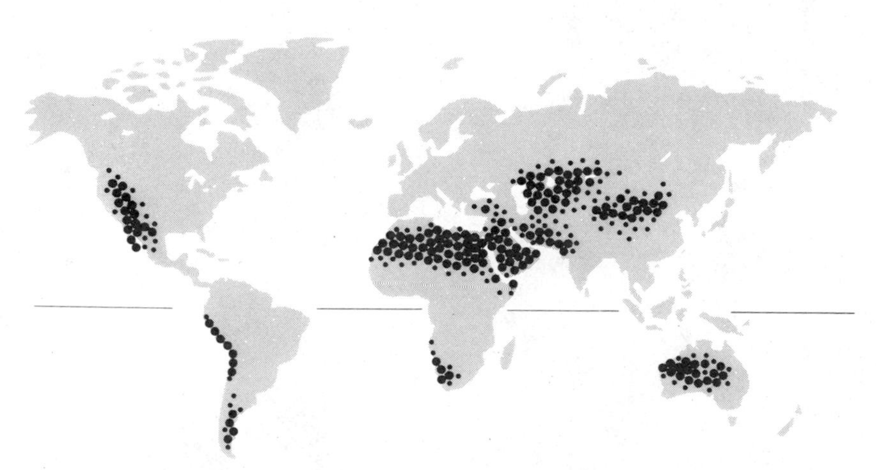

Large dots mark areas of widespread dominance; smaller dots indicate occasional occurrence, transitional mixtures, or scattered stands.

Most deserts have fairly many different species of plants and animals; annual productivity usually is between 10 and 200 grams per square meter.

CHAPARRAL (Mediterranean shrub, maquis, garrigue, sclerophyll forest)

The mild winter is the productive season in the dry-summer climate between the subsidence and the maritime frontal regions. Tall trees cannot tolerate the dry summer. The dominant shrubs have small leathery or waxy leaves that prevent excessive water loss during the dry season. Most of these shrubs are able to sprout up again after fires, which occur frequently in the dry months. Dead and dry plant material accumulates during a fire-free year and increases the likelihood of intense fires the next year. Therefore, allowing frequent fires is preferable to fire suppression in many parts of the chaparral biome.

Chaparral occupies a relatively small part of the world but is of considerable economic significance. The dry summer interrupts many disease and insect cycles. Places like California, southern Europe, South Africa, or the Black Sea coast of Russia have reputations as health resorts. The mild and moist winters permit crops to grow when most mid-latitude countries are snowbound. Like the grasslands and their grain crops, the chaparral zones gave the world a variety of domesticated plants typical of their environment. Among the products of cultivated shrubs and small trees from the chaparral are olives, figs, citrus fruits, cork, and grapes.

Permanent animal inhabitants of the chaparral have many of the characteristics of desert animals. Burrowing seedeaters, such as chipmunks and ground squirrels, are common. Many of them obtain needed water from the food they eat. They are adapted to conserve moisture, instead of indulging in wasteful humid-country habits, such as perspiration, dilute urination, or respiration through moist noses and mouths. Nocturnal habits and aestivation (resting) during the dry summer are other ways to avoid excessive loss of moisture.

On the dry margins of the chaparral, competition for water becomes fierce. Many plants wage chemical warfare by exuding toxic substances into the air or soil to keep competitors from getting too close. The result of this chemical activity is the "aromatic chaparral," a unique collection of odor-bearing shrubs. Many of these shrubs have uses as sources of raw materials for the chemical industry.

Westerly winds off the ocean cause the dry season to shrink on the poleward side of the typical chaparral. Redwoods and giant sequoias inhabit this productive zone in the United States; it is a mild region with a short dry season, an occasional fire, and a short cold season. The combination of minor stresses is just enough to exclude many species, while the reduced competition and the general warmth and moisture of the environment allow the remaining plants to thrive.

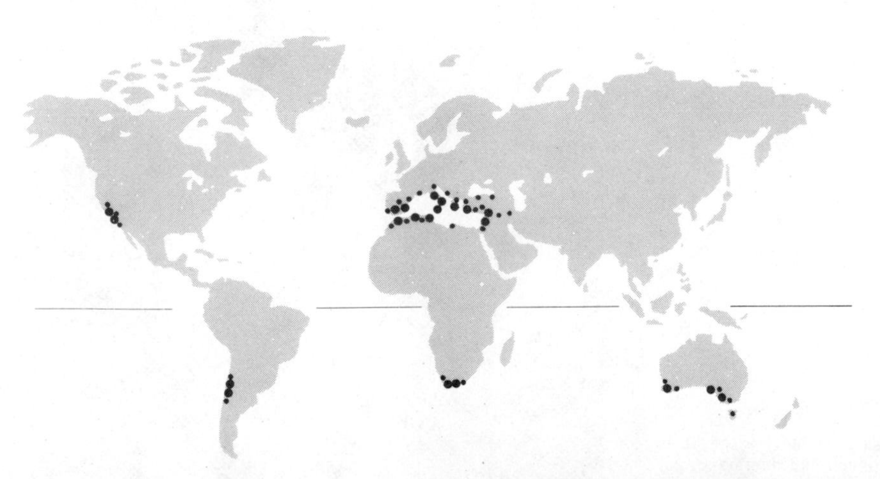

Large dots mark areas of widespread dominance; smaller dots indicate occasional occurrence, transitional mixtures, or scattered stands.

Most chaparral has fairly few different species of plants and animals; annual productivity usually is between **200 and 1600 grams** per square meter.

GRASSLAND (prairie, steppe, mid-latitude grassland, bunch grassland, pampas)

An endlessly rolling sea of grass stretches away to the distant horizon. The grasses are green and lush for a short time in late spring. On a year-long basis, however, the total productivity of mid-latitude grasslands is low. They occupy a frontal climate with a long cold season. Furthermore, they are often in the rainshadow of mountains and far from sources of moisture. Heat and scarcity of moisture limit summertime productivity, while the frozen soil allows no plant growth in winter.

Even this severe combination of weather conditions does not exclude trees and other woody plants. Trees in wet river valleys stay green and fairly fireproof, but the existence of moisture-loving trees on dry sites is not so easy to explain. The best available clue to the mystery seems to be the position of dormant winter buds on various plants. The buds of trees and other woody plants are exposed on the ends of twigs, whereas many grasses bury their growth structures underground. Thus fire is likely to kill woody plants while grasses survive. Rocky hills or steep slopes do not produce enough plant material to support a hot fire, and woody plants that cannot tolerate fires often find refuge on the dry hillsides. One form of backhand evidence for the idea that fire is partly responsible for the formation of natural grasslands is the fact that woody plants quickly invade many grassy areas when fires are suppressed.

Most grassland animals are either migratory, burrowing, or short-lived. Large herbivores such as elk, bison, and antelope roam the humid margins of the open area, where the grass is tall and nearby trees afford shelter from wolves and other migratory carnivores. Gophers, prairie dogs, weasels, snakes, and many other small animals live part of their lives underground, protected from the heat, cold, and grass fires in the heart of the steppe. Other animals live their entire life cycle during the favorable seasons. The most notorious of these short-lived creatures is the locust, a particular curse near the dry margins of the grassland.

It is difficult to study the complex interactions of climate, fire, grasses, migrants, burrowers, and insects on natural grasslands. Fences, grainfields, and pastures for domestic cattle occupy most of the former steppe. Farmers are drawn into the grasslands by the rich soil and occasionally driven away by the unreliable rainfall. The few remnants of natural grassland are isolated, cut off from the animal migrations and fires that formerly were such crucial parts of their essence.

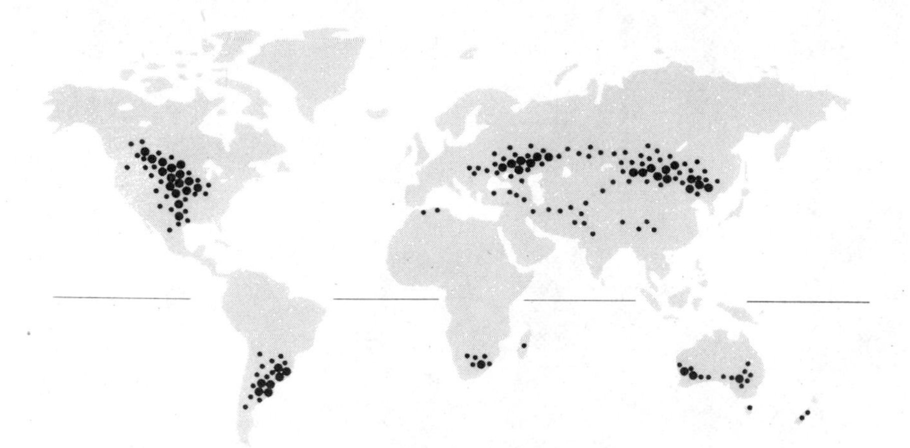

Large dots mark areas of widespread dominance; smaller dots indicate occasional occurrence, transitional mixtures, or scattered stands.

Most grassland has fairly many different species of plants and animals; annual productivity usually is between 100 and 1600 grams per square meter.

HARDWOODS (deciduous forest, mid-latitude broad-leaf forest, summergreen forest)

A single mountain range in western North Carolina contains more tree species than are found in all of Europe. The mountains are far enough from the pole so that they were not covered during the Ice Age. The southern Appalachians enjoy a humid frontal climate with moderate temperature and abundant rainfall. The summer sun is more direct and summer days are longer in North Carolina than at the equator. As a result, summer productivity is greater in the hardwoods than in the selva. Wintertime, however, brings a nearly total interruption of plant activity.

Most of the plants in the deciduous forest use a sophisticated procedure to measure the length of time the sun stays above the horizon each day. When days become shorter than a critical length, many trees interrupt the flow of moisture and nutrients to their leaves. The leaves, in turn, dry out and often fall to the ground before subfreezing weather can damage the trees. The autumnal display of colored leaves is one of the striking characteristics of the mid-latitude deciduous forest.

The seasonal synchrony of leaf-set, flowering, fruiting, and leaf fall imposes its rhythm on the other inhabitants of the forest. Ground vegetation flourishes mainly in the spring and autumn, when sunlight can pass through the leafless tree canopy. Seed-eating animals must employ a storage system to provide for the months when seed fall is scanty or nonexistent. Deer and other large animals do better near the edges of the forest, where trees provide shelter but also where edible ground vegetation is available all year.

The core of the deciduous forest is quite diverse and nearly as productive as the equatorial selva. Productivity and diversity decrease where the growing season is short or moisture is scanty. Toward the pole, the deciduous forest grades into the evergreen taiga. The efficiency of broad leaves is sacrificed for the durability of short needles. Toward the centers of the continents, a progressively decreasing supply of moisture eliminates one tree species after another, until only a few drought-resistant and fire-tolerant trees remain scattered among the grasses.

The transition zone between forest and grassland in the middle latitudes is one of the most valuable natural environments from an agricultural point of view. Grain crops thrive on the abundant moisture at the wet margins of the grasslands. The soil is more fertile in the transition zone than in the rainy forest. The long freezing season interrupts biological activity, aids in food storage, and keeps pest populations low. The intense sun and long days of summer provide more energy for surplus plant production than is available at any other place on earth.

Large dots mark areas of widespread dominance; smaller dots indicate occasional occurrence, transitional mixtures, or scattered stands.

Most hardwoods have many different species of plants and animals; annual productivity usually is between 800 and 2400 grams per square meter.

TAIGA (boreal forest, coniferous forest, needle-leaf forest, sub-Arctic forest, subalpine forest)

Viewed from the air, the vast boreal forest is a monotonous expanse of dark green. A few species of needleleaf trees dominate the canopy, which is often dense and tall. The evergreen trees capture most of the incoming light, and few species can tolerate the gloom near the ground. Inside the forest canopy is a dense cathedral stillness: wind barely penetrates the trees, and stray noises are hushed by a deep carpet of mosses and partly decayed needles on the forest floor. The relative humidity of the air is high, mainly because the moisture-holding capacity of the cool atmosphere is low.

The long winter of its frontal-polar climate makes conventional agriculture very difficult and limits the total productivity of the needleleaf forest. Animals are scarce because undergrowth is sparse, the trees are old and generally unpalatable, and the forest as a whole is unproductive. The transition between the taiga and tundra usually has more animal inhabitants than either biome alone. Moose and caribou live in the shelter of the scattered clumps of trees and eat the low plants in the open area. The shrubs of the tundra may produce less plant matter per year than the needleleaf forest, but a greater proportion of the annual growth is accessible to animals on the ground.

The sheer immensity of the needleleaf forest gives the impression of great productivity, but the slow rate of annual growth exposes the illusory nature of that wealth. From an economic point of view, the trees in much of the taiga can be mined but not farmed. If the money it costs to plant trees was deposited in a savings account in a bank, the interest on the savings account would be worth more than the trees by the time they are finally ready to harvest.

The low diversity of the taiga poses still another ecological problem with economic overtones. The food supply lacks variety since there are only a few tree species and little undergrowth. The scanty menu, in turn, restricts the number of plant-eating animals. An ecosystem with a limited number of species is subject to wide fluctuations in the populations of individual species. For example, a beetle outbreak may reach epidemic proportions and defoliate millions of trees before natural forces can bring it under control. The infrequent but apparently all-consuming fires, windthrows, insect attacks, or diseases often look like absolute catastrophes to the untrained observer. However, the ecologist recognizes them as part of the essence of the taiga because they are necessary for the long-term health of the forest and are indispensable for the survival of the ground animals.

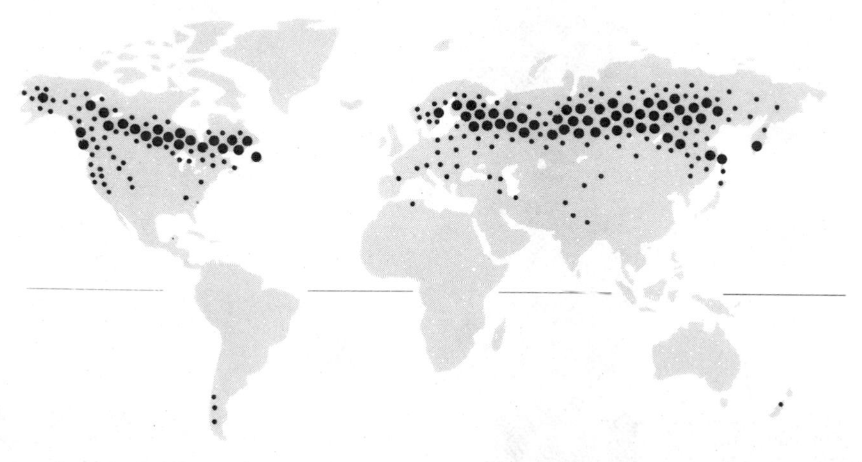

Large dots mark areas of widespread dominance; smaller dots indicate occasional occurrence, transitional mixtures, or scattered stands.

Most taiga has few different species of plants and animals; annual productivity usually is between 200 and 1600 grams per square meter.

TUNDRA (Arctic grassland, heathland, moor, willow scrub, Arctic barrens, cold desert)

Intense stresses are the major facts of life in the Arctic tundra. Temperatures remain well below the freezing point for most of the year. Occasional frost can occur even during the short summer, although daytime temperatures may go above 30°C (86°F). Precipitation is scanty, limited by low air temperature subsiding air, and great distance from sources of moisture. Winds, often armed with crystals of ice or sand, sweep across the unprotected ground. The soil itself remains permanently frozen except for a shallow surface layer that may thaw for a few weeks. The combination of bitter winter cold, occasional summer heat, lack of rain, abrasive wind, and frozen soil makes life impossible for most organisms.

The most successful strategy for coping with such an inhospitable environment appears to be avoidance. Many plants avoid the rigors of winter by completing their life cycles quickly: they go from emergence through flowering to death within a few days or weeks. Insects appear in immense numbers during the short period when the surface soil thaws and the flat areas of the tundra are transformed into vast swamps.

Other animals prefer migration as a means of avoiding the unbearable seasons. Birds fly into the tundra by the millions during the summer, when they can feast on the abundant plant and insect life. Land animals move from the forest to the tundra and back as the seasons change. Even the coastal mammals, seals and walruses, move out into more favorable parts of the ocean during severe seasons.

A few animals and plants have made special provisions to tolerate the winter. Most of the half-meter high "trees" and other long-lived plants hover near the ground and hide beneath the meager snow cover for protection from the icy winter winds. Thick coats of fur protect some animals against a variety of threats: cold, wind, water loss, and insects. Elaborate social arrangements ensure the continuation of some species, particularly the carnivores that must roam great distances to find their prey. Small animals construct extensive tunnel systems that link their sleeping quarters with food storage areas beneath the snow.

Equatorward-facing slopes in the tundra are often islands of greater-than-average plant productivity. Poleward-facing hillsides, on the other hand, are characteristically rocky and sparsely vegetated. Where life teeters on the brink of existence, minor differences in energy, moisture, and nutrients become very significant. The tundra as a whole is moderately productive during the long summer days. The total annual production of this essentially uninhabited region, however, is very low because the summer is too short.

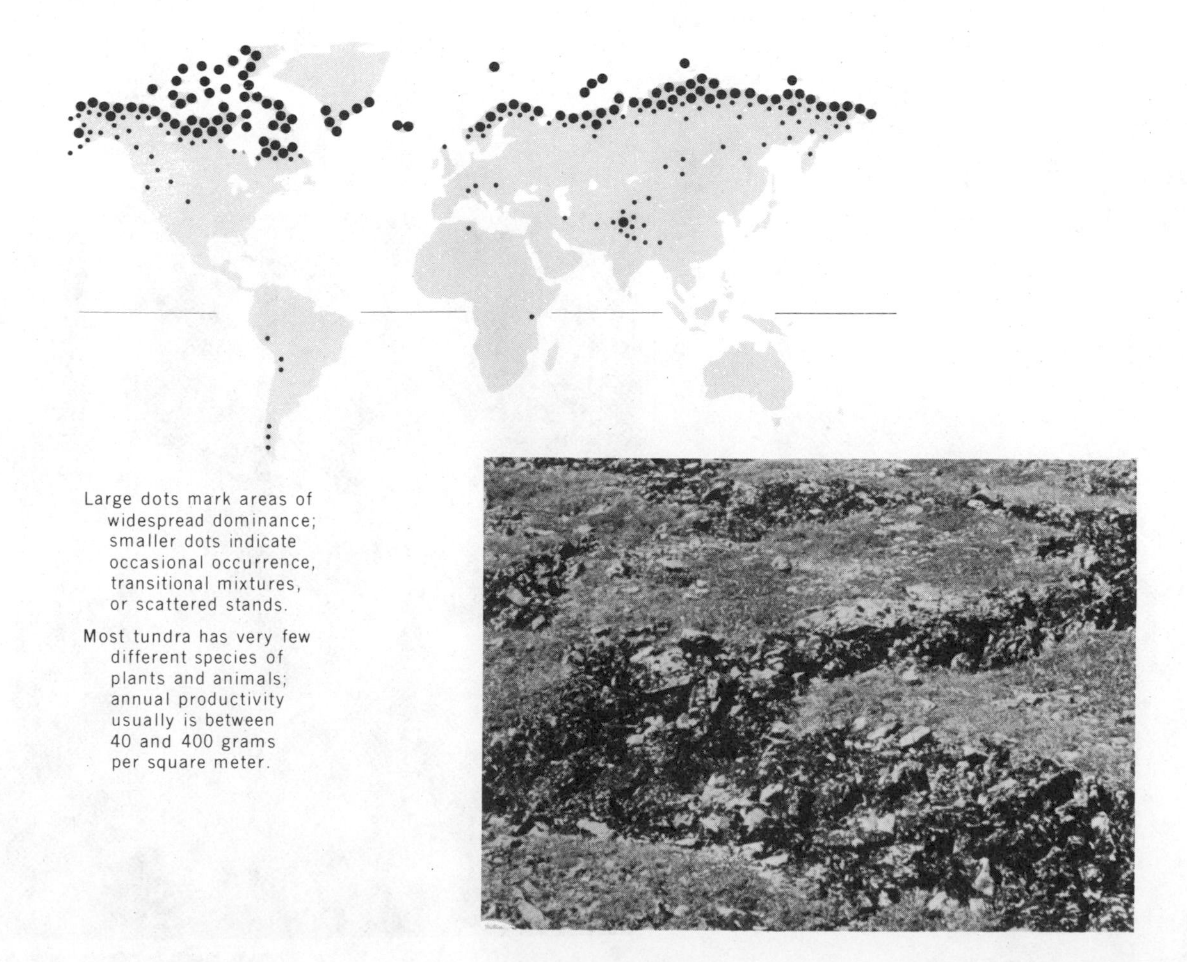

Large dots mark areas of widespread dominance; smaller dots indicate occasional occurrence, transitional mixtures, or scattered stands.

Most tundra has very few different species of plants and animals; annual productivity usually is between 40 and 400 grams per square meter.

Chapter 12

Biogeochemical Cycles: The Balance of Nature

You cannot make a nutrient; you can only try to move it around to where it will be most useful to you.

John W. Busch, Soil Conservation Service

The dry season wore on and the landscape became parched. Thin and listless cattle lay in the dust near the shrinking waterholes. On what seemed to be the hottest and driest day, the villagers carried coals from the council fire to the base of a nearby hill. After a brief incantation, they ignited the dry grass, and soon the entire hillside was engulfed by crackling flames and billowing smoke. This seemingly destructive ritual, performed by hundreds of "primitive" societies for thousands of years, illustrates a facet of environmental knowledge that technological societies have "discovered" only in the last few decades: fire is one of many natural ways of moving plant nutrients back to the soil where plants can reuse them.

Each region on the earth has a finite amount of energy, moisture, and nutrients to support life. The wise use of the three necessities involves three different kinds of budgets:

ecosystem a local environment and its living inhabitants, linked together by transfers of energy, water, and nutrients

1. Energy is difficult to store and therefore easily lost. Fortunately, the sun provides a daily income for most places on earth. The task of an energy user is to try to make efficient use of a limited but reliable allowance (as discussed in Chapter 1).
2. Some water can be stored in the soil and plants. The storage is like a moderate savings account that can tide an area over short dry periods between paychecks (rainfalls). The management problem is to arrange the use of water so that the savings account is not prematurely depleted (to be discussed in Chapter 14).
3. Nutrient wealth of a place is nearly all in the form of savings, because the annual nutrient income of most places is very small compared to the amount of nutrient used every year. The circulation of nutrients in nature is an exchanging of existing wealth rather than a budgeting of steady income. The rules of the exchange, however, are different in different places; they form the subject of this chapter.

THE NUTRIENT CYCLE AS A CLOSED SYSTEM

Picture a grove of trees as a closed box and carefully count all of the plant nutrient material in the box at different times. Most of the individual plant nutrients seem to be constantly in motion, yet the total quantity of nutrients in the box remains fairly constant. The realization that mature forests are in a kind of long-term nutrient equilibrium is one of the cornerstones of modern ecology.

Arbitrarily putting some interior walls in the boxed forest makes it possible to study nutrient movements in a more detailed manner. A particularly rewarding way to subdivide a forest is to separate it into three parts: inanimate soil, living plants and animals, and dead organic matter lying on the ground (Fig. 12-1).

compartment arbitrary subdivision of a system

compartmental modeling mentally picturing a complex system as a set of distinct compartments connected by pathways for movement

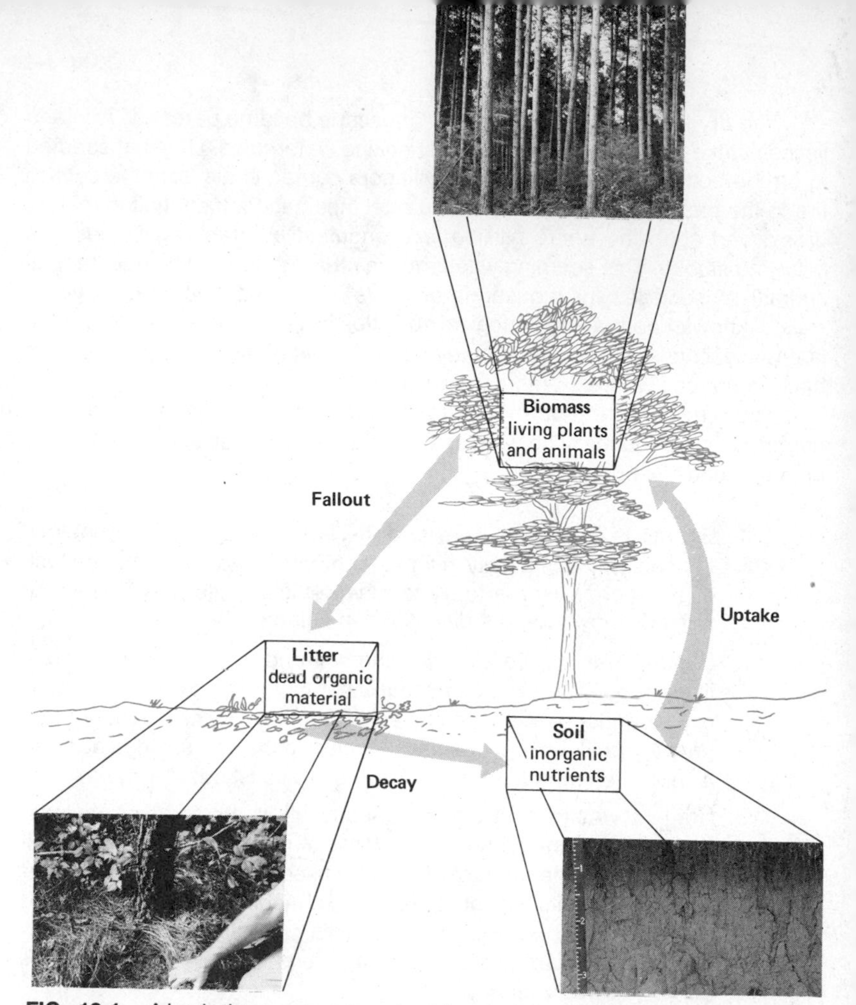

FIG. 12-1. A basic three-compartment nutrient cycle model. The division of a natural system into three compartments is an arbitrary choice, not necessarily more valid or "real" than any other carefully thought out separation of the world. An ecosystem is like an automobile; it consists of a number of parts that all function as the car moves along. Each part operates at its own pace and does its own thing, yet all are connected to one another by the materials or energy they exchange. One mechanic may choose to separate the fuel system of a car into four parts: the tank, pump, fuel line, and carburetor or injectors. Another may combine fuel pump and line into a single transportation unit but prefer to recognize the vapor recovery devices as a separate part of the fuel system. The usefulness of the results is the best judge of the validity of a particular selection of subdivisions for a model of a system. Three compartments appear to be adequate for the general model under consideration in this chapter. Some analysts separate plants and animals into discrete compartments, but as far as the authors are concerned, a squirrel is simply a nut on its way to the ground. Likewise, distinctions between woody stem and green leaves, between undecomposed litter and humus, between exchangeable and nonexchangeable soil nutrients, or between different kinds of nutrients with different behavior are all valid for certain purposes, but they make the basic model too complex for the purpose of this chapter. (Trees, courtesy of B. W. Muir, U.S. Forest Service.) (Litter, courtesy of C. J. DeLand, Soil Conservation Service.) (Soil, courtesy of B. C. McLean, Soil Conservation Service.)

This three-compartment forest has a straightforward internal cycle.

- ***Uptake.*** Nutrients rise from the soil and enter the living part of the system (the biomass) as plants grow.
- ***Fallout.*** Nutrients fall from the biomass to the litter when leaves fall or when plants and animals die.
- ***Decay.*** Nutrients pass from the litter into the soil as organic matter decomposes.

biomass living organic part of an ecosystem

litter dead organic part of an ecosystem

The movement of an individual bit of nutrient is a sequential migration from soil to plant to litter and back to soil. From the perspective of the entire ecosystem, however, the nutrient cycle consists of three distinct nutrient movements that occur more or less simultaneously. Each flow of nutrients apparently depends on the quantity of nutrients in the source compartment. A rich soil can support many plants that take up many nutrients. By contrast, the few scraggly plants in a nutrient-poor soil receive only a meager allowance from the impoverished land. The amount of nutrients in falling leaves depends on the number and size of the trees; big trees drop more leaves in autumn than tiny ones do. Finally, a large volume of litter supports a large number of decay microorganisms, which in turn consume litter and release nutrients into the soil rapidly.

transfer measurable movement of material or energy from one system compartment to another

To see how a system of compartments and transfer rules works, imagine a time interval during which half of the nutrients in the litter move down into the soil compartment. The plants in this example take up 30 percent of the soil nutrients in the same time period, and 60 percent of the nutrients in the plant compartment fall to the litter. (These numbers are arbitrary, chosen to make the illustration easy to follow. A later section will explain how the numbers are determined for a real ecosystem.) As will be shown, this combination of movement rules has an inevitable outcome. No matter how many nutrients are put into the boxes, or where they are placed at the beginning, the system invariably ends up with about half of the nutrients stored in the soil, about one-fourth in the litter, and about one-fourth in the biomass. The outcome is determined as long as the percentages of nutrients forced to leave each compartment during a time period remain unchanged. The only way to alter the outcome is to change the rules that govern the movement of nutrients.

equilibrium state of balance; further observation reveals no significant change
(see p. 15)

Actually, this principle is an extension of an old idea. Chapter 1 interpreted the temperature of an object as a consequence of the rules governing the movement of energy. Chapter 8 noted that the level of the water table in a soil depends on the input and output of water. Chapter 9 emphasized that the organic content of a soil is a consequence of the rules governing production and decomposition. The nutrient cycle, however, differs from these simple flows in one key respect: the output from one nutrient compartment becomes the input to another. Each compartment is thus partly responsible for the nutrient balance of its neighbors. Such a system of mandatory mutual exchange guarantees that any compartment with more than its proportional share of the wealth will have to give away more nutrients than it receives. As long as the rules for movement stay the same, the system always goes to the same equilibrium (Fig. 12-2).

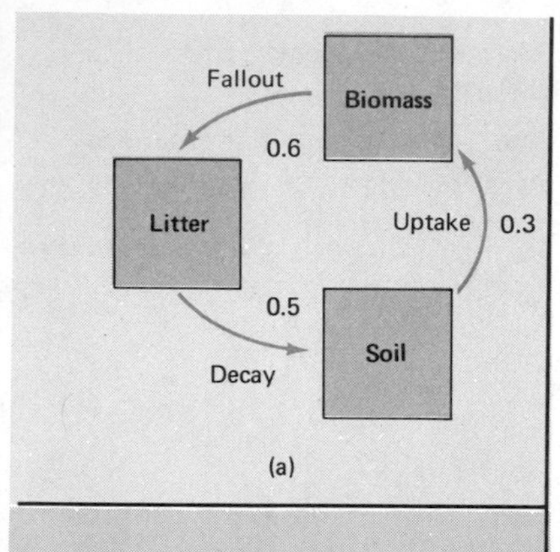

(a)

FIG. 12-2. Equilibrium in a closed system: a poker-chip simulation. To help visualize the workings of a closed three-compartment model of the nutrient cycle, make three boxes and three arrows as shown (a). Write the transfer rules by the arrows. These numbers show how much of the amount stored in a box will move out in each time period. Set up a data table (b) with four headings: Time Period, Soil Storage, Biomass Storage, Litter Storage. Put five chips into each box (c). From this point on, act like a low-memory computer and truncate all calculations at the decimal point. For the purposes of this simulation, 0.3 times 5 equals 1; 0.5 times 5 equals 2; and 0.6 times 5 equals 3 (as does 0.6 times 6, or 0.9 times 4, or 0.7 times 5). Use of whole numbers is essential for the health of the poker chips (d). Count the chips in each box and record the number in the appropriate column of the data sheet (e). Determine how many chips will be forced to leave each box during the next time period. This is done by taking the amount stored in each box and multiplying it by the corresponding transfer rule. Truncate the calculations as described above and put the appropriate number of chips onto the arrow going out of the box (f). Do this for each box. Then, move the chips from all of the arrows into the receiving compartments (g). Repeat the process of counting, determining the number of chips to move, and making the transfers, until further repetition produces no more change (h). This state is called the equilibrium of the system. Test the equilibrium by putting all 15 chips into any one compartment and repeating the simulation (i). Test it again, if you wish, by putting any combination of chips initially into the various compartments. The only way to change the equilibrium proportions is to change the transfer rules. (Note: as long as the simulations use only whole chips, chance variations of a single chip at equilibrium may occur during repeated trials. This problem disappears when a computer is taught to carry the calculations out to many decimal places. Computer simulations proceed inexorably to the same equilibrium as long as the transfer rules are unchanged.)

Time period	Soil storage	Biomass storage	Litter storage

(b)

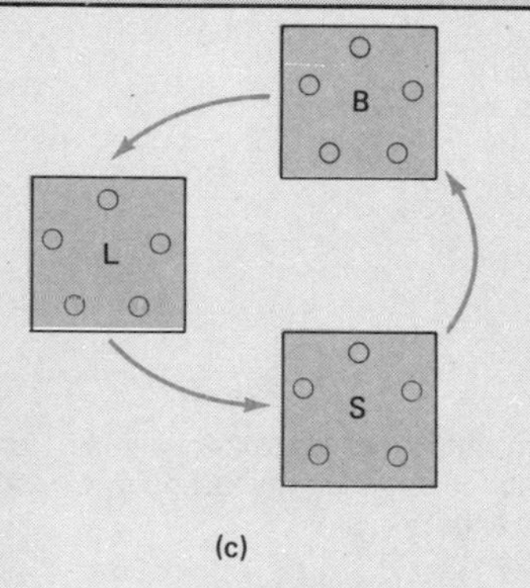

(c)

First time transfers

Uptake = .3 × 5 = 1

Fallout = .6 × 5 = 3

Decay = .5 × 5 = 2

(d)

Time period	Soil storage	Biomass storage	Litter storage
(1)	5	5	5

(e)

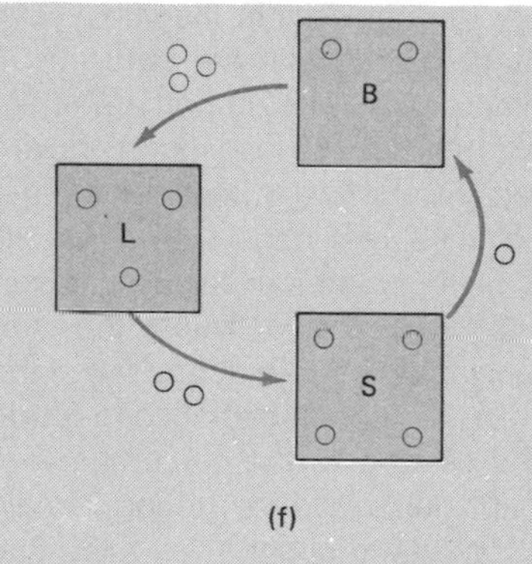

(f)

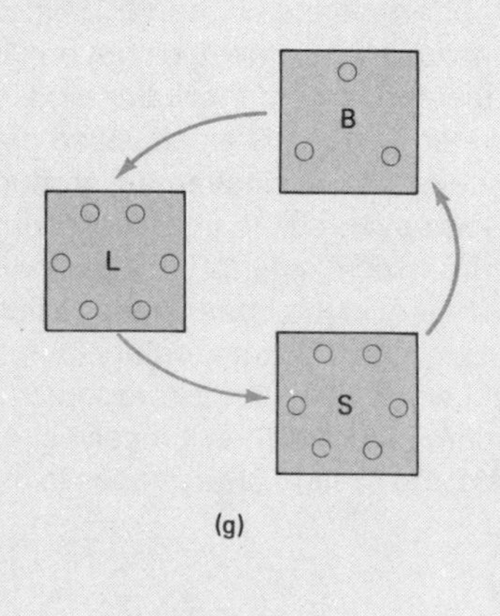

(g)

T	S	B	L
(1)	5	5	5
(2)	6	3	6
(3)	8	3	4
(4)	7	4	4
(5)	7	4	4
(6)	7	4	4

(h)

T	S	B	L
(1)	15	0	0
(2)	11	4	0
(3)	8	5	2
(4)	7	5	3
(5)	6	5	4
(6)	7	4	4

(i)

THE NUTRIENT CYCLE AS AN OPEN SYSTEM

A closed system is easy to analyze but not very realistic. In nature, plant foods are always entering and leaving local areas. Creeks remove nutrients from the land in the form of twigs, bits of soil, and dissolved substances. Nutrients enter as rocks weather and as rain brings impurities. The wind can blow leaves or dust across the arbitrary border around the forest. Animals may graze in nearby meadows and then leave nutrient proof of their existence on the forest soil.

condensation nucleus bit of ice, dust, or salt that serves as a base for raindrop formation

One way to make the model more realistic is to enlarge it by including rocks and clouds and rivers within the system. The enlargement complicates the analysis, but it still does not produce a closed system. The ocean serves as a reservoir for rainwater, a destination for streamflow, and a setting for its own nutrient circulation. Grand convection systems within the earth bring rocks to the surface in some places and bury them in others. The sheer enormity of the system needed to create a closed nutrient cycle has forced most environmental scientists to use two different models:

An immense closed geosystem. This large model of the nutrient cycle involves continental drift and ocean circulation as well as biological processes (Fig. 12-3). The geosystem takes a staggering amount of time to make

closed system system that has no inputs or outputs; all processes are internal

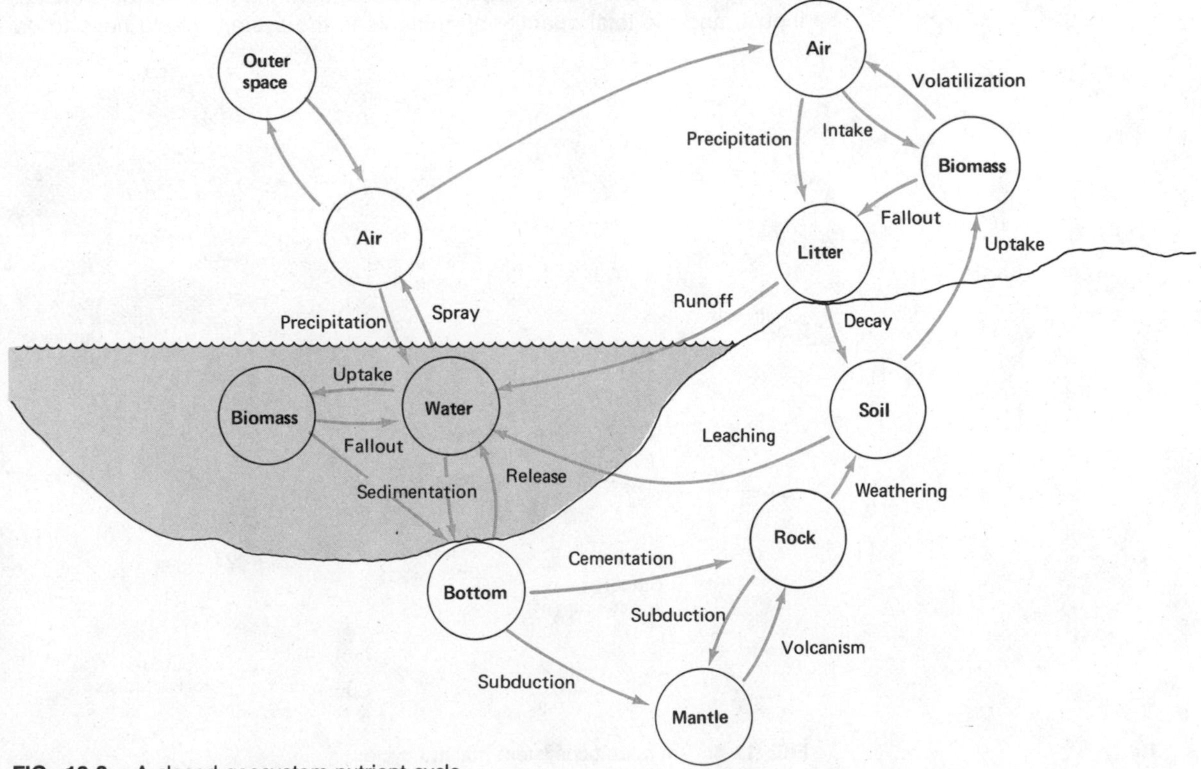

FIG. 12-3. A closed geosystem nutrient cycle.

open system system that has recognizable inputs and outputs as well as an internal circulation

source input to a system

sink output from a system

source-forced transfer a movement whose magnitude depends on the amount of movable substance in the source compartment

its rounds and therefore is understood only in a highly generalized manner. Furthermore, the biologically significant portion of the geosystem is only a minor part of the entire circulation.

An open ecosystem. This small model of the nutrient cycle admits the existence of identifiable and measurable but not necessarily traceable inputs and outputs (Fig. 12-4). Users of this model are content to accept the fact that a certain quantity of nutrients enters the system every time the rain falls. To ask where the wind originated, what field the dust left, or what ocean contributed the salt, would make the problem so big that it would never be solved.

Outputs from an open system, like the internal movements, are governed by their sources within the system (Fig. 12-5). For example, more nutrients can leach out of a nutrient-rich soil than from a poor one. Inputs are also governed by their sources, but the sources are outside of the system and thus beyond its control. The rain falls and contributes its load of nutrients regardless of the size of the plants or the fertility of the soil. Of course, the more nutrients the rain brings in, the richer the system will be.

A local nutrient system cannot become too rich; if it did, runoff from the litter and leaching from the soil would remove more nutrients than rainfall and rock weathering could replace. Outputs from the system would exceed inputs, and the total quantity of nutrients in the system would have to de-

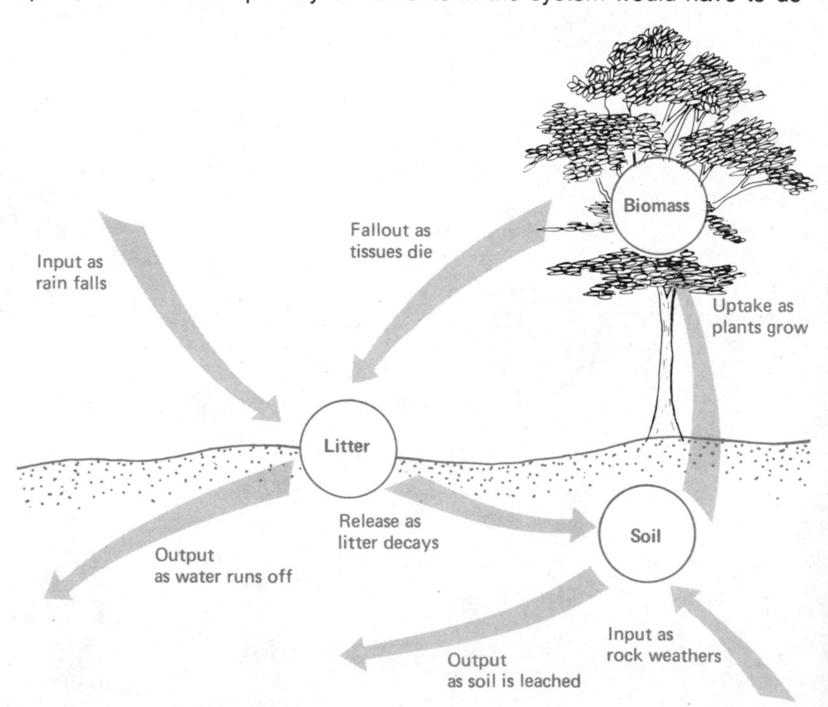

FIG. 12-4. An open ecosystem nutrient cycle.

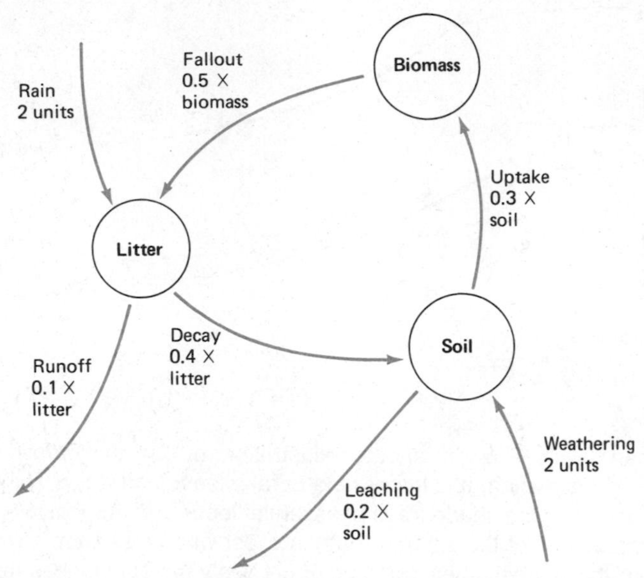

FIG. 12-5. A poker-chip simulation of an open system. Only three modifications are needed to adapt the procedures of Figure 12-2 to an open system. First, an external "bank" must be created as a source for inputs and destination for outputs. Second, the inputs during each time period are fixed amounts that enter at the same time the internal movements are made. Third, movements along several pathways out of a single compartment are calculated simultaneously, not sequentially. In other words, do *not* compute decay losses from the litter and then figure runoff losses on the basis of the amount of remaining litter. Rather, compute both movements on the basis of the total storage amount.

crease. The rate of loss by runoff and leaching would then decline as the amount of nutrients in the system decreased. Eventually, the system would adjust itself so that its outputs exactly balance its inputs.

HARVESTS AND FERTILIZERS

Suppose that the owner of an orchard creates an additional nutrient output by harvesting plums from the trees. A reduction in the biomass is not the only result of the new output from the system. Harvested plums cannot fall to the ground and become inputs to the litter compartment. (Whether the falling is direct or through a bird is irrelevant for now.) The litter compartment of the nutrient cycle must shift toward a lower equilibrium because it has a lower input. A reduction in the amount stored in the litter decreases the amount of nutrient that can be transferred to the soil by litter decay and thereby lowers the amount of nutrient in the soil. The impoverished soil cannot send as many nutrients up to the trees, so that they do not grow as well and cannot yield as many plums as formerly. In summary, the inevitable consequence of regular harvesting is a reduction in the nutrient wealth of the entire ecosystem, not just the part being harvested (Fig. 12-6).

compensation adjustment of an entire system to a change in one part

system sensitivity estimate of the amount of adjustment made by other parts of a system to change already made in one part

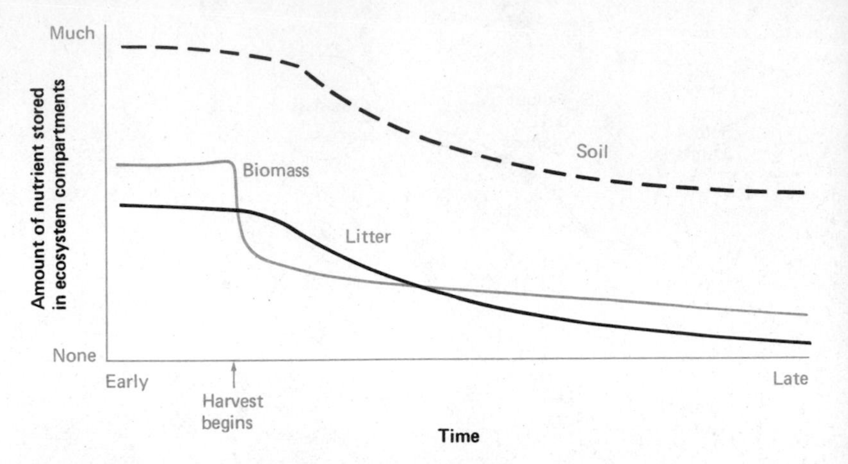

FIG. 12-6. Response of ecosystem compartments to a regular harvest from the biomass.

Urban lawns provide another illustration of the systematic consequences of harvesting. It is impossible to rake leaves off a lawn every year without depriving the shade trees of essential food. Standard lawn fertilizers do not replace all of the different nutrients lost when tree litter is removed. Therefore the soil becomes depleted of precisely the substances that trees need. A close look at the entire system suggests a link between leaf-raking and the susceptibility of urban trees to disease and insect attack (Fig. 12-7).

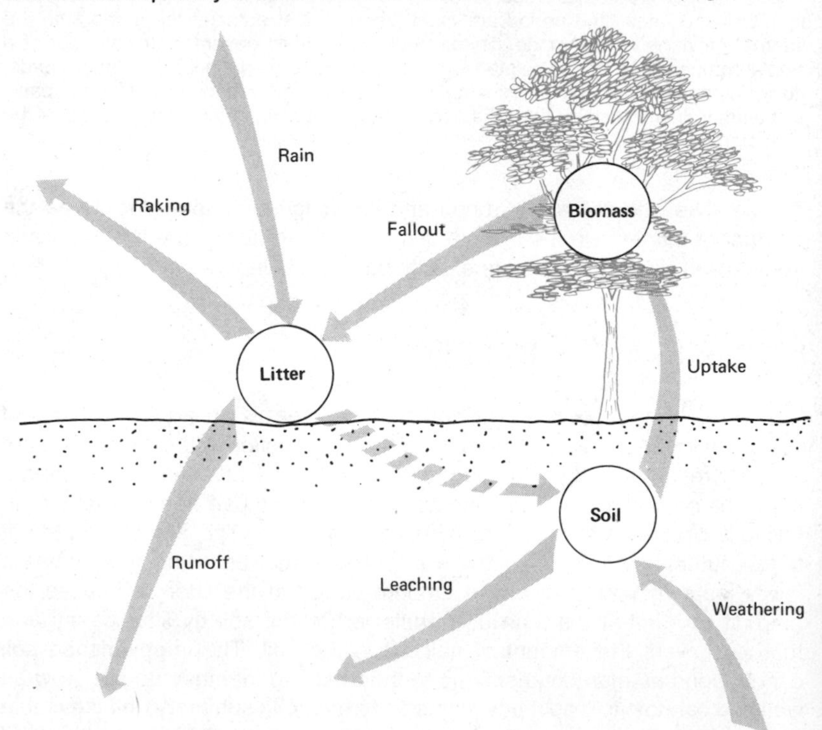

FIG. 12-7. The interrupted nutrient cycle of a suburban tree. Leaf raking is in effect a mining of the soil, because raking allows few nutrients to return to the soil compartment.

The straightforward response of modern farmers or suburbanites is to try to replace the harvested nutrients by applying fertilizers to the ground. Unfortunately, this solution also has drawbacks when the entire system is analyzed. A plant must increase its uptake of nutrients from the soil in order to tolerate two outputs (fallout and harvest). As long as the rules of nutrient movement stay unchanged, the only way to increase uptake is to add to the amount of nutrient stored in the soil. The soil must receive more input from the litter in order to stay rich enough to supply the plant, and the only convenient source of additional nutrients for the soil is an increase in litter storage. Enriched litter and soil, however, also lose more nutrients by runoff and leaching than the unharvested system did. Fertilizer applied to the ground is forced to run a gauntlet of potential losses before it arrives at the crop and becomes available for harvest (Fig. 12-8). No harvester can hope to recover all of the fertilizer that must be applied in order to maintain good yields. A typical by-product of crop fertilization is an increased output of nutrients through runoff and leaching to the groundwater and to nearby ponds, lakes, and rivers. (Chapter 13 will illustrate how enrichment of a water ecosystem may be harmful.)

eutrophication nutrient enrichment of an aquatic ecosystem (see p. 280)

This discussion is not a plea to turn back the clock on agricultural progress. Rather, it is a request to consider all of the systematic consequences of human actions. Present techniques of fertilization do not merely boost harvests; they enrich entire ecosystems and therefore have consequences that go far beyond the harvested plants. In the past, people have looked only at the biomass part of the system and considered only fertilizer cost and yield increase when they made their decisions. Now it is increasingly important to look at the entire system and to try to maximize yields while, at the same time, minimizing both the economic cost of fertilizer and the ecologic cost of nutrient outputs to other ecosystems.

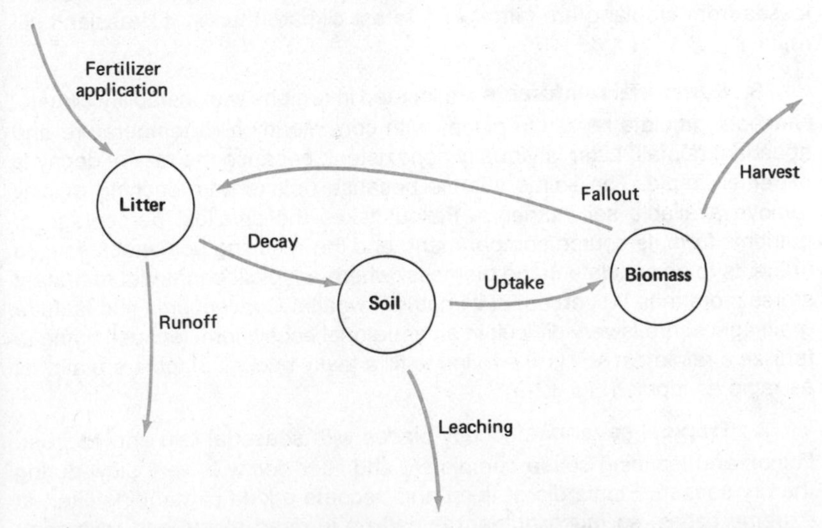

FIG. 12-8. Nutrient cycle of a fertilized crop. Fertilizer has several places where it may stray from the productive path before it gets to the crop.

NUTRIENT CYCLES AS AN EXTERNALLY GOVERNED SYSTEM

biogeochemical cycle the transfer of nutrients among rock, soil, water, air, and life

Different environments have different rules for the movements of minerals from one ecosystem compartment to another. Decomposer organisms work at a rate that depends on temperature and moisture supply (Chapter 9). Heat and water control the rate of rock weathering (Chapter 7). Runoff and leaching increase in rainy climates (Chapters 8 and 14). Climate affects plant vigor and nutrient uptake (Chapter 11). It follows, then, that different climatic regimes should produce different sets of transfer rules and therefore create distinctive patterns of nutrient circulation:

frontal climate weather dominated by air-mass collisions that produce occasional rain and changes in temperature (see Fig. 4-14)

prairie grassland with tall plants

steppe grassland with short plants (see Appendix II for information on all of these biomes)

1. **Mid-latitude grasslands** are inland places with frontal climate, frozen for half the year and not very rainy. Few nutrients come into the system, but most of the rainwater evaporates and there is not much surplus water to cause runoff and leaching. Only fallout and decay, aided by regular fires of the dried grass, remove a high percentage of nutrients from their source compartments. At equilibrium, a grassland stores few nutrients in its biomass, a moderate quantity in the litter, and vast amounts in the soil compartment. In turn, the rich soil becomes the focus of agricultural technology: tractors, plows, harrows, fertilizers, and grain crops are all merely different ways of manipulating soil.

deciduous forest forest with trees that do not have leaves in adverse seasons

2. **Mid-latitude forests** occupy places with rainy frontal climate. The surplus of water enables uptake, decay, runoff, and leaching to take nutrients from the soil and litter more rapidly than in a grassland. A forest has more biomass than a grassland, but its soil usually contains fewer nutrients. Clearing a forest and planting crops on the land condemns the farmer to a constant struggle to maintain soil fertility, because the potential nutrient losses from cropland are larger in a forest climate than in a grassland climate.

selva evergreen rainforest of broad-leaved trees

instability climate weather dominated by warm air that tends to rise and produce showery rains (see Fig. 2-16)

savanna scattered trees and grasses

3. **Equatorial rainforests** are located in regions with instability climate. Nutrients circulate rapidly in places with consistently high temperature and abundant rainfall. Litter is virtually nonexistent, because the rate of decay is extremely rapid. The soil is infertile because uptake and leaching quickly remove available soil nutrients. Fallout takes the smallest percentage of nutrients from its source compartment, and the resulting bottleneck causes nutrients to accumulate in the biomass, where a typical equatorial rainforest stores more than 80 percent of its nutrient wealth. Conventional mid-latitude grain agriculture is very difficult in an equatorial ecosystem because trying to fertilize a rainforest soil is like trying to fill a leaky bucket: outputs are almost as rapid as inputs (Fig. 12-9).

complex or hybrid climate weather seasonally dominated by two or more distinct climate types (see Fig. 3-9)

4. **Tropical savannas** occupy places with seasonal rain and no frost. Runoff and leaching cease completely and litter decay is very slow during the dry season. Plants die of thirst and become a kind of standing litter. In extreme cases, so much nutrient is tied up in dead plants and undecomposed litter that the next generation of plants may be malnourished. Fire is

one way to transfer nutrients from the standing litter to the soil and thus make nutrients available for new plants at the start of the next rainy season. Natural fires are common in savannas, and most people there deliberately use fire as a tool in manipulating the environment for their benefit.

5. **Deserts** occur in subsidence zones and rainshadow areas. Leaching and runoff cannot occur without water; therefore, deserts tend to accumulate nutrients. Uptake into the sparse vegetation is very slow; consequently, the soil is the major storage area for the nutrient wealth of a desert. The ground in some desert valleys may be so rich in nutrients that it needs intentional leaching before it can be used for crops. Extremely dry areas, however, may not have enough moisture for rock weathering and thus may remain poor in nutrients.

subsidence climate weather dominated by air moving downward and becoming warmer and less likely to produce rain
(see Fig. 3-8)

rainshadow dry area on the leeward side of a mountain in a region of prevailing winds
(see p. 95)

6. **Chaparral areas** have dry summers and wet winters. The rainfall seems too scanty to be very effective in removing nutrients, but it arrives in the cool season, when the slow growth of the plants allows the rain to concentrate on filling the soil and running off to the rivers. The result is an unexpectedly rapid loss of nutrients by leaching and runoff. At the same time, rainfall brings few nutrients and rock weathering is hindered by low temperature in winter and lack of moisture in summer. As a result, many chaparral soils are shallow, rocky, gullied, and nutrient-poor.

chaparral (maquis) low shrubs, usually evergreen

7. **Needleleaf forests** occupy areas with polar climate during part of the year. Dense shade, a short growing season, acid litter, and persistent snow-cover conspire to make life miserable for decomposer microorganisms. They respond by working very slowly and therefore allowing nutrients to accumulate in the litter compartment. The soil is leached and infertile, and

taiga evergreen forest of needleleaf trees

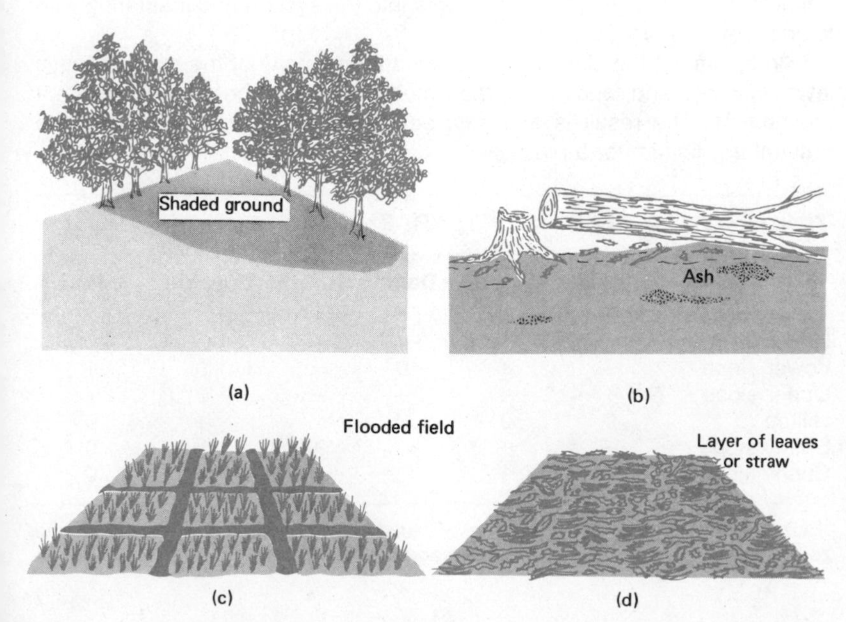

FIG. 12-9. Techniques of handling the tropical nutrient cycle: (a) Tree-crop farming, which directs attention at the major storage area for nutrients in the biomass and shows little concern for soil or litter. (b) Slash-and-burn practices, which transfer nutrients from the biomass to the litter and soil, where they remain to support crops for only a short period of time. (c) Flooding of fields, which retards litter decay, prevents runoff, and thus helps keep nutrients in the litter and soil. (d) Mulching techniques, which enrich the soil in a field by gathering and applying litter from large tracts of nearby forest land.

plants often show symptoms of nutrient deficiency because they are unable to extract the elements they need from the undecayed litter. The rapid rate of leaching makes it wiser to use the land as a tree farm rather than as a cornfield.

tundra low shrubs, herbs, and lichens

inversion climate weather dominated by cold and stable air that rarely produces precipitation
(see Fig. 2-13)

8. **Tundras** occur in the cold climates of high mountains or polar areas. Slow weathering and scanty precipitation bring few nutrients into a tundra. Cold weather shuts the nutrient cycle down for much of the year by locking water up in solid form. The slow nutrient circulation of the tundra hampers its ability to recover from stresses. Accidental devegetation, oil spills, overhunting, or toxic pollution simply cannot be repaired as rapidly in the tundra as in an ecosystem with a more speedy nutrient cycle.

The tour of world ecosystems emphasizes the importance of climate as a traffic controller directing the flow of plant foods along various pathways of an ecosystem. Thus the global location of a place has a powerful influence on the nutrient wealth of various parts of the local environment.

simatic or mafic material magma or rock rich in calcium and magnesium silicates
(see pp. 121 and 138)

The geologic foundation is another external factor with an effect on local nutrient cycles. Soft limestones have many nutrients, decay readily, and release nutrients into an ecosystem rapidly. Some volcanic rocks are so rich in nutrients that they can build fertile soils even in an equatorial climate. Coarse sandstones, on the other hand, yield soils that are infertile at the outset and that are also prone to leaching.

The topographic setting of a place is a third external influence on its mineral cycle (Table 12-1). South-facing slopes in the United States are warmer than north-facing ones. The warmth hastens rock weathering and litter decay but hinders leaching and runoff because it increases the rate of evaporation. Steep slopes usually lose nutrients because they allow water to run off more rapidly than from a gentle slope. Floods of water enrich the soil and litter in valleys by bringing nutrients into the system and interfering with litter decay and plant growth.

In summary, the climate of a place creates many of the nutrient traffic laws. Geology and topography then modify some of the rules for nutrient movements. The result is a unique set of transfer rules that govern the nutrient equilibrium for a place.

TABLE 12-1 EFFECTS OF TERRAIN ON NUTRIENT MOVEMENTS[a]

	Uptake	Fallout	Decay	Runoff	Leaching	Inflow
Bottomland	–	–	– –	–	– –	++
Lower slope	–	+	0	+	0	+
Upper slope	+	+	+	++	0	–
Hilltop	0	0	0	–	+	0
Shady slope	–	–	–	++	+	0
Sunny slope	+	+	++	–	–	0

[a]Pluses indicate increases in the rate of transfer; minuses denote decreases; and zeros represent no change from the average rate of transfer in the region.

THE NUTRIENT CYCLE AS AN INTERNALLY LINKED SYSTEM

External factors do not have absolute control over the nutrient movements in an ecosystem. The plant inhabitants, feeble though they may seem, can change some of the nutrient rules by subtly altering the effect of climate and geology. For example, plants consume energy and moisture. If plants are removed, the unused energy and water become available for nutrient-moving purposes (Fig. 12-10). Decomposer organisms, stimulated by the additional heat and water, become more active than previously. Runoff and leaching take larger shares out of the soil and litter than before. Uptake of plant food dwindles to zero when plants are removed.

system linkage mechanism that allows a system compartment to govern the flow between two other compartments

Different kinds of plants have different effects on local nutrient cycles. Some plants consume more of particular nutrients than others do. Deciduous trees and evergreens have different seasonal patterns of water use, nutrient uptake, and fallout. Trees and grasses have different effects on the temperature and moisture characteristics of the ground. As long as two different plants can survive in the same climate, substituting one for the other can alter the nutrient cycle. Like opposing political candidates, the alternative plants must obey the general laws of the realm, but they can make their own interpretations of many minor statutes.

plant exudate chemical released by plant root into surrounding soil, where it aids weathering
(see p. 145)

The role of plants as adjusters of the rules of nutrient movement becomes especially significant when humans begin to alter the natural vegetation. Replacing a forest with a cornfield does not merely add a harvest pathway and a need for fertilizer. The removal of the trees also increases the rates of runoff, leaching, and litter decay. The more drastic the change from natural vegetation, the more radical the alteration of the normal nutrient cycle.

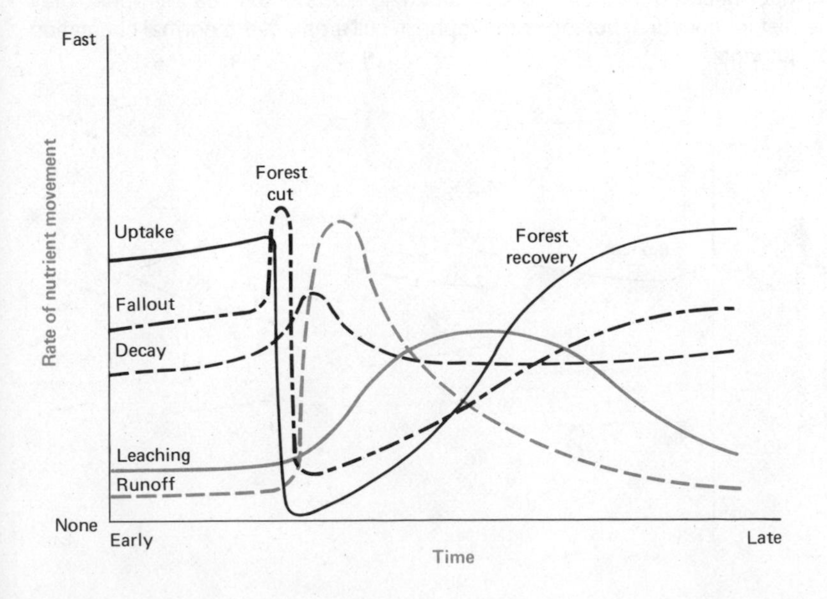

FIG. 12-10. Effects of vegetation removal on nutrient movements. Removal of plants changes many of the rules of nutrient movements. Energy and water formerly used by the plants now become available to hasten decay of litter and to carry nutrients away. The return of the same kind of vegetation restores the original transfer rules and the system goes back to its former equilibrium.

THE NUTRIENT CYCLE AS A SYSTEM TENDING TOWARD EQUILIBRIUM

dynamic equilibrium fluctuation around an equilibrium condition; the system spends more time reacting to frequent disturbances than remaining in balance

In its abstract form, the model of the nutrient cycle assumes that a nutrient system always comes to equilibrium if given enough time. In the real world, however, ecosystems rarely seem to be in balance, even though nutrient movements and energy transfer are continually trying to bring the ecosystem toward equilibrium. In effect, ecosystems are like slightly uncoordinated bicycle riders in a gusty wind. They must always be compensating for external and internal changes—irregularities of the inputs of energy and moisture, and changes in the size and composition of the living population.

For example, trees that die and come crashing down to earth furnish an important part of the total litterfall of a forest. The hundred-year lifespan of a typical tree suggests that one out of a hundred trees ought to die each year, but natural causes of death are not nearly so precise and regular. It is far more common for a large number of trees to die during an occasional severe storm, dry year, fire, or insect attack. Thus a typical forest accumulates minerals in the biomass for a number of years, only to release the nutrients in an occasional mass death (Fig. 12-11).

recovery time amount of time it takes to return to equilibrium after a disturbance

Catastrophes such as fires, storms, or diseases are therefore significant components of the normal nutrient cycle in most places. The impact of catastrophes is not very noticeable in places such as the equatorial rainforest where minerals circulate rapidly. The rainforest can recover from occasional stress quickly, so that its nutrient cycle at any given instant is likely to be quite close to equilibrium. Compensation for stress is less rapid in mid-latitude forests than in the selva, and the slow-growing northern forests take even longer to react to occasional catastrophes. On some severe polar or mountain sites, centuries of biomass accumulation may be followed by an abrupt transfer of nutrients to the litter. The scars of the "catastrophe" may persist for decades, but the catastrophe is still a part of the normal circulation of nutrients.

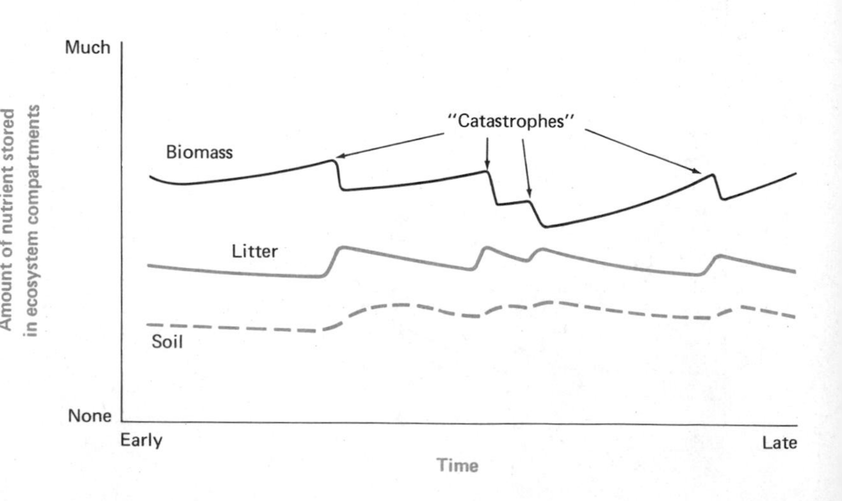

FIG. 12-11. Equilibrium as a dynamic average of stress and recovery. Because the deaths of individual large organisms occur at irregular intervals, the nutrients stored in the biomass and litter and the amount of fallout vary around average equilibrium figures.

CONCLUSION

The idea of equilibrium in nutrient circulation has become a very abstract notion, but its abstractness does not detract from its usefulness. The bank account of a person who gets a single monthly paycheck but has expenses every day is hardly ever at its average level. Nevertheless, it is still possible to budget expenditures on the basis of average wealth. Paychecks that are big and frequent allow a person to live in luxury. Small paychecks depress the standard of living for the individual. Paychecks at long intervals demand careful budgeting and saving if the person is to avoid bankruptcy before the next payday. Likewise, the rules of nutrient circulation have an effect on the kinds of things people can do in a particular region.

The villagers who opened this chapter were using a "tool" adapted to the rules of nutrient cycling in their environment. Repeated fires transfer nutrients from the biomass and standing litter to the soil, where they become available for new plant growth. The villagers gain a reliable source of food income at the expense of a slight increase in nutrient losses by runoff and leaching. A plow is a more useful tool than fire in places that store nutrients in the soil rather than in the standing litter. An orchard is more sensible than either fires or plows in regions where nutrients accumulate in the biomass. Every place on the earth has a unique pattern of nutrient circulation that influences, and is influenced by, its living inhabitants.

Chapter 13

Oxygen and Biochemical Demand: Life in Water

Things have to be getting mighty bad when you see fish coming up for air.

Minnesota Department of Natural Resources

Water usually remains low on the list of environmental concerns as long as faucets deliver an acceptably clear liquid and streams have few bad odors. Lack of attention, like ignoring a disease until symptoms are obvious, is especially questionable in the case of water pollution, since the final result of water contamination is a serious health hazard. The initial steps to that end, however, are deceptively benign. The change from a life-supporting body of water to a sterile one is not a smooth downward slope of increasing contamination. Rather, there is a gradual rise in productivity followed by fairly abrupt drops. Furthermore, the initial improvement and the later declines in water quality are both results of the same basic processes. The history of a typical polluted lake helps to unravel the tangle of causes and effects.

FOOD AND OXYGEN IN WATER

The present condition of the lake was, of course, completely predictable. The stench of rotting fish is merely an unpleasant exclamation point for a sentence that was passed long ago. An examination of one of the dying fish seems to center around the fact that animals always need four things: a safe environment, food, water, and oxygen. Land animals are not likely to think of all four. They tend to take oxygen for granted, because they can breathe the abundant supply in the air. Animals that spend much time underwater, however, must be concerned about the adequacy of their oxygen supply. Many aquatic animals have special mechanisms for obtaining oxygen. Whales and skindivers are obliged to come to the surface periodically. Some insects extend a long tube up to the surface for air. Fish "breathe" the minute amounts of oxygen dissolved in the water.

dissolved oxygen gaseous oxygen in liquid water

A few years ago, conditions in the hypothetical lake could not have been better for the fish. The temperature was comfortable, the water contained no poisons, and the oxygen concentration in the water was a favorable 8 ppm (eight units of oxygen dissolved in each million units of water). In addition, a new meat-packing factory on the shore was contributing a welcome supply of food for the inhabitants of the lake.

ppm parts per million

Any piece of food, however, can be viewed through chemist's glasses as a demand for oxygen (Fig. 13-1). Food must combine with oxygen inside

FIG. 13-1. Apparatus used to determine biochemical oxygen demand. The recipe for a BOD test is simple: take a sample of the water to be tested and put it in an airtight bottle with a known amount of oxygen. Allow the mixture to simmer for 120 hours at an ideal temperature for some bacteria. Then measure the amount of oxygen consumed and express as ppm (units per million units of sample). (Courtesy of Hach Chemical Company, Ames, Iowa, U.S.A.)

an animal in order to be useful. The actual amount of oxygen in the water at any given time depends on the balance between consumption and replacement. The amount of dissolved oxygen must decrease if oxygen is consumed more rapidly than it can enter the water.

positive feedback reaction that tends to increase the original cause and thus to promote a still greater reaction

Aquatic organisms differ in their oxygen requirements. Most trout need about 6 ppm of dissolved oxygen in their surroundings. A whitefish wants at least five, while carp seem to get by with only four. The sensitive trout are the first to sicken and die when oxygen becomes scarce. Ironically, the death of a sensitive fish increases the demand for oxygen in the lake, because the corpse becomes food for other creatures. Most fish are doomed when the oxygen concentration goes below 3 ppm, though sludgeworms, rat-tail maggots, and bloodworms can still survive.

anaerobic (anoxic) in the absence of free oxygen
(see p. 180)

Some tiny organisms are able to function even in an environment that has no free oxygen; they are able to obtain oxygen from other chemical compounds. One common source for combined oxygen is sulfate (SO_4 to the chemist), which yields four oxygen atoms to any microorganism that knows where to look for them. After the oxygens are removed, the divorced sulfur looks for someone to remarry, settles for two hydrogen atoms, and forms the gas hydrogen sulfide. This gas is the reason for the characteristic rotten-egg odor that rises from some swamps and other places where organic matter is consumed in the absence of free oxygen.

oxygen sag decline of dissolved oxygen that accompanies an input of food

In summary, the addition of food to a body of water causes the amount of available oxygen to decrease. An intelligent polluter should be able to predict the oxygen decrease that will accompany a given input. A body of water can assimilate a certain amount of pollution without serious consequences, because the concentration of dissolved oxygen stays high enough for most forms of life. Vast numbers of cesspool lakes and sewerlike streams, however, testify to the difficulty of making reliable forecasts of oxygen depletion and then enforcing the measures needed to stay on the safe side of the critical amount of pollution.

WATER CHARACTERISTICS AND OXYGEN SUPPLY

The size of the receiving body of water is one of the first factors to consider in evaluating the probable consequences of a food addition. The ecological contribution of a single toilet might overwhelm a tiny creek but would have no noticeable effect on the Mississippi River.

specific surface amount of surface area per unit of volume
(see Fig. 8-3)

oxygen recharge replenishment of dissolved oxygen by diffusion from the air

turbulent flow chaotic movement in many directions as well as forward, like a surging crowd

The depth of the water is also significant in an indirect way. Deep ponds and streams modestly expose only a small amount of surface to the air. Surface exposure is significant because only a limited amount of oxygen can pass from the air into each square centimeter of water surface. The air can replenish a wide and shallow lake more easily than a deep one with the same amount of water but with less surface area.

Recharge of oxygen from the air is more rapid when water is moving than when it is standing still. Fast streams tumble along, and the turbulence brings water up from the bottom levels and replaces it with oxygen-rich surface water. Slow water tends to move in layers, and only the uppermost layer of a slow stream or stagnant pond has easy access to the vital oxygen in the air. A quiet pool just below a rapids or waterfall is probably ideal for most fish. They get the advantage of aerated water without having to struggle with fast currents.

laminar flow layered movement in one direction, like a parade band

Moving water also aids the oxygen balance by spreading an input of food throughout a large volume of water. Spreading the food reduces the local use of oxygen because the food is not consumed in any one place. Dilution, of course, will fail if too many polluters rely on the lake or river to carry their oxygen demands away from where they put them into the water.

A film of oil or a layer of leaves on the surface of a body of water can block oxygen recharge from the air. Thus an oil slick is a double hazard. On the one hand, it provides food for some organisms and therefore takes oxygen out of the water. At the same time, it interferes with oxygen replenishment.

monomolecular film thin layer of oil or wax on a water surface; the film interferes with evaporation from the water and oxygen recharge into the water

Even inorganic nutrients exert an oxygen demand when they are added to a body of water, because they stimulate the growth of algae and other water plants. Green plants have a positive effect on the oxygen balance during the day and a negative one at night. When the sun shines, plants give off more oxygen than they consume. By night, however, they must take in oxygen for their own metabolic processes. Furthermore, water plants are a major source of food for animals in the water. In theory, plants should give off exactly as much oxygen as would be needed to consume their bodies (Table 13-1). In practice, however, much of the daytime production of oxygen escapes into the air. Water has a limited ability to store oxygen, and the storage limit makes it impossible to save all of the daytime surplus to satisfy the demands of the night. Moreover, surface algae interfere with oxygen recharge from the air.

TABLE 13-1 A BARE-ESSENTIALS VIEW OF THE OXYGEN CYCLE

	Inputs	Outputs
Photosynthesis by green plants in daytime	6 CO_2 (carbon dioxide) 6 H_2O (water) sun energy	1 $C_6H_{12}O_6$ (carbohydrate) 6 0_2 (oxygen) chemical energy
Respiration by all living things at all times	1 $C_6H_{12}O_6$ (carbohydrate) 6 O_2 (oxygen) chemical energy	6 CO_2 (carbon dioxide) 6 H_2O (water) mechanical or heat energy

The temperature of a body of water is the final item on the list of major influences on water's ability to supply oxygen. Temperature affects the rates of oxygen use and recharge in a variety of ways:

poikilotherm cold-blooded animal (e.g., fish, frog, insect) whose body temperature varies with the outside temperature; poikilotherms become more active as the temperature rises

homoiotherm warm-blooded animal (e.g., bird, cat, human) whose body temperature stays essentially constant; homoiotherms may become less active as the temperature rises

stability tendency for air or water to resist vertical movement
(see Fig. 2-2)

1. Hot water cannot hold as much oxygen as cold water (Fig. 13-2). Trout and other oxygen-loving fish are common in the cold streams and lakes of Maine or Michigan. The warm waters of Mississippi are more likely to contain carp or catfish, which can tolerate low oxygen levels.
2. Changing temperatures may kill some organisms outright, and their corpses are added to the food supply.
3. Animals tend to be more active, to eat more food, and to consume oxygen more rapidly in warm water than in cold water.
4. Poisons act more quickly at high temperatures. The carcasses of the victims, of course, become food for other organisms.
5. Warm water is not quite as heavy as cold water. The difference in density causes hot water to stay near the surface, where its inability to hold much oxygen prevents oxygen from penetrating to the cool layers below. In this sense, stable water is just as harmful for polluters as is stable air.

Several of these conditions can be combined to produce a real polluter's nightmare: a small, hot, slow-moving stream with an oil slick and a mat of algae on the surface. An unfavorable situation can occur as a result of a number of independent actions, each seemingly safe in its own right. For

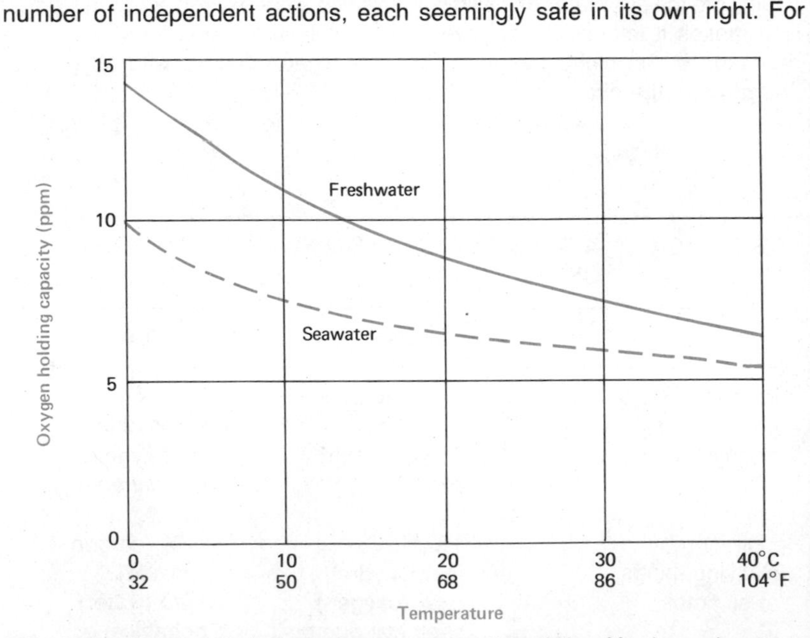

FIG. 13-2. Relationship between temperature and the ability of fresh or salt water to hold oxygen.

example, a dam may slow a stream. Meanwhile, the discharge of heated water from a power plant warms the stream, a small town removes some water, and another city pours partially treated sewage into the rest. The river may be able to absorb any two or three of these actions without undue strain, but the third or fourth may push it over the brink. Even worse, a river that is usually a fairly reliable absorber of pollution may become temporarily unsuitable during the late hours of a hot night after a long dry spell. At that time, the stream is small and warm, unable to hold as much oxygen as it normally could, whereas the plants and animals are exceptionally active. The oxygen concentration may dip below the critical level for some animals for a few hours, but even a short period of oxygen stress may be enough to cause death. If casualties are numerous, the stream may not recover from the burden of dead animals for weeks, months, or even years. The loss of a large number of sensitive animals also removes a vital check on the populations of other organisms and on the nutrient-cycling processes in the stream.

synergism the combined effect of two or more factors, which is different from the sum of their individual effects

OXYGEN SAG IN A FLOWING RIVER

A stream with a single big source of pollution will have several distinct zones of stream response (Fig. 13-3). Upstream of the pollution source, the stream probably contains nearly as much oxygen as it can hold at its temperature. A small amount of food continually enters the stream from natural sources—leaves, bits of soil, substances dissolved in rainwater, and so on. However, the oxygen demand that is exerted as these natural inputs are consumed is usually much less than the recharge ability of the stream. The oxygen concentration therefore remains fairly high, and the animals in the stream are not threatened with suffocation.

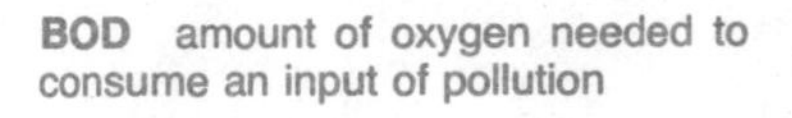

BOD amount of oxygen needed to consume an input of pollution

Just downstream from a big source of pollution, stream organisms consume the edible parts of the pollutants, and the amount of dissolved oxygen decreases. If anything in the pollution is poisonous to some of the inhabitants of the stream, the toxic effects usually show up in this zone, and the corpses of the dead animals or plants are added to the food supply.

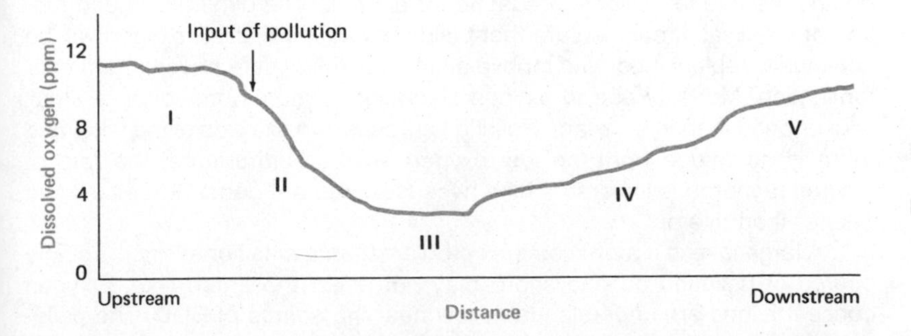

I Zone of clean water
II Zone of stream degradation
III Zone of reduced species diversity
IV Zone of stream recovery
V Zone of enriched water

FIG. 13-3. Oxygen sag in a typical flowing stream with a moderate amount of pollution. Zone III may not occur when the biochemical oxygen demand is small; Zones IV and V are absent when the pollution is too great for the stream to absorb.

diversity number of important different kinds of organisms (see p. 241)

anaerobic lacking free oxygen (see p. 274)

An oxygen demand that is very great will cause the concentration of oxygen in the water to go below the critical level for some species of animals. Mobile animals rarely die of oxygen starvation in a flowing stream; they simply migrate upstream or downstream until they find an area with abundant oxygen. The number of species in the low-oxygen zone decreases, but the total mass of living things may be considerably higher in the area with little oxygen than in cleaner parts of the stream. Few species can tolerate low oxygen levels, but those that do survive may become very numerous because of the abundant food supply.

Water pollution becomes most threatening to humans when the oxygen concentration drops nearly to zero. The virtual absence of oxygen allows only a few small anaerobic organisms to survive, but anaerobic organisms are responsible for a wide variety of human diseases. In clean streams, however, other bacteria and larger animals usually kill disease organisms quickly.

Still farther downstream from the source of pollution, much of the oxygen demand has already been exerted. The rate of oxygen consumption decreases, and natural recharge may be able to increase the concentration of dissolved oxygen. If so, clean-stream organisms will return, and the plants and animals that thrived in the zone of reduced oxygen tend to decline in numbers.

The final zone in the hypothetical stream often is more productive than the first because the pollution made some nutrients more abundant than they were upstream. Decomposition of animals and plants releases nutrients into the water, where green plants can use them again. The continual recycling of nutrients, however, may demand more oxygen than the air can provide; in that case the stream remains low in oxygen and hazardous to fish and human health.

OXYGEN SAGS IN LAKES AND RIVERS

The difference between polluting a flowing stream and contaminating a lake is apparent in the autopsies of the animal corpses. In the stream, most deaths are due to poisons, predators, or old age. The dimensions and motion of a typical stream assure that pollutants will be diluted, oxygen will be continually replenished, and mobile animals can find safe havens from contamination. Mortality due to oxygen starvation is more common in a small lake or pond than in a stream. Pollution spreads into all parts of the lake, and there is no refuge from the low oxygen levels. Furthermore, the rate of oxygen recharge is lower in a lake because lakes are generally calmer and deeper than rivers.

A large lake is a more complex problem than a small one, since a single source of pollution on one shore may not affect the entire lake. Oxygen concentrations are understandably low near the source of BOD. The pollutant, however, is consumed as it drifts away from the input, and thus the demand for oxygen is not so great on the opposite side of the lake (Fig. 13-4). Fish soon learn what parts of the lake are safe to occupy.

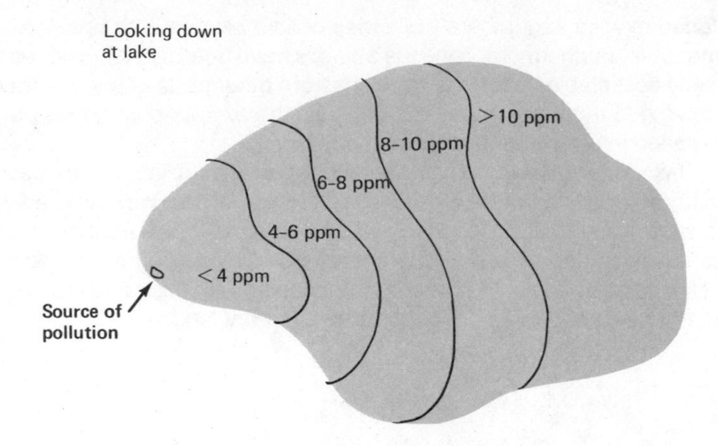

FIG. 13-4. Oxygen sag in a hypothetical lake. Organisms in the lake consume the pollutant as it drifts away from its source. Therefore the greatest consumption of dissolved oxygen is near the source of the oxygen demand (unless the pollution is so great that the area near the source becomes uninhabitable for most living things).

The simplified river and lake models need some modifications before they can handle many of the complications brought in when a body of water has several nutrient inputs. Figure 13-5 shows multiple oxygen sags caused by several sources of pollution along a stream. Several areas of very low oxygen concentration act as fences that restrict the movement of oxygen-demanding organisms up and down the stream. A big new input of pollution on such a segmented stream is more dangerous than in a stream with no areas of low oxygen concentration. Clean-stream organisms cannot escape

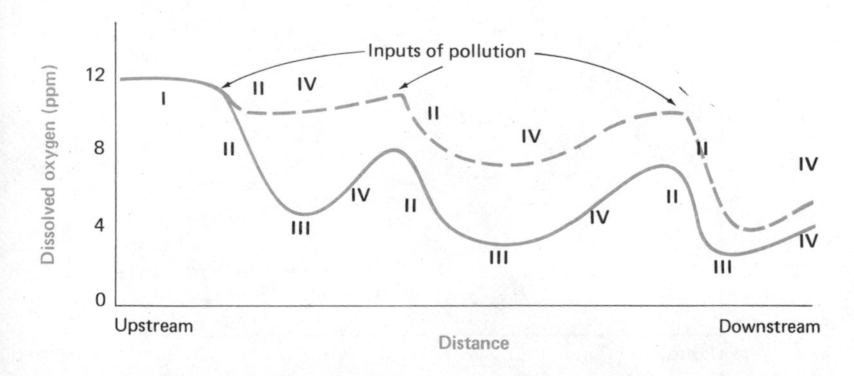

FIG. 13-5. Multiple inputs of pollution along a single stream. Each source of pollution produces its own oxygen sag; the result may be several distinct zones of reduced species diversity along the stream. The Roman numerals indicate stream zones as defined in Figure 13-3. The dashed line shows the oxygen profile if the first polluter builds a treatment plant. A sewage treatment plant is simply a controlled environment, rich in oxygen and favorable for the microorganisms that consume many kinds of pollutants. A properly designed treatment plant of sufficient size is odorless and inoffensive, but an overloaded one can undergo an oxygen sag, complete with reduced species diversity and the risk of anaerobic conditions.

a sudden oxygen sag if there are zones of little oxygen both upstream and downstream. Furthermore, once the animals have been killed in one reach of normally acceptable water, their cousins from other parts of the stream may not be able to repopulate the area, because the oxygen-poor zones prevent immigration into a clean reach of the stream.

A lake is less likely than a river to have clearcut internal fences that prevent the migration of organisms toward clean areas. However, a lake is more susceptible to occasional severe oxygen sags, which may exterminate some species of animals from the entire lake. Obviously, natural repopulation of a species-depleted lake may be difficult. Most fish find that even a polluted stretch of river is easier to cross than dry land.

EUTROPHICATION

The interaction between food and oxygen is very obvious in the simplified world of a dramatic pollution event. The same basic relationships, however, characterize all natural waters. Oxygen sags of various significance appear in nearly all flowing streams. Small pieces of organic debris and soil nutrients enter and demand oxygen from the supplies stored in the stream. Algae and other plants grow in response to the increasing supply of nutrients and, in turn, the plants furnish food for animals. Plants and animals die all along the way, and decomposition releases their nutrients into the water. Most rivers are carrying a significant burden of biochemical oxygen demand by the time they reach the ocean. Understandably, dissolved oxygen usually is less abundant near the ocean than in the upper reaches of a stream (Fig. 13-6).

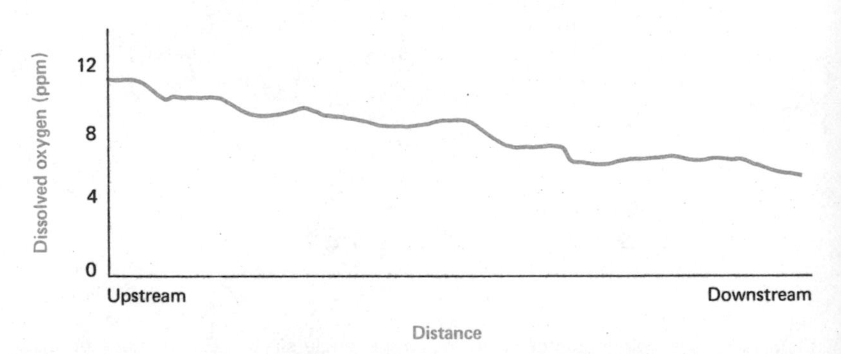

FIG. 13-6. Oxygen profile of a typical river. The general decline in oxygen concentration as water flows from the stream source at high elevations toward its mouth at sealevel is a result of three factors: (1) increasing productivity of plants because the nutrient content of the stream is increasing; (2) decreasing capacity to hold oxygen because the temperature of the water is increasing; and (3) decreasing ability to recharge oxygen because the stream becomes deeper and less turbulent.

Ponds and lakes go through a similar sequence of events as they age. A lake that has just been formed is clear and blue, rich in oxygen but poor in nutrients. As time passes, nutrients enter the water and stimulate the growth of plants and animals. Meanwhile, the lake begins to fill with dust from the air and sediment from inflowing streams, together with an increasing quantity of dead organic remains. The oxygen concentration declines as the amount of food in the system increases. The water becomes shallow, green, and murky, and eventually the lake becomes so shallow and choked with plants that it is more properly termed a marsh or swamp. As the centuries pass, the marsh fills in, dries, and hardens. Finally, land plants come in and take the place of water-loving plants. There is an inevitable conclusion to this discussion: all lakes are ephemeral features on the landscape, doomed to become dry land in time (Table 13-2).

marsh soft, wet, treeless land

swamp soft, wet, forested land

bog soft, wet area with acid organic soil

TABLE 13-2 STAGES IN THE BIOTIC SUCCESSION OF A HYPOTHETICAL POND[a]

Stage	Oligotrphic	Mesotrophic	Eutrophic	Marsh	Dry land
Clarity	Clear	Cloudy	Murky	Muddy	Opaque
Color	Blue	Gray	Green	Black	Brown
Nutrient	Little	Moderate	Much	Much	Moderate
Oxygen	Much	Moderate	Little	Little	Moderate
Temperature	Cold	Cool	Warm	Variable	Variable
Inhabitants	A few aerobic organisms and very few fish	Many fish and other organisms	Very many organisms but few fish	Bottom-dwelling organisms	Trees, grass, or other land organisms

[a]These diagrams represent arbitrary snapshots along what is really a continuous change from a lifeless body of water to solid ground.

The aging of a lake is a succession process. A particular plant or animal species enters the lake or marsh when conditions there become tolerable. In turn, each organism alters its local environment, changing the concentration of oxygen and food and other characteristics of the water. The changes may cause the extinction of some organisms, while paving the way for other living things to settle there. For this reason, the populations of living things in a lake are useful clues to the apparent age of the lake. Moreover, efforts to eliminate "undesirable" organisms such as carp or algae with poisons are often futile because a surplus of nutrients is the ultimate reason for the existence of these life forms. The poisons simply produce a lot of corpses that become food for the survivors.

succession progressive modification of environment by successive groups of inhabitants
(see Fig. 11-8)

POLLUTION AS ACCELERATED EUTROPHICATION

cultural eutrophication rapid increase of nutrients as a result of human activity

eutrophication nutrient enrichment of a body of water

- oligotrophic—containing little nutrient
- mesotrophic—containing some nutrient
- eutrophic—containing much nutrient

In a sense, most water pollution could be called a speeded-up natural process. Before rejoicing in their essential normality, however, polluters should be aware that there is some deception in this model of pollution as premature aging of a body of water. The natural aging of a typical lake consists of nutrient enrichment, oxygen sag, and depth reduction, all occurring at approximately the same rate. Natural regulators usually prevent lakes from getting too rich in nutrients for their depth. In most polluted lakes, on the other hand, nutrient accumulation and oxygen depletion occur more rapidly than the filling of the lake basin. The result of imbalanced aging can be a sort of cesspool, too deep for many of the shallow-water organisms that would occupy a marsh but too low in oxygen for most of the free-swimming inhabitants of clean lakes.

The fish and other organisms that were poisoned or incapacitated by artificial substances are another reason that water pollution should not be dismissed as merely premature aging. Natural aging usually involves materials that the occupants of the water are able to eat. Through the centuries, many plants have developed chemicals which make them less palatable to the animals that feed on them. Fortunately for the animals of the world, plants have been fairly slow in their efforts to make themselves inedible. Animals generally could develop a taste for each new natural poison shortly after it was invented by the plants, but the ability of animals to become resistant to toxic compounds is simply no match for the recent, sudden, and widespread onslaught of thousands of complex and often lethal artificial chemicals.

flushing time amount of time it takes for all the water in a lake to be replaced with new water from rain or incoming rivers

Nutrient enrichment and artificial poisoning are not very threatening in fast rivers, because moving water can dilute pollution and carry it away from the source. The danger is greater in small lakes near metropolitan areas; newspapers frequently headline the "death" of such water bodies. The ecological threat is most insidious and ultimately most serious in the case of the ocean basins, whose future productivity may be seriously compromised by present contamination.

OCEAN PRODUCTIVITY

At this time in history it is a cruel joke to tell a starving world that the human race can turn to the boundless ocean for food. The open ocean, with a few striking exceptions, is a biological desert. Plant productivity is very low, and large populations of animals cannot survive without plants to eat. The reason for the limited ability to support life is simple: the essentials for life rarely appear together in the same area. Water is available everywhere, but sunlight, oxygen, and nutrients are unevenly distributed.

Poems and photographs to the contrary, the dominant color of the sea is black. A glass of seawater may look transparent, but sunlight is able to penetrate only the top few hundred meters of water in the ocean. The energy is absorbed in the upper layers, and more than 95 percent of the ocean volume is permanently beyond the reach of the sun. Waters with abundant plant life or suspended sediment may have perpetual darkness within 15 m (50 ft) of the surface. Green plant life obviously can grow only in the lighted part of the ocean. Most of the production of organic material in the oceans occurs within 20 m of the surface, in less than one percent of the total volume of water.

euphotic zone lighted surface layer of an ocean or deep lake

The behavior of nutrients in deep water tends to diminish the productivity of even the tiny lighted portion of the ocean. Suspended plants take nutrients from seawater and thus inwittingly rob their descendants of a food supply. The corpses of plants or their predators usually are heavier than water; their weight causes them to sink into the lightless depths of the sea. Ecosystems on land sustain their productivity by reusing the same nutrients many times. The option to recycle nutrients is simply not available for most places in the ocean, because sunken nutrients rarely return to the surface (Fig. 13-7).

nutrient sink nutrient output that has no return flow and no apparent pathway for recycling

In the depths beyond the reach of the sun, life depends on the slow downward drift of plants, animals, and dead organic matter from the productive zone above. This material exerts a strong biochemical oxygen demand as it sinks. Consumption and decay release nutrients but take oxygen from the deep water, where oxygen recharge is extremely slow. As a result of these processes, the surface layers of the ocean are lighted and oxygenated but poor in nutrients. The depths are rich in nutrients but dark and frequently depleted of oxygen.

pelagic in the lighted part of the open ocean

abyssal in the dark part of the open ocean

benthic on the dark bottom of the open ocean

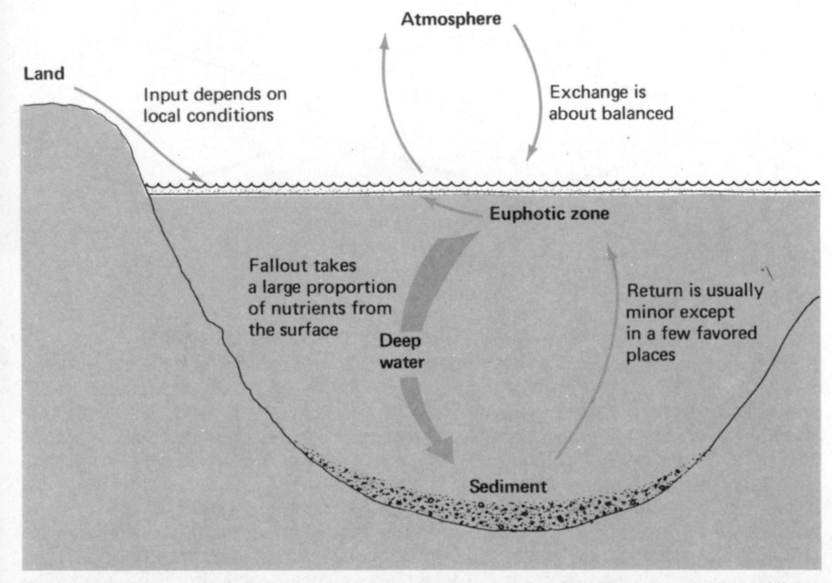

FIG. 13-7. The cycle of nutrients in the ocean. Incorporation into the bodies of living things is usually the first step in the eventual removal of nutrients from the biologically productive part of the ocean. The thickness of the lighted layer at the top of the ocean is exaggerated to make it possible to be printed.

UPWELLINGS AND OCEAN CURRENTS

Small areas just off the west coasts of tropical landmasses are striking exceptions to the general rule of ocean unproductivity. The rich tropical fisheries owe their existence to the upward movement of water from deep in the ocean. The ocean upwelling is a product of the wind and thus has its ultimate origin in the surplus of solar energy at the equator.

wave wind-raised moving mound of water

Most waves are born when winds blow across the water (Fig. 13-8).

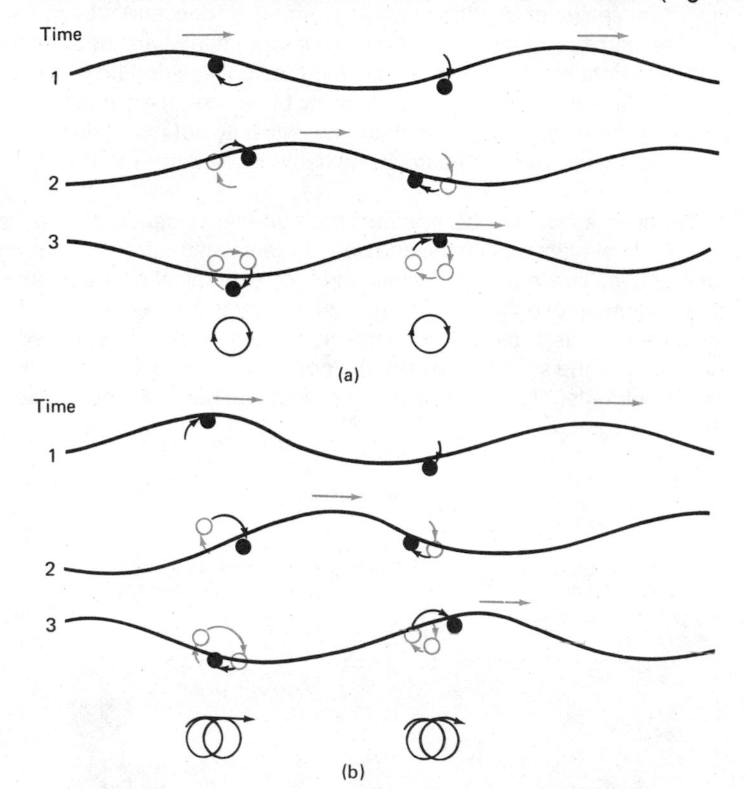

FIG. 13-8. Two kinds of waves on water: (a) Shock waves occur as a result of sudden movement, such as an earthquake or an object falling into the water. Waves radiate outward from the point of impact, carry energy away from the source of the shock, and dissipate the energy over a large area. An individual particle of water in a shock wave does not migrate. It merely moves in a vertical circle and passes energy to its neighbors without going far itself. (b) Current waves are produced by a steady force, such as a prevailing wind. Much of the movement of particles of water is still circular, but the surface water receives a continual input of energy from the wind, so that the waves advance in the direction of the wind. The stronger the wind and the farther it blows across the water, the larger the waves become. Water beneath the surface receives its energy from the water above, not directly from the wind. The subsurface water does not flow in the same direction as the wind, because all things in motion on the rotating earth tend to be deflected from a straight path.

Once started, waves may persist long after the wind has changed its mind and gone elsewhere. It is easy to see how a confused wave pattern can develop as a result of several winds blowing in different directions at different speeds in different places. The pattern becomes even more complicated because moving water, like the winds described in Chapter 3, is deflected from a straight path by the rotation of the earth.

Coriolis effect apparent deflection of moving objects on the surface of the rotating earth
(see Fig. 3-6)

An examination of the North Atlantic Ocean shows how prevailing winds and the deflection of ocean waves cause water to accumulate in the center of the ocean basin. At about 15°N, the general atmospheric circulation pushes air southward toward the equator, but the wind turns to the right and blows toward the west. This wind pushes surface ocean water to the west, but the subsurface water turns to the right and flows northward. Meanwhile, about 45°N, the surface ocean water receives an eastward shove from the wind, and the water immediately below the surface turns to the right and flows southward. It is not easy to sort out the actual movements of water from the maze of pushes and turns. The wind tends to mask the true motion of most of the water by continually pushing on the surface water and keeping it in line. Immediately beneath the thin surface waves, however, water flows nearly at right angles to the winds and moves toward the thirty-degree parallel from both north and south. These converging currents involve only a thin layer of water just beneath the surface, but they pile up a considerable quantity of water in the middle of the ocean.

trade wind wind blowing toward the west about 15°N or 15°S
(see Fig. 3-13)

westerly wind blowing toward the east about 45°N or 45°S
(see Fig. 3-14)

The subtropical mound of water acts like the subtropical high in the atmosphere. It pushes down on the water below and causes deepwater currents to flow north and south away from the high. Like a wind in the lower atmosphere, a deep ocean current is deflected from a straight path. The deflection of deepwater currents gives the ocean a second set of easterly and westerly flows. The first set consists of currents of surface water that get their push directly from the wind. The second set of currents flow far beneath the waves and are driven in a very indirect fashion by the wind. Currents at both levels move toward the west in tropical waters and toward the east in the middle latitudes.

subtropical high zone of high air pressure about 30°N or 30°S
(see p. 54)

At this point, the continents arise and make an essential contribution to the theory: no current in the North Atlantic Ocean can flow indefinitely toward the west or east. Eventually the currents run into the continents and have to stop or change direction. Tropical currents take water away from the coast of Africa, carry it across the ocean basin, and pile it up near the opposite shore. Mid-latitude ocean currents flow toward the east and push water against the European shore. Thus the North Atlantic Ocean has three areas in which water piles up: a subtropical mound in the center of the ocean, a tropical accumulation near Cuba on the west side of the ocean, and a mid-latitude accumulation near Ireland on the east side of the ocean. Some of the excess mid-latitude water drifts toward the equator along the coasts of Europe and Africa. Meanwhile, surplus tropical water flows northward along the American shore. The result is an immense whirl of water around a calm high area in the center of the subtropical ocean (Fig. 13-9).

countercurrent narrow current flowing eastward, against the wind, near the equator

FIG. 13-9. Circulation of water in a hypothetical ocean. Deepwater currents are more complex and less understood than surface currents. (Compare this idealized diagram with the actual ocean circulation depicted in Fig. 13-10.)

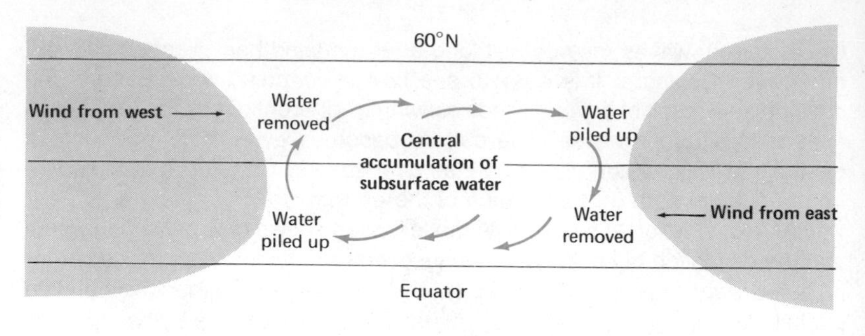

East-west movements of ocean water are less significant for life than the north-south flows they cause. Along the east coasts of the continents, the poleward flow of warm tropical water carries heat away from the equator and brings rainier conditions than would normally be expected. Fish consider the warm currents undesirable, however, because warm water cannot hold much dissolved oxygen.

upwelling upward movement of ocean bottom water; upwellings occur off the west coasts of tropical continents and off the east coasts of high-latitude continents

The equatorward currents along the west coasts of the continents are more favorable for fish. The water in these currents is cold and capable of holding much oxygen. Moreover, the surface water turns away from the land when it enters the latitude of the trade winds. The resulting "emptiness" near the surface draws water up from the ocean bottom, and the rising water is rich in nutrients lifted from the lightless floor of the ocean.

The combination of nutrients from the bottom, tropical sunshine, and abundant oxygen makes the areas of rising water very productive (Fig. 13-10). The rich upwellings occupy only a few percent of the ocean area, yet they produce more than half of the ocean fish catch. They are, in effect, small oases in a vast desert; the average productivity of the open ocean is little more than that of the central Sahara.

ESTUARIES, SHELVES, AND SALT MARSHES

littoral in the shallow margin of the ocean

continental shelf undersea margins of the continents

Downward movement of nutrients is largely responsible for the unproductivity of the open ocean but poses no threat in the shallow areas around the margins of the sea. The floor of a shallow sea is lighted by the sun; plants can grow directly on the bottom and can use the nutrients there. Shallow areas of the ocean are productive because their nutrient movements resemble the cycling processes on land, rather than the downward escalator that characterizes deep ocean basins. Rivers bring nutrients from the land to the

FIG. 13-10. (a) Ocean currents and areas of upwelling. (b) Average annual primary productivity of the oceans. More than half of the annual deep-ocean fish catch comes from the upwellings, where productivity is great enough to warrant trying to harvest it.

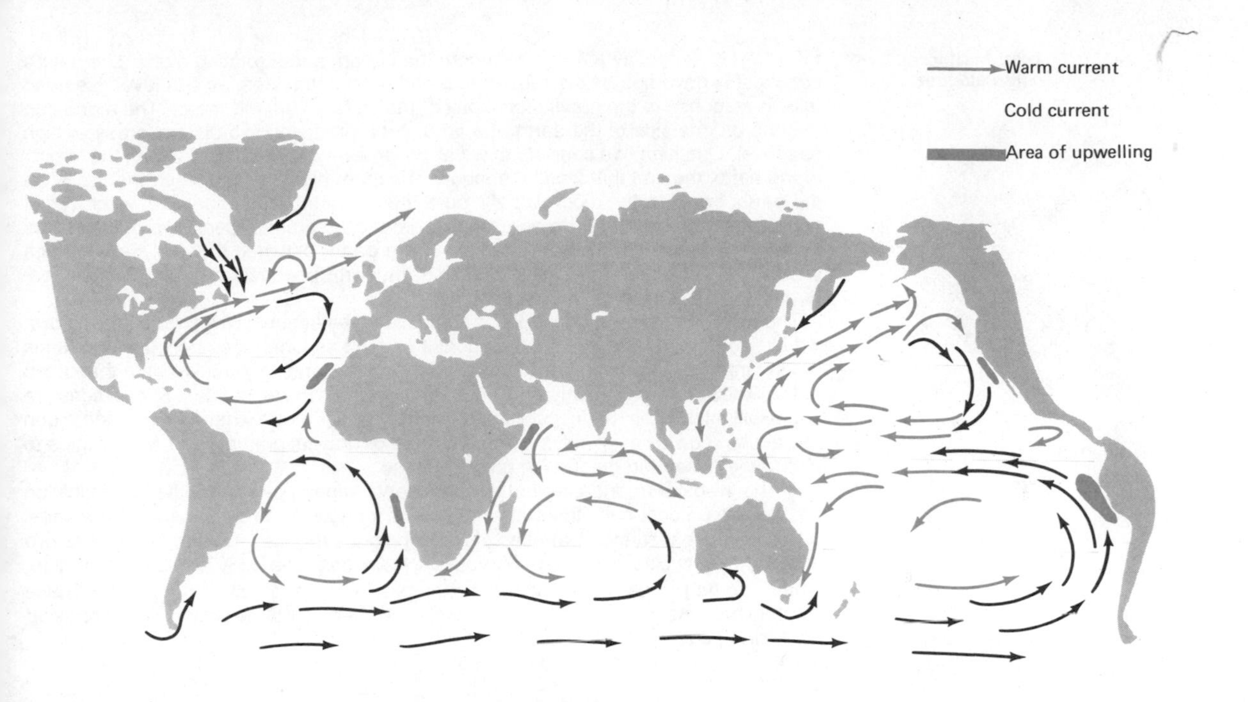
Warm current
Cold current
Area of upwelling

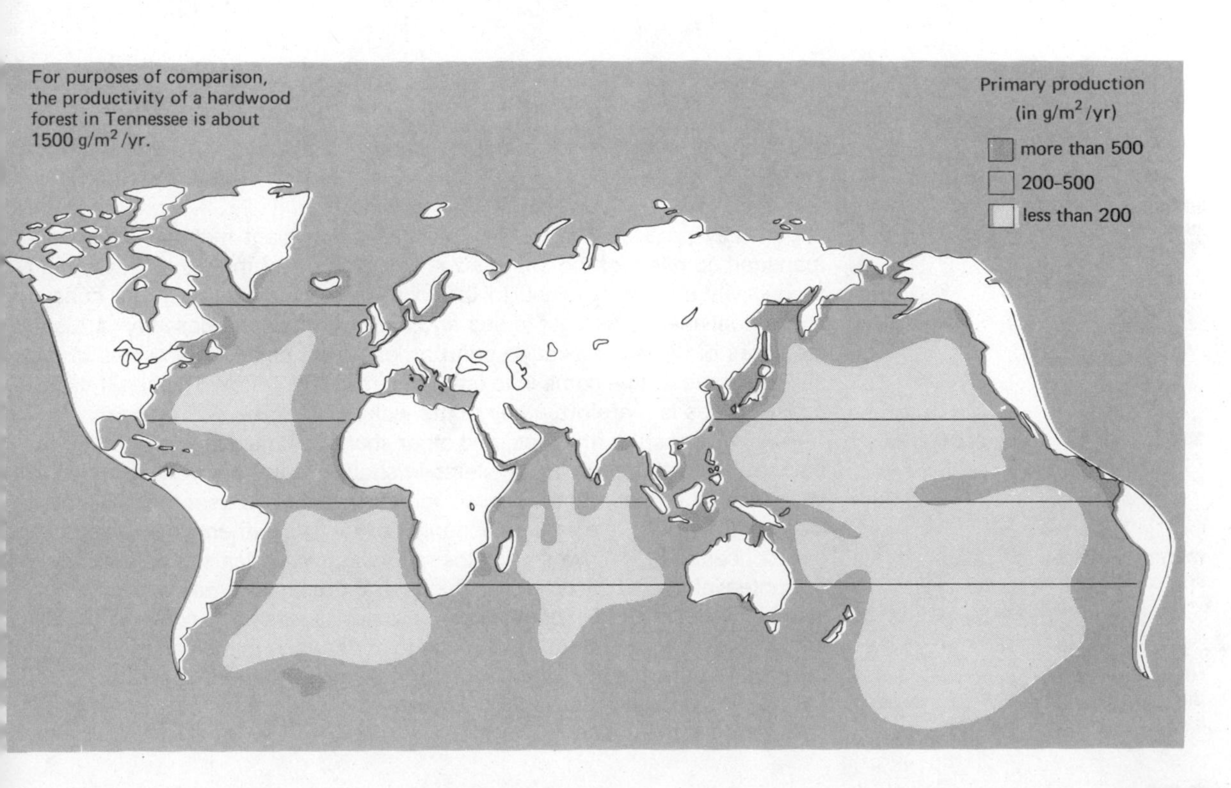
For purposes of comparison, the productivity of a hardwood forest in Tennessee is about 1500 g/m2/yr.
Primary production (in g/m2/yr)
more than 500
200-500
less than 200

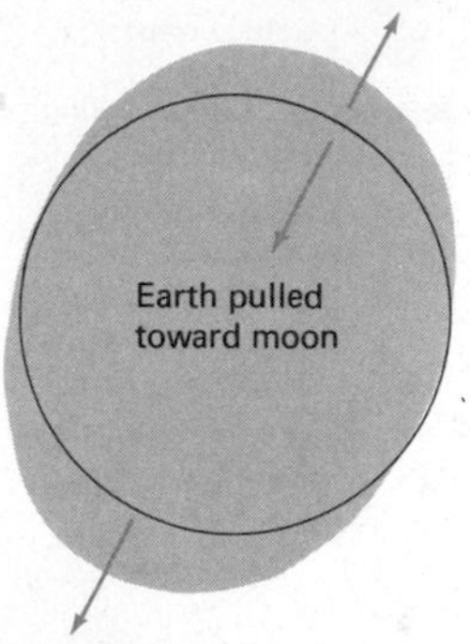

Centrifugal force pulls water out

Earth pulled toward moon

Gravity pulls water toward moon

FIG. 13-11. Interplay among the moon, the ocean, and a rotating globe. Life near a seacoast is governed by the regular ebb and flow of the tides, as sea level rises and falls in response to the gravitational pull of the distant sun and moon. The moon can only be on one side of the earth at a time, yet it produces two distinct areas of high sea level. One high tide consists of water pulled away from the surface on the "front" of the earth, the part that faces the moon. The other high tide occurs on the "back" of the earth, because the moon literally pulls the earth out from under the water. (Or, if you prefer, both earth and moon revolve around a common center of gravity.) The level of the sea is lower on the "sides" of the earth, between the two areas of high water. The tides appear to sweep around the earth because the earth rotates "beneath" the two zones of high water.

The regular pattern of lunar tides is complicated by the sun, which produces its own set of tides. Solar tides are only about half as high as lunar tides because the sun is much farther away and its gravitational pull is correspondingly weaker than that of the moon. The sun serves mainly to strengthen or weaken the lunar tides. *Spring* tides are especially high ones that occur during the new or full moon when the sun and moon are in line. Low or *neap* tides are the results of gravitational pulls at right angles to each other when the moon is in quarter phase.

The irregular arrangement of the continents superimposes another complication on the tides. Local conditions also affect the periodicity and intensity of the tides. Some places experience two tides of equal height every day. Other places have two unequal tidal cycles each day. Still others have only one daily period of high tide. Finally, some places have different patterns at different times of the month. These different patterns are evident on the monthly tide charts for three United States ports on the facing page.

shore. Tides flow into the marshes along the shore, stir up the soil, and aerate the water (Fig. 13-11). Plants in the shallow water respond to the nutrient inflow and tidal mixing by growing bountifully. A typical coastal marsh may produce five to ten times as much plant material as a well-managed cornfield of the same size. The benefits of this fantastically high productivity are not confined to the local area. Nutrients from the coastal zones continually wash out to sea and enrich the ocean ecosystems. Many creatures of the deep ocean spend at least part of their life cycles in the shallow areas. Some come in to feed, others to reproduce. The health of the coastal areas is therefore a key to the well-being of the entire ocean.

Reefs, saltwater marshes, and other shore communities are productive, but they are also fragile. Ecosystems in shallow water are very sensitive to changes in the depth, temperature, volume, nutrient content, and sediment loads of the water. The same mechanisms that trap nutrients near the coast also concentrate pesticides and other poisonous chemicals, thus increasing the potential impact of pollution. Because the coastal areas are linked so intimately with the health of the ocean, it is not necessary to contaminate the entire ocean basin to do great damage to the sea. As in the case of the lake that opened this chapter, the ocean shows few symptoms of pollution until the disease is quite far advanced. By that time the cure may be very expensive, perhaps impossible.

spring tide high tide that is above average

neap tide high tide that is below average

salt marsh grassy areas in the tidal zones near shore

mangrove forested areas in the tidal zone near shore

reef rocklike shelf or barrier made of the "skeletons" of living and dead coral

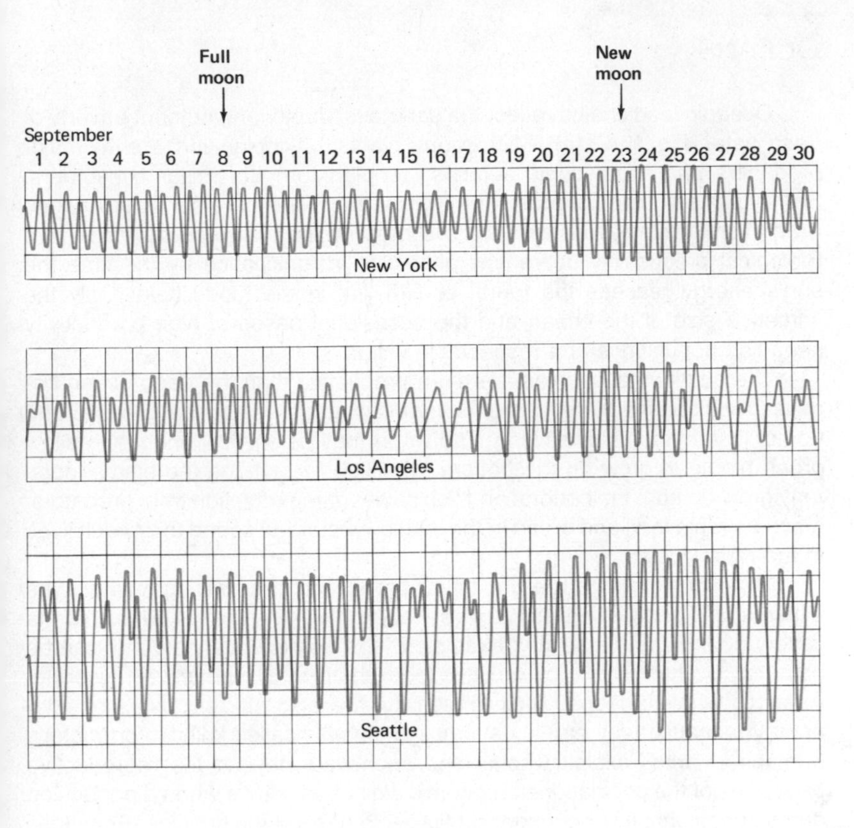

APPENDIX TO CHAPTER 13
Aquatic Biomes

Watery environments offer a variety of lifestyles to their potential inhabitants. Differences in nutrient supply, oxygen concentration, depth, temperature, light intensity, and salinity all evoke different associations of plants and animals. Like the terrestrial biomes in the Appendix to Chapter 11, aquatic biomes are characterized by internal variations, transitions between types, and migratory linkages. The figures show the typical inhabitants of each biome and the flows of nutrients among them; the widths of the flow lines indicate the comparative importance of various pathways and groups of organisms.

OCEANS

Oceanic food chains reflect the darkness, depth, and nutrient poverty of deep water [Fig. 13A-1(a)]. Microscopic plants (phytoplankton) are the major producers and microscopic animals (zooplankton) the major consumers. Some of the zooplankton are carnivores; they feed on other zooplankton. Small fish or shrimplike animals are next in line, followed by large fish. Biological productivity decreases almost to insignificance by the time the sun's energy reaches the fourth or fifth link in the food chain. Only the immense size of the ocean and the occasional oases of high productivity keep fish populations at a respectable level.

The continental shelves are more productive than the deep ocean because they receive nutrients from rivers flowing off nearby landmasses and do not lose nutrients by sinking. The water may be shallow enough to allow green plants to grow on the bottom. Shellfish live among the plants; coral may grow up from the bottom; and fish have some protection from predators. Many animals feed and breed in the shallow water but spend their adult lives in the deep ocean.

Microscopic plants are the major producers in upwellings, where the water is cold and nutrient-rich but very deep [Fig. 13A-1(b)]. The anchoveta fishery near Peru is a striking example of the tremendous yield that can be obtained by harvesting low on the food chain in a productive upwelling. Antarctic whales, despite their immensity, are also efficient users of solar energy, because they eat tiny shrimplike animals called krill. Unfortunately, humans at times exploit whales and anchoveta beyond the reproductive capacities of the populations. Economic extinction occurs when a population drops so low that it is no longer profitable to harvest the species. Economic extinction may be followed by physical extinction if the species does not have a high reproductive rate. Anchoveta are less likely than whales to go extinct, because anchoveta are prolific breeders with short lifespans. Whales, by contrast, are like human beings in many respects: they have long lifespans, do not reproduce until late in life, and bear few offspring per adult. Physical extinction therefore looms as a real possibility if whaling is not controlled.

REEFS

Reefs owe their existence and distinctive character to a unique alliance between algae (noncellular plants) and coral (rigid-walled animals). The animal skeletons provide a sturdy foundation that can grow upward to adjust to gradual changes of sea level. The coral hold the algae up in the bright sunlight near the ocean surface. The algae, in turn, produce abundantly in the warm water and provide food for the coral. Both are nourished by the back-and-forth movement of the tides [Fig. 13A-1(c)].

Coral reefs can form only in shallow warm water. Relict reefs in the middle latitudes are symptoms of climatic change or crustal movement. Tall

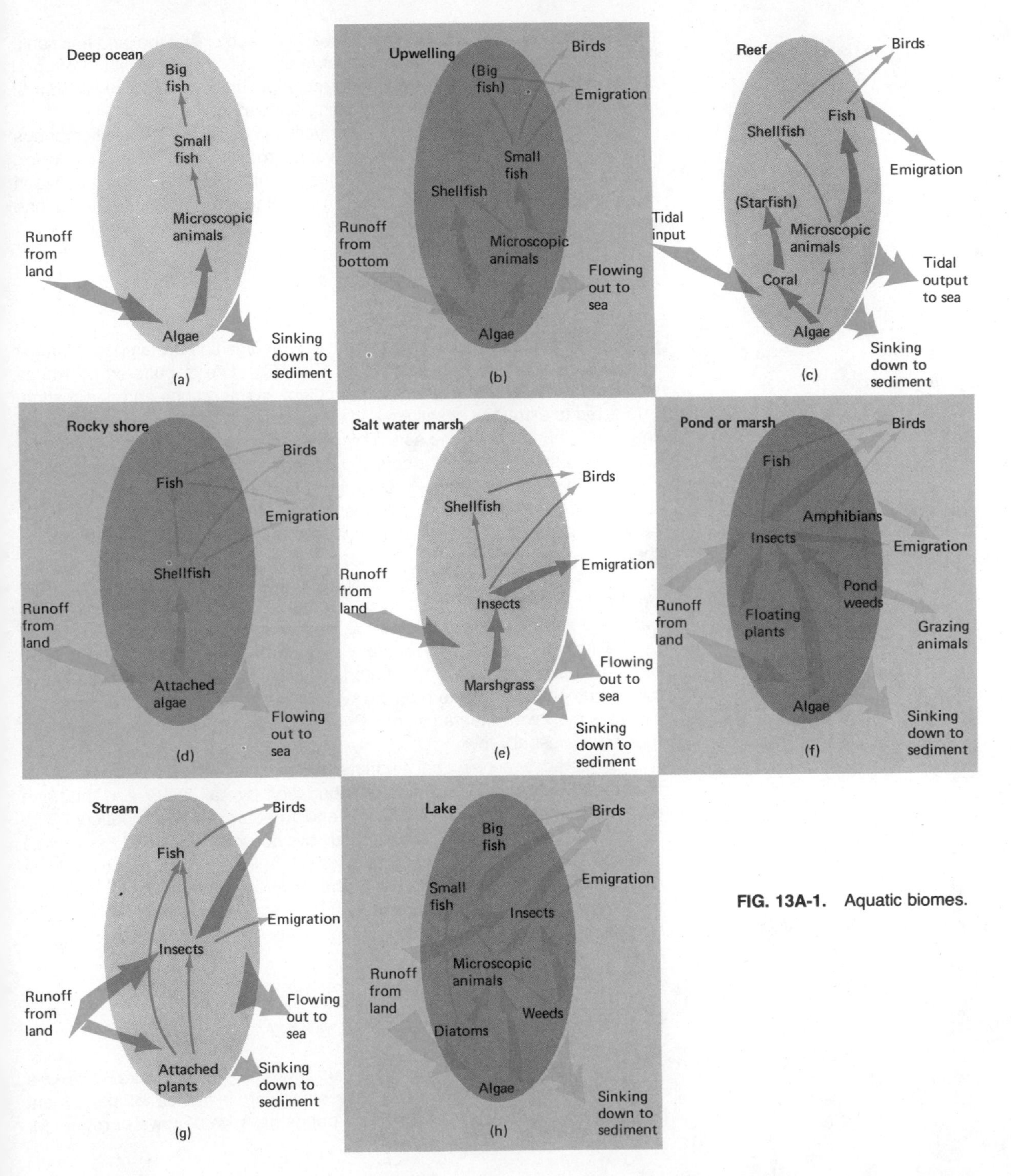

FIG. 13A-1. Aquatic biomes.

reefs in deep water indicate that sea level has risen or that the land has sunk. The classic example of a growing reef is an atoll, a circular island of coral with a hollow center. The hole was formerly occupied by a rocky island that slowly sank while the coral kept growing upward.

A coral reef is an exceedingly productive ecosystem. Tidal action brings nutrients into the area, removes toxic waste products, and aerates the water. The reef supports a complex assemblage of higher forms of life that live in the shelter of the coral and subsist on the algae, the tidal detritus, or one another.

TIDAL ZONES

Rocky shores that are under water only at high tide are among the most rigorous environments on earth [Fig. 13A-1(d)]. Life is buffeted by waves and wind, threatened by alternate periods of submergence and desiccation, subjected to extremes of salinity and temperature, and exposed to predators from both the land and the sea. The ability to cling tightly to the rock is the most successful strategy for coping with the stresses of a rocky tidal zone. Snails and slugs creep slowly over the rocks and consume the masses of sticky algae (seaweed) that grow there. Other shellfish simply attach themselves to the rock in a convenient place and let the tide bring their food to them.

Sandy beaches offer their animal inhabitants more protection from predators and the elements than rocky shores do. Burrowing into the loose sand is a common adaptive strategy, both for permanent inhabitants, such as shellfish and insects, and for many occasional visitors, such as turtles and birds. Plants, on the other hand, consider shifting sand a poor foundation, low in nutrients and apt to blow away from the roots or cover the leaves of even the most well-situated plants. Primary production on an exposed beach therefore is usually low.

Sheltered beaches and mudflats, by contrast, are favorable sites for plants that can tolerate occasional inundation by salt water and thus can reap the benefits of tidal circulation and freedom from competition [Fig. 13A-1(e)]. Salt marshes (dominated by cordgrasses) and mangrove swamps (dominated by trees) achieve high levels of productivity. The grasses and trees furnish food and shelter for a variety of consumers, although the basic stresses imposed by tides and changes in salinity keep the number of different kinds of organisms lower than in most other aquatic ecosystems.

PONDS

Pond life resembles life in the shallow margin of a lake, with one significant exception: ponds lack the stabilizing influence of permanent populations of large animals, because ponds have no deepwater refuge for animals to use during unfavorably cold or dry seasons. Ponds may freeze during a cold winter or dry up completely during long rainless periods. Sunlight usually penetrates to the bottom and allows plants to grow there, so that a pond does not have distinct temperature layers or a dark region occupied solely by decomposers and consumer organisms.

Ponds usually produce more plant material than lakes because plants can grow on the bottom and nutrients cannot sink out of the lighted zone [Fig. 13A-1(f)]. Algae and other microscopic plants are less significant in a pond than in a lake. Large plants such as marsh grasses, lily pads, or cattails are favored by the availability of nutrients in the bottom mud of a pond. Insects, both adult and larvae, are the major consumers of plant material in a pond. Frogs, other amphibians, and reptiles are important carnivores. Amphibious animals spend only part of their lives in water, so that they are uniquely adapted to the occasional disappearance of ponds.

STREAMS

Motion of the water is a major influence on life in a flowing stream [Fig. 13A-1(g)]. An individual drop of water usually describes a corkscrew path as it moves along, and the helical flow scoops up earth material and deposits it in alternate areas along the bed of the stream. The result is a sequence of deep pools and shallow riffles rather than a smooth course with uniform depth.

Water moves swiftly in the riffles; plants must attach themselves to the coarse bottom in order to stay in place. Clams and other attached filter feeders strain edible material out of the flowing water. Some insects are strong enough to move against the current and thus can select their residence; others are able to choose only between attachment to the bottom or drift with the current, depending on the adequacy of the food supply in a particular place. Microscopic floating plants and animals are rare in riffles but common in stagnant pools. The pools also contain many typical pond insects, fish, and mudworms.

Local production of green plant material is less significant in a stream than in most other aquatic environments. More than half of the diet of stream animals usually consists of leaves and other debris washed into the water from adjacent land. Therefore a typical stream contains more scavengers and decomposers than consumers of live plants and animals.

Streams that enter large lakes or oceans often serve as breeding grounds for fish that spend their adult lives in deep water. Minute differences in the odors of different streams enable many of these fish to return to the exact stream in which they were hatched. They produce large numbers of young in the relative safety of a tiny brook. The young fish, however, are too numerous for the limited productive capacity of the stream; they migrate out into deep water and feed on microscopic plants and animals or other fish in the larger body of water.

LAKES

Water is heavier at 4°C than at any other temperature. This unusual response to temperature is the reason for one of the key characteristics of deep lakes in the middle latitudes. The temperature of water near the bottom remains near 4°C, whereas the temperature of water near the top fluctuates from winter to summer. Seasonal temperature changes are important for fish, because they provide a means for oxygen recharge of the entire lake.

A lake is stable in the summer because surface water is warm and light and the dense cold water on the bottom has little tendency to rise. The oxygen concentration in the bottom water decreases since oxygen is difficult to replace once it is removed by fish or microorganisms.

In the autumn the surface water cools and becomes more dense. Lake water has a uniform density when surface temperature reaches 4°C, and even a slight breeze can set up a circulation that mixes surface and bottom water. The mixing, called the fall overturn, carries great quantities of vital oxygen down into the depths and brings nutrients to the surface.

The mixing slows when water near the surface gets colder than 4°C and therefore becomes lighter than the water below. Formation of ice on the surface stops the mixing. Ice rarely gets thicker than a meter or two because frozen water cannot conduct heat out of the lake very efficiently. Fish survive beneath the protective cover of ice until spring comes. Another overturn and vertical mixing occurs in the spring when the surface temperature again climbs to 4°C.

The productivity of a lake parallels its overturn cycle. The spring overturn often is the most productive time of the year. During the summer, nutrients are tied up in plant tissues or lost in the lightless depths. The fall overturn brings nutrients to the surface and induces another surge in plant growth, but low temperature and weak light depress production again in the winter.

Lake food chains resemble ocean chains rather than pond chains [Fig. 13A-1(h)]. Most plants in a lake are microscopic, although nutrient abundance during the overturns often produces a visible surface scum of algae. In an unpolluted lake, algae are eaten by microscopic animals. Small fish are next in line, followed by large carnivorous fish. Pollution stimulates algae while it hampers large animals by reducing the oxygen supply. The alternating periods of oxygen recharge and stagnation in the annual cycle of overturns and stratifications impose a seasonal rhythm in the ability of a lake to tolerate pollution.

Lakes in hot climates may not experience overturns because the top layers stay warm and buoyant. Scarcity of nutrients limits plant productivity while oxygen depletion hinders animal activity. The productivity of lakes in cold regions is depressed by weak sunlight and low temperature. In extreme cases the water is virtually sterile, since it may not get warm enough to melt the ice cap and cause an overturn.

Chapter 14

Water Budgets: Floods and Droughts

When Bailey's Crick gits up, ya can't get ta Suger Holler.

Philip Gersmehl

Ask any person on the streets of Kansas City why rivers are full in spring and dry up in late summer, and the likely response is a discourse about snow melting. The meltwater idea sounds reasonable, but like many lacy theories, it cannot fly against the cold wind of measurable fact. True, places like the Colorado Rockies receive several meters of annual snowfall, and the water released when this snow melts is a major resource for dry plains nearby. Nevertheless, more than one-third of the yearly precipitation on these mountains falls during the three summer months, whereas only one-eighth arrives in the form of winter snow. Central Indiana experiences many spring floods, but the annual snow accumulation of 60 cm (2 ft) represents less than 7 cm of water, about as much as may fall during a single summer thunderstorm. Georgia rivers also flood in early spring and shrink in summer, but the average annual snowfall is a mere 4 cm, the equivalent of half a centimeter of rain. In all of these cases, streams are large during seasons with comparatively little precipitation and small during the rainy summers.

precipitation water falling from the atmosphere; includes drizzle, rain, sleet, snow, hail, and dew

THE WATER BUDGET

hydrologic cycle circulation of water from sea to air to land and back to sea

Apparently, there is no simple connection between cloud and river. In fact, most of the precipitation that falls on the ground never reaches the rivers (Fig. 14-1). A recognition of alternative destinations for rainfall is the key to understanding spring floods.

The temperature of the air and other characteristics of the energy budget at a place have considerable control over the water budget there. A weak winter sun does not provide much energy to evaporate water, and cold winter air cannot hold much water vapor. Moreover, many plants are dormant during the winter. The lack of evaporation keeps the soil filled with water, and excess water runs off to the ocean.

Kansas City receives five times as much precipitation in the summer as in the winter, but the intense summer sun can evaporate even more water than the clouds deliver. Evaporation and plant growth deplete the supply of

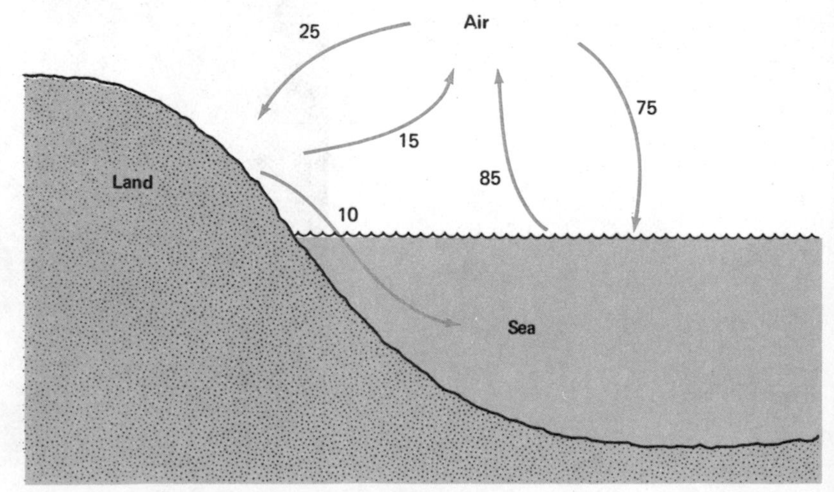

FIG. 14-1. The hydrologic cycle of the earth. The total amount of water in motion during a given year is less than one percent of the amount stored in the ocean. The figures on the arrows indicate volumes of water expressed as percentages of the total annual precipitation over the entire world. Thus, 25 percent of the total falls on the land, but only 15 percent of the total evaporates from the land, so 10 percent of total rain becomes runoff from land to sea.

water in the soil, and the thirsty soil absorbs a large share of the summer rainfall. The result is a decrease in the size of rivers during the season with the most precipitation.

The fundamental idea of a tradeoff between precipitation and evaporation was presented in Chapter 5 as the basis for a specialized classification of climate. Air temperature, sun intensity, daylength, wind speed, relative humidity, and cloudiness are combined in order to determine the potential evapotranspiration for a given period of time. The primary purpose of the Thornthwaite procedure is to estimate the adequacy of rainfall for agricultural purposes, but the method also does a good job of predicting seasonal river volume at different places (Fig. 14-2). The framework of a water budget, like that of an energy budget, is a way of connecting a variety of observations and experiences. The value of the approach can be illustrated by examining water budgets in a number of places.

potential evapotranspiration amount of water that could evaporate from a surface if there was an unlimited supply of available water

Thornthwaite method procedure for estimating potential evapotranspiration and for comparing it with available precipitation in order to work out a water budget
(see Table 5-1)

FIG. 14-2. Annual water balance at a number of places on the earth. Water surpluses do not necessarily flow off the land during the season in which they occur. Some of the water surplus may be stored as snow or soil water and released later. For this reason, rivers may flow even during a dry season, if the water surplus is sufficient and the dry season is not too long.

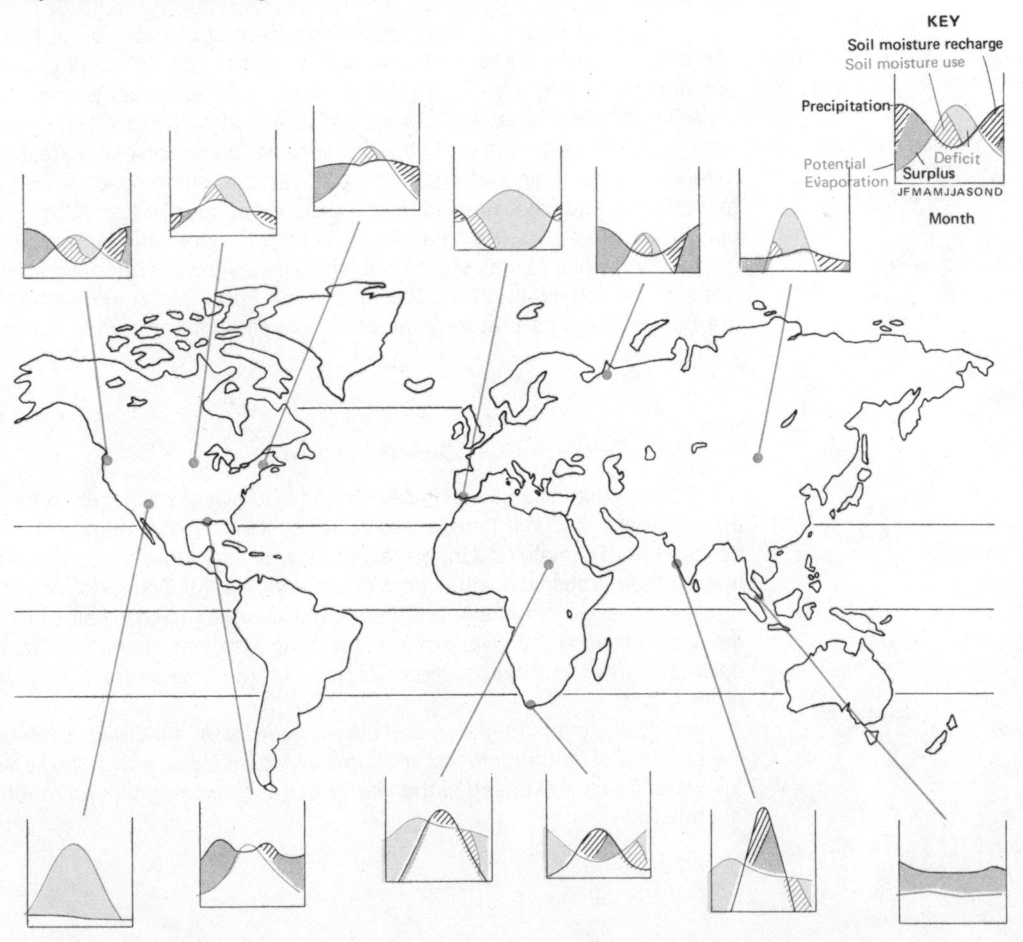

A Water Budget for a Forest Region

megalopolis popular term for the interconnected urban area extending from Boston to Washington, D.C.; also called the Northeast Urban Corridor

The idea of isolated cities is no longer appropriate for the region around New York City. Rather, there is a complex of interconnected urban areas stretching along the coast from Maine to Virginia. Some of the cities outgrew their local water supplies a long time ago. Their drinking water now comes through aqueducts that extend far into the hilly interior of the continent. The pipes and canals originate in reservoirs in the mountains, where water is clear and clean.

siltation filling a reservoir with sediment
(see Figs. 7-6 and 15-11)

Some of these reservoirs were not always so clear. Half a century ago, land around these lakes was used for farming. Rains beat down on the exposed soil between crop rows and carried earth into the reservoirs. The resulting cloudiness of the water was a minor nuisance compared to the threat of mud filling the reservoirs and rendering them useless. Sediment removal from a reservoir is prohibitively costly, forcing the engineers to find another solution. The most obvious alternative was to stop the sediment flow at its source by keeping soil on the land, and a good way to do this was to establish a forest or other permanent vegetation cover on the slopes around the reservoirs.

gauging station place fitted with a weir or other device for measuring streamflow

watershed area drained by a single river and its tributaries; a large watershed may contain several smaller ones

transpiration evaporation of water from tiny openings on plant leaves

A forest of pine trees may effectively control erosion, but it has other impacts on nearby lakes. Pines are able to survive on 60 cm (24 in.) of rain per year, but if they are given twice as much, spread reasonably throughout the year, they are capable of consuming nearly all of it (Fig. 14-3). The ability of pine trees to evaporate water has dire consequences for reservoirs, because reservoirs owe their existence to the favorable outcome of a very basic hydrological equation: precipitation equals runoff plus evaporation. Planting pine trees increases the total evaporation of water, and any increase in evaporation while rainfall stays constant will inevitably reduce the amount of water in the creeks that feed the reservoirs. Fortunately, the eastern cities are not yet so large that they face the ugly dilemma of muddy water or no water at all.

A Water Budget for a Desert Region

subsidence climate dryness that is caused by sinking air
(see Fig. 3-8)

The irrigation districts of the American Southwest are not as fortunate as the eastern cities. The land is arid because it lies in the heart of the subsidence type of climate and in the rainshadow of lofty mountains. The farmers irrigate their crops with water from rivers such as the Colorado, which originates in the snowy mountains. Spreading river water on crop fields increases the loss of water by evaporation and crop use; the inverse relationship between runoff and evaporation dictates that the size of the river must decrease.

rainshadow dry area on lee side of mountain
(see p. 86)

Ironically, even the dams and other water storage facilities contribute to the problem of inadequate water supplies. A dam slows a river and enlarges the water surface exposed to the air. Evaporation increases, and runoff must decrease.

A

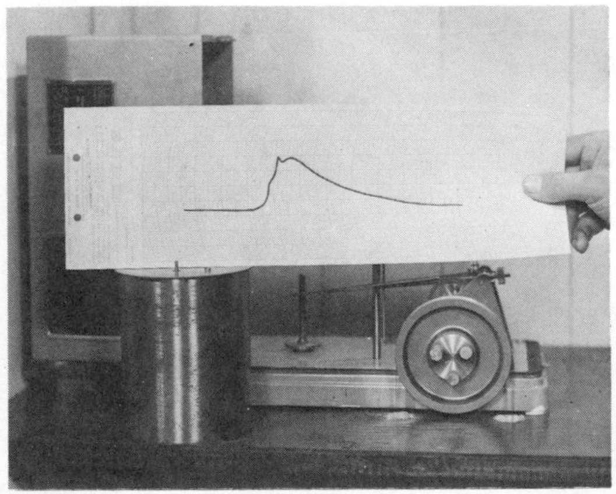

B

FIG. 14-3. The white pine watershed at the hydrologic experiment station near Coweeta, North Carolina. White pine trees were planted on all of the land drained by one small creek, and a gauging station was constructed to monitor the changes in stream volume as the pines grew. Inset A shows a weir, a place where stream flow can be measured very accurately. The tiny building beside the weir contains a depth gauge that prints a record of river stage at all times. Inset B shows a graph of stream response to a typical storm. It is impractical to build weirs over large streams and rivers, but rectangular openings through bridges can serve the same purpose of providing a regular cross section for the depth measurements. Therefore tiny depth-gauge shelters are common sights near highway bridges on many streams. River flow is presently measured at more than 3000 gauging stations in the United States. (Courtesy of A. Hubbert (main photo) and L. J. Prater (insets), U.S. Forest Service.)

Long ago the states along the Colorado River signed a complicated water-allocation agreement that specified how the measured annual flow of the river should be divided. Subsequent construction of a number of reservoirs and diversion projects has increased total evaporation and seepage losses. Annual riverflow is no longer as great as it was, and someone must make do with less water than expected. Not surprisingly, water rights may be just as important as malpractice and divorce are as generators of lawsuits in some arid regions.

A Water Budget for a Suburban Area

floodplain frequently flooded area of lowland near a river

The problem in suburban Chicago is hardly one of too little water. Rivers and creeks overflow their banks and inundate suburban houses nearly every spring (Fig. 14-4). Old residents of these towns can remember when the rivers were well-behaved and rarely rose "beyond the third brick on that bridge abutment."

The reason for the recent delinquency of the creeks is not hard to find. All around the formerly small towns, new housing developments are replacing fields, forests, hills, and swamps. Streets and roofs keep the rain from sinking into the soil, and ditches and sewers carry water away before plants can use it. The energy budget demonstrated how temperature must rise as evaporation decreases; the water budget now asserts that stream runoff also must increase as evaporation decreases. The additional flow is most noticeable in the spring, when the rivers in this region ordinarily would be full anyway. The resulting floods, however, are emphatically not "natural catastrophes."

FIG. 14-4. Flood near Chicago. A flood such as this may actually be a consequence of the process of suburbanization. (Courtesy of E. E. Offerman, Soil Conservation Service.)

REFINED WATER BUDGETS

The hydrologic equation outlined above (precipitation equals runoff plus evaporation) is valuable as an interpreter of general characteristics of river flow. Accurate flood prediction, however, demands some refinements in the basic hydrologic equation. One way to improve the basic equation is to consider it as a number of separate parts, each treating a different aspect of the water budget (Fig. 14-5):

1. Precipitation must be allocated among evaporation, runoff, or temporary storage. The difference between this equation and the basic water budget is the inclusion of the idea of storage. Measurements of rainfall, evaporation, and runoff for a given month rarely yield a balanced equation, because the soil is capable of hiding an appreciable quantity of water from

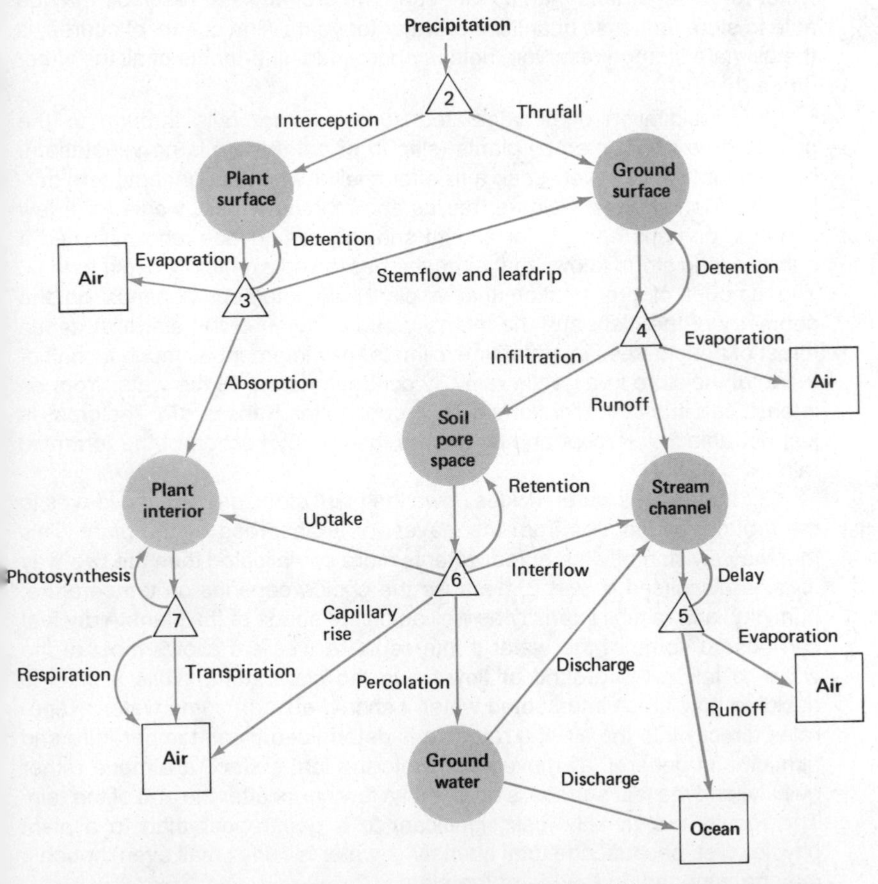

FIG. 14-5. Compartmental model of a local water budget. The circles represent temporary storage areas. The squares are final destinations. The triangles indicate decision points, where incoming water must separate and go to alternative destinations. Various characteristics of the environment influence each decision, as described in the correspondingly numbered sections in the text.

observers on the surface. The delayed discharge of soil water often produces the puzzle of more riverflow than rainfall in a dry month. Over the long term, of course, deposits and withdrawals from the soil storage bank should be balanced. Over the short term, however, changes in rainfall or water use generally provoke changes in the storage part of the hydrologic equation. Like the nutrient cycle, the water cycle takes awhile to return to its equilibrium.

The idea of storage appears first in the list of modifications of the basic water budget for an important reason: all other parts of the basic equation also have an implied storage term. Water can delay awhile at every one of the junctions where it has to choose between alternative destinations. For some of the decision points, such as the surface of the ground or a plant, the amount of storage is small and storage time is short. Soil can hold more water for a longer time than a plant can. The groundwater reservoir may be able to store immense quantities of water for years. The ocean, of course, is the ultimate storage reservoir, holding more than nine-tenths of all the water on earth.

groundwater table top of the saturated zone in the soil. The water table is a sensitive indicator of storage trends; if more water is removed than is replaced, the water table goes downward
(see Fig. 8-9)

2. Precipitation on a vegetated surface either falls through to the ground or is intercepted by plants (skip to part 4 if there is no vegetation). Many people instinctively seek a tree for shelter when caught outdoors during a sudden shower. Using a tree as an umbrella usually works for a few minutes, perhaps enough for a brief shower, but anyone who has tried it during a long storm knows that even the best trees eventually begin to leak. The amount of precipitation that a plant can intercept depends on the geometry of the plant and the intensity and duration of the storm. A dense forest of needle trees in a maritime climate may intercept as much as half of the total moisture in a gentle rain. By contrast, nearly all the water from an intense cold-front thunderstorm gets through to the Kansas soil. The grass is just not able to intercept and hold more than a tiny fraction of the torrential rain.

interception precipitation that is caught by plant structures

throughfall precipitation that is not caught by plants

3. Intercepted water trickles down the plant stem, drips off the leaves to the ground, evaporates from the leaves, or is absorbed by the plant. This four-way division of intercepted water is more complicated than the two-way division discussed in part 2, because the choice depends on temperature, humidity, and rainfall intensity as well as on the shape of the plant. A dry leaf can absorb some of the water it intercepts. A wet leaf allows most of the water to fall to the ground or flow down the plant stem. While a plant is deciding how much intercepted water it should absorb, some water evaporates directly into the air at a rate that is determined by air temperature and humidity. In general, all movements from the leaf surface are made rather quickly, and the leaf surface is dry within a few hours after the end of the rain. The results are usually less significant to a geographer than to a plant physiologist, because the total quantity of water is fairly small even though it may be important in the life of the plant.

stemflow intercepted water that subsequently flows down the plant to the ground

leaf drip intercepted water that subsequently falls directly to the ground

4. Throughfall (from part 2) plus stemflow and leaf drip (from part 3) combine to become water on the ground surface, and the surface water must then infiltrate, evaporate, or flow off as overland runoff. Infiltration is like a child eating candy: it starts fast and then declines to a slower rate as the

infiltration water that enters the ground

overland runoff water that flows away on top of the ground

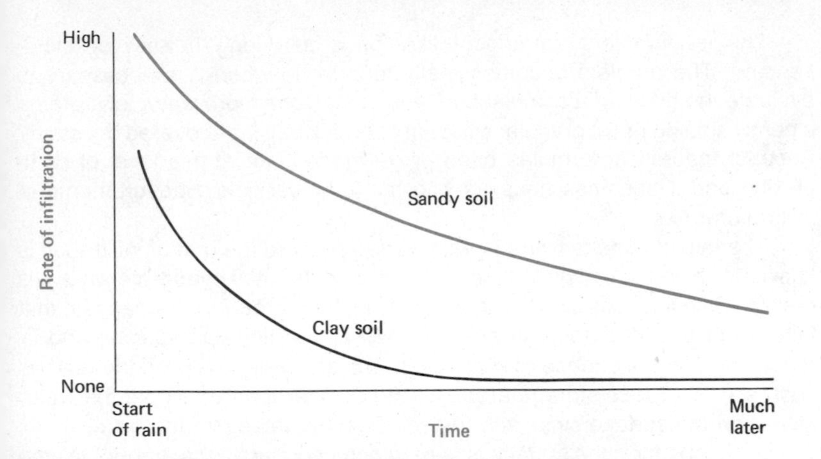

FIG. 14-6. Infiltration curves for two different soils. The rate of infiltration usually declines as soil becomes wetter. One scientist suggests that the process is like people going into a theater. The first ones inside can find seats easily, and therefore they enter quickly. Latecomers enter a crowded theater and must stand in the aisles while looking for seats. Many soils delay the entrance of water even further, because the soil particles enlarge as they are moistened, and the expansion makes the "aisles" narrower than they were when the soil was dry.

emptiness inside is filled (Fig. 14-6). The entry of water into a soil is a very complex process, but its importance can hardly be overstated: infiltration stands at a crucial place in any study of temperature, groundwater, plant health, soil fertility, erosion, flooding, or water quality.

It is easy to look at the multitude of factors that affect infiltration and to evaluate the probable effect of each factor by itself. A dry soil can absorb more moisture than a wet one. A coarse-textured soil can take water faster than a fine-textured one (Fig. 14-7). A deep soil can accept more water than a shallow one. Water has less time to sink in on a steep slope than on a flat area. Granular soil structure promotes infiltration, while massive structure inhibits it. A leaf cover on a soil aids infiltration, whereas an unprotected soil is usually hardened and sealed by sun and raindrop impact. Finally, gentle drizzles have more chance to infiltrate than torrential rains.

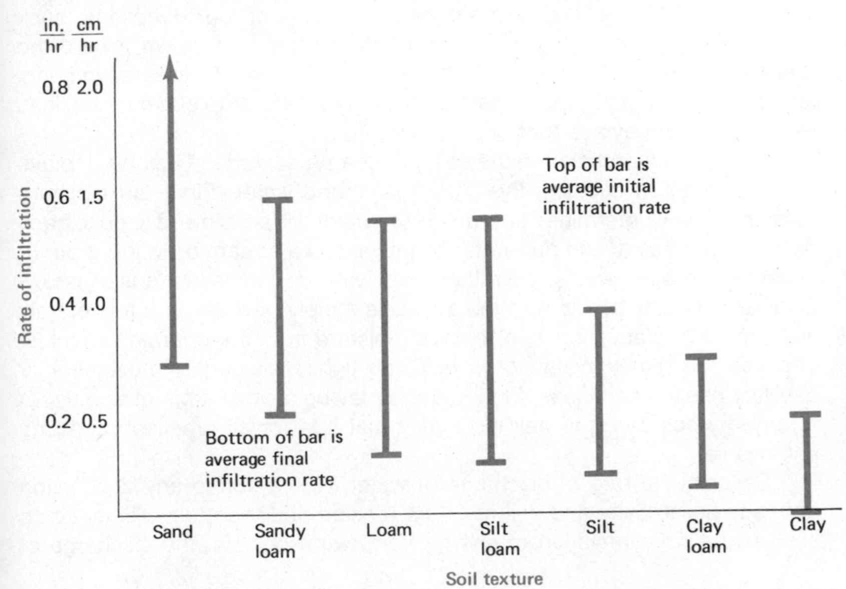

FIG. 14-7. Ranges of infiltration rates for a variety of soils. The initial infiltration rate is the amount of water a dry soil can absorb in an hour. The final infiltration rate is the amount of water that can pass through a completely saturated soil in an hour.

The list of factors that affect infiltration is quite long, but not too mind-bending. The problem becomes really formidable when it is necessary to evaluate the effect of a combination of factors. Does more water infiltrate on a gently sloping moist granular silt loam or on a steep leaf-covered dry sand? Abstract theoretical formulas often prove inadequate at this level of complexity, and it becomes necessary to make actual field measurements of infiltration rates.

infiltrometer cylindrical device for measuring the rate of infiltration

The fate of surface water is easy to decide once the rate of infiltration is known for a particular soil. A small amount of water may evaporate while it is raining. Rain that falls slowly usually sinks into the ground, whereas rain that falls faster than the soil can absorb it will fill the tiny depressions and irregularities on the surface of the ground. Excess water flows off the land as soon as the surface storage areas are filled. After the rain stops, the water stored on the surface sinks into the soil or evaporates into the air.

surface detention water that is temporarily stored in minor depressions of the ground surface

5. Runoff from the surface is separated according to the amount of time it takes to reach the main river. Mountain folk are acutely aware of the "flashiness" of streams on steep slopes. Creeks rise shortly after the rain falls because sloping land carries runoff to the streams quickly. Gentle slopes and hillsides cluttered with debris give water more time to get sidetracked by filling a small hole here, evaporating a little there, or sinking into a patch of coarse soil in another place.

sheetwash runoff that is not organized into channels

rill erosion runoff that is organized into small channels
[see Fig. 15-6(a)]

The effect of slope is complicated by the size and shape of the area drained by a creek. Water from all sectors of a circular drainage area arrives at the bottom at the same time and produces a higher flood than if the watershed were elongaged (Fig. 14-8). Moreover, a storm is not likely to be equally intense in all parts of a large watershed. Different parts of the drainage area usually send their biggest floods toward the main river at different times (Fig. 14-9).

6. Water that infiltrates a soil can go in one of four directions: back toward the surface by capillary action, into a plant root, down toward the groundwater storage area, or off into a stream or river. All four movements take place underground and therefore are more difficult to observe and less well-known than events that occur aboveground.

capillary rise upward movement of soil water from a layer of little tension toward a layer with great soil moisture tension
(see Fig. 8-12)

The moisture tension in the soil, an idea presented in Chapter 8, plays two vital roles in deciding the fate of infiltrated water. First, the moisture tension in a soil determines how much soil water will be allowed to go to each destination. Gravity can pull water downward to a stream or to the groundwater table only when a soil is rather full of water and moisture tension is low. Capillary rise and plants split the available supply of water in a moderately wet soil. Only plant roots can remove moisture from a moderately dry soil, because the remaining water is held too tightly for gravity movement or capillary rise to pry it loose. Finally, there may be a small amount of water in a very dry soil, but it is held so tightly that it is for all practical purposes immovable.

field capacity amount of water in a soil when gravity has removed all it can

wilting point amount of water in a soil when plants have removed all they can
(see Fig. 8-8)

Second, the rate of discharge of water from a soil to any destination depends on porosity and soil moisture tension and therefore ultimately on soil texture. Like radiation of energy from warm objects, the discharge of

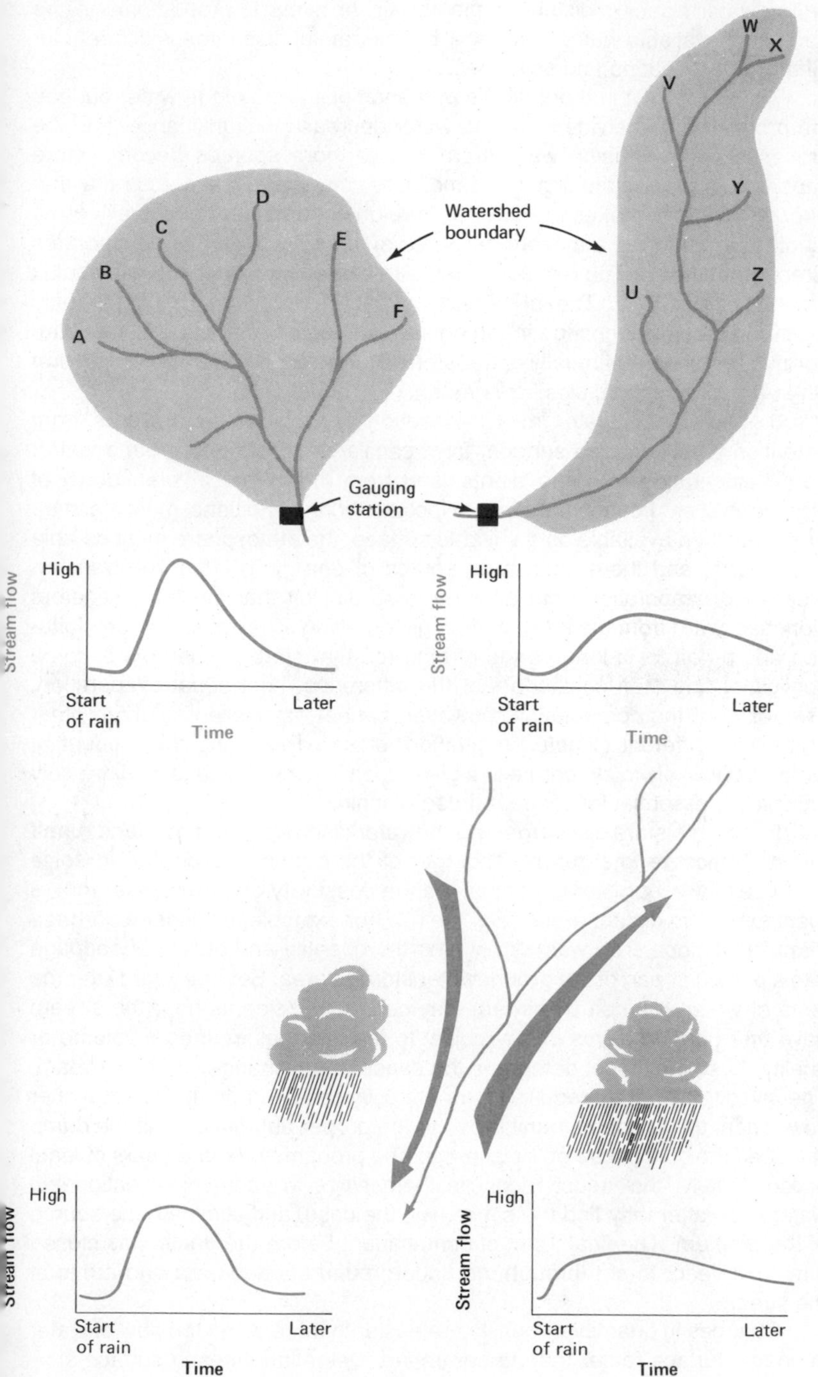

FIG. 14-8. Effect of watershed shape on river flow. Runoff from places A through E of a round watershed goes by the gauging station about the same time. Runoff from places W and X on the elongated watershed must travel farther to the gauging station and therefore arrives later than water from places U and Z. The result is a lower flood that lasts longer on the elongated watershed than on the round one.

FIG. 14-9. Effect of storm direction on river flow. The graph on the left shows a flood caused by a southward moving storm on the Chattahoochee River. The flood is large because runoff from the land in the southern part of the watershed is synchronized with the arrival of floodwater from the earlier rain up north. A northward moving storm produces a much less severe flood, because southern floodwater is long gone by the time runoff arrives from the delayed rain in the northern part of the watershed.

discharge rate percentage of soil water that is discharged into a stream or spring each day

photosynthesis production of food by a green plant

respiration combustion of food; conversion of hydrocarbons and oxygen into water and carbon dioxide and energy
(see Table 13-1)

latent heat energy absorbed when a substance evaporates and released when it condenses; the latent heat of water is about 600 calories per gram (see p. 10)

interflow water that flows through the soil quickly to the stream; the significance of this kind of intermediate flow is debated

flushing time amount of time for all of the water in a lake or other storage area to be replaced by raw inflowing water

groundwater or soil moisture to plants, air, or rivers is proportional to the amount of available water in storage, but the rate of discharge is different for different destinations and soil types.

7. At this point in the analysis of the various parts of the water budget, the processes that divide incoming water decrease in significance, and the processes that combine water from two or more sources become more important. A plant is the first of the moisture combiners: it absorbs rainwater through its leaves, takes in soilwater through its roots, and eventually gives all of its water to the atmosphere. Some of the water in a plant evaporates directly into the air. The rest combines with carbon dioxide to make complex organic compounds. The storage of water in organic form is the only significant chemical change in the entire hydrologic cycle, but it is only temporary. The water eventually is transferred to the atmosphere in vapor form when the plant grows, dies, or is eaten.

8. The air gets water from interception by the surfaces of plants, from detention on the ground surface, from capillary rise of underground water, from transpiration by living plants, and from the chemical breakdown of organic matter. For evaporation to occur, three conditions must be met. Water must be available on a suitable surface, the atmosphere must be able to accept it, and there must be a source of energy to effect the transfer. Measuring evaporation from an area is so difficult that most investigators work backward from the basic hydrologic equation. They measure precipitation and runoff for a long period of time (so that storage changes become negligible) and then conclude that the difference must equal evaporation. The result of the computation, however, cannot be applied safely to other areas with different climate, vegetation, or soil. Predicting the amount of water that is likely to evaporate in a given area is a frustrating task, extremely difficult yet essential for wise land-use planning.

9. Soil moisture discharge, groundwater discharge, and overland runoff join and become total runoff. This part of the equation is difficult to solve because of the long storage times and the possibility of groundwater movement to or from distant areas (Fig. 14-10). For example, consider a garbage dump on a slope. The waste may alter the quantity and quality of seepage into a particular part of the groundwater storage area. Several years later the taste of water in a nearby stream may change. Residents near the stream have three big problems if they object to the changes in stream volume or quality. First, they must determine the cause of the changes in their stream. The investigation may require years of costly research on the groundwater flow. Then, the residents must prove to the proper authorities that the dump is indeed the sole cause of the change. The proof may require years of legal action. Finally, they must reconcile themselves to years of objectionable water even after they find the cause, win the case, and eliminate the source of the problem. The final input of contaminant before the dump was closed may take years to get through the underground storage area and arrive at the stream.

Changes in quantity or quality are easier to trace when the flow of water is on the surface rather than underground. Detention times in surface storage areas usually are short, and responses follow soon after changes. Moreover, the flow is visible and can be monitored anywhere along its way. Even so, it is often difficult to obtain the legal proof needed to compel a polluter to stop.

10. The results of parts 8 and 9 join to yield the basic hydrologic equation again: total evaporation plus total runoff equals precipitation. The refined water budget provides a framework for the analysis of most hydrological problems. For practical planning purposes, however, a general theoretical model is less valuable than the specific characteristics of a particular river basin. The translation of general principles into practical information is the job of graphic tools such as output graphs, rating curves, hydrographs, routing maps, and diagrams of river regimes.

GROUNDWATER GRAPHS

Depth-yield graphs for wells are essential wherever people rely on groundwater storage for a significant part of their water supply. Any measurement of groundwater is inevitably uncertain because the subject is hidden from easy observation. Ultimately it is impossible for a groundwater storage area to yield any more water than it receives. However, the extremely long detention times in some groundwater storage ares make it possible for communities to pump water out of the ground for years before realizing that their annual withdrawals are greater than the annual deposits.

aquifer layer of soil or rock that is porous enough to allow water to flow (see Fig. 8-11)

FIG. 14-10. Separation of surface and subsurface flow. A sloping aquifer (water-carrying layer) may cause groundwater in the soil to go to stream X, while overland runoff water goes to a different stream (W).

recharge area area of surface that allows water to infiltrate and enter an aquifer

connate water water that was put into an underground layer at the time the rock or soil itself was deposited (see p. 381)

Prolonged and complex storage systems are most common in precisely the kind of environments in which groundwater is a vital source of water for human inhabitants. For example, the dry plains from Texas to Alberta are underlain by geologic formations that bring groundwater from wet areas far away. Some Nebraska towns and farms have wells that tap groundwater that entered the earth near the Rocky Mountains of Colorado. The problem of overwithdrawal of groundwater is extremely acute in parts of Texas, where wells are taking water out of layers that have no source of present recharge. Groundwater entered the storage area long ago, when the local climate was rainier. In effect, the people there are mining a nonrenewable resource.

Failure to recognize the depletable nature of the water supply can be triply disastrous. First, the population of the local area becomes far larger than the resource base can support. Second, people often turn to complex engineering solutions in order to prevent economic hardship when the resource is depleted. An elaborate water transfer plan often becomes merely an expensive way to perpetuate an unwise distribution of population. Third, the transfer itself has environmental consequences that also may be delayed by the complex nature of environmental systems. For these three reasons, groundwater yield graphs must be carefully interpreted (Fig. 14-11).

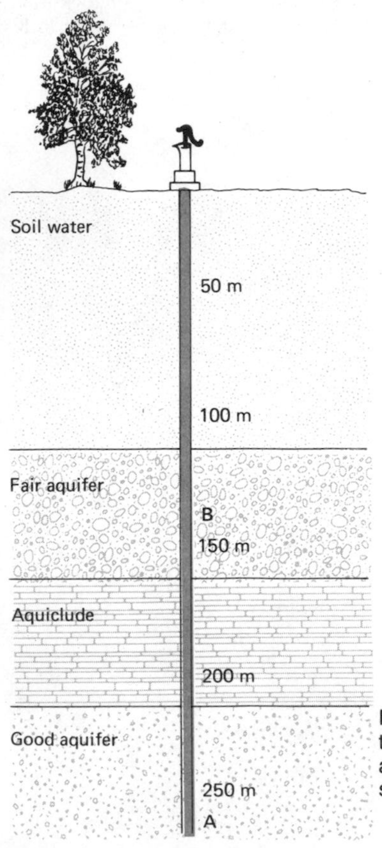

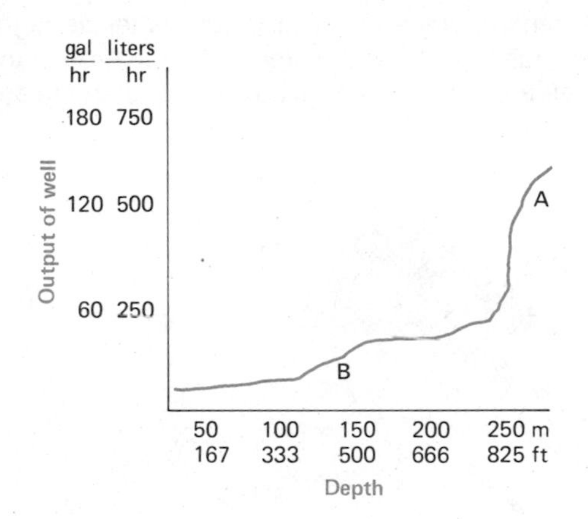

If the well were only as deep as B, the output would be 130 l/hr

If the well were as deep as A, the output would be 550 l/hr

FIG. 14-11. Depth-output graph for a well in the Great Plains. The deeper a well is, the larger the volume of soil it intercepts and the more water it can deliver. However, adjacent wells interfere with each other's recharge and yield less water than this graph suggests (see Fig. 8-10).

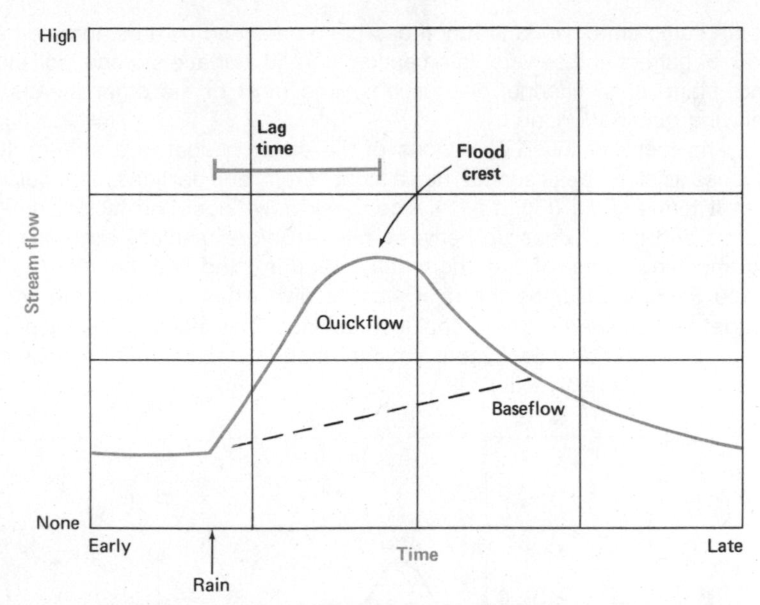

FIG. 14-12. Hydrograph of a typical stream. The dashed line separates the water into baseflow (water that originates as discharge from the ground) and quickflow or stormflow (water that originates primarily as overland runoff).

STREAM HYDROGRAPHS

The normal hydrograph of a stream is a graph of the typical response of the stream to a storm of average size (Fig. 14-12). The amount of delay between the storm and the flood crest, the height of the flood peak, and the rate of decline after the crest are all unique features of a particular watershed.

hydrograph graph of river response to an input of precipitation

stage water level in a river at a given time

flood crest highest level a river reaches following a storm

The owner of the "Hypothetical Creek Applesauce Factory" is acutely aware of the relationship between watershed characteristics and stream hydrographs. Nestled on the banks of Hypothetical Creek, the factory takes water from the gentle stream to cool its gleaming machines. The placid scene was recently disturbed by the construction of a large shopping center upstream. Thousands of square meters of forest were converted into paved parking lots and presumably waterproof roofs. Predictably, the removal of grass and trees drastically reduced the total evaporation of water. Elaborate root systems no longer draw moisture up to be transpired from millions of leaves, and the basic water budget allows little room for negotiation: if evaporation decreases, runoff must increase.

The additional runoff, however, does not merely raise the river level a proportionate amount at all times. Quite the contrary; the stream is now actually lower most of the time. The shopping center prevents infiltration of water over much of the former watershed. The soil receives less moisture than it did and therefore discharges less water to the river between storms.

baseflow part of a river that comes mainly from groundwater or soil moisture discharge

storm flow or quickflow part of a river that originates mainly as overland runoff

At the same time, overland runoff has been hastened by an elaborate network of gutters and sewers. Interception storage, surface storage, soil storage, plant roots, channel obstructions, and most of the other means of delaying runoff are gone.

The result of these alterations of the water budget is a stream that reaches its storm peak sooner, has a higher crest, and declines more quickly than it formerly did (Fig. 14-13). The creek now floods the factory during storms and nearly dries up between rains. (Before mentally accusing the big shopping center of extreme cruelty, keep in mind that the little Applesauce Factory assumed the river was its own when it altered the water budget and diverted water to cool its machines. The effect of the shopping center is greater only because it is bigger than the factory, not because big business is inherently evil.)

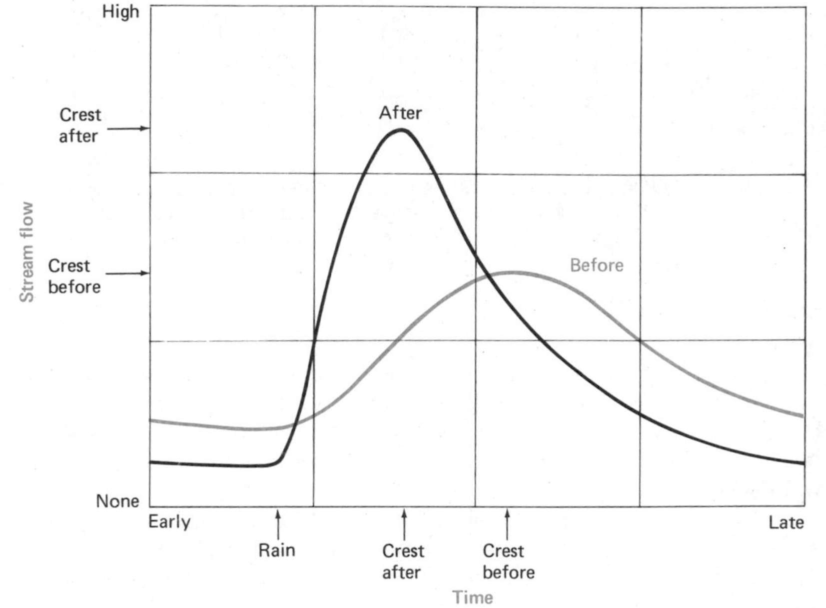

FIG. 14-13. Hydrograph of a stream before and after its watershed was partly paved.

REGIMES, STAGES, AND FREQUENCIES

A graph of the annual regime of a river shows the normal depth of the river for different months of the year. Rivers in most climates have distinctly seasonal regimes with times of high and low flow determined by the seasonal patterns of precipitation and potential evaporation (Fig. 14-14). Responses to individual storms in particular years, however, may deviate significantly from the average annual river regime.

The stage-frequency graph shows how often a river exceeds a particular depth of flow (Fig. 14-15). Contrary to a popular opinion, very few rivers stay in their channels all the time. A normal river will create a channel that is just large enough to contain all except the biggest flood of an average year.

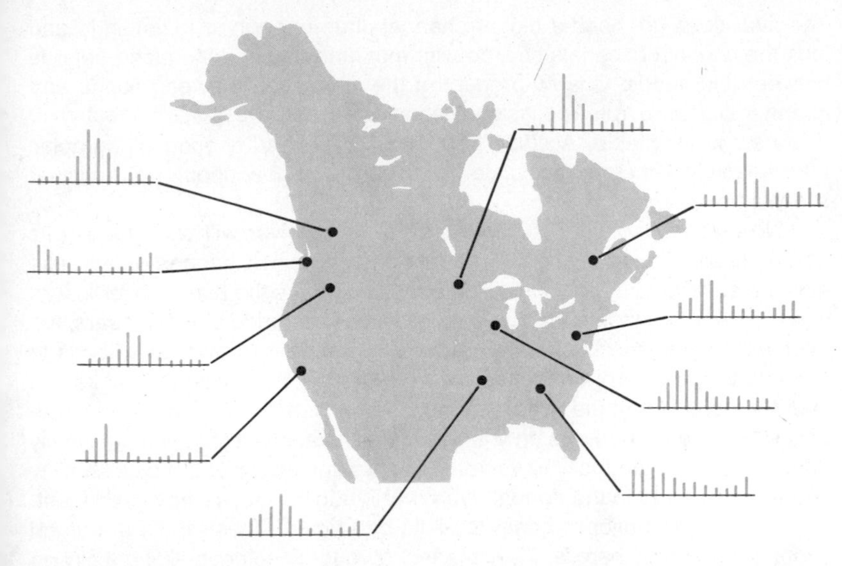

river regime normal pattern of river stage throughout a year

FIG. 14-14. Annual regimes of North American rivers. Heights of the bars indicate the volumes of river flow during the 12 months of the year from January to December. High flows are correlated with seasons of greater rainfall and/or lower evaporation, unless some means of water storage delays the runoff.

FIG. 14-15. Stage-frequency diagram for the Susquehanna River at Harrisburg, Pennsylvania. Like any expression of recurrence interval (see p. 100), this graph is based on years of observation, but it becomes obsolete whenever there are major changes in land use or other characteristics of a watershed.

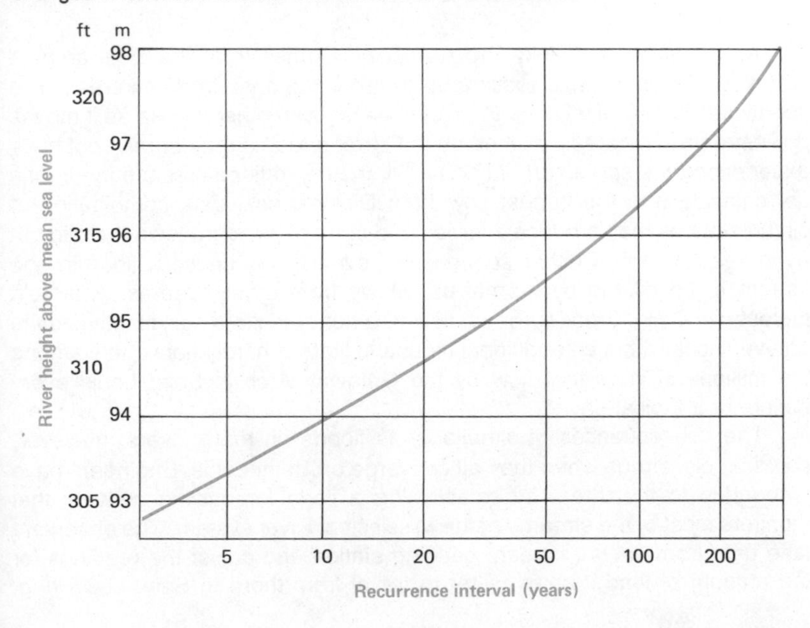

stage-discharge rating a graph showing the volume of flow associated with a given river stage

stage-frequency graph graph of return periods of different river stages

The river does not need a bigger channel often enough to maintain it, and thus the channel becomes choked with mud and debris during long periods between big floods. One by-product of the interaction between floods and channel size is a rule: occasional flooding of low areas near most river channels is inevitable. A stage-frequency graph and a good topographic map enable planners to calculate how often (and how deeply) a particular site will be flooded.

bankfull stage depth of water that fills the natural channel of a stream

No matter how high its banks are, a typical river will work toward its normal habit of overtopping them about once a year. Changes of land use may increase the frequency of large flows and cause the river to overflow its banks more often than once a year. Slowly, over a period of many years, the river will enlarge its channel and reduce the number of overbank floods to one every year or two. Sandbags or concrete walls may raise the banks of a river temporarily, but the river responds to confinement by depositing material in the unconfined areas downstream, which causes it to flow more slowly than it did and to gradually fill its bed with mud and debris until it gets shallow enough to indulge in the annual overflow. Dredging may remove sediment, but the river will patiently begin to fill its bed again. These traits of natural rivers suggest that there is no permanent cure for flooding; building a city on a floodplain inevitably condemns the occupants to a lifetime of dredging riverbeds, building levees, or reconstructing after floods. The modern practice of zoning floodplains for recreational or other noncritical uses is one of the best solutions to the problem of flooding, because the floodplain is actually a part of the river, not the land.

flood stage any stage higher than bankfull

FLOOD-ROUTING GRAPHS

All the water in "Dinky Brook" usually comes from the local area. A storm of a particular size elicits a predictable response from the creek. The floodwater from Dinky Brook then joins water from other creeks as it moves downstream. The areas drained by the other streams may or may not have experienced the same storm. If they did, their own flood crests may or may not coincide with the largest flow from Dinky Brook. One combination of circumstances may produce a large flood in a downstream town that did not even see a cloud. Another sequence of events may cause a flood in one stream to be diluted by normal baseflows from other streams. A tenfold increase in Dinky Brook's normal flow of a hundred liters per minute is quite an event locally, but an additional thousand liters is hardly noticeable among the millions of liters that flow by the Gateway Arch in Saint Louis every minute of a typical day.

tributary stream that flows into a larger river (see Fig. 15-6)

The consequences of simultaneous floods on many rivers, however, could be disastrous when they all converge on Saint Louis. Engineers have responded to the threat by establishing a flood forecasting network that monitors most of the streams of the Mississippi River system. The observers take data from each upstream gauging station and adjust the readings for the amount of time it takes water to travel from there to Saint Louis (Fig.

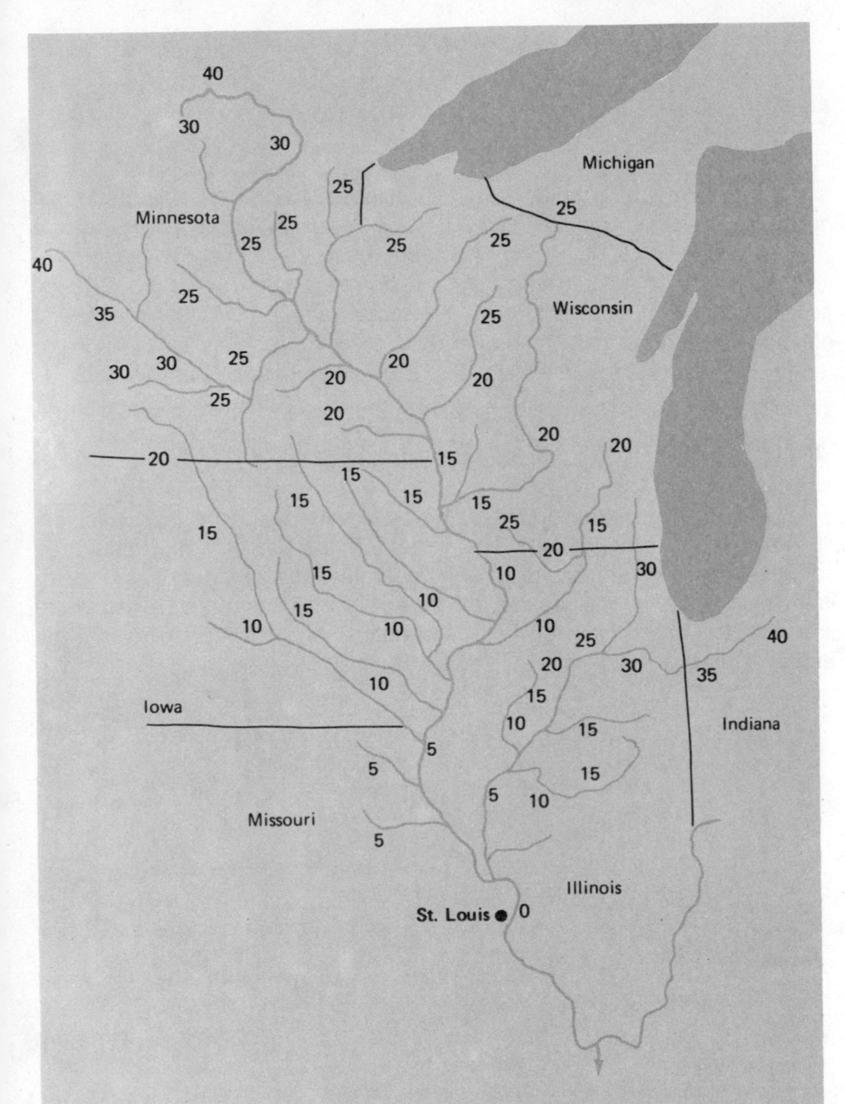

FIG. 14-16. Travel time for floods in the Upper Mississippi River Basin. The numbers indicate the days needed for stream water to reach St. Louis from any part of the watershed.

14-16). Then, it is a fairly simple procedure to add the expected inflow from all of the upstream watersheds and to predict the future depth of the river.

Flood forecasts of this kind for large watersheds are quite accurate but do not provide the kind of control some people prefer. By the time a flood is forecast, it is too late to do anything except leave or shore up the defenses. For this reason, many researchers are going back into the small upstream watersheds and are studying local water budgets. They believe that the way to control the movements of water is to analyze and manipulate the detailed water budgets of the small areas around the tributary streams. Adapting the use of upstream land to promote infiltration and evaporation of water near the sources may be the best way to prevent floods downstream.

APPENDIX TO CHAPTER 14
Estimating Runoff, Erosion, and Sediment Yield

Forecasting the hydrological consequences of a change of land use is a necessary part of sound planning. The Soil Conservation Service, Geologic Survey, and Army Corps of Engineers have developed a number of techniques for estimating runoff, soil erosion, and sediment yield. This appendix contains three simple techniques that use crude data to give general estimates. Accurate predictions demand complex procedures that in turn require sophisticated measurements, far beyond the scope of this book.

URBANIZATION AND FLOODS

The hydrologic effect of urbanization of a small watershed is the subject of the first graph (Fig. 14A-1). The *P* factor is the percentage of total area occupied by pavement, rooftops, or other impervious surfaces. The *S* factor is the percentage of total area served by storm sewers or gutters. The curved line near the intersection of the appropriate *P* and *S* lines represents the expected increase in the size of floods when a city replaces a typical forest or

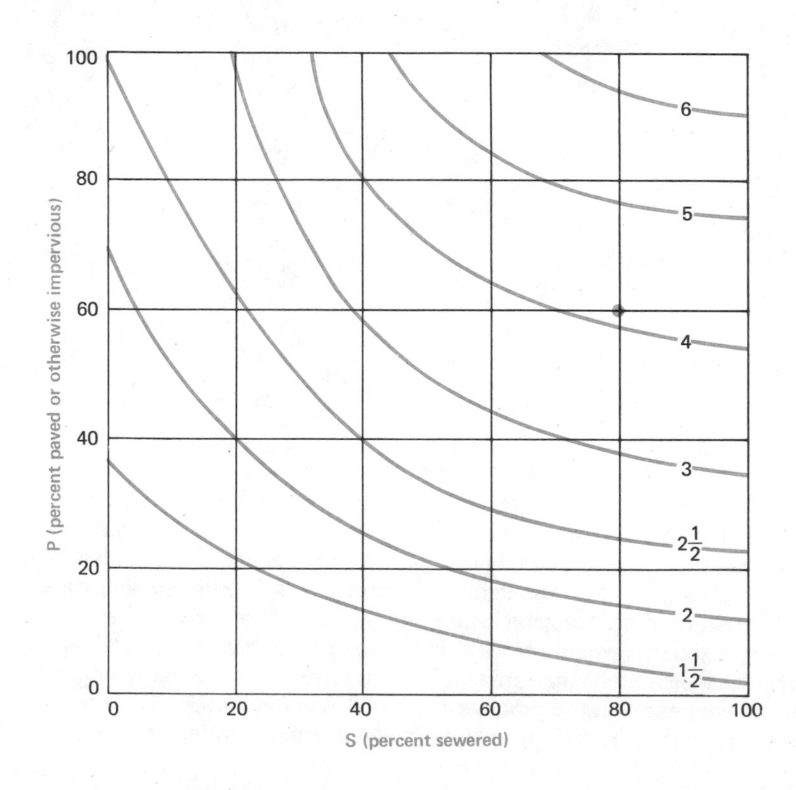

FIG. 14A-1. Flood multiplier graph. (Adapted from Leopold.)

pasture. For example, the dot shows a 60 percent impervious industrial area that is 80 percent sewered and will have floods about four times as large as those in areas with natural vegetation. The estimate is very crude because it does not consider climate or other watershed characteristics. Industrialization in different regions may double average flood size or multiply it by 10, depending on soil, slope, climate, and watershed characteristics.

LAND COVER AND FLOODS

The curve-number or cover-index procedure is used to predict the amount of storm runoff from a watershed after a rain of a particular size. Table 14A-1 shows the cover index that corresponds to certain combinations of soil type and land use. Figure 14A-2 shows the centimeters of storm runoff to be expected for any particular cover index and storm size. For example, a thin pasture (grass, poor cover) on sandy loam soil will have a cover index number of 73 and will deliver about 2 cm of runoff from a 7-cm storm (dot on graph). The total volume of runoff is equal to the depth of runoff multiplied by the area of the watershed. A more complex series of graphs

TABLE 14A-1 COVER INDEX NUMBERS

	Soil Type				
Cover Type	**Clay**	**Silty Clay**	**Loam**	**Sandy Loam**	**Sand**
Pavement	98	98	98	98	98
Business district	96	95	94	93	91
Industrial area	94	93	91	88	84
Dense residential	92	90	87	83	77
Crops, poor cover	90	88	84	78	70
Grass, poor cover	88	85	80	73	63
Medium residential	86	83	77	68	56
Sparse residential	84	80	73	63	49
Woods, poor cover	82	78	70	58	42
Crops, good cover	80	75	66	53	35
Grass, good cover	78	73	63	48	28
Woods, good cover	76	70	59	43	21

FIG. 14A-2. Storm runoff graph.

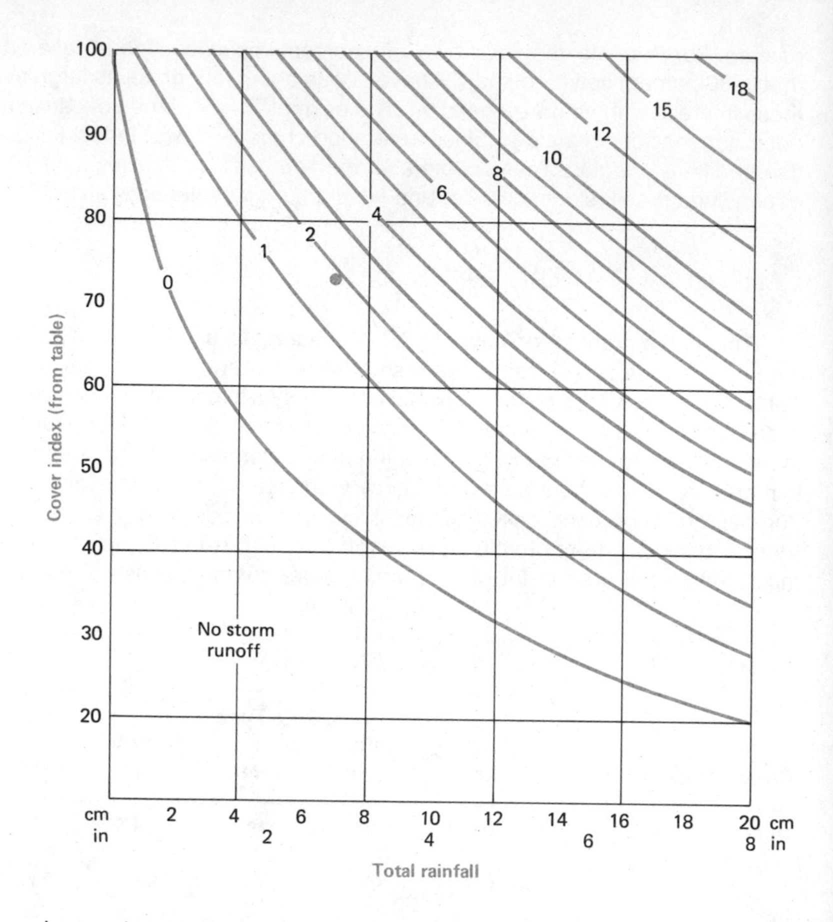

can be used to predict peak flows and lag times. Even so, the predictions fail to consider such factors as moisture content of the soil or seasonal variation in temperature and plant activity. These variables must be included in a detailed analysis of watershed response.

RUNOFF AND SOIL LOSS

The Universal Soil Loss Equation is a method of predicting soil removal by measuring four factors that govern the rate of erosion from a plot of land:

> *K*—the soil factor (Table 14A-2). Soils with high *K* hinder infiltration, promote runoff, and erode easily. The simplified table shows *K* factors based only on texture and organic matter, although structure, depth, profile development, and cover are also significant.
>
> *R*—the rainfall factor (Fig. 14A-3). Climates with long growing seasons and intense storms are able to exert more erosive force on a soil than climates with long freezing seasons and gentle rain.

TABLE 14A-2 *K*-FACTORS

Texture	Percent Organic Matter 0	2	5
Coarse sand	.05	.03	.01
Loamy sand	.20	.15	.10
Sandy loam	.30	.25	.20
Loam	.40	.35	.30
Silt loam	.50	.40	.30
Silt	.60	.50	.40
Sandy clay	.15	.12	.10
Clay loam	.30	.25	.20
Clay	.25	.20	.15

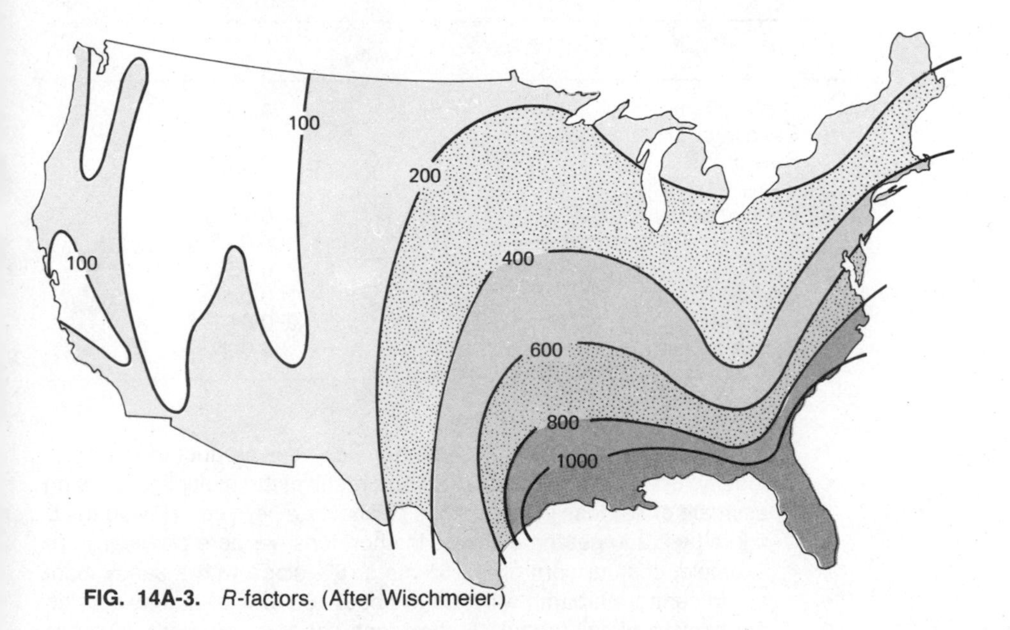

FIG. 14A-3. *R*-factors. (After Wischmeier.)

S—the slope factor (Table 14A-3). Slope angle has a direct effect on erosion: water is more likely to run off a steep slope than off a gentle one. Slope length has a cumulative effect: water that starts down a slope exerts an erosive effect all the way to the foot of the hill.

L—the land use factor (Table 14A-4). A dense cover of vegetation hinders runoff and erosion. The land use factor depends on the density and duration of plant cover. The table of land use factors gives at best a general idea of the effect of land use, because matching the life cycles of plants with local rainfall patterns is a difficult procedure that is further complicated by differences in land management and crop productivity.

TABLE 14A-3 *S*-FACTORS

Slope Angle in Percent[a]	Slope Length 50 / 167	100 / 333	150 / 500	200 / 667	250 / 833	300 / 1000	meters / feet
4	0.5	0.7	0.9	1.1	1.2	1.3	
8	1.3	1.8	2.2	2.5	2.8	3.1	
12	2.3	3.3	4.0	4.7	5.2	5.7	
16	3.7	5.2	6.4	7.4	8.3	9.1	
20	5.4	7.6	9.3	10.7	12.0	13.2	

[a]Meters of rise per 100 meters of horizontal movement.

TABLE 14A-4 *L*-FACTORS

Bare soil	1.00
Seedlings	0.50
Row crops	0.30
Straw mulch	0.20
Row crops on contour	0.15
Row crops on terraces	0.05
Grains	0.05
Grazed forest	0.02
Grass sod	0.01
Dense forest	0.005

E—the erosion loss or sediment yield, the product of the other four factors: $E = K \times R \times S \times L$. The result of the multiplication is an estimate of soil loss in metric tons per hectare per year. (Divide the *E* figure by 2.2 to get the estimate in short tons per acre per year.) For example, contour corn on a 150-meter 8% slope with a sandy loam soil in central Alabama will lead to about 100 tons of annual erosion per hectare of soil (about 45 short tons per acre, or nearly 10 times the amount that many soil scientists consider an acceptable level of erosion). The same land use on a South Dakota loam with 5% organic matter will cause only about one-fourth as much erosion.

The importance of estimating runoff and sedimentation is increasing. Damages from uncontrolled runoff amount to many millions of dollars annually. Many states now have laws that require landowners to control runoff and sediment, because the victims of mismanagement include not only the landowners and their children but also people that live downstream, perhaps far away.

Chapter 15

Erosion and Isostasy: Gravity at Work

Every valley shall be exalted, every mountain shall be abased, and the rough places made plain.

J. Higgins, U.S. Forest Service

Whenever part of the earth aspires to rise above its surroundings, it becomes a victim of the "Great Terrain Robbery." No rock can remain intact when exposed to the elements, and a disintegrated rock is easy prey for the forces of gravity, which work incessantly to humble the high places and bring everything back to level.

Gravity does not work alone in this effort to lower the terrain. A whole host of agents enlist as allies—snow, ice, wind, wave, soil moisture, rivers, plants, animals, and the exploitable weaknesses of the rocks. Each of these allies leaves its own imprint on the surface of the earth. The unique terrain features that identify the work of certain agents are subjects of the next chapter. This chapter uses rivers and mass movements as examples, but its emphasis is on the traits that all landforming processes have in common. An investigation of the basic principles of land shaping leads inevitably to a discussion of energy, the ability to do work.

PRINCIPLES OF LANDFORMING ENERGY

energy ability to do work (see p. 8)

first law of thermodynamics energy cannot be created or destroyed; it can only be transformed from one form to another.

friction resistance to motion; friction converts energy of motion into heat

meteorite metallic or rocky particle that enters the atmosphere from outer space

second law of thermodynamics energy transformations always leave energy in a less concentrated form; it is easy to turn all of the energy of motion in an object into heat but impossible to convert all heat in an object into organized motion

An object has an energy advantage over its neighbors if it is higher than they are. This advantage occurs because an elevated object must have expended some effort (or had effort spent on it) to lift it to a higher level. For example, a water droplet in a cloud represents many calories of sun energy, first to evaporate it (Chapter 1), then to heat the lower air and to make it unstable enough to lift the drop to its high altitude (Chapter 2), and finally, to create the horizontal pressure differences needed to drive the wind that carried the water drop to its present position (Chapter 3). Physicists assert that it is impossible to create or destroy energy; thus the tremendous input of work must somehow be stored within each water droplet.

The stored energy is released as soon as raindrops begin to plummet toward the ground. Fortunately for the inhabitants of the earth's surface, friction with the air consumes energy and prevents a continual increase in speed. If friction did not intervene, a raindrop falling from the top of a 10-kilometer thundercloud would impact the earth at a speed of 450 m/sec (1500 ft/sec, about the velocity of a pistol bullet).

A small meteorite plunging into the atmosphere from outer space gives a glowing report on how friction takes energy of motion and converts it to heat. Raindrops are also warmed by friction; they do not get as hot as meteorites because they do not fall as far and would evaporate long before they glowed visibly. The history of raindrops and meteorites suggests that speed and height and heat are merely different disguises for the same basic entity (Fig. 15-1).

Landforms are shaped by the direct application of one of these forms of energy. Pieces of weathered rock and soil are picked up and moved around by any agent that has enough energy to dislodge them. Most earth-forming agents gain speed energy by trading height for it. Therefore landforming should proceed most rapidly wherever large objects move quickly down steep hills.

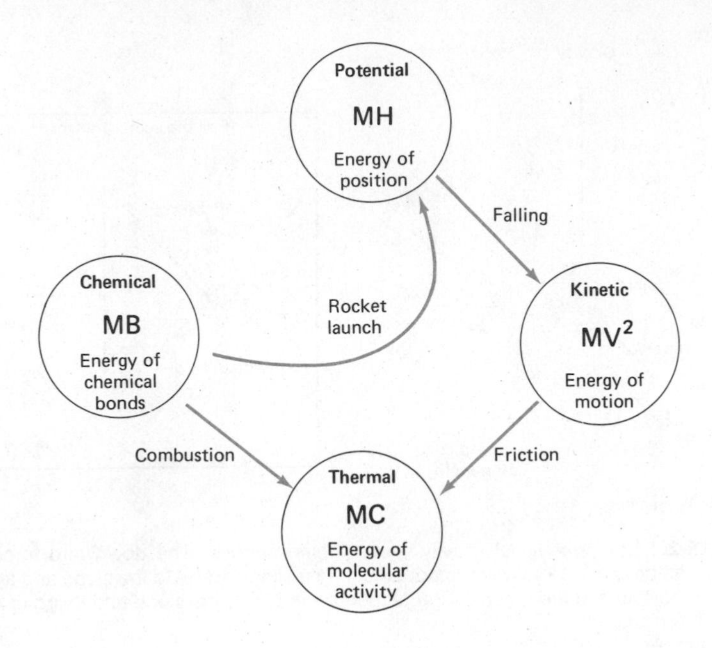

Symbol	Concept	**Result** (all other things being equal)
C	Specific heat	hot objects have more energy than cool ones
B	Bond strength	complex bonds have more energy than simple ones
H	Height	high objects have more energy than low ones
M	Mass	heavy objects have more energy than light ones
V	Velocity	fast objects have more energy than slow ones

FIG. 15-1. Relationship among various forms of energy.

DOWNHILL MOVEMENTS AND GRAVITY COMPONENTS

Every few years some movie writer discovers a convenient way to eliminate a villain without sullying the image of the movie with excessive violence. The writer makes the villain an unwitting suicide by positioning the evil one at the base of an appropriate slope before the final shoot out. The impact of the first shot (of necessity from the villain's gun) triggers a landslide, avalanche, mine cave-in, or other dusty disaster that buries the evil one under tons of rock or snow. The "happy" ending of the movie also illustrates why many hills have the shapes they do.

mass movement downhill movement of earth material under the force of gravity

Consider a hillside from the viewpoint of a rock lying on it. Loose rock will slide downhill only when the forces trying to make it move are greater than those acting to keep it in place. Paradoxically, the pull of gravity is the major force on both sides of the balance (Fig. 15-2).

component of gravity proportion of gravity force that acts in a particular direction

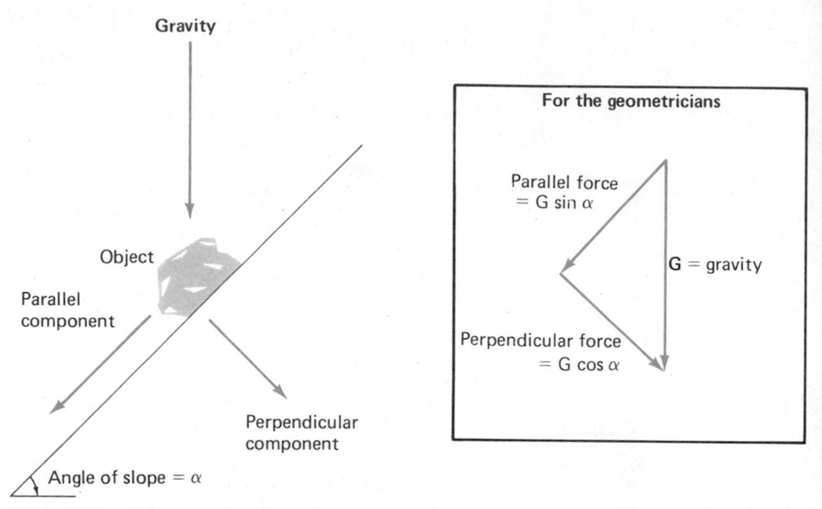

FIG. 15-2. Components of gravity on a sloping surface. The downward force of gravity can be divided into two components, one pulling parallel to the slope and trying to move objects and the other pulling perpendicularly into the slope and trying to hold objects in place.

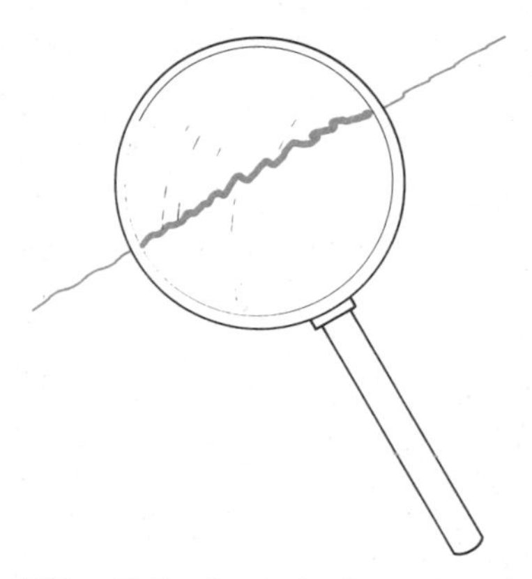

FIG. 15-3. A rough slope viewed as a combination of smaller slopes of varying angles of inclination. Roughness usually aids friction in resisting motion along the surface.

angle of repose steepest slope that a given kind of material can form without sliding downhill

coefficient of friction measure of the frictional resistance to motion; rough surfaces have high coefficients of friction, whereas smooth surfaces have low coefficients

effective surface amount of frictional contact with a surface

The angle of slope determines which side of the balance will receive the most force. All of the force of gravity pulls directly into a horizontal surface. An object has no urge to move, whether or not it is physically fastened to the floor. The situation is reversed on a vertical wall. Gravity cannot hold an object against the wall; all of the force of gravity is used to pull toward the floor. The forces into and along a surface must compromise when the surface slopes at an angle somewhere between horizontal and vertical. On steep slopes, the downward force is stronger than the one into the slope. On gentle slopes the willingness to stay is greater than the urge to move.

The roughness of a surface helps to determine how much force is needed to start downhill motion. Humans can climb a rocky hill with an angle of 40° or a varnished ramp at an angle of 20° from horizontal, but trying to ascend a 5° icy slope usually provides more amusement than progress. A rough surface could be interpreted as a combination of small slopes, some steeper and some gentler than the average slope (Fig. 15-3). The gentle portions, acting like treads on a staircase, provide resistance to slipping. For each particular type of surface, there is a critical angle of slope that is related to roughness of the surface. Material can remain on the surface only when the slope is more gentle than this critical angle.

An object resting on a slope at nearly the critical angle will begin to slide if anything alters the rather delicate balance between the downhill component of gravity and frictional resistance. Adding more weight on top of the object increases both pulls of gravity, the one into and the one along the slope. However, the additional weight does not increase the surface area of contact between the object and the slope. A new snowfall may start an avalanche, because the frictional resistance to motion usually does not increase as much as the urge to move when the depth of snow increases.

A more common way to start avalanches is to alter the friction of the surface. Water from melting snow acts as a lubricant and decreases the frictional resistance holding a snowmass in place. Loud noises or the rumble of a moving train can induce slight vibrations that may trigger avalanches by momentarily breaking some of the frictional contact with a surface. Railroad companies, ski resorts, and mining communities sometimes use explosives to cause small avalanches, rather than wait for nature to pile up so much snow that an avalanche could be a catastrophe.

avalanche mass movement consisting primarily of snow (see Fig. 16-6)

substrate liquefaction temporary weakening of earth support by vibration, a major factor in earthquake damage

THE DIVISIONS OF A MASS MOVEMENT SLOPE

Avalanches and landslides are like riots; once started, they frequently involve bystanders. Much of the weight of a typical avalanche is material that would otherwise have been content to remain in place indefinitely. Thus there is another distinct kind of slope, one which is not steep enough to start material moving by itself, yet is too steep to stop a mass movement that is underway. Friction and gravity normally can keep surface material in place, but the additional downslope force imparted by material already in motion tips the balance in favor of movement along the slope. Like the critical angle for starting mass movements, the angle for stopping them depends on the nature of the surface material and the size of the sliding objects.

erosion all forms of downhill earth movement; also called leveling, denudation, or gradation

These two critical angles are the dividing lines for a convenient three-part division of slopes (Fig. 15-4):

1. Initiating slopes, which are steep enough so that gravity movements of loose material start of their own accord. Initiating slopes generally are free of loose material, because gravity usually moves material off the slope soon after weathering loosens it.
2. Maintaining slopes, which are not steep enough to initiate mass movements but not gentle enough to stop them once they are underway. Maintaining slopes often consist of loose debris at the foot of a steep cliff.
3. Absorbing slopes, which are so close to level that friction with the surface not only keeps local material in place but also absorbs the additional energy of material moving on to the slope from above.

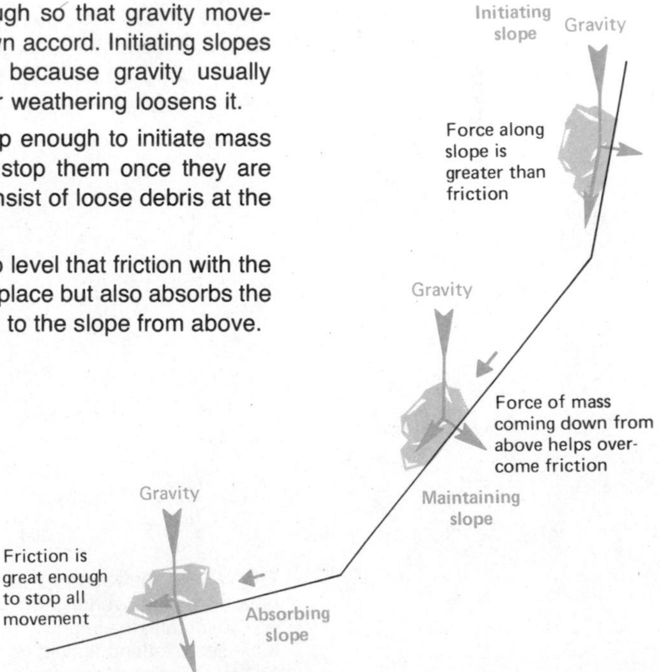

FIG. 15-4. Components of gravity on initiating, maintaining, and absorbing slopes. Steep slopes start mass movements, gentle slopes stop them, and intermediate slopes are unable to alter whatever is happening at the moment.

The three kinds of slopes are quite distinct on some cliffs. Other hills have maintaining and absorbing slopes that merge together into a long gradual curve, the consequence of mass movements of different sizes. Large mass movements have sufficient energy to travel far downhill and to overpower material on gentle slopes. Small landslides usually stop on steep slopes because they have little force and little tendency to involve other materials in the slide.

The matter is further complicated by the absence of initiating slopes on some kinds of terrain. Rock cliffs are rare in rainy regions, and other slopes are too refined and gentle for something as crude as a landslide. Gravity seems to recognize that catastrophic mass movements will not work, and it resorts to other means of lowering the land.

RIVER PROFILES AND GRADED STREAMS

A raindrop on a hillside above sea level represents a unit of potential energy, a tool that gravity can use in its effort to lower the terrain. Water is a fluid, able to flow readily down a slope (Fig. 15-5). Frictional resistance to river flow is great near the source, where a river is just a tiny trickle of water during storms. A single twig may make an effective dam across the creek. A leaf is a major burden, a pebble a significant obstacle. The stream needs energy to overcome frictional resistance and can get the needed power only by trading height.

overland runoff downhill movement of rainwater on top of the ground surface (see Fig. 14-12)

Once underway, channeled water begins to carry a load of whatever soil material it can lift. Tiny rivulets move tiny pieces of soil and create tiny valleys. Steepened slopes on the sides of the valleys attract other drops of water toward the stream. The result is an enlarged stream whose increased energy enables it to cut a deeper valley (Fig. 15-6).

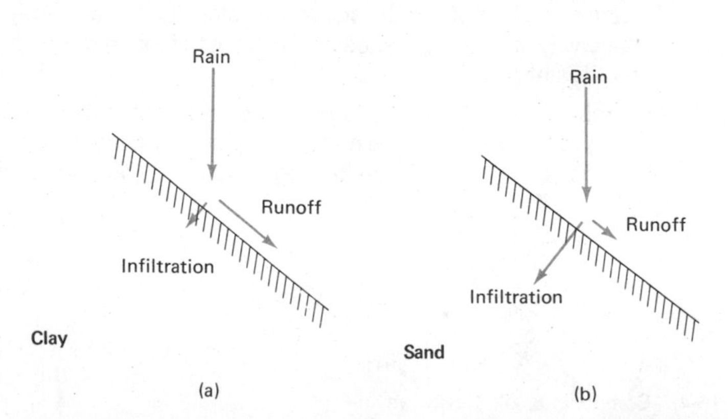

FIG. 15-5. Relationship of slope and overland runoff. The critical slope angle needed to start overland runoff of water depends on soil texture, soil moisture, vegetation cover, and the intensity of the rain. A wet clay soil (a) cannot absorb water rapidly; it causes overland flow at gentle angles, whereas dry sand (b) allows much water to infiltrate and thus minimizes runoff.

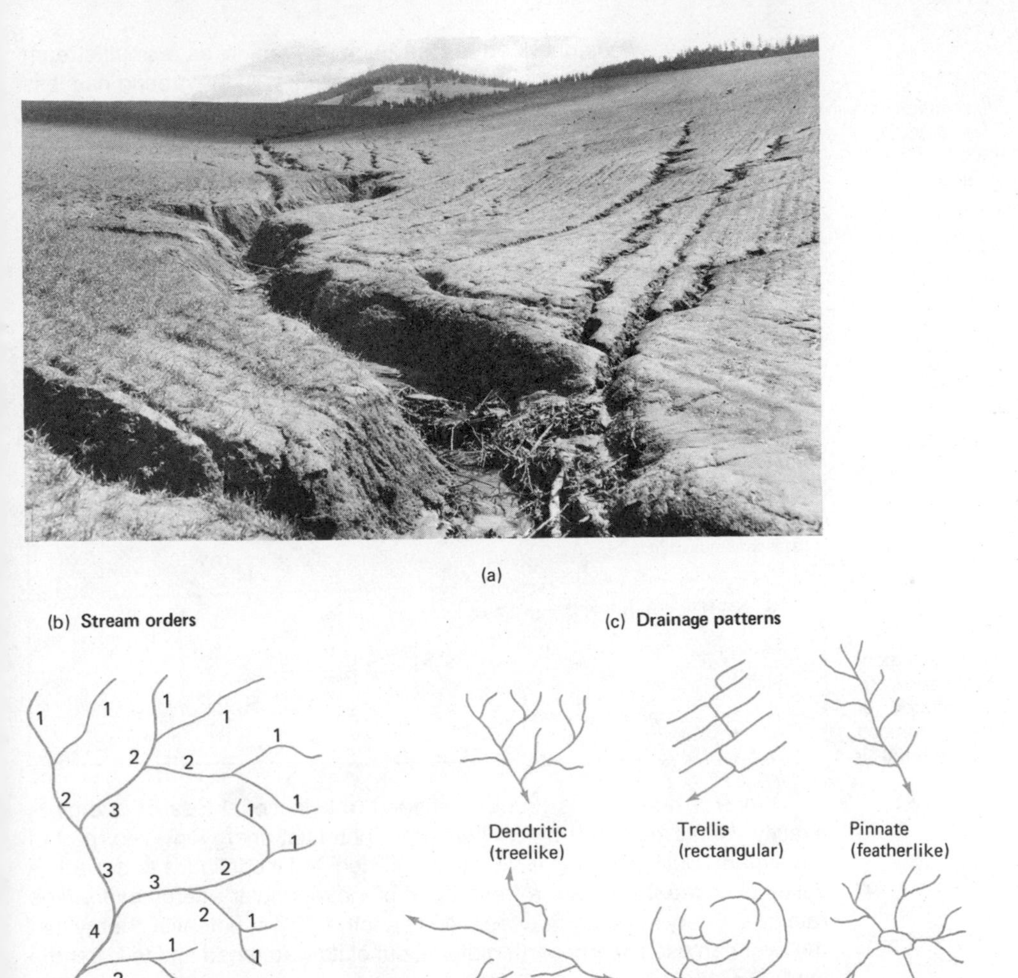

FIG. 15-6. Beginning of a stream system. (a) Small rills have begun to unite into larger valleys. (b) A measure called *Stream Order* has been devised to make precise comparisons between rivers or among various parts of the same river. Every source stream in a drainage system is classified as Order I or First Order. A second-order stream is formed when two first-order streams come together. Whenever two rivers of the same order unite, the combined river is one order higher. When two streams of different orders merge, the combined stream is of the same order as the higher of the two tributaries. (c) Stream networks can have a variety of spatial arrangements. All other things being equal, a dendritic system will produce a higher order stream than a trellis drainage pattern; a trellis higher than a pinnate. Stream order also has implications for flood prediction. High order streams are more likely to carry great floods of storm water, but low order streams that drain the same amount of land usually receive storm water in little pushes all along their length. When this happens, the flood is spread over a long period, and the peak flow is quite low. (Courtesy of C. I. Tyer, Soil Conservation Service.)

wetted perimeter, hydraulic radius expressions of the amount of bank and bottom surface touched by a stream of a given volume

A large river does not need to lose height as rapidly as a small stream does in order to maintain the same speed. Friction with the ground has less effect on the large river, because most of the water can flow along without touching the bottom or sides of the riverbed. The diminishing importance of friction helps to explain a seemingly unbelievable fact: broad stately rivers usually flow at least as fast as the tumbling mountain brooks from whence they came. The situation exists because a stream is always working to shape its valley and to adjust its slope so that its speed stays uniform. If one part of the valley is too steep for the size of the river there, the water speeds up, and its greater kinetic energy allows it to remove more material from the riverbed, which of course lowers the high spot. A low spot, on the other hand, slows the water and induces deposition, which builds up the slope (Fig. 15-7).

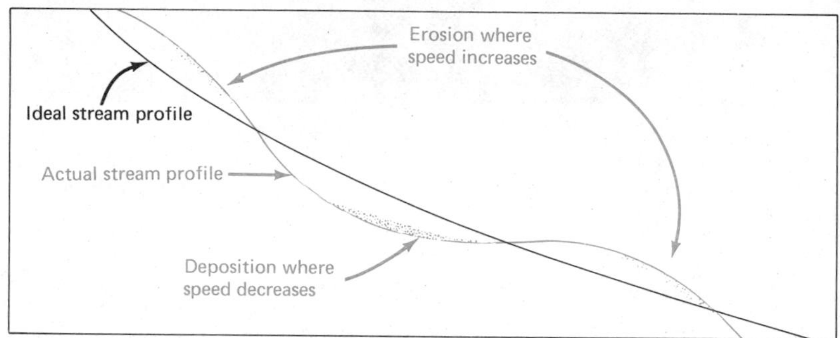

FIG. 15-7. Force relationships along a stream with an irregular bed. Erosional energy is concentrated on the high places of the streambed, whereas deposition of material is more likely.

When a river bed is properly shaped, the frictional loss of energy is exactly counterbalanced by conversion of potential energy in every part of the stream. Such a stream is said to be graded, but it continues to dig out its valley because it receives a fresh input of indirect solar energy every time rain falls. Every new raindrop represents another unit of potential energy that the river can use to carry earth material out of its watershed and to lower the land (Fig. 15-8).

graded stream stream with a bed that slopes properly so that the frictional loss of energy is compensated by fall along its length

watershed area drained by a single stream and its tributaries (see Fig. 14-3)

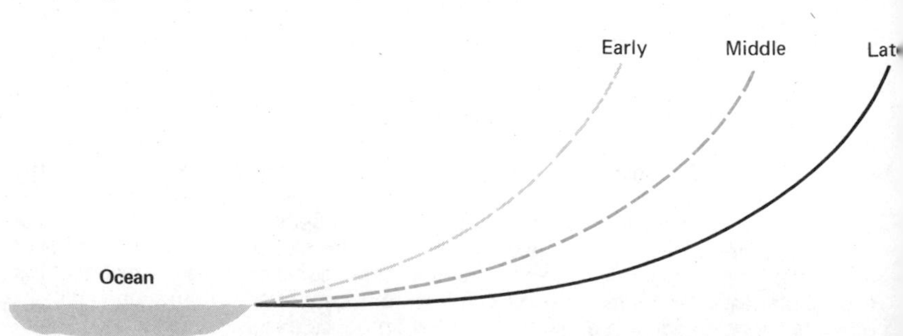

FIG. 15-8. Longitudinal profile of a typical stream at three times in its history. Steep slopes near the source and gentle gradients near the mouth reflect differences in the power of the river and the friction of its bed. As time passes the stream carries material out to sea, the source moves back away from the coast, and the slopes all along the river become more gentle than before.

BASELEVEL AND THE CYCLE OF EROSION

There is a limit to the land-lowering ability of a stream. Drops of water on a hillside have energy as long as they have somewhere to fall, but the water in a typical stream cannot fall any lower than the surface of the ocean. Deep water causes a river to slow, lose its energy of motion, and drop its load of earth material. In this way, places beneath the ocean surface are built up while places above sea level are worn down. The elevation of the ocean surface is the zero line or baselevel, the elevation of terrain that rivers try to produce by wearing down high places and by filling up low places. Rivers could retire when the continents were finally worn down to sea level. They would have nothing left to do, and no energy with which to do it anyway.

baselevel elevation of zero stream energy, usually mean sea level

The idea of an earned retirement after finishing a finite task inspired an almost allegorical conception of landforming called the Geographical Cycle of Erosion (Fig. 15-9). Now more than three-quarters of a century old, this grand analogy suggests that the stages in the development of a landscape resemble the phases of life:

Davisian Cycle theory that terrain passes through a recognizable sequence of stages—"Youth," "Maturity," and "Old Age"—as it is eroded down

1. At "Birth" some mysterious force lifts a block of land out of the sea and raises it to some suitable height. The forces responsible for the uplift were unknown at the time the theory was first proposed, but plate motion seems to fit the job description.

plate block of rigid crustal material that moves sideways across the melted rock beneath the crust
(see Chapter 6)

2. In "Early Youth" the terrain is a vast reservoir of potential energy. Each drop of rain that falls on to high land has a long way to go before it reaches the sea. The water is capable of doing much work, but since the land is mainly flat, only the water near the edges of the block actually moves downhill. Consequently, the edges of the block of land are marked by waterfalls and tumbling rapids and spectacular gorges cut by energetic streams. The rest of the block is still an undrained upland, an unfulfilled promise of greater activity in the future.

3. In "Late Youth" the block of land has realized some of its potential. Streams are longer than they were and can tap larger watersheds, carry more water, and make deeper and wider valleys than before. Large streams often "capture" other streams by enlarging their valleys and diverting additional water into them. The landscape loses its level monotony and becomes organized into drainage networks. The process of valley deepening proceeds until the main river valleys are graded smoothly down to sea level, at which time the landscape is said to have spent its youth and entered maturity.

stream piracy large stream deepening its valley sufficiently to divert water from another stream
(see Fig. 17-11)

drainage pattern spatial arrangement of tributaries
(see Fig. 15-6)

FIG. 15-9. The Geographical Cycle of Erosion, also called the Davisian Cycle, after its inventor, W. M. Davis (continued on next page).

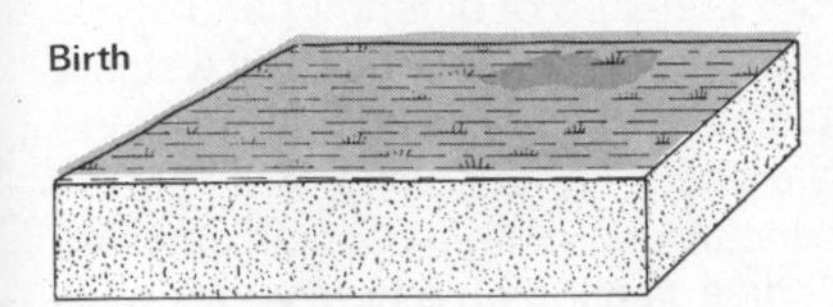

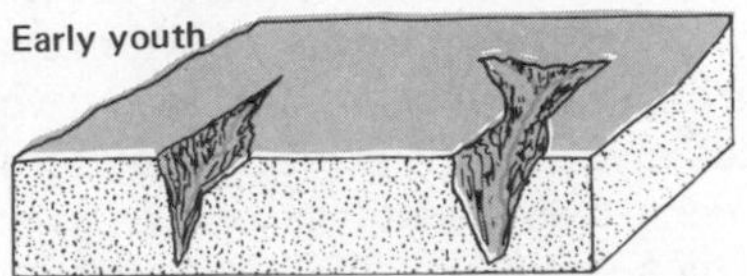

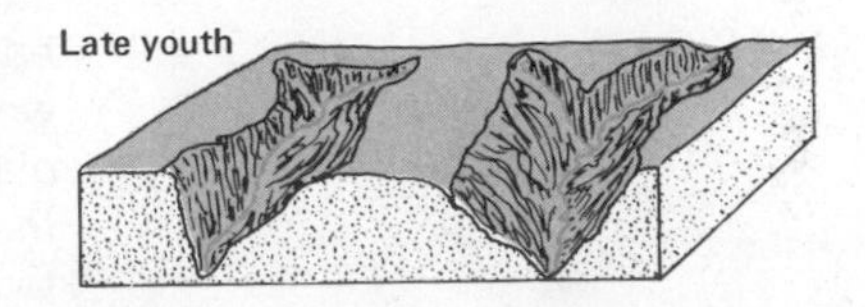

Early maturity

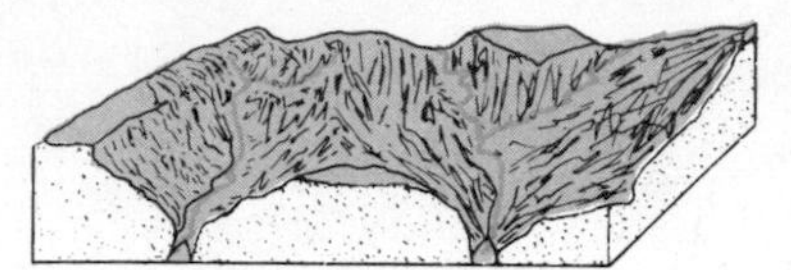

Late maturity

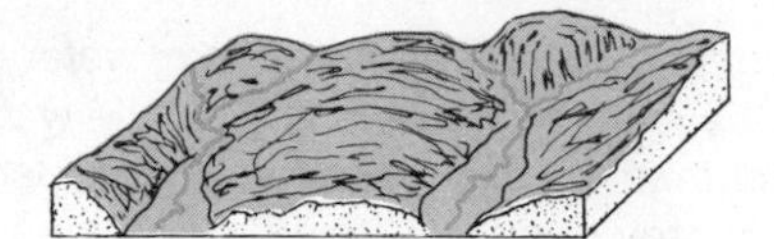

Old age

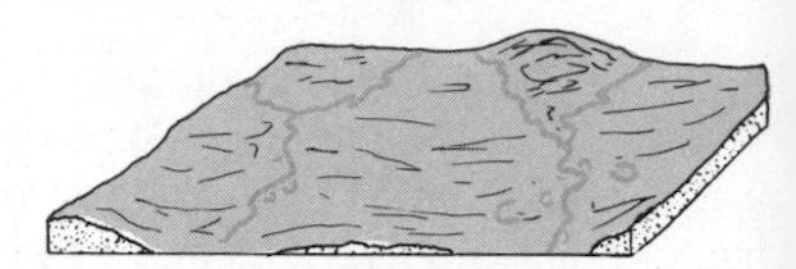

drainage divide highland between watersheds

4. In "Early Maturity" most of the original flat upland is gone. Tiny level areas along the drainage divides between rivers are all that is left to show where the original surface was. Most of the terrain is sloping, so that the work of removing earth material and of carrying it downhill takes place nearly everywhere. In terms of sheer quantities of earth moved, this stage is the most productive.

5. By "Late Maturity" no hint of the original upland remains. It is difficult to guess the age of mature terrain, because the last remnant of youth may have just disappeared or may be long gone. In any event, the incorporation of all areas into the drainage network means that no new source of energy remains to be tapped. Slowly, imperceptibly, the divides wear down. The potential energy available for earthmoving decreases as the hills shrink. The valleys cannot be deepened because they are already at sea level. From this time on, valleys can only get wider, while small branch streams continue to carry material away from the remaining uplands. Maturity passes quietly into "Old Age" when most of the terrain has been lowered to near sea level.

peneplain, old-age landform a plain at sea level

6. "Old Age" means that the average elevation of the land is low. Most of the landscape energy is gone and little earth movement can take place. Streams meander lazily across broad floodplains as if dreamily remembering their youth, when there were mountains to conquer and great earth-moving tasks to be done.

ISOSTASY

isostasy condition of equal downward pressure upon the mantle
(see p. 128)

mantle plastic layer of molten rock beneath the rigid crust
(see Fig. 6-1)

plate tectonics theory about movements of rigid crustal plates across the plastic mantle
(see Chapter 6)

The biggest problem with the idea of an erosion cycle is that terrain just does not age along a logical sequence of stages, for reasons that are also related to gravity and potential energy. Imagine an enormous block of rock suddenly placed on the surface of the earth. The material underneath the earth's crust is not rigid; the added weight causes the crust under the block to sink as mantle material slowly shifts beneath it. Meanwhile, the subterranean movement pushes surrounding areas upward. In this respect the earth is like a giant rubber balloon that absorbs a push in one place by bulging out somewhere else. The weight or downward pressure of different parts of the earth's crust must be nearly uniform everywhere around the globe. If weights were not equal, subsurface material would move out from under the "heavy" places and push up beneath the "light" areas. The balancing of weight on the crust thus underlines one of the conclusions of the chapter on plate tectonics—continents are like light-rock boats floating on a heavy-rock pool.

Consider now a hypothetical situation in which half a continent is abruptly "eroded" away and deposited in the adjacent ocean. The added mass increases the amount of downward pressure on the mantle beneath the ocean bottom. At the same time, the part of the continent that lost material is less weighty than it was. The region of deposition sinks and the eroded section is pushed up (Fig. 15-10). The resulting uplift of the continent sets the stage for a new cycle of erosion, which in turn produces another adjustment beneath the surface. In practice, both erosion and uplift of the continents are usually going on simultaneously, although rarely at exactly the same rate of speed.

The eventual goal of this slow interaction between erosion and uplift is to spread continental material horizontally on top of the heavy material of the ocean floor. The time needed to lower the entire continent to sea level, however, is so incredibly long that today's students probably need not warn their children, or even their grandchildren, about a future drowning of the entire human population. Long before the continents disappear under the sea, the sideways movements of crustal plates will probably push continental blocks together and will create new mountains.

In effect, then, plate motion is a geological fountain of youth that gives new height to the land and new energy to the rivers. Isostatic uplifting is more like a medicine that delays aging by counteracting some of the energy-robbing effects of long-term erosion.

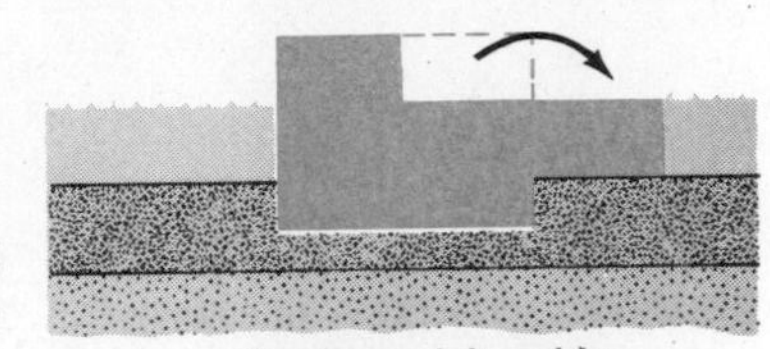

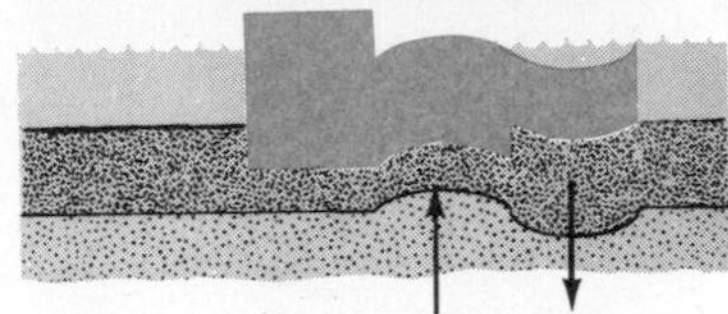

FIG. 15-10. An illustration of the principle of isostasy (equal weight on the mantle beneath the crust). Removal of a block of lighter continental rock and subsequent deposition in the adjacent ocean causes the ocean floor to sink and the land area to rise.

rejuvenation addition of potential energy to a river system by an uplift of the land or lowering of sea level

SPATIAL DIFFERENCES IN LANDFORMING ENERGY

A dam makes landform analysis more complex by rejuvenating one part of a landscape and causing the premature aging of another (Fig. 15-11). A drop of water upstream from the dam is not able to "see" the ocean as its goal. A stream that enters the reservoir behind a dam behaves exactly as if it were joining the ocean: it slows, stops, and drops the burden of sediment it was carrying. Thus the dam reservoir becomes an energy baselevel for the

temporary (or local) baselevel lake, dam, hard rock layer, or other feature that reduces stream velocity to zero

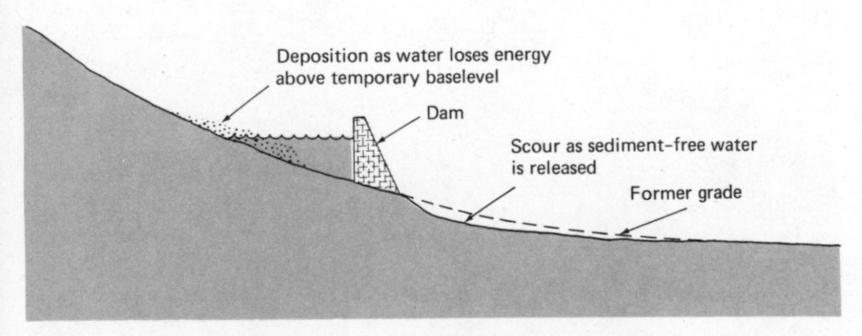

FIG. 15-11. Energy relationships on either side of an artificial dam.

streams flowing into it. Terrain upstream from the reservoir develops as if it were going to be lowered only to the level of the reservoir. Streams begin to flood and meander across valleys that look like old age, even though they may be hundreds or thousands of meters above true sea level. Many farmers and townspeople along a river above a reservoir have experienced floods that show how far upstream the effects of a dam can be felt.

Water that is released through the dam suddenly regains its youthful vigor. A full-fledged river should not be empty of sediment on a slope a thousand meters above the sea; the released water quickly picks up a load of earth material to replace the sediment lost in the reservoir. Erosion deepens the channel below the dam and causes tributary streams to flow more quickly than before. In turn, the tributaries excavate deeper gorges for themselves, and the entire landscape takes on more of the rugged grandeur of youth.

Niagara falls is a natural dam, an extreme example of the effect of a hard rock layer across the valley of a stream (Fig. 15-12). Lake Erie is its reservoir, and rivers such as the Maumee and Sandusky cannot flow below the level of the lake and therefore appear old before their time. Below the falls, the rejuvenated stream churns through its gorge, foaming over rapids on the resistant rock. Slowly the river wears the hard rock down. If engineers did not intervene, the Niagara River would eventually obliterate Niagara Falls, empty Lake Erie, and thus force the honeymoon industry to seek a new location.

nickpoint (knickpunte) waterfall or rapids where a stream crosses resistant material; nickpoints "retreat" upslope as the material is worn down by the stream

Rock differences do not negate the basic principle of a graded stream. Hard rocks offer more frictional resistance than soft rocks, and therefore the river must flow downhill more steeply over hard rocks in order to convert enough potential energy to offset the greater loss of energy to friction. The effect of rock differences is to make the slope of a graded stream a variable that depends on rock type as well as distance from the ocean. A river in geologically complex terrain may look old in some places, youthful in others,

rock hardness has two meanings: geologic—able to resist weathering and erosion; mineralogic—able to scratch other softer rocks

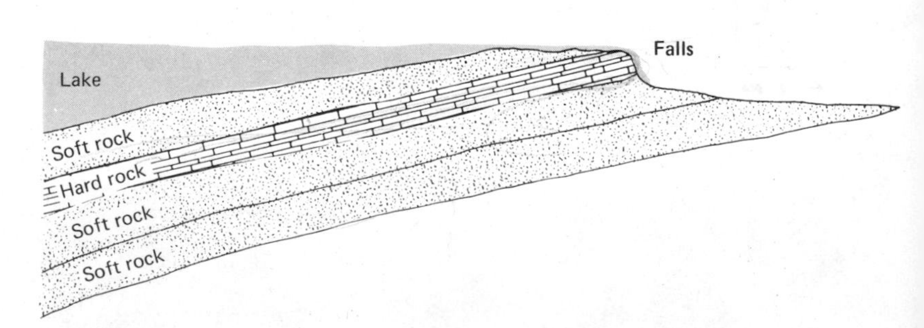

FIG. 15-12. Geologic setting of Niagara Falls. The sloping layers of hard rock do not weather as rapidly as the softer rocks both above and below the Falls (see also Fig. 17-4).

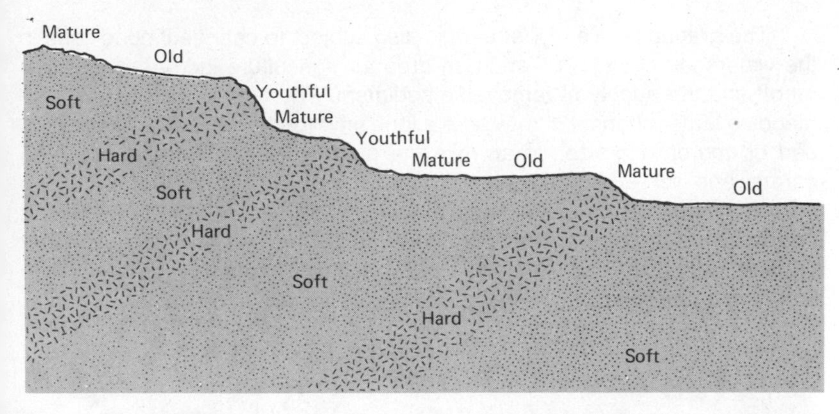

FIG. 15-13. A typical long profile of a river over geologically mixed surfaces. Hard rocks resist erosion and remain youthful, whereas soft rocks allow terrain to proceed toward old age quickly.

and mature in still other reaches along its length (Fig. 15-13). Given enough time and no interruptions, a river will breach every dam, level all the falls, wear down the hard rocks, and lower the land to the level of the sea. In this sense, intermediate baselevels are only temporary.

TEMPORAL VARIATIONS OF LANDFORMING ENERGY

load-discharge ratio relationship between the amount of material a stream can carry and the volume of water it has; the ratio increases as the river enlarges

Most of the time, a river is too weak to do much besides carry itself to the ocean, but it becomes a potent earthmover when it reacts to a big storm or regional snowmelt. Its volume increases greatly, its speed increases moderately, and its ability to carry sediment skyrockets. In most cases the river promptly exercises its right to a heavier load by picking up some loose material from its bed (Fig. 15-14). Channel scouring alone rarely satisfies the river's newfound capacity for work. Tributary streams may help by bringing sediment in from nearby hills. Collapse of the riverbanks also adds material to the load. In all respects, rivers do a disproportionately large share of their earth-moving work during the short periods of time that they are in flood. Friction slows streams when they are small and causes them to deposit material in their channels.

One major result of temporal variations in size is that a river is almost never graded at any instant in time. It usually is adjusting its channel to accommodate a new level of flow different from that of yesterday. Thus the grade of a stream is an average condition that can be determined only after a long period of observation.

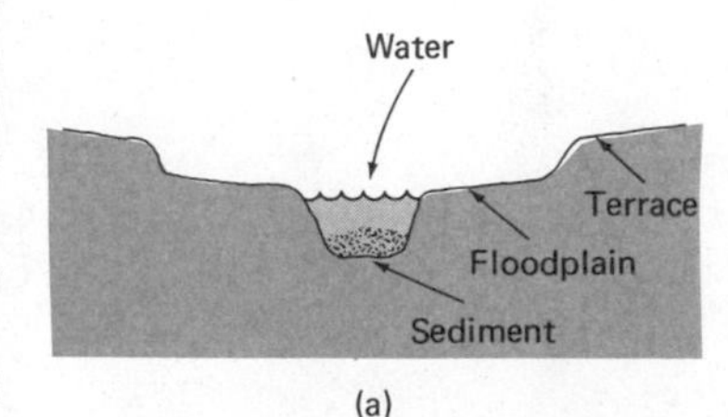

(a)

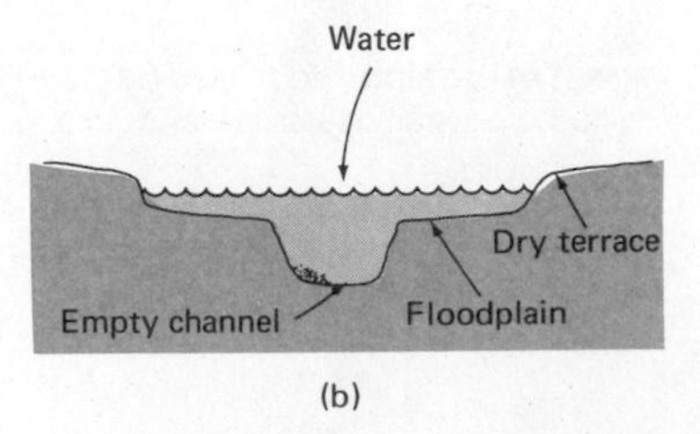

(b)

FIG. 15-14. Cross-sectional area of a river during (a) normal stage and (b) flood. Increasing the energy of a stream enables it to excavate a deeper channel by removing some of the material dropped in its bed during times of low flow and little energy.

bankfull stage depth of water that fills the natural channel of a stream (see p. 312)

The graded profile of a stream is also subject to change if conditions in the watershed change. Clearing an area for agriculture increases the total runoff and the supply of removable sediment. The larger river, in turn, demands a larger channel. It may take a long time for the stream to excavate a bed of appropriate size. Meanwhile the river floods more frequently than normal and spreads sediment from the crop fields across the floodplain. Former floodplains in some parts of the United States have been buried under several meters of new sediment during the first few decades after farmers settled the land (Fig. 15-15). Logging in mountainous areas can have the same effect, as many a silt-choked salmon can testify.

FIG. 15-15. Second-generation bridge over a small creek in Wisconsin. Beneath the present bridge are the supports for an old bridge, now almost completely buried by recently deposited sediment

CLIMATIC CONTROL OF LANDFORMING PROCESSES

All erosional processes have the same force (gravity) and the same goal (sea level). Furthermore, all are constrained somewhat by the underlying rock, and all would pass through a recognizable sequence of developmental stages if the land stood still long enough. Increasing the height of a landform increases the amount of energy available for the agents of erosion. The effects of landforming processes are more noticeable on soft rocks than on resistant ones.

agent of erosion tool used by gravity to help wear down the land

These principles of landforming have been illustrated by referring to the erosional work of landslides and rivers. However, these two landforming agents are not the only ones that help to lower high places. Gravity may use frost or ice to do its work in cold places. Mass movements in wet places take forms that are far more subtle than a landslide. Wind and occasional torrential rains are more important landformers than permanent rivers in arid places. The next chapter outlines the domains of the various landforming agents and the signatures they leave on the surface of the earth.

Chapter 16

Terrain Shapers: Landforming Processes

Landformers are careless burglars; they leave their fingerprints all over the place.

U.S. Forest Service

A person trying to trace the development of a landform is like a detective trying to identify the perpetrator of a crime but faced with a complex assemblage of meaningful clues and irrelevant detail. Each agent of terrain formation has its own unique mode of operation, its own preferred region of activity, and its own characteristic set of marks left on the surface of the earth. Clues about the origin of some landforms may be overwhelmingly obvious. Other terrain features may have been shaped by several different processes that partially obliterated the distinctive signatures of one another. This chapter stresses the unique features produced by the major landforming processes. The appendix outlines one theory about the relationships among landforming processes in a particular place. The next chapter examines the interplay of geologic structure, landforming processes, and history in some typical landscapes.

morphogenetic region place in which certain processes operate to produce particular landforms

TERRAIN SHAPED BY MASS MOVEMENT

The term mass movement includes all ways in which gravity shapes the earth's surface without the assistance of transportation agents such as rivers, glaciers, or waves. Mass movements occur in a wide variety of guises that include the almost imperceptible downhill creep of an Indiana hillside soil, the implacable advance of a California mudflow, the giddy flight of a rock falling off an Arizona cliff, and the awesome suddenness of a Colorado avalanche. Spectacular landslides and avalanches, however, do not transport as much earth as the rarely noticed movements of individual particles of rock and soil when they are splashed by the rain, trampled or burrowed by animals, wedged by frost or plants, plowed by farmers, or otherwise dislocated. Disruptive forces push soil in many directions, but the relentless pull of gravity guarantees that most of the resulting motion will be downhill. Nevertheless, different kinds of mass movements shape distinctive landforms in different environments.

Rockfall. An arid climate cannot support a continuous cover of vegetation to protect the land. Soil often is thin or absent because chemical weathering operates slowly when water is scarce. Frost, plant roots, intense sunlight, and subzero nights inflict tremendous mechanical stresses on the exposed rock. The widely scattered plants cannot hold loosened material in place: the rocks fall downhill if the slope is steep enough. Vertical cliffs flanked by aprons of broken rock are characteristic signatures of mechanical weathering and rockfall in an arid environment (Fig. 16-1).

talus slope sloping apron of broken rock at the base of a cliff, usually in dry regions

Exfoliation. Rounded rock outcrops occur whenever exceptionally dense rocks resist weathering more effectively than their neighbors can (Fig. 16-2). Occasional rockfalls may occur as sheets of rock break loose, but much of the rock disintegrates slowly and is removed grain by grain, usually by a transporting agent such as rainwater. Accumulations of debris at the base of hills are rarely noticeable in a rainy region because the results of rockfalls are carried away by a river or incorporated into the general mass of soil and vegetation at the foot of the resistant outcrop.

exfoliation or spalling thin sheets breaking off a convex surface, like layers of an onion

bornhardt dome-shaped outcrop of rock in tropical regions

FIG. 16-1. Talus slopes. Accumulations of angular rock debris at the bases of cliffs are typical products of mechanical weathering and rockfall in arid regions. (Courtesy of E. A. Busek, Soil Conservation Service.)

FIG. 16-2. Rounded rock outcrops, natural consequences of chemical attack on all sides of an exposed mass of dense rock. (Remember Fig. 7-11.) (Courtesy of J. D. Love, Geological Survey.)

Soil creep. An observer cannot watch the kind of mass movement that is most important in humid environments, yet the results of earth movement are obvious: tilted fence posts, broken foundations, displaced rocks, and accumulations of dirt on the uphill sides of buildings or rocks. Any expansion of the soil on a nonlevel surface pushes earth directly away from the surface, and therefore sideward as well as upward, while gravity still pulls straight downward (Fig. 16-3). A wide variety of processes can cause the soil to expand: frost, earthworms, plant roots, gophers, expanding clays, cattle, thunderclaps, sprouting seeds, or earthquakes. Together with gravity, these disturbances can move several tons of soil downhill each year from a typical hectare of hill land (about 2½ acres or two football fields). One result of this slow downhill creeping of the soil is a reduction in average slope and a decrease in the likelihood of dramatic mass movements.

frost-heaving expansion and contraction caused by water freezing and melting in tiny pore spaces (see Fig. 10-8)

soil creep slow downhill movement of soil

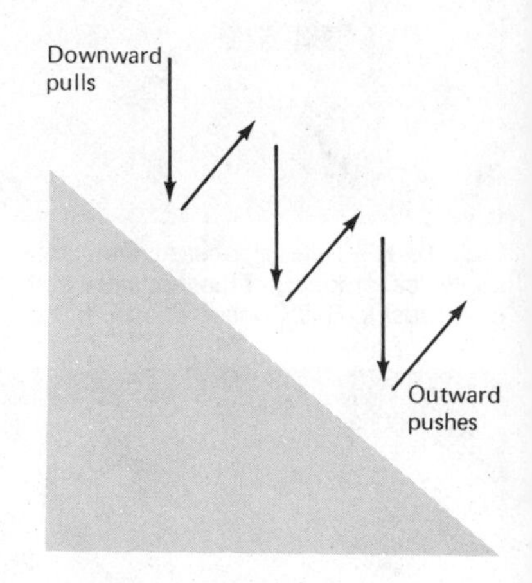

FIG. 16-3. Soil creep. An almost imperceptible downhill motion of soil is the result of a stepwise alternation of downward and outward motions.

Mudflow. Soil water promotes downhill motion by lessening the internal friction of soil, but vegetation interferes with earth flows by binding soil together. The most likely site for a mudflow is therefore a place that has a wet season to soak the soil and a cold or dry season to kill vegetation. The texture of the surface material is also significant. Coarse or angular material resists flowage; rounded sands and fine particles are more willing to slide downhill. Some kinds of waterlogged clay can ooze down a very gentle slope and produce a nearly level surface. Slowly flowing mud may be the reason for the extreme flatness of some terrain near the equator. Mudflows are even more significant near the poles, where permanently frozen soil provides an ideal skidway for gravity movements of surface material that thaws and becomes waterlogged during the brief summers (Fig. 16-4). Waterlogged soil on mid-latitude slopes is bound together by plant roots and less likely to flow downhill slowly than to slump with an abrupt backward rotation resembling a person accidentally but suddenly sitting down on a slippery surface (Fig. 16-5).

solifluction downhill flowage of waterlogged soil over rock or permanently frozen soil

slump fairly rapid downhill slippage of earth material

FIG. 16-4. Solifluction, a downhill movement of waterlogged surface soil lying on top of permanently frozen subsoil. The white strip on the right is a road in this view from an airplane. Inactive solifluction lobes often indicate episodes of cold climate in the past. (Courtesy of R. B. Colton, Geological Survey.)

FIG. 16-5. An earth slump, a moderately rapid mass movement characterized by a backward rotation of the sliding surface material. (Courtesy of McDole Soil Conservation Service.)

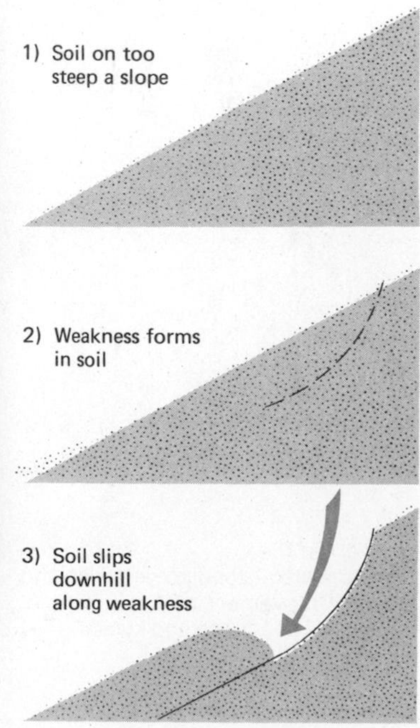

Avalanche. Soil creep, mudflows, and slumps are also common on high mountains, but the symptoms may be overshadowed by the consequences of spectacular mass movements such as avalanches and landslides. Snow avalanches are so common on some mountains that they leave a recognizable signature in the form of well-defined avalanche chutes (Fig. 16-6). Rock slides or debris avalanches occur less frequently than snow avalanches but carry more earth material downhill. Like mudflows, debris avalanches and landslides occur in certain favorable environments. They are rare in dry areas because slopes often are so steep that rocks fall off as soon as they weather loose. Landslides also are rare in warm and humid areas, since weathering and soil creep produce gentle slopes that do not permit rapid mass movements.

avalanche chute track cleared by frequent snow avalanches

FIG. 16-6. Avalanche chute. Frequent avalances keep the surface smooth and free of vegetation, which in turn promotes avalanching. (Courtesy of F. E. Matthes, Geological Survey.)

Treefall. Trees cause their own distinctive form of mass movement on forested slopes. A dead tree usually falls downhill when it finally collapses. Its roots lift a mass of soil, which then settles back to earth in a pile some distance downhill. A few centuries of intermittent "treefall" produces a slope marked by pits and mounds, and the irregularity of the slope prevents most forms of smooth downhill motion. Treefall movement of earth material is much less regular than soil creep, but the two mechanical processes move about the same amount of soil per year in their respective environments.

silvaturbate slope irregularities and downhill movement of earth material caused by soil "uprooted" when trees fall

In summary, the shapes of slopes in a particular place depend on the nature of the surface material and on the kinds of mass movements that occur in the region.

TERRAIN SHAPED BY WATER—PEDIMENTS

Raindrops hit an unprotected surface with considerable force. A centimeter of rain on a football field represents forty tons of water whose impact may splash many tons of dirt into the air. Most of the earth splashed from a sloping surface lands downhill of its takeoff point. Rain splash is a significant form of erosion in normally dry places, on recently plowed fields, or on other exposed soils.

Raindrop impact also plugs soil pores and interferes with infiltration. Surplus water flows downhill, carrying bits of earth with it. Runoff tries to form a channel whenever the volume of water becomes great enough to remove sizeable amounts of earth. In a normally dry region, however, the incipient channel is the wettest part of the landscape. Plants naturally prefer to grow on the moist areas rather than on the dry slopes. They fill the channel with a dense tangle of roots and stems, which in turn obstructs the flow of water and hinders further erosion by the stream. Consequently, the runoff water often chooses a different path downhill after the next storm. The erosive effect of running water is spread over the surface rather than being concentrated in a narrow channel, and the terrain is lowered uniformly. The inevitable result of this tendency for plants to plug valleys as soon as they begin to form is a gently sloping but monotonously flat surface that resembles a tilted tabletop. Much of the land in states like Nevada and Wyoming consists of these gently sloping pediment surfaces.

sheetwash runoff that is not confined in a channel
(see p. 304)

pediment (bajada) surface planed by sheetwash
(see Fig. 16-9)

Occasional peaks may stand out above the erosion surface for a variety of reasons. Exceptionally hard rocks resist weathering and therefore survive as islands of high land. Cliffs retreat as rocks weather loose and fall off, but the apron of loose rock protects the base of the cliff from erosion. Fortuitously located areas escape erosion for a short time and then find themselves isolated above neighboring areas, saved from further erosion by their height above normal flood levels. However, these remnant areas only serve to emphasize the flatness of the surrounding surface.

TERRAIN SHAPED BY WATER—VALLEYS

Large desert rivers and most streams in humid regions erode the land in a manner that is strikingly different from the way small intermittent streams plane the land in dry regions. The channel of a more-or-less permanent river is usually waterlogged and unfavorable for plant growth, while the surrounding land is protected by a cover of vegetation. Gravity concentrates the energy of the stream on the area immediately beneath or beside the water. The concentration of transportation energy on the lowest available path produces the most characteristic feature of a river-eroded landscape, the V-shaped valley (Fig. 16-7).

FIG. 16-7. Valley with a V-shaped cross section, an indicator of active erosion by running water. (Courtesy of Soil Conservation Service.)

However, there is a limit to the depth of valley that a river can carve. As the valley gets deeper, its sides become steeper and mass movements become more likely. Most rivers function primarily as transporters, not as eroders. They spend most of their energy carrying away material after it has been loosened by weathering and brought to the stream by mass movement, rain splash, or unchanneled runoff.

A river that is strong enough to carry all of the available sediment will usually occupy the entire bottom of a narrow valley. The land slopes upward immediately beside the banks of the stream. The hills of West Virginia, Alabama, Missouri, or western Oregon are typical examples of mature river-carved landscapes.

canyon narrow V-shaped river valley

arroyo valley of intermittent stream in dry area

A river needs surplus energy and abrasive tools in order to maintain a classic V-shaped valley. A river that changes volume seasonally or after big storms often has a fairly flat bed that is usually choked with debris. The need to accommodate occasional floods makes the valley much wider than the stream, which may not even be present at all times. Occasional thundershowers or meltwaters from nearby mountains turn dry streams into raging torrents and thus make it necessary to build substantial highway bridges across seemingly insignificant creeks. A flooding river usually carries a heavy load of sediment taken from adjacent land surfaces, but the stream weakens as it flows along, because water evaporates or sinks into the dry soil. As the flood subsides, the river fills its valley with debris rather than gouging it out.

The reliable rains of an instability climate guarantee that equatorial rivers will not lack water. However, intense weathering in a hot and humid place usually reduces rocks to very fine clays. Rivers may carry a substantial load of sediment and still be unable to gouge their valleys downward, because the particles of sediment are not heavy or hard enough to act as abrasive tools. The terrain around an equatorial river often resembles a staircase in cross section, with level areas formed as consequences of the inability of the river to wear down the rapids or waterfalls that form where resistant layers of rock cross the valley. Moreover, the efficiency of mass movement in wet clay soils soon chokes an overly ambitious river: a temporarily deepened valley only serves as an invitation for soil on the slopes to move toward the stream valley and to fill it with sediment.

cataract waterfall or rapids in a zone of hard rock

Rivers in cool regions are more effective than equatorial rivers because they can use sharp quartz sands to help them overcome geological resistance and create a smooth profile. In still colder regions, however, frozen subsoils impede downcutting and encourage widening of the rivers that flow in summertime. Many Arctic rivers are shapeless swamps that ooze gently toward the sea. Those that flow northward have an additional obstacle to efficient valley formation: their sources thaw out long before the ice downstream near the Arctic Ocean begins to melt. The ice dam across the river encourages it to spread out even farther. Winter refreezes the landscape and obscures the river valleys. Another summer may bring a slightly different pattern of melting and therefore put the rivers in different places for a year. Of course, rivers on steep terrain may have enough energy to produce permanent V-shaped valleys even in a frozen soil.

permafrost permanently frozen layer of soil in cold area

TERRAIN SHAPED BY WATER—PLAINS

A river rushing down a steep slope has energy to pick up bits of rock and soil. Friction slows the water when it enters a lake, ocean, or level stretch of land and has no more altitude to trade for speed. A river that loses speed must drop material that it can no longer carry. Even in its depositional role, running water retains its sensitivity to load size and tendency to seek the lowest possible path. However, as a depositer, a river does not enlarge the lowest path; it destroys it. Water finds a low place, slides down into it, and then loses speed and drops its load of debris. One by one the low places are filled with sediment and the landscape is leveled.

alluvium sorted and layered material deposited by moving water when it slows
(see Fig. 16-18)

A flood is a river's way of forsaking the mad race in its channel and taking a more leisurely route to the ocean. A river usually loses speed as it moves farther from the main channel. The largest and heaviest particles in its load drop out first. Sandbars are left in the stream and alongside the channel, while silt and clay can be carried farther away from the channel and deposited in the backwater swamps.

floodplain alluvial plain caused by occasional river floods

natural levee alluvial deposit of sand or coarse silt immediately adjacent to a river channel

The same sorting of material occurs when a river enters a calm body of water or leaves a steep valley to venture forth onto a plain. The river loses energy and drops material in a sequence from coarse to fine. Deposition chokes the riverbed with debris and forces the river to shift sideways and to seek another path. Eventually the river creates a broad, fan-shaped deposit

delta alluvial deposit where a river enters a lake, ocean, or larger river

FIG. 16-8. Alluvial fan, formed when a river flows out of a steep valley onto a level area. (Courtesy of W. C. Alden, Geological Survey.)

that consists of concentric arcs of different sizes of sediment (Fig. 16-8). The fan often has a relatively steep slope of coarse debris near the mouth of the valley. Gentler slopes and finer sediments may merge imperceptibly into the level plain beyond. Adjacent fans frequently coalesce into a broad plain in their lower reaches (Fig. 16-9). The level expanses of western Kansas and Oklahoma are products of long-term river deposition of material from the western mountains. Some grainfields lie on top of hundreds of meters of deposits that were meticulously sorted and leveled by ancient streams.

alluvial fan fan-shaped deposit where a river passes from a steep valley onto a plain

FIG. 16-9. Coalescing alluvial fans. Many mountains in arid regions are surrounded by flat pediments and broad aprons of debris deposited by intermittent streams. (Courtesy of J. R. Balsley, Geological Survey.)

FIG. 16-10. Lateral erosion by a small stream. When this stream floods, it undercuts the steep bank on the left side of the photo and deposits material on the low bar to the right. (Courtesy of R. J. Turner, Soil Conservation Service.)

Uplift of the land or additional water in the channel may allow a stream to erode into its own floodplain deposits. Where a stream curves, it pushes against the outside bank, undercuts the soil there, causes the bank to collapse, and deposits material at a lower elevation on the inside of the curve (Fig. 16-10). Several generations of streams, each operating at a different elevation, can shape an old floodplain into an intricate assemblage of flat benches and curving slopes. The benches represent former floodplain levels, while the slopes are abandoned riverbanks that mark the positions of the river at different times in its past (Fig. 16-11). The soil pattern may be even more complicated than the topography, because different parts of the floodplain were near the river at different times in history, and nearness to the river governs the speed of floodwaters and thus influences the kind of material deposited in a particular place.

meander sinuous curve in a river

terrace flat area above the present floodplain or shoreline

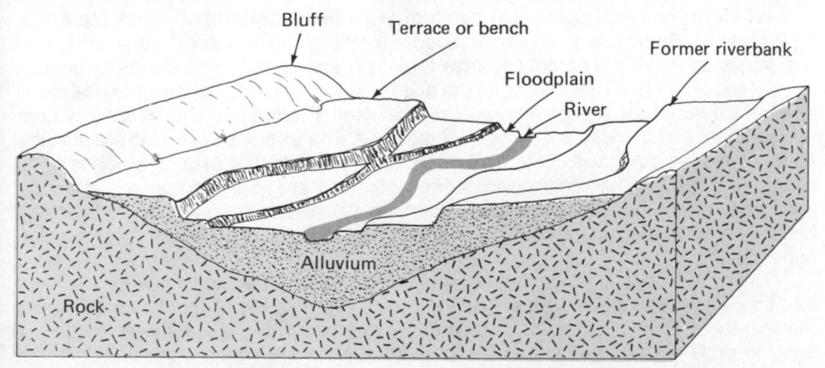

FIG. 16-11. River terraces. These flat areas at different heights above the stream indicate the elevation of the river at different times in the past.

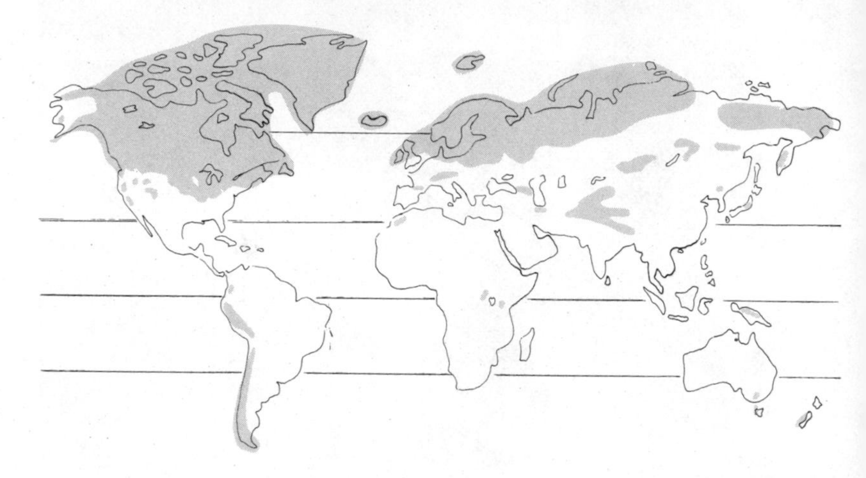

FIG. 16-12. Area covered by glaciers during the Ice Age. The modern world has only two large ice sheets, one on Greenland and the other on Antarctica. A vast array of geological, topographical, biological, and oceanographic evidence indicates that glaciers were far more extensive ten thousand years ago. The mechanisms by which large amounts of water were taken from the oceans and frozen on the continents are simply not known. Perhaps an excess of carbon dioxide in the air hindered outgoing radiation, raised surface temperatures, increased the rate of evaporation, and thus caused more snow to fall. Maybe a melting of Arctic Ice exposed more ocean surface, therefore increasing the supply of moisture in polar air masses. On the other hand, a thickened cloud or a weakened sun could have helped to preserve ice through the summer. There is considerable agreement that a change of temperature of only a few degrees may be capable of beginning a glacial age; whether the earth must become warmer or cooler first is still a matter of debate.

Whatever its initial cause, ice can further its own growth once it has started to form on land. Snow has a very high albedo; it reflects solar energy, stays cool, and postpones its melting. The cold surface may increase short-term snowfall by cooling air masses laden with moisture from the ocean. In the long run, however, an ice mass creates a thermal cell of high air pressure over itself. Prevailing winds blow away from the ice sheet and hinder the approach of moist air. The existence of an ice sheet in the middle latitudes may thus set in motion a chain of events that eventually leads to its disappearance. On the other hand, storm enhancement near the margins of an ice sheet may cause the glacier to expand even though the center no longer receives any new snow. Once again, a discussion of continental glaciation leads to a puzzling contradiction.

The record does indicate that glaciers advanced into the middle latitudes at least four times in recent geologic time. The ice sheets melted away all four times, but some modern theorists add an ominous footnote: if any one of these ice sheets had managed to grow another few hundred kilometers toward the equator, the albedo effect would have become dominant, the snow would have been unable to absorb enough sun energy to melt, and the ice would have continued to grow until it covered the earth.

TERRAIN SHAPED BY ICE—SCOUR ZONES

Glaciers presently occur only on high mountains or in the icy fastnesses of inversion climate in Greenland and Antarctica, but nearly half of the people in the United States live on land formerly covered by kilometer-thick masses of ice (Fig. 16-12). The accumulation of great masses of snow requires an excess of snowfall over snowmelt and evaporation. The bottom of a snowpack becomes ice when the weight of overlying snow is great enough to recrystallize the water. In turn, the ice becomes a glacier when it begins to move under the influence of gravity.

inversion climate cold and dry weather near the poles (see Fig. 2-13)

nevé snow that has been compressed into ice

glacier moving mass of ice

Ice moves when gravity can overcome friction with the ground and the internal resistance of the crystalline water. An ice sheet on flat land must be quite thick before outward flow begins. The thickness of ice is not so important in mountainous areas, because the ice already has a slope to follow. In either case, the speed of glacial movement is slow, generally on the order of a centimeter to a meter per day.

ice sheet glacier that covers most of the land

A valley glacier begins life as a snowfield high on a mountain. Years of snow accumulation, freezing and thawing, and perhaps an occasional avalanche create a sizeable pile of snow in a depression on the side of the hill (Fig. 16-13). When enough ice builds up, the mass begins to move downhill.

alpine glacier glacier confined to a valley on a mountainside

cirque bowl-shaped depression at the source of a valley glacier

FIG. 16-13. Cirques, bowl-shaped depressions where snow accumulated at the heads of former valley glaciers. (Courtesy of T. S. Lovering, Geological Survey.)

arête sharp ridge between two valley glaciers

horn pointed mountain produced by valley glaciers on three or more sides

A valley glacier may start down an old stream valley but refuses to be confined by a narrow river canyon. A V-shaped valley may be efficient for water, but ice moves several thousands of times more slowly and therefore must be several thousands of times larger than a river in order to carry the same amount of water downhill. Furthermore, glacial ice is too rigid to follow the curving path of a typical river. Therefore a valley glacier reorganizes a river valley into a flat-bottomed "U" shape, which is more convenient for the movement of ice (Fig. 16-14).

FIG. 16-14. U-shaped trench made by a valley glacier. The dark stripes on the glacier in the back part of the valley are medial moraines, accumulations of debris on the ice where separate valley glaciers merged. (Courtesy of L. J. Prater, U.S. Forest Service.)

medial moraine stripe of debris in the middle of a glacier

striation groove in rock caused by glacial scour

roche moutonnée rounded and grooved rock exposed and scoured by a glacier

paternoster lakes chain of rock-floored lakes in scoured part of formerly glaciated valley

Sliding ice pushes a great deal of earth material along. Most of the debris carried by a valley glacier came from the sides of the valley and is held in the ice near the side edges of the glacier. A ribbon of debris appears in the middle of the combined ice flow when several glaciers from different valleys join together.

A vigorously advancing glacier removes loose material from the ground and then uses the loosened rocks as tools to scrape and gouge the material that it passes over. The characteristic signatures of glacially scoured terrain are parallel grooves, bare rounded rocks, and lake-filled basins scooped out of the earth. The orientation of grooves and lakes indicates the direction the ice was moving, while their depth and evenness are clues to the thickness and speed of the glacier (Fig. 16-15).

FIG. 16-15. Glacially scoured rock outcrops. A glacier moved across the terrain, cut grooves in the rocks, and occasionally shattered a rock under the immense weight of the ice. (Courtesy of F. E. Matthes, Geological Survey.)

A powerful valley glacier may deepen its valley rapidly, while the side valleys that enter into the main valley remain relatively high. Some spectacular scenery emerges when the glacier melts away and a river in the side valley has no easy way to get to the floor of the main valley. Finally, as if in desperation, it simply dives off the side as a waterfall and cascades down the steep side of the glacially scoured U-shaped valley.

hanging valley tributary valley whose floor is much higher than that of the deep glacial valley it enters

TERRAIN SHAPED BY ICE—MORAINES

A glacier begins to lose mass when it moves too far from its frigid birthplace and enters the hostile warmth of other environments or when the environment near its origin changes and becomes warmer or drier. In either case, ice melts more rapidly than new snow can accumulate (Fig. 16-16).

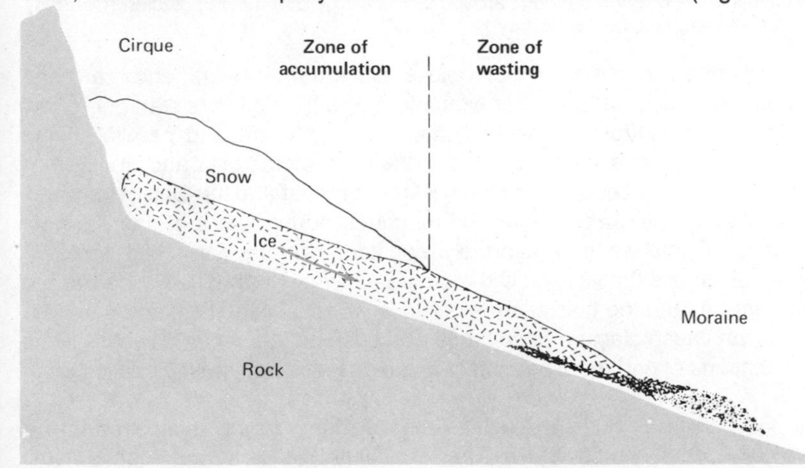

FIG. 16-16. Mass budget of a glacier. A glacier loses some of its earth-moving ability wherever losses by evaporation or melting exceed additions by snowfall or avalanching.

The glacier becomes thinner and the ratio of debris (passengers) to ice (vehicle) becomes greater. Eventually the weight of debris may become great enough to halt the outward motion of the glacier. Ice near the margins of the glacier melts completely away during warm times. The melting causes the edge of the glacier to retreat, although the ice does not physically move backward. The load of earth carried in the ice falls to the ground as a jumbled accumulation of boulders, gravel, sand, and fine particles.

A retreating glacier may advance again if it is forced to move by the added weight of new snow. Like a colossal bulldozer, advancing ice pushes previous deposits before it. Repeated advances and meltbacks can pile up a considerable heap of debris at the outer edge of the farthest advance. The debris is an end moraine, one of the characteristic signatures of glacial deposition. Lines of irregular hills across many midwestern states mark the end moraines of various glacial advances during the Ice Age (Fig. 16-17).

end moraine mass of till left at the outer edge of a glacier

FIG. 16-17. An end moraine. The ribbon of hilly land behind the barn in this photograph is the pile of earth material left at the outer edge of a melting glacier. (Courtesy of W. C. Alden, Geological Survey.)

The hills often lie across former valleys, block former rivers, and rearrange previous drainage patterns. For example, some Illinois rivers start only a few kilometers away from Lake Michigan, flow southward and parallel to the shore but behind a moraine for hundreds of kilometers, and finally turn southwest and proceed toward the Mississippi River and the Gulf of Mexico.

Valley glaciers also leave end moraines at the points of their farthest advance. Centuries of reasonably constant climate can build sizeable moraines across the valleys. Old moraines may be abandoned if the climate gets warmer and the glacier shrinks. Snowy or cool conditions, on the other hand, can cause glaciers to advance and push old moraines downhill. Thus the locations of end moraines offer valuable clues about past climatic conditions.

ground moraine till left by a melting glacier

till plain mosaic of till, moraine hills, swamps, and lakes

Ground moraines are glacial deposits that were not bulldozed into a heap by later glacial advances. Year after year, the ice margin melted back, leaving a sequence of seasonal moraines that blend together into a disorganized mass of low hills and closed swampy valleys.

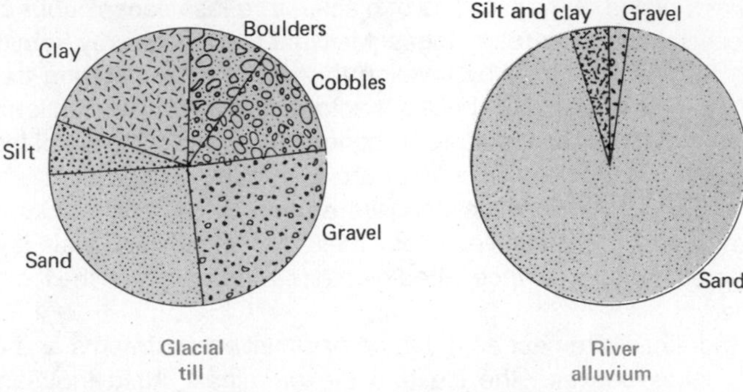

FIG. 16-18. Textural composition of samples of unsorted material deposited by a glacier (left) and sorted material deposited by a river (right).

The material in a moraine is invariably poorly sorted; unlike water or wind, moving ice has little sensitivity to particle sizes (Fig. 16-18). Water from melting ice may rework a glacial deposit by smoothing the terrain and sorting the debris. Many formerly glaciated areas are confusing mixtures of unsorted glacial moraine and well-sorted water or wind deposits.

till or boulder-clay unsorted mixture of material deposited by a glacier

TERRAIN SHAPED BY MELTING ICE—OUTWASH DEPOSITS

When glaciers melt, they release vast quantities of water in a fairly short period of time. The water flows over land that often was recently covered by ice and now lies bare of vegetation and littered with debris deposited by the glacier itself. The loose and exposed material is easily eroded, so that meltwater streams usually carry a heavy load of earth. In fact, their loads are often far too much for them to carry any distance. Meltwaters frequently spread out, slow down, or sink into the soil soon after they leave their icy sources. As the streams lose their ability to carry earth, they drop their loads of material.

outwash material deposited by meltwater flowing off a glacier

Meltwater deposits are responsible for three distinct features of glaciated terrain. The least extensive but most conspicuous meltwater landforms are formed within the ice mass itself. Water droplets join and form streams that begin within the glacier and flow toward the edge. These streams erode down through the ice and form valleys or tunnels. As the drainage system becomes more extensive, the glacial streams begin to deposit material in their beds, just as ordinary rivers do on the floodplains that form around streams as they age. Unlike a river on land, however, a subglacial stream cannot spread over a vast floodplain. The ice around the stream confines it to a fairly narrow channel, where it deposits a thick layer of sediment in its bed rather than spreading a thin layer across a wide valley. When the ice finally melts, the "streambed" stands above the surrounding terrain and forms a long but narrow ridge of well-sorted deposits. The ridge may be tens of meters high and many kilometers long, occasionally straight but more commonly winding, and often superimposed across the terrain with little regard for the hills and valleys below.

esker ridge formed by alluvial deposits in the bed of a subglacial stream

Outwash plains at the margins of glaciers are less conspicuous but far more extensive than the stream ridges. Meltwaters flowing away from the ice rearrange the glacial debris by leveling the deposits and sorting them by particle size. The result may be a monotonous plain with a uniform soil, usually sandy or gravelly in texture. Frequently, however, blocks of ice float away from the glacier-like icebergs or are left on the ground as the glacier melts. Meltwaters pile outwash deposits around these relic blocks of ice. When the ice blocks themselves finally melt, the flat outwash plain is transformed into an irregular surface pitted by depressions in which the ice blocks were buried.

pitted outwash irregular surface formed as outwash is deposited over ice blocks that subsequently melt away

End moraines often act as dams across meltwater streams and create temporary lakes between the ice and the moraines. Amid thousands of nameless and short-lived glacial lakes, a few achieved immense sizes and lasted for a considerable time during the Ice Age. Glacial Lake Erie extended all the way to Fort Wayne, Indiana. Glacial Lake Bonneville was many times larger than its shrunken remnant, the Great Salt Lake. Other ice-fed lakes covered large portions of Nevada, the Dakotas, Michigan, Connecticut, and many other states during the Ice Age.

Pleistocene period in geologic history when ice sheets covered much land not presently under glaciers

Erosion and deposition of material by Ice Age rivers and lakes obeyed the rules of water-formed landscapes, but the scale of activity was many times greater than at present. Meltwater rivers often carved valleys that are much larger than needed by present streams. Ice Age winds took materials from moraines and floodplain deposits and reworked them according to the laws of wind activity. Thus the past Ice Age produced many terrain features that are related to glaciers, could not have occurred without them, but were not shaped directly by them.

TERRAIN SHAPED BY WIND—PAVEMENTS

Ice has little concern, water is moderately sensitive, but wind is very fussy about the sizes of particles it transports (Fig. 16-19). Gravel is essentially immovable; sand is too heavy to do more than bounce along near the

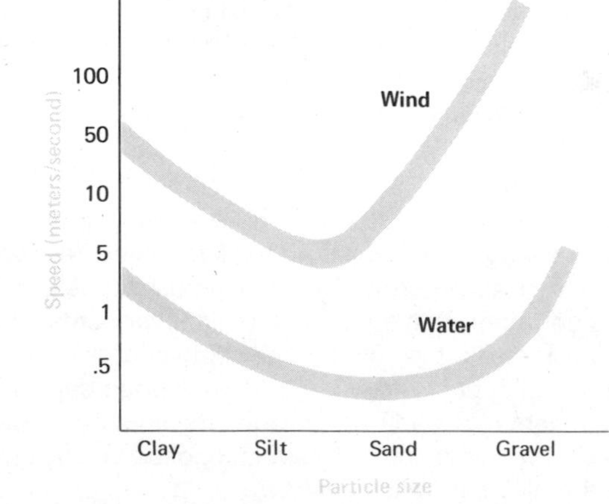

FIG. 16-19. Minimum speed of air or water needed to move particles of different sizes. Rivers are strong and able to carry sand-sized particles easily; winds are most efficient at carrying silt-sized material, but air is relatively weak and must be going at least 5 m/sec before it can move a significant amount of soil.

ground. Tiny clay particles are easily carried, but clay usually clumps together in large aggregates that resist motion by the wind. A medium-sized silt grain, on the other hand, is a natural hitchiker, easily picked up and able to remain suspended as long as the wind maintains speed. If the wind slows, the silt drops out and waits for another ride, unless the soil or vegetation is able to trap the silt and make it settle down.

ped clump of soil particles (see Fig. 8-14)

The wind can leave recognizable erosion scars on the land only where other landforming processes are very weak and the surface is unprotected by vegetation. In these favored environments the wind carries away whatever particles it can. The larger particles are left behind as a "desert pavement," a rock-armored surface that resists further wind action.

ablation or deflation removal of material by the wind

desert pavement surface cover of gravel and rock left after wind removes fine material

The wind is not content with merely removing all loose, fine material. As if sensing that the supply of silt and sand is finite, the wind seems to use the available supply of sand to make additional fine material. Rocks or plants unfortunate enough to stand above the surrounding terrain are sandblasted; loose material near the ground is chipped away by the sand-armed wind. Wind removal of partially weathered rock is a delicate landforming process, especially sensitive to minor differences in rock hardness (Fig. 16-20).

ventifact stone with flat faces polished by wind abrasion

FIG. 16-20. Pedestal rocks. Partially weathered rock is removed by wind abrasion. Softer rocks are more vulnerable to weathering and abrasion, especially near the ground, where the wind is armed with sand and weathering is aided by somewhat more moisture than elsewhere. (Courtesy of M. E. Gregory, Geological Survey.)

Rocks protruding a short distance above the level of the pavement are especially vulnerable, because sandblasting concentrates its energy in a narrow band near the surface. Closer to the ground, friction keeps the wind too slow to be effective. Farther away from the ground, the wind blows rapidly but carries only dust. The rock-carving sand particles bounce along within a few meters of the surface and exert their force at precisely the level that weathering is most active. Monuments, pedestals, and balanced rocks are characteristic features of a process that erodes the weathered base of an isolated rock more rapidly than the top.

monument, pedestal rock isolated, wind-abraded rock

TERRAIN SHAPED BY WIND—SAND DUNES

Wind is unique among transport agents because it pays little attention to local differences in elevation. Water and ice can flow only downhill, although ice occasionally creates its own slope. Wind, however, gets its impetus from large-scale differences in air pressure and thus can move uphill, downhill, or across hill.

saltation bouncing advance of wind-driven sand or water-driven pebbles

A ground level view (with protective goggles, please) of a sand dune on a windy day shows sand grains moving in hops and jumps as they bounce up a slope made entirely of sand. Eventually the sand grains reach the top, where the next puff of wind pushes them over the crest and on to the sheltered slope on the far side. In time, every sand grain on the windward side will be driven up the hill and shoved over the brink; the entire dune advances in the direction of the wind (Fig. 16-21).

leeward sheltered side of a hill or other object; side opposite the direction from which the wind is blowing

The shape of a sand dune is a good indicator of wind conditions. Undependable winds produce disorganized masses of sand, but dunes in regions with reliable winds usually have gentle windward slopes and steep leeward slopes. The windward side of a dune can be no steeper than the slope a sand grain can be forced to ascend. The wind keeps pushing sand over the top, so that the leeward side usually slopes at the critical angle for maintaining mass movements (between 31° and 36° for different kinds of sand).

maintaining slope or angle of repose slope that is not steep enough to start a mass movement but still not gentle enough to stop one once under way
(see Fig. 15-4)

barchan crescent-shaped dune in region of reliable wind

seif elongated dune

Sand dunes form only in areas that have a number of key characteristics. One requirement, a source of sand, seems too apparent to mention, but it does place some restrictions on the location. The right kind of rock material must be present and properly weathered. A second requirement for dune formation is an occasional dry period. Even hurricanes do not have the strength to move a sodden mass of wet sand covered by tall trees and dense grass. Finally, it is necessary to have winds with sufficient strength and directional reliability to collect the sand and pile it up in one area.

Contrary to what we see on television, less than 10 percent of the deserts of the world are covered by dunes. Blowing sand is rare near the

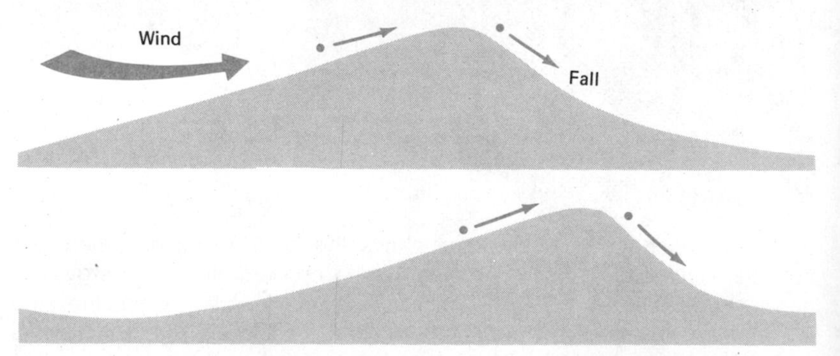

FIG. 16-21. Advancing sand dune. (top) Wind drives sand grains up the windward slope and pushes them over the crest. (bottom) Slowly, the entire dune moves in the direction the wind is blowing. The wavelike shape of a sand dune is no coincidence; however, sand waves advance much more slowly than water waves.

margins of hot deserts and in most cool deserts, because vegetation protects the ground. Dunes seldom appear in the cores of the tropical deserts, where subsiding air does not produce winds of sufficient strength and reliability. Limestone or shale bedrock does not yield enough sand for dunes to form. Consequently, most major dune fields occur downwind from outcrops of granite or sandstone in hot deserts with prevailing winds and harsh conditions for vegetation.

The second major dune environment is a lake or ocean coast, where rivers and onshore waves bring a regular supply of fresh sand onto the beach. Strong winds must be ready to move the sand inland when the top layer dries out. Offshore winds must be weaker, moister, or rarer than onshore winds. Large coastal dune fields are found on the West Coast of the United States, in the path of the oceanic westerlies, and on the south and east shores of the Great Lakes, where Canadian air masses provide the wind. Smaller dune fields dot the East Coast, where winds are less regular than on the West Coast. Coastal dunes rarely move far inland, because vegetation stabilizes the sand soon after the wind moves it away from its source on the beach.

dune stabilization a succession process covering a sand dune with vegetation that holds the sand and protects it from blowing
(see Fig. 11-8)

TERRAIN SHAPED BY WIND—DUST CAPS

Another kind of wind deposit is apparent whenever someone rubs a white cloth over a textbook or other piece of seldom-used furniture. Ordinarily, the wind merely rearranges material within a local area. Any building, tree, or other obstacle produces a local slowing of the wind, and piles of wind-carried material obediently accumulate in the sheltered area behind the obstacle (Fig. 16-22). Many a homeowner has sullenly cursed fate (or the

boundary layer layer of calm air near a surface
(see p. 106)

FIG. 16-22. Deposition of wind-carried dust where obstacles slow the wind. This scene from the American "Dust Bowl" days of the 1930s shows accumulations of dust behind fence posts and farm buildings. (Courtesy of Soil Conservation Service.)

architect) while shoveling drifts of snow away from an ill-located door. Snow fences in the mountains or shelterbelts on the plains are deliberate attempts to interrupt the wind pattern and thereby decrease wind erosion or collect wind deposits in particular places (Fig. 16-23).

particulate load material suspended in the air

In certain regions of the world, however, the wind produces a noticeable "rain" of fine particles. Except for the obvious requirement of wind, the climatic conditions for silt deposition are different from those for sand dunes. Weathering in most dry regions is too inefficient to produce many silt-sized particles. Therefore, sandstorms are more common than dust storms in true deserts. A protective cover of vegetation prevents wind erosion in humid environments. The ideal moisture environment for wind transportation has a distinct and prolonged dry season but enough water at other times for efficient weathering. Warm soils weather rapidly to fine clays and do not have the silty texture needed for efficient wind transport. Cold soils are usually coarse and frozen for long periods. The best combination of temperature, moisture, and wind for a dust storm occurs in a rainshadow frontal climate. Oklahoma and Kansas once earned the nickname "Dust Bowl," but similar storms also occur in central Russia, China, and Argentina.

rainshadow dry area on the leeward side of a mountain
(see p. 86)

exotic river river that flows from a wet region through a dry region

Immense dust storms, however, are rare even in a dust bowl climate unless something disturbs the protective cover of vegetation or otherwise exposes an unusually large supply of loose silt. One extremely common source of silt is a large muddy river that occasionally overtops its banks and spreads sediment over the adjacent floodplain. Grassland rivers such as the Missouri or the Rio Grande of North America or the Yellow River of China are fruitful sources of wind-blowable material.

glacial flour silty material produced by glacial scour

loess silty material deposited by wind
(see p. 192)

The all-time champion silt source, however, is a continental glacier. Ice scouring often produces a silty debris with excellent soaring characteristics. Melting of the glacier exposes the debris and provides floodwater to sort it and spread it out. The icy glacier creates a local area of high air pressure that causes strong winds to blow away from the ice. The winds scoop up exposed silt, carry it away, and drop it when the airflow weakens. Many parts of the world have thick layers of wind-deposited silt (loess), a signature of the former cooperative union of ice, water, and wind (Fig. 16-24).

ped clump of soil particles held together by a combination of electrical, chemical, and biological forces
(see Fig. 8-16)

dry farming special techniques to conserve soil moisture

Cows and plows have recently joined natural floods and glaciers on the list of agents that expose soils to wind erosion. Overgrazing destroys the protective cover of vegetation, and plowing and cultivating pulverize the soil. Fertilizers may compound the problem by further disrupting soil structure. Farmers in susceptible areas must use techniques such as minimum tillage, mulching, stripcropping, or shelter-belt construction in order to reduce wind removal of silt.

albedo reflectivity of a surface

Thus the seemingly inconsequential event of dusting furniture is tied to the rich loess soils of the Midwest and the constant hazard of dust bowls. Dust deposition has even helped to confirm theories of crustal movement: differences in the thickness of the dust layer on the ocean floor indicate differences in the ages of different parts of the ocean. The great dust storms of the Ice Ages also raise questions concerning the role of particles in the whole atmospheric energy budget. One theory contends that the absence of reflective dust in the atmosphere triggers a series of complex changes that

FIG. 16-23. Shelterbelts in a region of potential wind erosion. The lines of trees reduce erosion by keeping the wind near the ground from achieving the speed needed to pick up soil efficiently. Furthermore, they increase wind deposition by interrupting the wind and causing it to drop some of the dust it may have already collected. (Courtesy of H. Postlethwaite, Soil Conservation Service.)

FIG. 16-24. Deposits of windblown silt (loess).

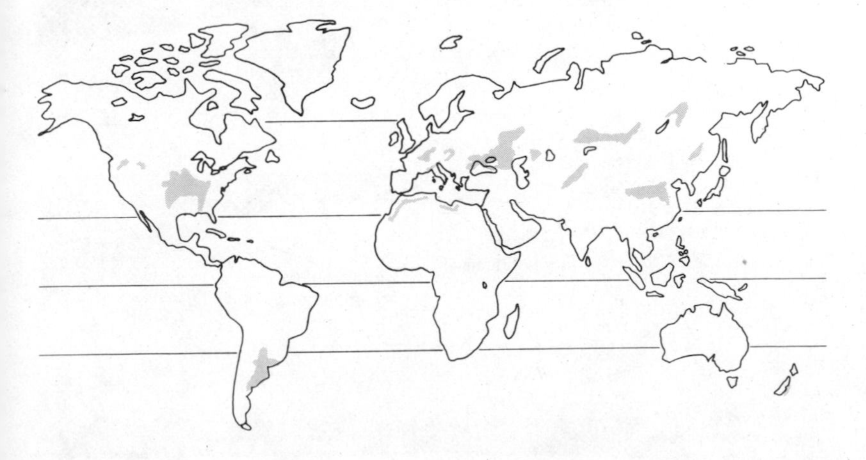

enhance precipitation. The precipitation produces vast snowfields that reflect radiation, fail to melt, and become glaciers. The masses of ice in turn create dust storms that block solar radiation, lower the temperature of the air, decrease the amount of moisture in the air, and cut off the supply of snow for the growth of ice. Another theory suggests that airborne dust causes glaciers to grow because the dust reflects light and permits snow to remain longer without melting. The fact that two theories about atmospheric dust come to diametrically opposite conclusions indicates that theories about the origins of Ice Ages are still somewhat speculative—a thought worth considering when someone suggests that we know enough to intentionally modify the weather on a large scale.

TERRAIN SHAPED BY WAVES—BENCHES

sea cliff cliff produced by wave erosion

marine terrace level bench produced near average sea level by wave erosion

Waves concentrate their energy on a narrow strip of land within a few meters of sea level. A wave that approaches shallow water usually trips over its own "feet" and falls forward in a watery somersault onto the shore. The breaking of a wave increases the landward speed of surface water and arms it with material scooped off the bottom. A wind-driven wave is a formidable land-shaping tool, able to cut into hard rocks with an ease matched by few other geologic agents. The result is a wave-cut notch that consists of a steep cliff at the shoreline and a level bench formed near sea level (Fig. 16-25).

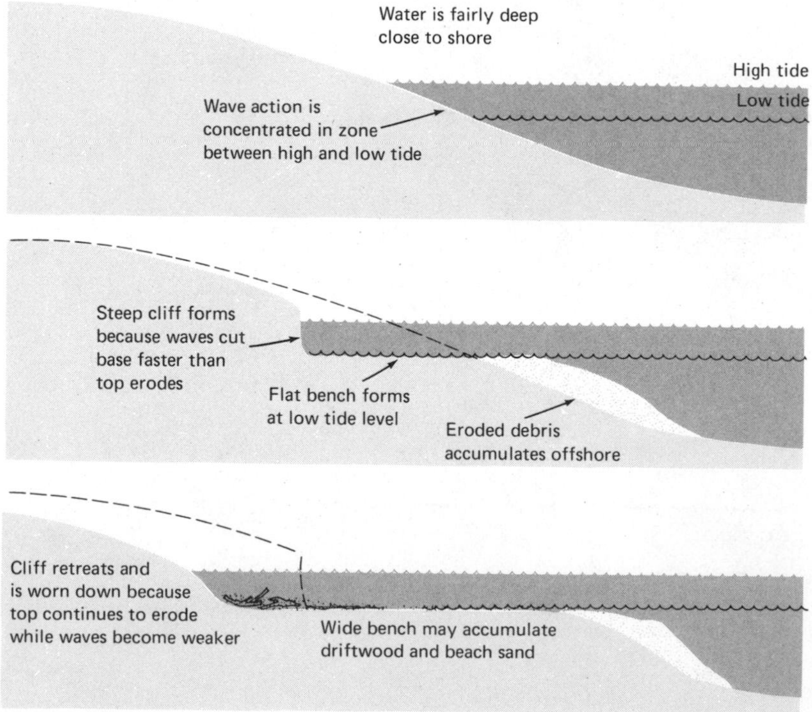

FIG. 16-25. Wave-cut bench and cliff along an exposed rocky shore. The cliff retreats under the impact of the waves until it is eroded back so far that the waves cannot cross the bench and still maintain enough strength to batter the cliff.

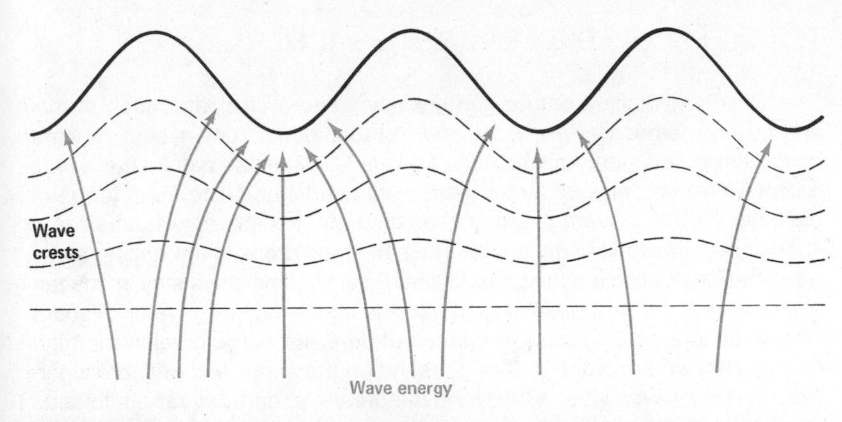

FIG. 16-26. Wave refraction. The bending concentrates the erosive force of the waves on the headlands and wears them back.

Wave-cut benches or platforms are usually quite straight when viewed from above, even though the coastline may have originally been hilly and irregular. Areas that protrude out into the sea are subjected to severe erosion, because the waves are bent as they approach the shore. The bending concentrates the energy of the waves on the points or headlands, while the sheltered bays escape with comparatively little erosion (Fig. 16-26). As time passes the points are eroded back faster than the bays, and the coastline becomes straighter.

headland point or promontory jutting out into the sea

sea stack island formed by erosion on both sides of a headland

The process of bench-cutting is self-limited, since each piece of material eroded causes the cliff to move farther away from where the waves break. The distance causes the rate of erosion to decrease: the waves fall harmlessly on their faces in shallow water during all except the biggest storms. Offshore reefs or deposits of sediment from rivers also cause waves to break too far from the shore to do much bench-cutting.

reef rocklike underwater feature formed by the "skeletons" of coral (see p. 290)

The relationship between slope and wave activity provides useful generalizations about the world's shorelines. Shores that face deep ocean trenches in regions of prevailing onshore winds are most likely to be cut by waves, because land surfaces are steep and zones of shallow water are narrow. To see a spectacular sea cliff, go to the West Coast. With the exception of parts of New England, most of the Atlantic and Gulf Coast of the United States is flanked by a gently sloping continental shelf with many depositional features but few symptoms of wave-cutting.

trench deep ocean where a crustal plate descends into the mantle (see Fig. 6-4)

continental shelf zone of shallow water near the margin of a continent (see p. 286)

Wave-cut cliffs and benches are sensitive indicators of changes in the elevation of the sea. A rise in sea level may inundate an old bench and move the cutting edge of the waves higher and farther inland. An uplift of the land may raise a bench safely above the waves and expose it to other landforming processes. Parts of the California coast have several distinct benches at different elevations, proof of many changes in the elevation of the land. Similar benches on the mountains around Salt Lake City are evidence for a much larger Great Salt Lake in the past. The degree of weathering and erosion of an elevated wave-cut bench gives a clue to the amount of time it has been above sea level.

TERRAIN SHAPED BY WAVES—BEACHES

strand line line of coarse material left high on a beach by storm waves

Waves advancing onto a gently sloping shore dissipate much of their energy long before they reach the land. A broad buffer zone of shallow water collects river sediment from the land and holds it up in the path of the waves. Storm waves can pick up larger pieces of material and fling them far up on the land. Waves of average size, however, prefer to carry only sand-sized or smaller particles toward the shore. The sand then drops to the bottom as the wave hesitates before ebbing out to sea. Fine silts and clays stay in suspension and ride the return waves out into the calm and deep water offshore. The result is a neatly sorted sequence of deposits, with gravel in the high-energy zone where storm waves go, sand on the shore and offshore where the returning waves slow, silt beyond the breakers, and clay far out to sea.

There is also a latitudinal sequence of particle sizes. Equatorial coasts are often muddy because rocks weather rapidly in a hot and wet climate. Sand is more common on mid-latitude beaches, while many polar shorelines are rocky or stony, because weathering processes operate slowly in cold climates.

swash incoming wave

backwash outgoing wave

soil creep slow downhill movement of loose soil (see Fig. 16-3)

longshore transport movement of sand along a beach by incoming waves and outgoing undercurrents

Waves that strike a shore at an angle often move material along the beach. Incoming surface waves wash loose material diagonally onto the shore, while outgoing undercurrents carry it directly out to sea. The result is a diagonal-in and straight-out motion that resembles the geometry of soil creep and carries material sideways along the shore. Obstructions such as rocks or piers lose material from one side and accumulate it on the other side because they interrupt the normal ebb and flow of the waves (Fig. 16-27).

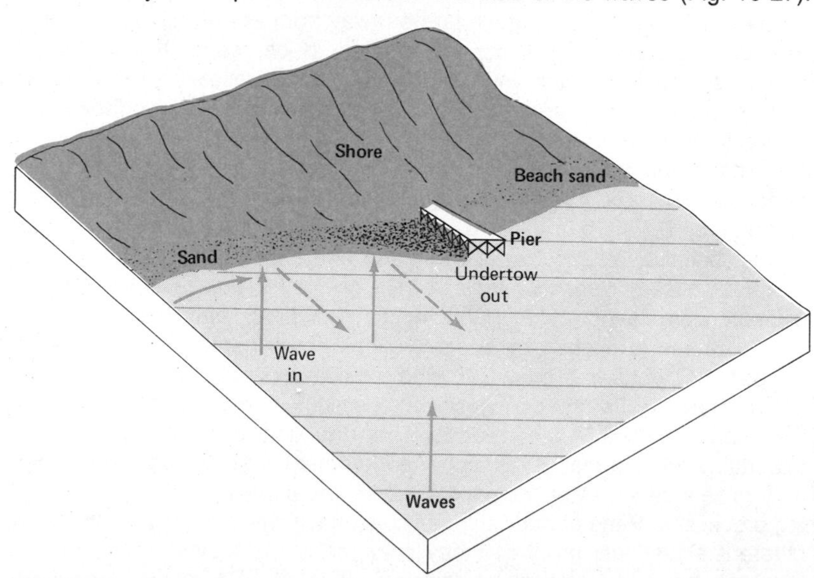

FIG. 16-27. Beaches forming in front of obstructions in the path of longshore currents. Sand is carried along the shore when surface waves come in diagonally and undercurrents flow out perpendicular to the coast. Obstructions break the rhythm of sideways flow and cause sand to accumulate.

The waves can even build beaches across river mouths or ports, if the supply of material exceeds the ability of local forces to carry it away. On other sites the waves may be able to remove material so rapidly that even an artificially augmented beach may last for only a short time. At still other places, winter waves may destroy beaches that were painstakingly built by summer waves.

The mouths of rivers are areas of complex interplay between the processes of river deposition and wave action. A symmetrical delta will form where a river enters and deposits its load of sediment in a shallow sea with fairly weak wave action. Variations in depth or wave forces will shape the delta deposits into a variety of forms that merge into the beach and bar deposits on the nearby shore (Fig. 16-28).

delta alluvial deposit where a river enters a larger body of water (see Figs. 7-6 and 16-8)

FIG. 16-28. Deltas, areas of land built up by rivers where they enter the ocean and drop their sediment loads. (Courtesy Gabler et al.)

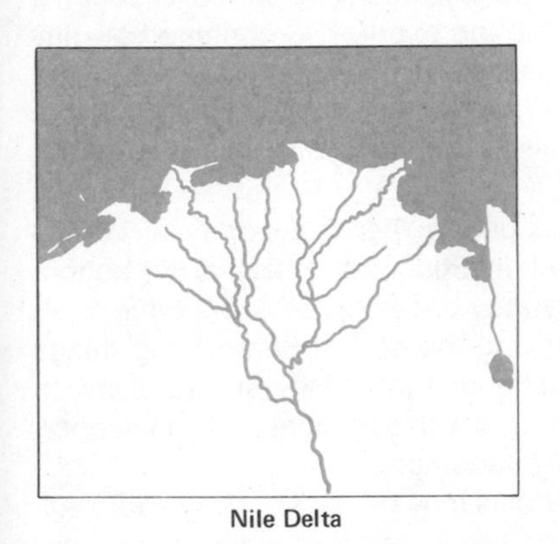

Nile Delta

Niger Delta

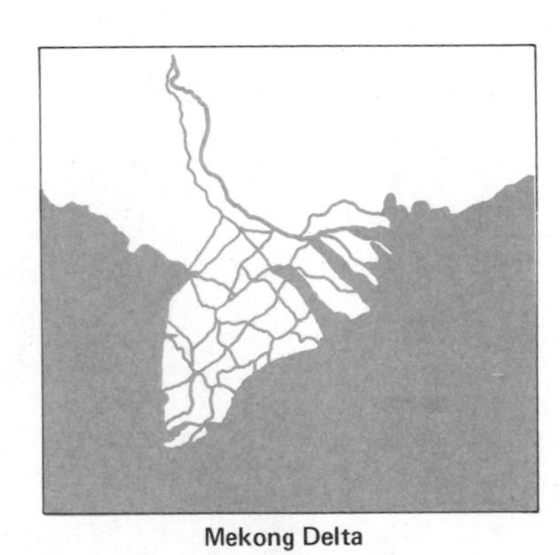

Mekong Delta

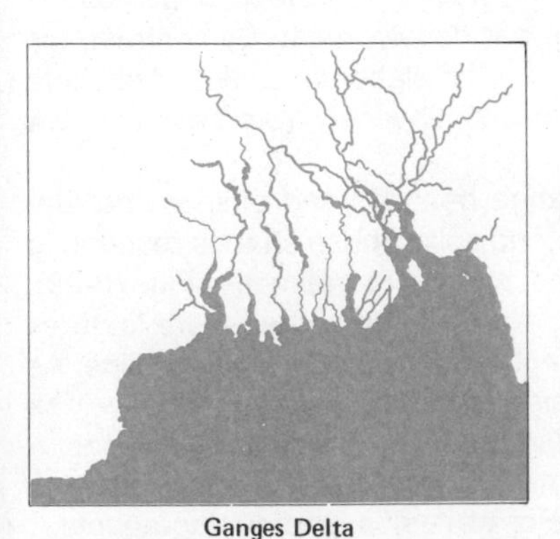

Ganges Delta

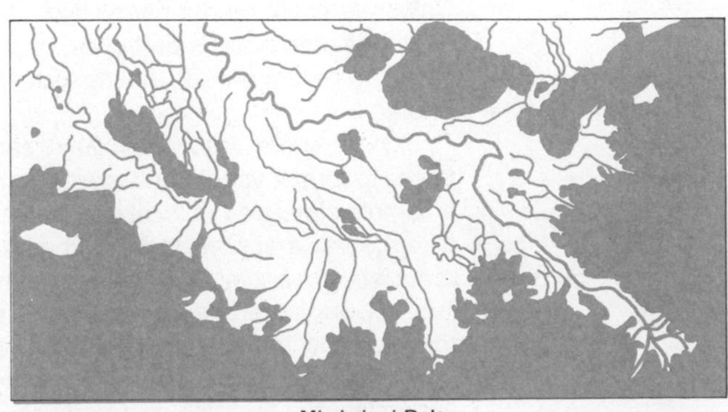

Mississippi Delta

Like wave-cut benches, relic beaches often mark the heights of past sea levels. Alternative sandy and clayey soils on the coastal plain of North Carolina indicate half a dozen former shorelines. Beach ridges surround many of the Great Lakes and demonstrate that some of the fertile fields of Michigan, Indiana, and northwest Ohio are really the bottoms of ancient lakes. Low sandy ridges in Minnesota, the Dakotas, Montana, and Nevada mark the margins of other Ice Age lakes that disappeared when the glaciers retreated and the climate changed. Most of the urban area of Salt Lake City is built on a coarse sandy or gravelly material that was piled up along the shores of Glacial Lake Bonneville, an enormous inland sea whose shrunken remains now bear the name of Great Salt Lake.

TERRAIN SHAPED BY WAVES—BARS

Waves also create characteristic features a short distance away from the shore, in the zone where onshore breakers and offshore undertows conspire to accumulate earth materials. The accumulation forms a zone of shallow water that runs roughly parallel to the shore. The offshore bar may emerge as a low island during times of low tide or offshore winds. Coral often grow on bars in tropical oceans and give the bar an armoring reef. Changes in sea level may submerge a bar or raise it safely out of the reach of the waves.

offshore bar sandbar made by wave action parallel to a gently sloping beach

An offshore bar has a marked effect on the shore environment. Waves are stilled by the physical barrier of the bar. Sediment settles to the bottom and the water becomes clearer. Salinity may decrease as rivers bring fresh water into the lagoon between the bar and the shore. Some living things disappear if they cannot adapt to the changes in salinity and turbidity. In time, the bottom of the lagoon becomes filled with sediment and can support dense stands of marsh grasses or other vegetation.

lagoon area of quiet water behind an offshore bar or coral reef

Meanwhile, the breakers and undertows may be constructing additional bars still farther out to sea. The number and spacing of offshore bars depends on the slope of the sea floor, the nature of the material on the bottom, and the strength and regularity of incoming waves. The steep slopes off the California coast do not permit extensive bar development. By contrast, the gentle Atlantic slope is partly responsible for the elaborate chains of offshore bars and barrier islands that stretch almost continuously from Texas to New England.

barrier island long island parallel to a coast, usually a former bar or coral reef

Other shore processes may rearrange beaches and offshore bar deposits to form a variety of capes, points, and spits whose shapes respond to a complex interplay of local winds, waves, and shore geometry (Fig. 16-29). Like the sea that formed them, the beaches, bars, and other shore features are always changing in detail but seemingly changeless in broad outline. As long as the general surroundings remain the same, spits and bars will be formed, modified, moved, and reformed in essentially the same locations. A variety of costly engineering problems have arisen out of human failure to appreciate the strength and persistence of waves as landforming agents.

FIG. 16-29. A complex shoreline, the result of an interplay between coastal shifting and shaping by ice, river, and wave action. (Courtesy of D. J. Miller, Geological Survey.)

TERRAIN SHAPED BY ANIMALS—PITS AND MOUNDS

Many animals spend at least part of their lives underground, sheltered from predators and the direct effects of the weather. The burrowing of these animals—earthworms, soil insects, moles, foxes, termites, and so on—moves a surprising amount of earth. Whether the burrowing was done by an ant or a gopher, the result is an extensive underground excavation and an associated mound of earth near the entrance. In time, the subsurface cavern may collapse. Meanwhile, the surface mound may be covered by vegetation, and in this way a formerly smooth slope becomes rolling and irregular.

The effect of burrowing animals may extend far beyond the pits and mounds they create. The entrances to underground burrows are breaks in the cover of vegetation, while the tunnels may act as channels for subsurface movement of water. Thus, burrowing animals may initiate wind or water erosion on a slope that would be stable if protected by a continuous plant cover. The writer's farm has four large gullies that began as foxholes or gopher mounds. Burrows of other animals may also initiate blowouts in sandy grasslands, terracettes on high mountains, or mass movements on steep forested slopes.

blowout deflation hollow where wind has removed loose soil (see p. 351)

Down in the valleys, beavers and other river-damming animals have a significant but often overlooked effect on landforms in the immediate vicinity.

temporary baselevel local area of zero stream energy (see Fig. 15-11)

bulldozer important agent of man-made landforms

The obstruction of free flow greatly reduces the sediment-carrying capacity of the river. The valley floor is raised and mass movements on the slope become less likely. Downstream of the dam, the river becomes a more effective eroder because it is able to pick up additional sediment to replace the load it had to deposit behind the dam.

Human activity likewise is responsible for many intentional and inadvertent changes in landforming processes. Man-made landforms—dams, leveled hills, graded highways, artificially drained areas, and harbor structures—are more conspicuous but often less significant in shaping the terrain than the subtle and often unintentional human alterations of the wind patterns, soil characteristics, or water budgets of the surface of the earth.

CONCLUSION

Each landforming process operates in a particular kind of environment and leaves a unique imprint on the terrain. The ability to read landforms does more than enable the reader to understand the origin of particular terrain features. It also provides a clue to the climate and vegetation history of an area, because many landforms were shaped by processes that cannot operate under present conditions. The next chapter takes a look at the complex history of typical landforms.

FIG. 16-30. Human activity as a landforming process. (Courtesy of B. W. Muir, U.S. Forest Service.)

APPENDIX TO CHAPTER 16
Morphogenetic Regions

Each of the landforming agents (wind, rivers, glaciers, etc.) has its own arena of pronounced activity, and each one leaves its imprint in the form of particular landforms. The various agents work cooperatively to shape the terrain in a particular place. The results of this cooperative effort are reasonably predictable, and this regularity has led some geographers to suggest that the world could be divided into a number of distinct landform areas ("morphogenetic regions"), in which certain sets of processes almost inevitably produce particular landforms.

Several different people have tried to produce maps of landform regions. The various maps are based on somewhat different criteria, so that they are not identical, but they all bear a strong similarity to a map of climate. The similarity is hardly surprising, since climate acts as the major control of landforming processes. Moreover, the effectiveness of various landforming agents is partly governed by surface material and plant cover, and both soil and vegetation are strongly influenced by climate. The regions discussed in this appendix have approximately the same boundaries as the vegetation regions in the Appendix to Chapter 11, with one exception: the desert environments have been separated into core deserts and semiarid margins because the landforming processes change significantly with the addition of even a few rainstorms every year.

A classification of landform regions, like any other system of classification, consists of generalizations with many local exceptions. Obviously, an extremely hard rock will resist the efforts of rivers or glaciers more effectively than a soft rock can. Climatic changes and the slow rearrangement of continents by plate motion are responsible for changes in landforming processes through time. The present terrain may therefore contain many relics of previous conditions. For these reasons, the actual distribution of landforms is far more complicated than the map of landforming regions would suggest. The regions described in this appendix are simplifications, in which rock structure, plate movements, and previous history are not considered.

TABLE 16 A-1 LANDFORM ENVIRONMENTS

Region	Key Environmental Characteristics	Dominant Landforming Processes
Selva	Abundant energy and moisture for rapid chemica weathering. Absence of frost or drought. Dense forest cover. Fine-textured surface material with little abrasive power.	Deep and rapid chemical weathering of soil-covered rock. Less rapid weathering of exposed rock. Slow flowage of waterlogged soil down gentle slopes. Rapid mass movements on steep slopes.
Savanna	Distinct rainy and dry seasons but no frost. Hard iron oxide layers in the soil (called cuirasses or laterite armor). Fine soil particles, poor abrasive tools for winds or rivers. Fires during the dry season.	Intense chemical weathering of soil-covered rock. Slow breakdown of exposed rock. Extensive flooding during rainy season. Wind transport of exposed soil in dry season. Formation of cliffs where resistant iron layer acts as a caprock.
Desert	Insufficient moisture for rapid chemical weathering. Virtual lack of runoff. Near absence of vegetative cover. Temperature extremes.	Slow chemical weathering. Rockfalls from steep cliffs. Wind removal of fine particles. Wind abrasion of exposed rock. Flash floods during rare storms.

Characteristic Landforms	Geomorphic Problems for Human Use of the Land
Smooth and gentle slopes. Dome-shaped outcrops of bare rock (tors). Rivers barely incised into plains. Rapids and waterfalls. Occasional steep slopes and sharp ridges.	Risk of rapid gullying if forest cover is removed. River transport hindered by rapids. Mass movements on steep slopes.
Braided streams that are seasonally dry. Flat expanses of land planed by sheetfloods. Isolated rocky prominences (bornhardts). Steep cliffs and waterfalls (escarpments).	Floods on flat land. Dust storms.
Desert pavements. Monuments and pedestal rocks ("hoodoos"). Sand dunes in favorable places. Canyons near streams that originate elsewhere.	Wind abrasion of structures. Transport hindered by canyons. Sand and dust storms.

TABLE 16A-1 LANDFORM ENVIRONMENTS *(Continued)*		
Region	**Key Environmental Characteristics**	**Dominant Landforming Processes**
Semiarid	Absence of continuous vegetation cover except near watercourses. Insufficient water for rapid chemical weathering. Occasional intense storms.	Slope retreat by rockfall and mass movement. Wind removal of fine particles. Flash floods after occasional storms. Blockage of stream valleys by vegetation.
Chaparral	Hot, dry summers and cool, humid winters. Soils that expand and contract as they are wetted and dried. Thin cover of drought-adapted vegetation. Fires during dry summer.	Chemical weathering hindered by seasonal dryness and cold. Soil saturation and flowage during cool season. Flash floods at beginning of rainy season. Gullying and mass movements of fire-exposed soil on steep slopes.
Mid-latitude Forest	Enough moisture and energy for adequate chemical weathering. Continuous cover of vegetation except in watercourses. Intense seasonal frost. Springtime runoff from snowmelt.	Rapid chemical and mechanical weathering of exposed rock. Soil creep down moderate slopes. Treefall disruption of smooth slopes. Slumping of saturated soils on steep slopes. River dissection of uplands. River transport and deposition of fine sediment.

Characteristic Landforms	Geomorphic Problems for Human Use of the Land
Mesas, buttes, and other angular rock outcrops. Arroyos—usually dry valleys. Smoothly sloping pediments (bajadas). Intermittent lakes in closed drainage basins (playas). Canyons, especially along streams that originate elsewhere.	Flash floods. Landslides. Dust storms.
Shallow soils on steep slopes. Mudflow scars and lobes. Gullies and alluvial fans.	Mudflows and flash floods. Soil erosion on exposed slopes.
Smoothly rounded slopes covered by soil and vegetation. Pits and mounds on slopes with frequent treefall. Slumps and landslide scars on steeper slopes. V-shaped valleys along eroding rivers. Flat floodplains along depositing rivers. Smooth river profile with few waterfalls (concave profile).	Soil erosion when vegetation is removed.

Region	Key Environmental Characteristics	Dominant Landforming Processes
Mid-latitude grassland	Conditions barely adequate for rock weathering. Nearly continuous cover of vegetation. Frequent periods of frost and drought.	Flooding and deposition by rivers. Wind deposition of silt picked up from floodplains. Severe gully erosion of exposed soil. Retreat of cliffs by gullying and mass movement.
Tundra	Permanently frozen ground except for seasonally thawed surface. Insufficient energy for rapid chemical weathering. Low, often discontinuous vegetation cover.	Frost-shattering of rock. Frost-heaving of loose material. Nivation—concentration of weathering beneath snow banks. Solifluction—slow downhill flowing of surface material above permafrost. Avalanching—rapid mass movement of snow and earth material. Wind abrasion and removal of fine earth materials. Lateral spreading of rivers carrying meltwaters across permafrost.
Ice cap	Excess of snowfall near source of glacier. Temperatures low enough to permit ice to persist.	Accelerated weathering beneath snowpacks. Scouring by sliding ice. Plucking of rocks by frost action. Deposition of unsorted material. Snow avalanching down steep slopes.

Table caption: TABLE 16 A-1 LANDFORM ENVIRONMENTS *(Continued)*

Characteristic Landforms	Geomorphic Problems for Human Use of the Land
Loess deposited on uplands. Smoothly rounded slopes covered by grass. Cliffs and plateaus on resistant rocks. Intricate erosion of steep exposed slopes (badlands). Braided streams and alluvial fans.	Severe wind and water erosion of exposed soil. Dust storms.
Patterned ground (see tundra photo in Appendix to Chapter 11. Boulderfields (felsenmeere). Solifluction lobes. Avalanche chutes. Nivation hollows. Wide swampy valleys.	Ground subsidence if permafrost thaws. Frost-heaving of buildings and transport structures. Avalanches. Seasonal floods.
Scoured and grooved rocks (Roches moutonnées). Ice-scoured rock basins (Paternoster lakes). Deep U-shaped valleys. Sharp ridges and peaks between valley glaciers (horns and arêtes). Moraines and outwash features near edge of ice. Avalanche chutes.	Avalanches. Crevasses in glacier. Seasonal meltwater floods.

Chapter 17

Landforms: Structure, Process, and Time

Geomorphology is like psychoanalysis; when you start probing, you uncover a tangled past.

Philip Gersmehl

The shape of the land in a particular place depends on four things:

1. Internal processes such as crustal movements and volcanic activities.
2. Arrangement and resistance of rocks.
3. External processes such as weathering and erosion.
4. Amount of time the terrain has been subject to a particular set of conditions.

These four factors have different degrees of importance in different places. Continual volcanic eruptions will obscure the effects of most agents of erosion and weathering. Resistant zones of rock guide rivers into distinctive paths. Dry climates build dunes on rock structures that humid climates carve into V-shaped valleys. Finally, even the most resistant rocks will yield to the weakest atmospheric forces, given enough time.

Previous chapters on plate tectonics (Chapter 6), weathering processes (Chapter 7), water budgets (Chapter 14), and erosional forces (Chapters 15 and 16) have discussed the major principles individually. This chapter will bring the principles together by applying them to a set of case studies that illustrate the interactions of structure and process through time.

plate tectonics theory that rigid crustal plates slide across liquid or plastic mantle
(see Chapter 6)

CENTRAL APPALACHIA

Many Appalachian mountains are ridges of uniform height and great length, arranged like parallel fences across the countryside. Place-names such as Harrisburg, Gettysburg, Chattanooga, Cumberland, Roanoke, Antietam, Harper's Ferry, and Mohawk testify to the strategic importance of gaps through the Appalachian ridges.

Construction of the stage for the drama of human use of the Appalachians began long ago, when a vast sea covered the middle of the present continent and a lofty range of mountains paralleled the present east coast. At times the weather attacked the mountains vigorously and carried sand and gravel into the adjacent seas. The periods of intense erosion alternated with quiet times, when rocks on land weathered to fine clay and marine organisms lived in shallow seas offshore. Layers of sandstone, limestone, shale, and coal lie atop one another in the sedimentary rocks formed on the flanks of the ancient Appalachians.

sedimentary rock rock formed from compressed and/or cemented deposits
(see Appendix 7)

Differences in rock resistance are not very important as long as rock layers remain flat and the uppermost layer is unbreached. The layered rocks on the west side of the ancient granite mountains, however, did not remain flat. The relentless forces of the shifting crust of the earth pushed inward from both sides and bent the rock layers into a series of elongated folds that resemble a gigantic rug shoved against a wall. Some of the rock layers broke under the strain. Like present-day California, ancient Appalachia was plagued by frequent earthquakes. After the era of crustal restlessness gave way to a long interval of relative calm, the forces of erosion concentrated their attack on the high ground and wore through the tops of the folded rock layers (Fig. 17-1).

fault break in rock, usually accompanied by slippage along break
(see Fig. 6-14)

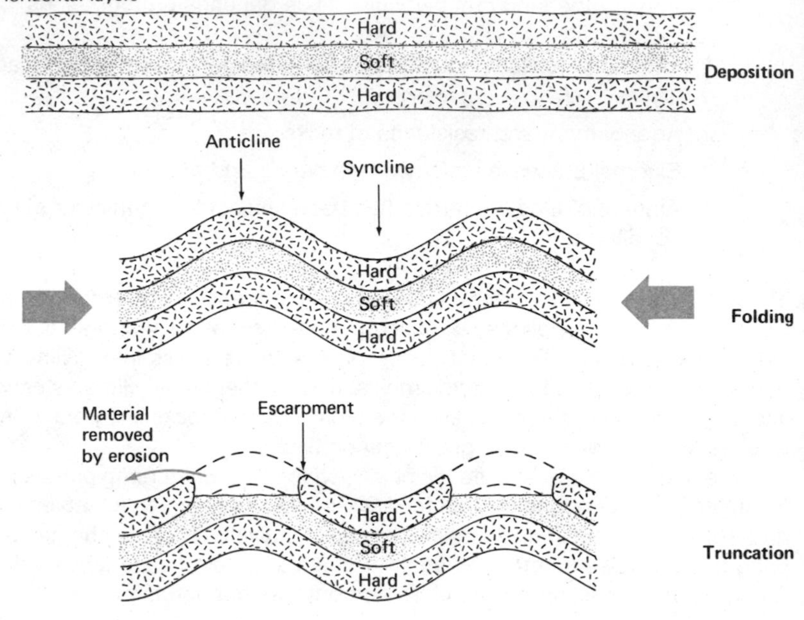

FIG. 17-1. Ridges and valleys on folded rocks. Sideways pressure compressed and folded the rock layers. Then, erosion removed material from the high places more rapidly than from the downward folds. Cutting the tops off the folds exposed rocks of different degrees of resistance to further erosion. (A geologically soft rock is one that succumbs readily to weathering and erosion, even though the individual particles may be mineralogically quite hard.)

truncated fold fold that has been leveled by removal of material from high places by erosion

plunging fold complex fold whose ridge is not horizontal

Erosion of this geologically complex surface is especially rapid wherever soft rocks such as shales or limestones are exposed. Resistant sandstones and conglomerates stand out as ridges above valleys carved into the softer rocks. The ridges are long and level where the rock folds are symmetrical, but they describe complex zig-zag patterns on the surface where the original folds were irregular (Fig. 17-2).

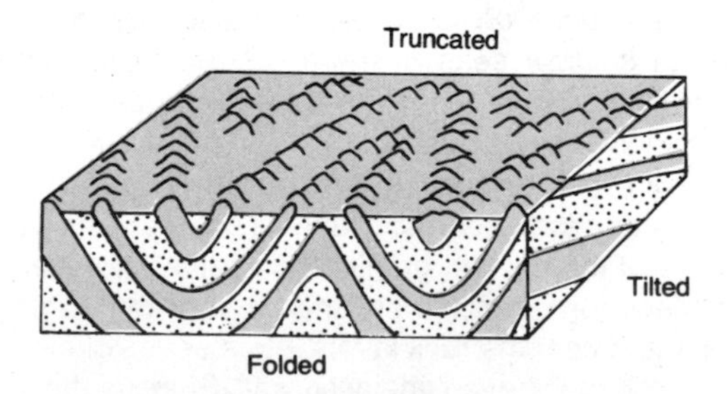

FIG. 17-2. Zig-zag mountains in Pennsylvania. The ridges represent surface exposures of resistant layers of rock in a complex nonlevel fold.

Sandstone ridges are biologically poor because the soil is coarse, acid, low in nutrient storage ability, and alternately subject to leaching during rainy periods and drought during dry seasons. Soils formed on soft limestone, on the other hand, are fertile and productive. Valleys with names like Shenandoah, Nolachucky, Holston, and Tennessee were magnets for early settlers, because the fertile soil promised good crops and an easy life.

leaching removal of nutrients by percolating water
(see p. 161)

Fortunately for the animals pulling the wagons for the settlers, the ancient geologic uplift was not very rapid. Large rivers often had enough energy to wear away the hard rock as fast as it was pushed up in their path. Early settlers, Civil War generals, and modern railroads all put a high value on gaps where rivers flow through the mountain ridges.

superimposed stream stream that cuts "across the grain" of the rock structure

A tour across the entire Appalachian Mountain region passes through three distinct zones. The eastern Piedmont and Blue Ridge are the rounded stumps of the ancient granite mountains. Soils are mostly clay, quite fertile but subject to rapid erosion when mismanaged. In the middle is the region of folded sedimentary rocks, long ridges, and fertile valleys. The western zone is a region of rugged hills formed on thick and nearly level layers of rock that weather slowly to form coarse and infertile soils. Transportation is difficult through the western zone because the stream valleys are steep and crooked, a striking contrast to the wide and straight valleys in the middle area.

dendritic drainage pattern treelike arrangement of tributaries

trellis drainage pattern rectangular arrangement of tributaries
(see Fig. 15-6)

In summary, the present Appalachian topography is the product of different rates of stream erosion on rocks with unequal resistance. Ancient cycles of erosion, sedimentation, and crustal movement produced complex arrangements of rock types that continue to guide the activity of modern erosional processes. The resulting landforms have a significant effect on agriculture and transportation, and thus on the course of human history in the Appalachians.

THE WISCONSIN GLACIERS

The topography of central Wisconsin is not spectacular by Colorado standards. Nevertheless, a perceptive observer views a trip from Milwaukee to La Crosse as a fascinating landforms textbook with half a dozen chapters of varying length. Some of the chapters are flashbacks to very ancient times, while others cover recent events. The terrain reader must keep one important fact in mind: very few Wisconsin scenes can be interpreted solely in terms of present-day processes.

The significance of past erosional processes becomes apparent with the first hour of travel. The last suburb of Milwaukee passes by, and the gently rolling terrain yields rather abruptly to a chaotic mass of lakes and forested hills. A map of the region shows that these low but rugged mounds are near the junction of two long arcs of hills, one enclosing most of eastern Wisconsin and the other outlining the shore of Lake Michigan (Fig. 17-3).

moraine landform made of material deposited by a melting glacier
(see Fig. 16-17)

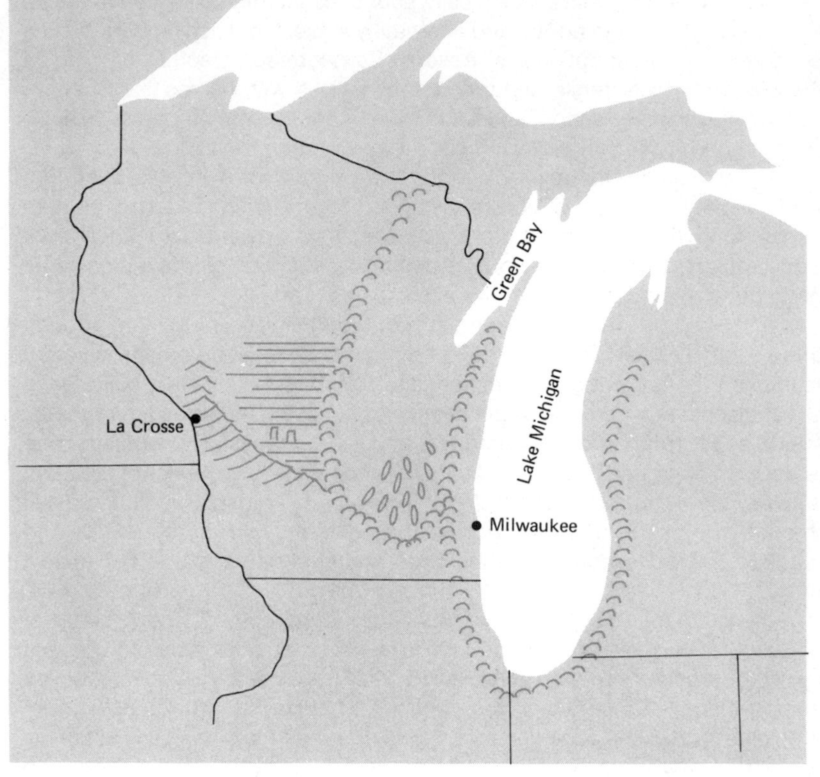

FIG. 17-3. Landforms associated with the Green Bay and Lake Michigan lobes of ice.

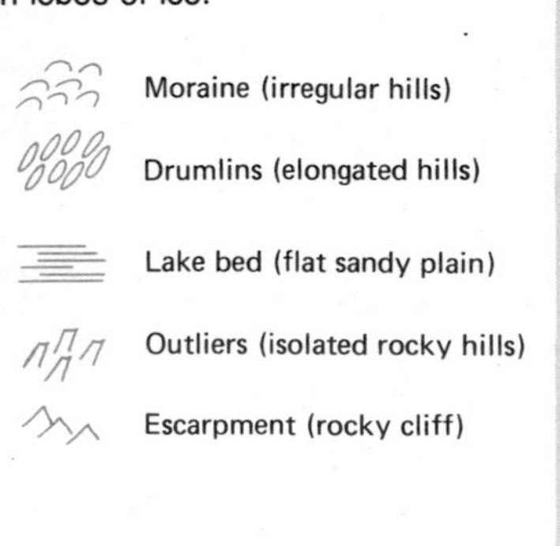

The similarity of the patterns of hills and shore provides a clue to the origin of both. The story begins with an accumulation of snow over eastern Canada a few tens of thousands of years ago. Eventually the snow became so thick that the lower layers hardened to ice, and the ice in turn began to slide outwards. A continental glacier usually ignores minor terrain features as it relentlessly bulldozes across the countryside, but a substantial ridge of hard rock under the Door Peninsula in northern Wisconsin acted like a traffic island on a highway and separated the flowing ice into two distinct tongues or lobes (Fig. 17-4). Ice on the Lake Michigan side of the ridge passed over low land on rocks that were soft and therefore easily moved. The Lake Michigan glacier gouged out a deep and wide hollow, carried the earth material toward the south, and deposited it over Indiana and Illinois. Ice on the west side of the ridge had to go over high land and resistant rock. Therefore the western or Green Bay tongue of ice moved more slowly, gouged less deeply, and could not penetrate as far to the south as the eastern lobe.

lobe earlobe-shaped extension of a glacier

ground moraine material deposited beneath a stagnant melting glacier

(see p. 348)

Both lobes advanced in an erratic fashion. Times of rapid extension alternated with intervals when the ice at the edge melted back faster than the

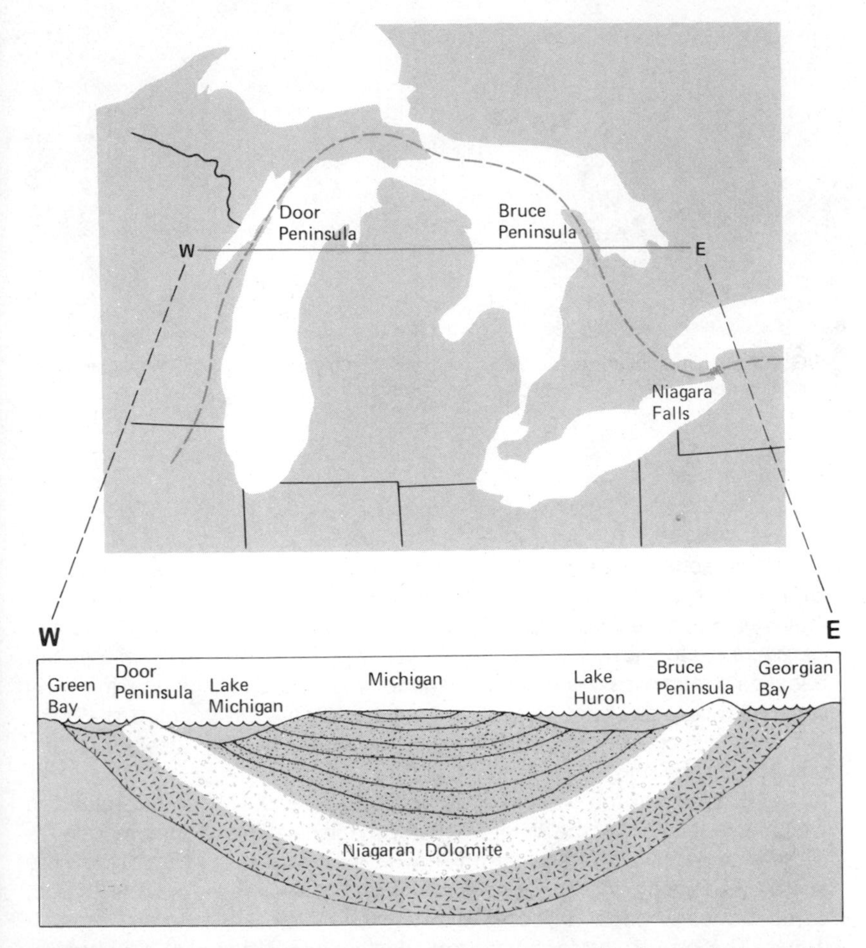

FIG. 17-4. The Niagaran Escarpment, a ridge of exceptionally resistant rock that surrounds Michigan.

glacier moved forward. A subsequent advance pushed the debris along and heaped it up at the outer edge of the advance. The long arc of hills in Wisconsin is the moraine that marks the outermost edge of the Green Bay glacier. The hills are especially rugged just west of Milwaukee because the Lake Michigan lobe of ice pushed from the east and contributed additional material to the deposits from the Green Bay glacier.

end moraine material deposited at the outermost edge of a glacial advance

interlobate moraine exceptionally rugged moraine built by deposition between two converging glacial lobes

To the west of the rugged moraine is another pleasant gently rolling landscape dotted with prim Wisconsin dairy barns. Half an hour of travel in this pastoral scene, however, raises the suspicion that the same force cannot have shaped terrain on both sides of the arc of rugged hills. The rolling land on the Milwaukee side appears patternless, whereas the hills on the west side are long and neatly aligned, like a school of enormous dolphins

FIG. 17-5. Drumlin in central Wisconsin, a smoothly elongated hill of glacial material. (Courtesy of W. C. Alden, Geological Survey.)

drumlin smoothly elongated hill built and shaped by a moving glacier

swimming southward (Fig. 17-5). The smoothly sculptured hills represent a different kind of glacial deposit, one formed beneath a mass of moving ice that was either melting or spreading outward as it moved along. The resulting decrease in thickness reduced the ability of the ice to carry material, but because the ice was still moving, it shaped the deposited material into long ridges. Present-day farm fields and highway patterns are oriented in a manner that reflects the terrain alignment imposed by the glaciers of ten thousand years ago.

outwash material carried from a glacier by meltwater streams (see p. 349)

lacustrine plain flat plain built by deposition of material in the still water of a lake

Still farther to the west is another line of irregular hills that mark the westernmost extent of the former Green Bay glacier. Beyond the moraine is a broad and monotonously flat region where signs of human habitation are scarce. The sandy swamp supports multitudes of pine trees, a military camp, some cranberry bogs, an occasional vacation home, and hordes of mosquitos. The plain is the former bed of Glacial Lake Wisconsin, a Delaware-sized body of water that was formed between the thick ice to the northeast and the highlands to the southwest (see Fig. 17-3). Meltwater streams flowed off the glacier, carried some of the sandy glacial debris with them, entered the still water of the lake, and deposited their loads of debris on the bottom. A few large stones were added by occasional icebergs that broke away from the glacier and carried their loads onto the lake. Year after year, sand and pebbles filled the former valleys beneath the lake. The flatness of the resulting terrain is most apparent when the land is viewed from one of the handful of isolated hills that stood as islands above the waters of the former lake.

retreating escarpment wall of hard rock that retreats as material erodes from the edge

Among the most prominent of these former islands are a cluster of rocky outcrops that look like Utah mesas transported to central Wisconsin (Fig. 17-6). These steep rocky prominences are remnants of an extensive layer of rock that once covered most of Wisconsin. Long before the Ice Age, erosion had removed most of this layer over northern Wisconsin. The outcrops are like the cracked and irregular cement at the edge of a pothole on a busy road. Farther to the southwest, the rock layer still exists and forms a rugged highland beyond the south shore of old Glacial Lake Wisconsin.

Southwestern Wisconsin has no natural lakes, morainal deposits, scoured rocks, or other clear evidence that the ice sheets ever penetrated the region. Rocky cliffs and V-shaped valleys reflect the work of rivers and mass movements. The terrain is mature: most of the land is sloping and the rivers flow in deep valleys. Even so, the landscape bears an unmistakable imprint of a different environment long ago. Roadcuts reveal a striking sequence of colors in the earth material, beginning with a deep red near the hilltop. Farther down into the ground is a mottled zone of red and white, underlain by a layer of pale grayish-white. All of these colors indicate different degrees of soil drainage (Chapter 8). The depth of the sequence suggests intense weathering, which only occurs under tropical conditions. Putting all of this evidence together leads to the conclusion that Wisconsin once had a tropical climate, with selva or savanna vegetation, no frost, and a deep red soil. Former tropical conditions in presently cold places offer a persuasive argument that climate patterns have changed or that the continents were not always in their present positions.

The tour of Wisconsin ends at the Mississippi River. A few small caves near the river indicate the presence of limestone beneath the surface. The river itself flows into a valley that seems much too wide for the present stream. The scenic valley is another remnant of past times, when melting glaciers made the river much larger than it is now and gave it energy to cut a deep valley. The present river occupies only a small part of the wide valley,

Driftless Area apparently unglaciated part of Wisconsin and adjoining states

maturity intermediate stage in the Davisian Cycle of Erosion (see Fig. 15-9)

deep-weathering profile sequence of soil colors and chemical composition that indicates intensive weathering

karst topography terrain with solution pits, caves, and sinkholes on soluble limestone (see Fig. 7-14)

spillway valley that was enlarged by glacial meltwater

underfit stream small stream occupying a large valley

FIG. 17-6. Castle Rock, a remnant of the retreating cliff at the left of the picture. The outlier was isolated when erosion removed the intervening rock material. The flat plain in the foreground is the bed of glacial Lake Wisconsin. (Courtesy of W. C. Alden, Geological Survey.)

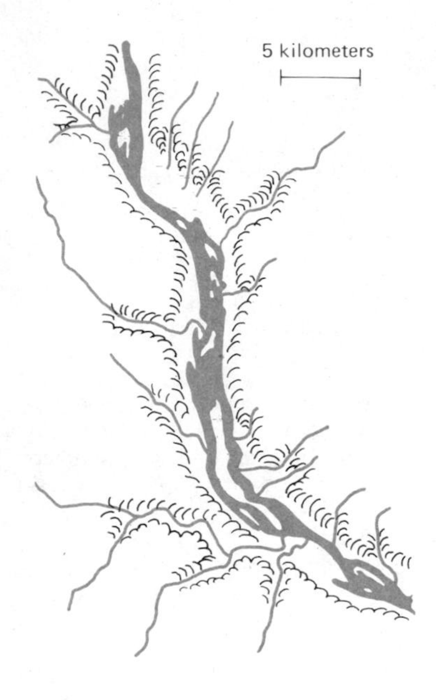

FIG. 17-7. The course of the Mississippi River between Wisconsin and Iowa. The river detours around alluvial fans at the mouths of tributary streams that enter the main valley from both sides.

but its pattern is not one of aimless wanderings. Rather, it is guided from side to side by partial dams built where tributary streams leave the surrounding hills, enter the flat valley floor, slow down, and drop their sediment loads (Fig. 17-7).

In summary, the perceptive traveler across Wisconsin can see moving masses of ice, huge inland seas, tropical forests, and gigantic rivers, all written into the pages of the present landscape. As an example of the origin of landforms, Wisconsin differs from Appalachia in one key respect. Appalachia inherited a distinctive folded rock structure that combines with present erosional processes to produce most of the present topography. Many Wisconsin landforms are direct results of erosional and depositional processes that no longer operate in the region. Given enough time, modern processes will obliterate most of the evidence of the Ice Age. For example, glaciers once covered Iowa and northern Missouri, but the ice departed so long ago that most of the lakes, moraines, and aligned hills have succumbed to extensive reshaping by modern rivers.

THE LOWER FORTY AND THE HIGH PLAINS

The "lower forty" introduced in Chapter 10 owes its present shape to a sequence of events that affected much of the American Midwest. Iowa rests on a deeply buried foundation of granite and other crystalline rocks. The detailed history of this "basement" complex is lost in the murky reaches of geologic time. On the top of the basement are layers of sedimentary rocks, evidence for a long period of submergence beneath the sea. There is reason to believe that the long submarine existence was broken by an interval of uplift and exposure to the air, followed by another submergence.

igneous rock rock formed from molten material
(see Appendix 7)

Then, fairly recently in its geologic history, Iowa emerged from beneath the sea. Time and the falling rain soon established a system of stream channels on the sedimentary rock. The streams converged on a large river flowing near the course of the present Mississippi River. In those days, however, the continent was lower than it is today, and the Gulf of Mexico extended nearly to Illinois. One of the old stream channels crossed just south of the lower forty in Iowa County, but river erosion was not very rapid because the low elevation of the land provided little energy for the rivers.

Davisian Cycle of Erosion theory that topography passes through recognizable stages of youth, maturity, and old age as uplifted land is eroded down to sea level
(see Fig. 16-9)

Pleistocene recent "Ice Age" in geologic history
(see Table 6A-1)

The gradual progression of mature terrain toward old age was interrupted when glaciers covered the northern part of the United States. There were four major periods of glacial activity, but it is difficult to find evidence of the first one in Iowa. The second ice sheet covered most of the state and left 50 to 75 m (150 to 200 ft) of debris over hills and valleys alike. The melting glacier then added vast quantities of water to the rivers, which, in turn, vigorously attacked the loose glacial deposits. Postglacial streams occasionally wore through the glacial debris and intercepted ridges on the former

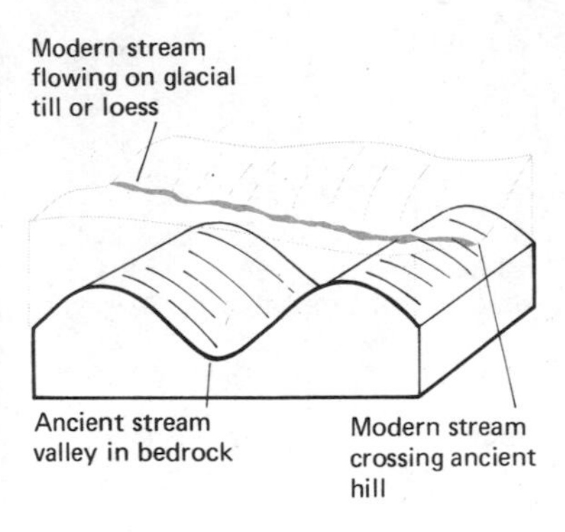

FIG. 17-8. The geologic setting of towns such as Cedar Rapids, Iowa Falls, and Cedar Falls, Iowa. Waterfalls or rapids often form where a modern river cuts down through the deposits of loess and glacial till on the surface and encounters a ridge on the buried bedrock surface.

bedrock surface (Fig. 17-8). Tributary streams from the debris-covered uplands rushed down steep slopes toward the deep meltwater river valleys.

The disappearance of the glacier, however, changed the energy relationship of the rivers and the surrounding terrain. The rivers were deprived of water from the glacier and could not maintain their big valleys. Sediment brought by tributary streams began to accumulate in the main valleys. Slowly but surely the terrain became adjusted to life without a neighboring glacier.

At least two more glaciers advanced into the Middle West after the one which covered most of Iowa, but neither sheet of ice reached Iowa County. Nevertheless, they had an impact on the lower forty. Meltwater-laden rivers once again scoured out their valleys and increased the energy disparity between valley and upland. Wind-borne dust settled over the land and imperceptibly raised the surface at the rate of a few millimeters per decade.

loess silt deposited by wind (see Fig. 16-22)

The lower forty is hilly because its streams flow directly to the deep channel of the Iowa River, a few kilometers away. By contrast, streams on the next farm to the north travel toward a valley that was not deepened by glacial meltwater. Therefore the north-flowing streams descend gentler slopes, have less energy, flow more slowly, and cannot carve the landscape as effectively as south-flowing streams (Fig. 17-9). Farms on the gentle terrain to the north are valuable because the land is nearly level and the silt that the wind brought from the glacier makes a fertile soil. The lower forty, by contrast, is steep and prone to erosion because meltwater from the glacier made the Iowa River deeper than it would otherwise be. A glacier that never entered the lower forty is thus responsible for the fact that the land there is worth less than half as much as land a kilometer to the north.

gradient slope, expressed as vertical drop per unit of horizontal distance.

drainage divide ridge between two river basins (watersheds)

land capability class group of soils with similar management problems (see Table 10-2)

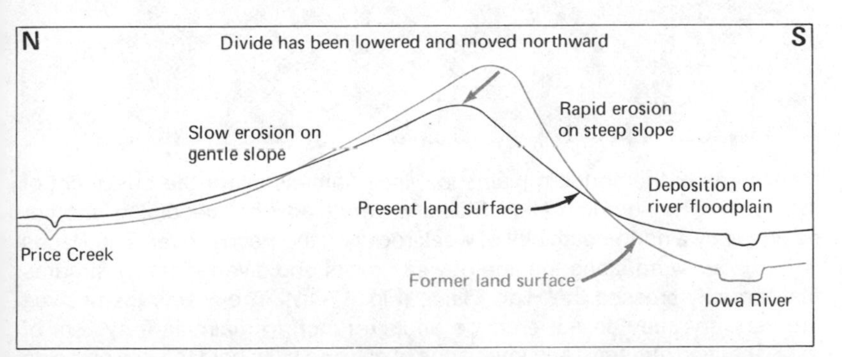

FIG. 17-9. Profile of the land near the lower forty. The rivers flowing southward go down steeper hills, have more energy, and therefore erode more rapidly than rivers flowing to the north. The different rate of erosion on the two sides of the drainage divide causes the ridge to move northward. Ultimately the imbalance of erosional force on the two sides of the ridge will be eliminated.

The High Plains are another example of terrain whose present form is largely due to forces originating and operating outside the local area. Long before glaciers invaded the northern states, internal forces raised a mountain range near the present Rockies. The mountains interrupted the flow of moist air from the west, gave rise to a multitude of energetic streams, and filled the streams with sand and silt. The streams, in turn, began to drop their loads as soon as they flowed onto the dry land in the rainshadow to the east of the mountains. Former valleys became choked with debris and the streams began to migrate across the flattened land. Over the centuries, ancient rivers built the plains out of weathered fragments from an ancient range of lofty mountains.

rainshadow dry region on the leeward side of a mountain in a region of prevailing winds
(see p. 86)

alluvial plain flat area created by river deposition
(see p. 341)

However, the High Plains needed another form of external assistance to enable them to persist into the present era. A period of crustal uplift gave new energy to the rivers and allowed them to dig deep valleys through the loose deposits of their ancestors. Tributary streams carved the land near many of these newly entrenched rivers into "badlands," intricate mazes of closely spaced rivers, steep V-shaped valleys, and sharp ridges (Fig. 17-10).

stream piracy capture of one stream by another that is downcutting more rapidly
(see p. 327)

FIG. 17-10. "Badlands" caused by accelerated erosion near the deepened valley of a river crossing the northern Plains. A small remnant of the original plain can be seen in the upper right corner of the picture. (Courtesy of Dwight Brown.)

Many parts of the northern plains lost their flatness under the onslaught of the rivers. The southern High Plains were saved from becoming another badlands by a north-south belt of weak rock and the Pecos River. The Pecos flowed southward along the line of weak rocks and diverted many streams that formerly crossed the High Plains (Fig. 17-11). The Plains themselves are very dry; they do not produce enough runoff to maintain a system of valley-carving streams. Accumulations of calcium salts hold the soil particles together and further decrease the likelihood of destruction of the flat surface.

caliche soil cemented by calcium carbonate

The High Plains are thus remnants of former times, preserved through the millennia by fortuitous circumstances. Human use of the region depends on another relic of the past. Modern irrigation systems rely on well water that

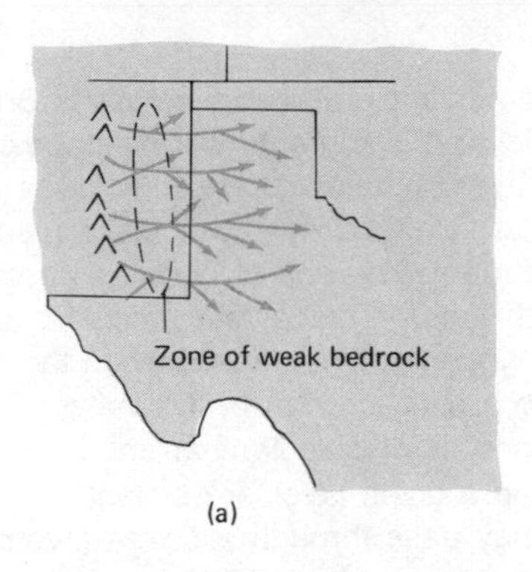

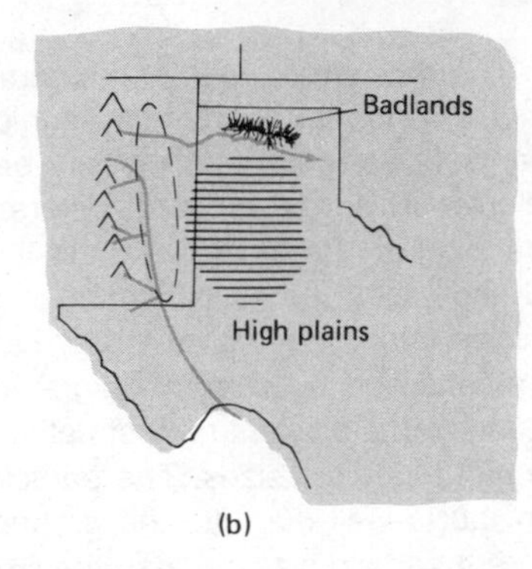

FIG. 17-11. Preservation of the High Plains of Texas. (a) Deposition of debris from mountains. (b) Uplift and erosion. The Pecos River cut into a zone of weak rock and diverted streams that otherwise would have crossed the plains and carved deep valleys in their level surface.

was stored in the ground when rivers flowed east from the mountains and crossed the Plains. Diversion of the ancient mountain streams by the modern Pecos River has cut off the supply of water. Therefore the wells are "mining" water that is not being replaced by any modern process.

connate water groundwater that entered its storage area under conditions that were different from the present
(see p. 308)

In summary, the landform of the lower forty is the legacy of a glacier that never entered the area. Wind from the ice piled the dust on the hills, and meltwater from the ice made the Iowa River a deep channel that gives extra energy to modern streams. The High Plains are a remnant flood plain, made of debris from eroded mountains and protected from destruction by the fortuitous diversion of rivers that otherwise would have cut across the region. Both areas are examples of regions whose landforms were influenced by events that happened far beyond the boundaries of the local area.

CANYONLANDS

Bryce Canyon in Utah is less than one-twentieth the size of its more famous cousin, the Grand Canyon of Arizona; but both have interesting geologic stories. Both canyons occupy dry regions where rainfall is too scanty to support a continuous plant cover. The exposure of the rock makes geologic structure easy to observe, but the intermittent landforming processes are difficult to analyze.

Bryce Canyon owes its general configuration to an old fault and the resulting energy relationships of usually dry streams. The canyon is carved into the side of a high ridge; streams flowing from the top of the ridge toward the north disappear in the desert sands of central Utah, whereas the south-flowing streams drop quickly into the canyon of the mighty Colorado River. In the first 2 km of travel, a south-flowing stream falls six times as far as a north-flowing one. The greater fall of south-flowing water gives it much more energy to use in carrying earth away. The north slope of the divide is gentle, nearly flat, while the south slope is deeply carved by falling water.

fault break in the rock, usually accompanied by sliding on one or both sides
(see Fig. 6-14)

intermittent stream stream that dries up between storms

The rock structure is at the same time simple and very complex. Bryce Canyon National Park rests on a thick layer of sedimentary rocks deposited by ancient streams flowing from nearby mountains. Stream deposits are meticulously sorted by size; different particles fall out at different places, depending on the local energy of the stream. The positions of the streams and shorelines, however, changed with time. A place that received large particles during one century may have gotten only fine clays when the sea encroached or the streams migrated to a different part of the valley. Consequently, the present rock structure consists of irregularly arranged layers of various particle sizes. The particles in a given layer are essentially uniform, but a well dug into the ground may pass through a dozen layers of different particle sizes in less than 10 meters.

lens layer of rock shaped like a magnifying glass, tapering off in all directions from the center

monument steep-sided, isolated hill, protected from erosion by a resistant cap layer
(see Figs. 7-13 and 16-20)

talus slope sloping pile of debris at the foot of a cliff
(see Fig. 16-1)

retreating escarpment edge of a flat plateau or sloping layer of hard rock; the edge retreats as erosion undercuts the steep slope
(see Fig. 17-6)

The exposed cliffs of Bryce Canyon are an intricate patchwork of rocks with different textures and hardnesses in different places. The bizarre rock formations in the canyons are consequences of dissimilar rocks weathering at unequal rates in a dry climate (Fig. 17-12). Hard rocks act as protective caps while soft layers weather away at the sides. Loosened debris slides down the steep cliffs and forms sloping piles of rubble at the bottom. Rivers flow too infrequently to carry away all of the broken rock, but the absence of protective vegetation gives the occasional storms an exceptional amount of landforming prowess.

The role of intermittent processes is even more apparent downstream in the Grand Canyon. Most desert rivers appear only during periods of intense local rainfall. The rain falls faster than the rock and sun-baked soil can absorb it, and some of the water flows away on the surface. The moving stream may leave the area of the storm and pass into regions that did not receive any rain. What begins as a sudden and often spectacular flood

FIG. 17-12. Monuments in Bryce Canyon. Resistant layers of rock act as protective caps and prevent erosion of weaker rocks beneath them. (Courtesy of E. S. Shipp, U.S. Forest Service.)

gradually diminishes in size as it moves along. A decrease in volume of water means that the river loses energy as it flows. For this reason, much desert terrain is depositional. Even the mighty Colorado contains dozens of boat-threatening rapids, formed where high-energy side streams bring debris down the steep canyon slopes and deposit it in the main channel, where it may remain for a long time before being removed by the main stream. (Fig. 17-13).

The man-made storage dams on the Colorado River may make future raft trips impossible in the Grand Canyon. The dams have three main effects on river flow:

1. Total flow is decreased because evaporation from the reservoir surface takes some of the river water.
2. Low flow is greater than it was before the dam was built. The river now can carry a greater load of tiny particles out of the Canyon in a year, and many beaches and sandbars in the Colorado have been destroyed by erosion.
3. Peak floods have been reduced drastically. Extremely high flows are the only way the river can roll large boulders out of the way and thus keep the channel open. Boat owners report a recent increase in the difficulty of the rapids, because the undammed side streams keep bringing in new supplies of debris, but the tamed river is not able to remove the large boulders.

stage-frequency graph graph of the relationship between river depth and the percentage of time that depth is exceeded
(see Fig. 14-15)

tributary stream flowing into a larger one
(see Fig. 15-6)

FIG. 17-13. Hance Rapids in the Grand Canyon. An energetic tributary entered the canyon from the left, dumped its load of coarse debris in the main channel, and caused the rapids at this place. (Courtesy of L. B. Leopold, Geological Survey.)

In summary, the stark, intricate, and rapidly changing landforms in the canyons are products of the essential weakness of many landforming processes in a desert. Bare rock remains near the surface because weathering is inefficient. Protective vegetation is scanty because rainfall is sporadic. Rapids are built when side streams flood and are destroyed when main rivers flood. The terrain bears the unmistakable imprint of infrequent events whose effects overshadow the meager results of impotent day-to-day processes. Intermittent processes are significant in all environments but often dominant in the canyonlands. The answer to the often-asked question—"Did that little river make this big canyon?"—is a resounding "NO." The Canyon was built by the Colorado River in flood, a rare sight before the dams were built and a memory now.

MOUNT RAINIER

Mount Rainier is by no means the tallest mountain on the continent but it seems to be because it rises far above a low plateau. It owes its imposing form to a rather dramatic interplay of fire and ice.

As mountains go, Rainier is a youngster, formed just before the Ice Age. The history of the mountain is marked by episodes of rapid eruption and of slow lava flow, interrupted by periods of dormancy and occasional explosive reawakenings. Successive layers of ash, cinders, and lava built the lofty cone over a period of many thousands of years. The symmetry of the mountain is due to the fact that a single vent dominated most of the eruptions (Fig. 17-14). Cinders and ashes spewed forth and landed near the vent, only to be covered by subsequent flows of liquid rock. A landslide or explosion removed the top of the cone within the past ten thousand years. At present, the volcano is dormant. Its crater is blocked by a "cork" of hardened lava, but it

cinder cone explosive volcano; ash and bombs of sialic material (see Fig. 6-12)

FIG. 17-14. Mt. Rainier, a symmetrical volcanic cone. (Courtesy of B. W. Muir, U.S. Forest Service.)

may erupt again if internal pressures become great enough to blow the cork out.

extinct volcano permanently plugged and inactive volcano

The second facet of Rainier's character is its mantle of snow and ice. The mountain looms up in the path of westerly winds blowing across the Pacific Ocean. The humid air is pushed up the slopes, cooled, and forced to drop some of its moisture. Air that reaches the summit is so cold that it no longer contains a significant amount of moisture. For this reason, the sides of the mountain get more rain and snow than the summit does.

westerly wind blowing from the west; common in middle latitudes (see Fig. 4-14)

The summit is an icy desert, smooth and covered by only a thin mantle of snow. Valley glaciers are active on the snowy side slopes. Meltwater lakes lie behind moraine dams in U-shaped valleys on all sides of the mountain. At still lower elevations, frigid streams rush down forested slopes and carve the deep V-shaped ravines that characterize the work of running water.

moraine debris left by a melting glacier (see Fig. 16-17)

Glaciers are fairly accurate indicators of climatic changes, because their size depends on the balance between snowfall and snowmelt. The glaciers on Mount Rainier were much more extensive a few centuries ago than they are today. From the mid-1800s to about 1950, ice melted more rapidly than new snow was added, and the glaciers shrunk. A general cooling since about 1940 is the apparent cause of a very recent extension of the glaciers. The 20-year lag between climatic change and glacial response is due to the sheer mass of ice and the inadequacy of length measurements as estimates of total size.

The mountain itself is warm, evidence for the recency of volcanic activity and the possibility of future eruptions. The internal heat is partly responsible for the frequent avalanches, landslides, and other mass movements on Rainier. Water percolates down into the rock and is warmed, occasionally forming steam and exerting great pressure on its surroundings. The bottom layers of ice and snow may melt, lubricate the slopes, and thus aid mass movements and glacier flow (Fig. 17-15).

mass movement downhill movement of earth material pulled by gravity, without the assistance of a transporting agent such as a river or glacier (see Fig. 15-4)

FIG. 17-15. Meltwater cavern beneath a glacier on the slope of Mt. Rainier. (Courtesy of U.S. Department of the Interior, National Park Service.)

There is a marked afternoon maximum in the rate of landforming activity. Streams that flow clear and placid in the early morning are swollen and debris-choked by midafternoon, when melting is most rapid. Landslides and avalanches are triggered during the times of minimum friction. Glaciers hasten their meter-a-week pace somewhat. Even the atmosphere cooperates by providing occasional storms, which, while unspectacular by midwestern hailstorm standards, nevertheless add a great amount of moisture.

In summary, three major factors are responsible for shaping the terrain of Mount Rainier. First, its birth as a composite volcano gave it a smooth, conical outline. Second, its height and resulting climate allowed glaciers to gouge deep U-shaped valleys on its slopes. Finally, the abundant precipitation and warm earth assist gravity in causing frequent avalanches and landslides. The present changes of the mountain shape are mainly erosional, but there is no proof that the volcano will not erupt again. Thus the modern landform is the work of a complex assemblage of activities that vary in intensity and effect. This characteristic of Mount Rainier is merely an exaggeration of the endless process of disturbance and adjustment among independent and often opposing forces that continually reshape the face of the earth.

CONCLUSION

Chapter 16 illustrated the many ways in which various landforming processes shape the surface of the earth. However, the case studies in this chapter added important qualifications to the general principles of terrain shaping. Appalachia highlighted the role of geologic structure in governing the rates of activity of present landforming processes. Wisconsin illustrated the persistent imprints of events that occurred long ago. The High Plains stated flatly that processes at work in distant regions can have an effect on a local area. The Canyonlands emphasized that rare events may be more significant than day-to-day processes. Finally, Mount Rainier pointed out the rather sensitive balance that may exist between opposing forces. In reality, all of these principles are applicable to every landform on the surface of the earth.

APPENDIX TO CHAPTER 17
Topographic Maps

Much of the United States has been mapped very accurately by the Geological Survey. This appendix contains parts of 13 contour maps, each part depicting an area of about 6 km² (about 2 mi²), carefully chosen to represent the characteristic features created by different landforming processes operating on different rock structures in different environments (Fig. 17A-1). The units on the maps are feet and miles because most of the maps of the United States have not yet been translated into the metric system.

A contour map is an isoline map of elevations above an arbitrary standard elevation (mean sea level is the usual standard). A single contour line remains at the same elevation as it traces its way across the terrain; it separates regions of higher elevation from lower places. Contour lines are usually drawn at equal differences in elevation (for example, 100, 200, 300, 400, etc. meters above sea level). The contour interval for a particular map is the elevation difference between contour lines on the map.

Closely spaced lines indicate a steep slope, a rapid change in elevation across a given horizontal distance. Smooth contour arcs depict a smooth slope, whereas the contour lines on a gullied slope are notched and ragged. A stream in the bottom of a valley is below the land on either side of its channel. A contour line that approaches the stream from one side must turn upstream in order to stay at a constant elevation as it crosses the valley. Upstream-pointing notches are therefore good clues to the locations of valleys.

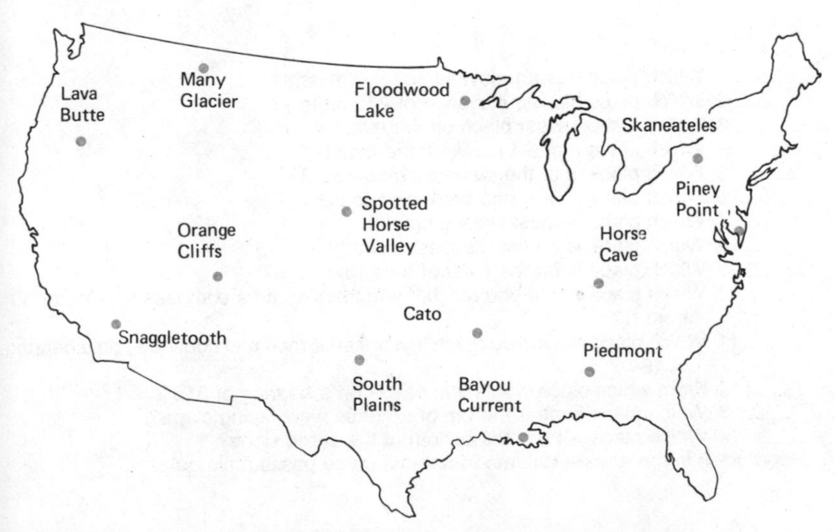

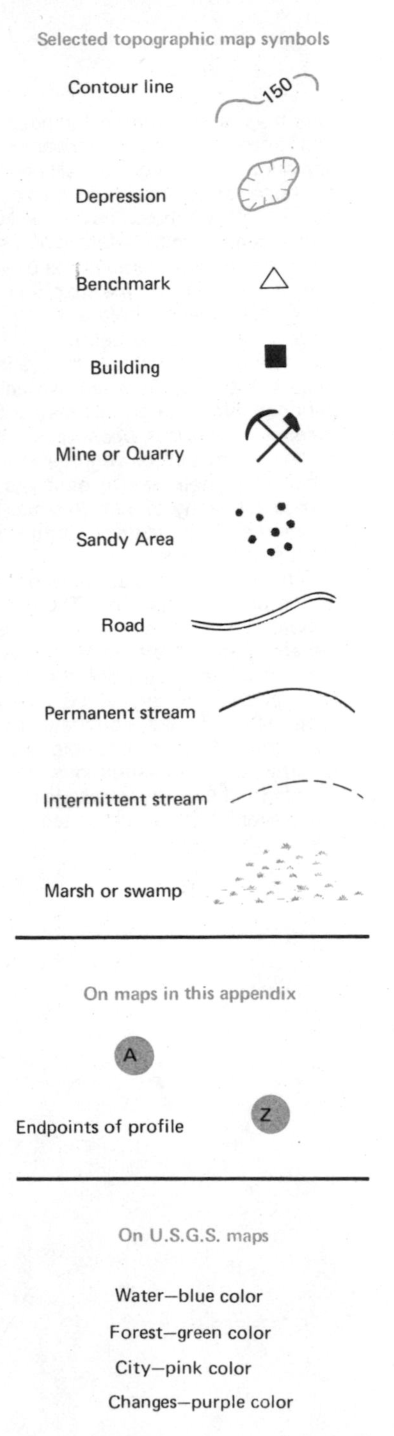

FIG. 17A-1. Locations of topographic maps and key to map symbols.

This map is an exercise intended to familiarize you with the mechanics of contour map reading. You will need a protractor or compass and a little ingenuity; if you have never worked with a contour map before, you may find it helpful to determine the elevation of every line on the map and to label them clearly before trying to answer the questions below.

The *contour interval* (the difference in elevation between two adjacent contour lines) on this map is 50 meters. A *compass bearing* is a direction expressed in degrees to the right of magnetic north; east has a compass bearing of 90°. This map is oriented so that magnetic north is at the top of the page.

A *profile* is a side view of the terrain along a given line. The graph beneath the map is a profile across the area between places M and R on the map. When you finish the questions below, you should be able to "see" hills and valleys on the contour map, and it should be possible for you to draw a not-too-strenuous hiking path from M to R (the profile tells you that a straight path would be too steep and tiring).

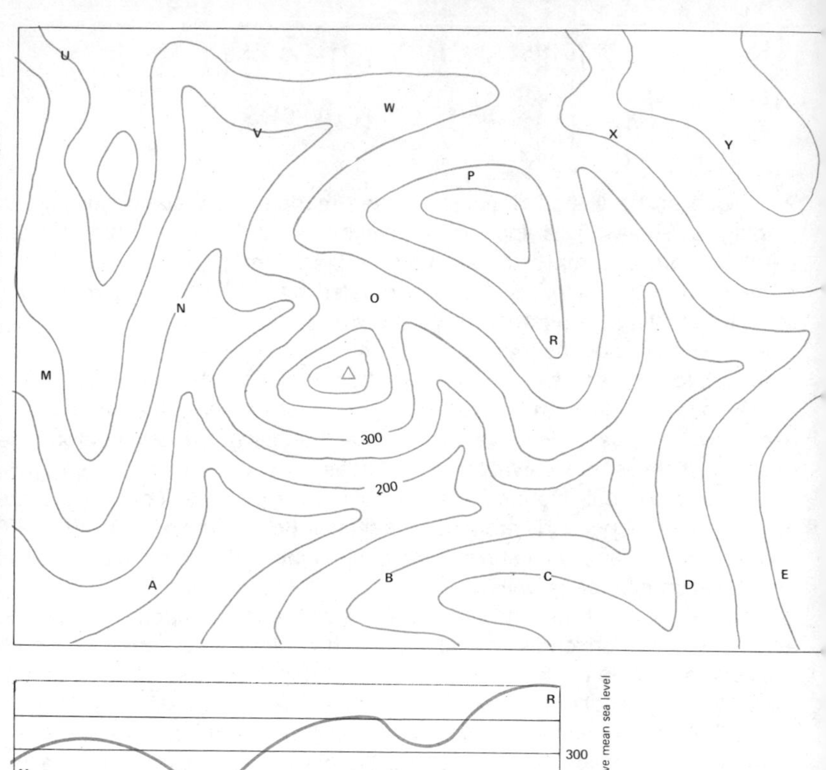

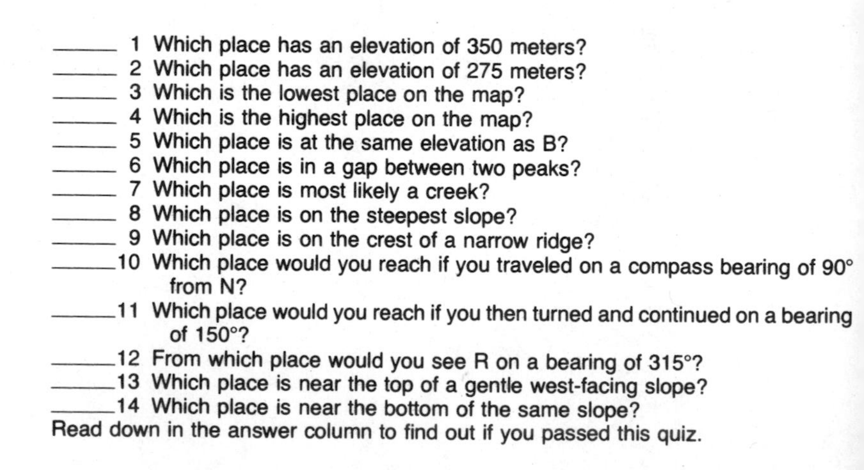

_____ 1 Which place has an elevation of 350 meters?
_____ 2 Which place has an elevation of 275 meters?
_____ 3 Which is the lowest place on the map?
_____ 4 Which is the highest place on the map?
_____ 5 Which place is at the same elevation as B?
_____ 6 Which place is in a gap between two peaks?
_____ 7 Which place is most likely a creek?
_____ 8 Which place is on the steepest slope?
_____ 9 Which place is on the crest of a narrow ridge?
_____10 Which place would you reach if you traveled on a compass bearing of 90° from N?
_____11 Which place would you reach if you then turned and continued on a bearing of 150°?
_____12 From which place would you see R on a bearing of 315°?
_____13 Which place is near the top of a gentle west-facing slope?
_____14 Which place is near the bottom of the same slope?

Read down in the answer column to find out if you passed this quiz.

Piedmont SE Quadrangle, Alabama

latitude: 33° 45′
longitude: 85° 30′
interval: 100 feet

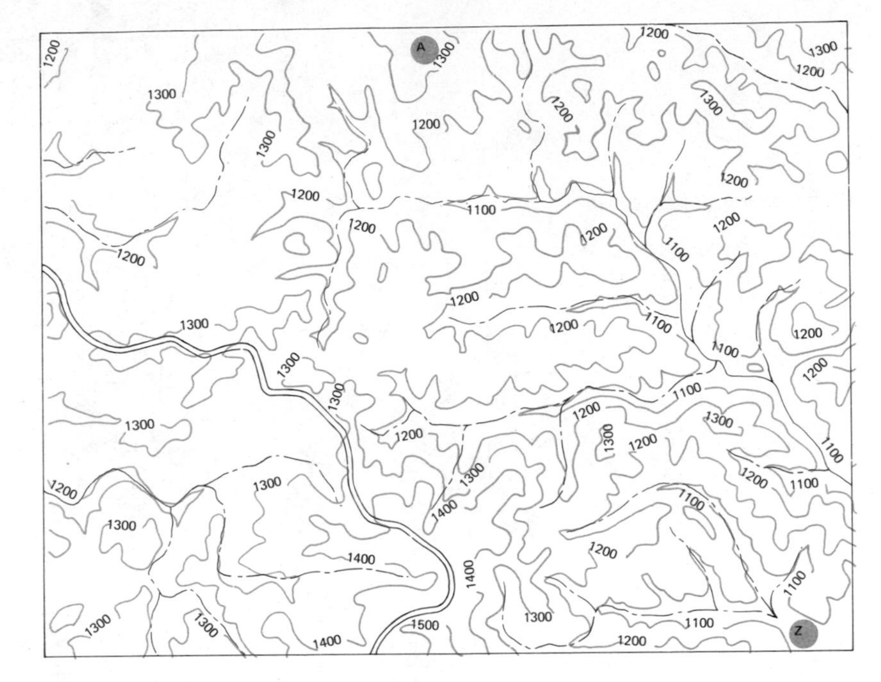

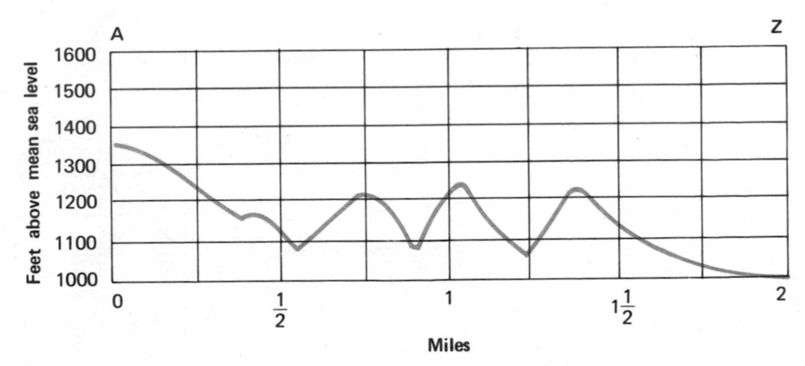

Stream erosion—uniform surfaces and humid climate

Treelike patterns of branches and main trunks are the rule when rivers drain rock materials that do not have significant differences in resistance. This map shows part of a maturely dissected plateau that flanks the Appalachian Mountains. Flat land is virtually nonexistent, so that transportation is difficult and the land has little agricultural value. Roads usually follow the ridges between the steeply sloping valleys. Most of the land remains in forest, a land use that protects slopes, regulates rivers, and provides recreational space for the inhabitants of nearby urban or agricultural areas. The central part of the profile crosses several characteristic V-shaped valleys. The right-hand part of the profile follows the typical concave path of a single stream as it tumbles down a steep hill and gradually moves on to more level land downstream.

Cato Quadrangle, Arkansas

latitude: 34° 52 ½′
longitude: 92° 15′
interval: 50 feet

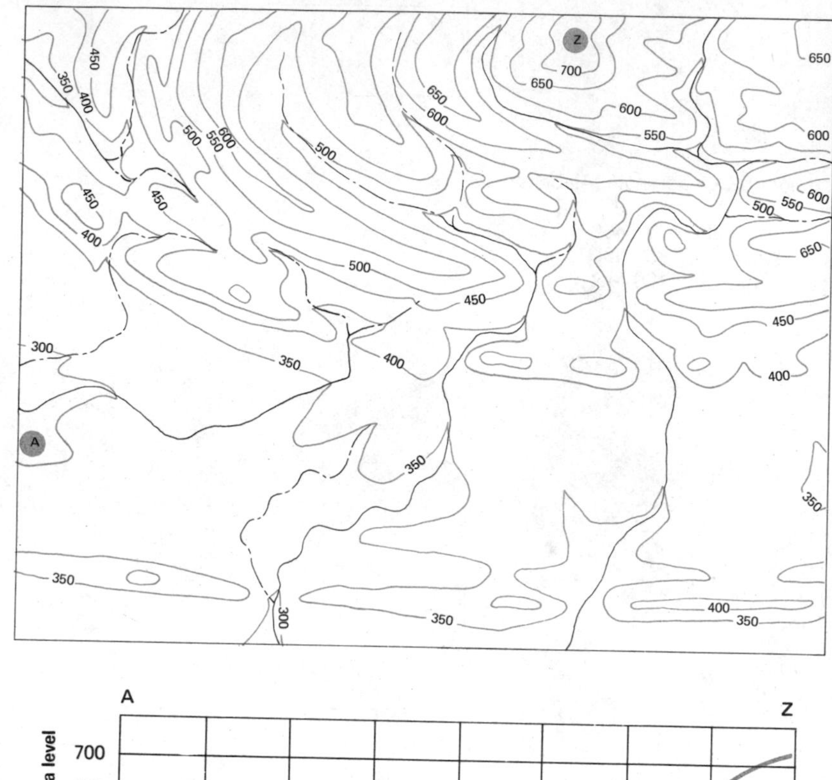

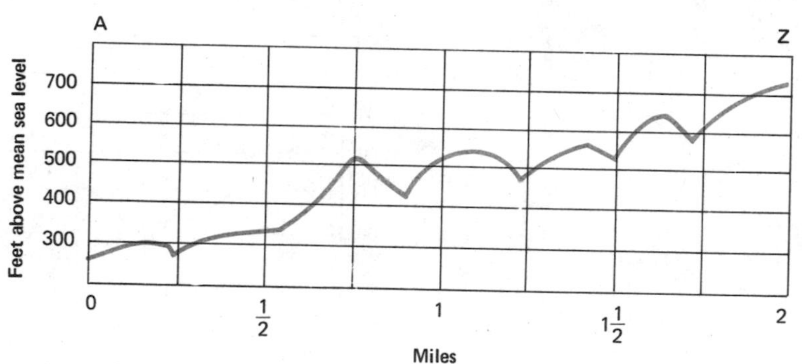

Stream erosion—nonuniform surfaces and humid climate

The connected network of V-shaped valleys indicates that flowing streams are the dominant landforming agents. The rock structure beneath the surface, however, frustrates the efforts of rivers to create the normal treelike pattern of an efficient drainage system. Long ago, geologic forces pushed flat rock layers upward and formed a dome. Erosion decapitated the dome and exposed alternate layers of resistant and weak rock that are arranged in concentric arcs. The small streams obediently follow the lines of least resistance and stay on the weak rocks. The combined volumes of several streams, however, may be sufficient to overcome the resistance of the hard rocks and make a gap through the arc of hills. The profile shows a transect inward from near the edge of the structural dome.

Piney Point Quadrangle, Maryland–Virginia

latitude: 38° 7½′
longitude: 76° 30′
interval: 20 feet

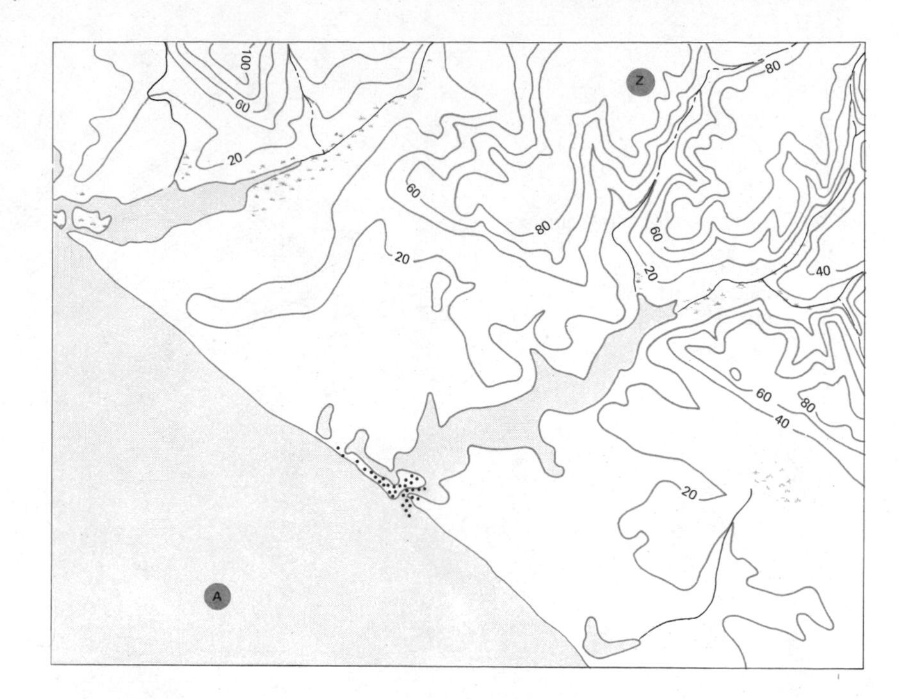

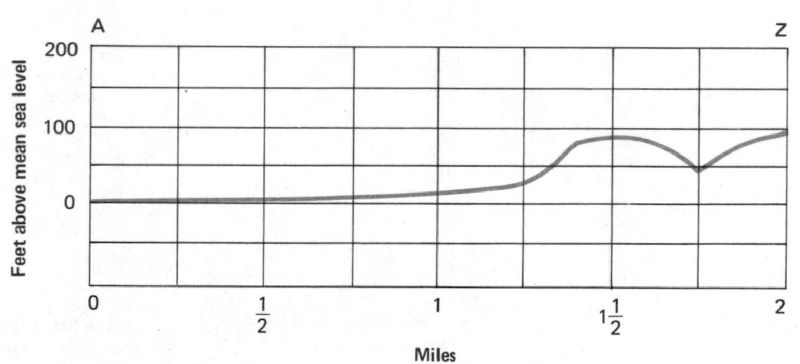

Wave erosion—sea level change

This part of the shoreline of Chesapeake Bay is a miniature model of the entire Middle Atlantic Coast. The closely spaced 40- and 60-foot contour lines mark an almost straight line of hills that stretches from the northwest to the southeast. This cliff represents a former shoreline. Raising the land or lowering the sea exposed the wide wave-cut bench that extends to the southwest of the line of cliffs. Streams cut across the bench and made characteristic V-shaped valleys in it. Then, sea level rose again, submerging the lower reaches of the valleys and forming the long inlets of the sea. Finally, waves moved sand along the shore and built sandbars and islands across the mouths of the former valleys. The slope of the valley beneath the sea is not apparent on the profile, which shows the present sea level, the broad low plain, the cliff along the former beach, and the irregular upland.

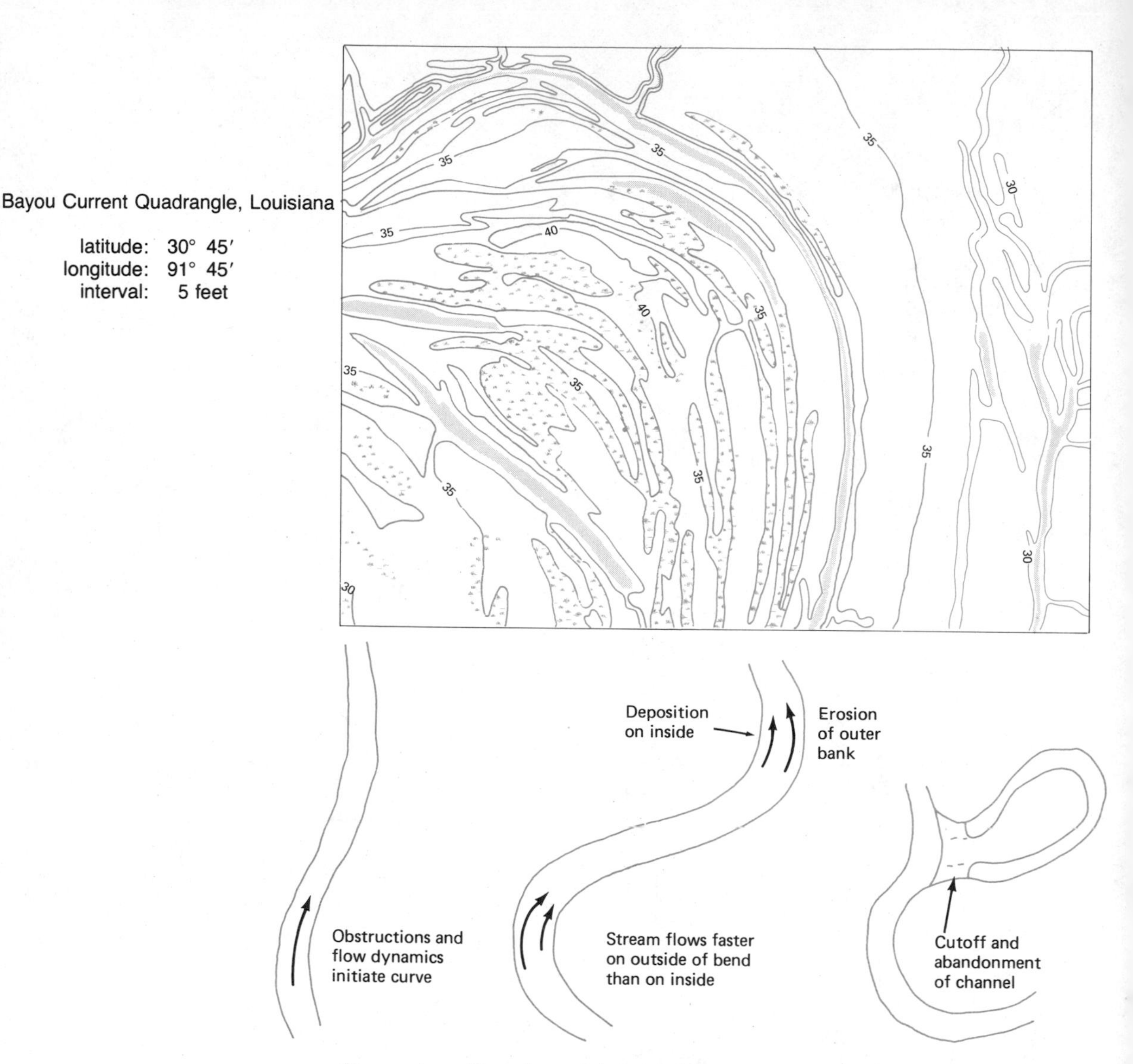

Stream deposition—flood plain in humid climate

Repeated flooding and deposition of material is responsible for the flat terrain on this map. The streams have no surplus energy because the plain is barely above sea level and river gradients are very gentle. The sluggish rivers cannot hold all the material that was eroded from the upper reaches of the immense area drained by the Mississippi River and all its tributaries. The lower reaches of the rivers frequently choke their own channels with sandbars and other deposits, particularly on the inside bank of a curve in the river. Meanwhile, the water continually undercuts the bank on the outside of the curve, as shown on the force diagram that substitutes for the monotonous profile of a level plain. Many centuries of deposition and sideways movement of the rivers produced the present valley: an ever-changing maze of meandering streams, abandoned channels, low depositional ridges, and stagnant swamps.

South Plains Quadrangle, Texas

latitude: 34° 7½′
longitude: 101° 15′
interval: 10 feet

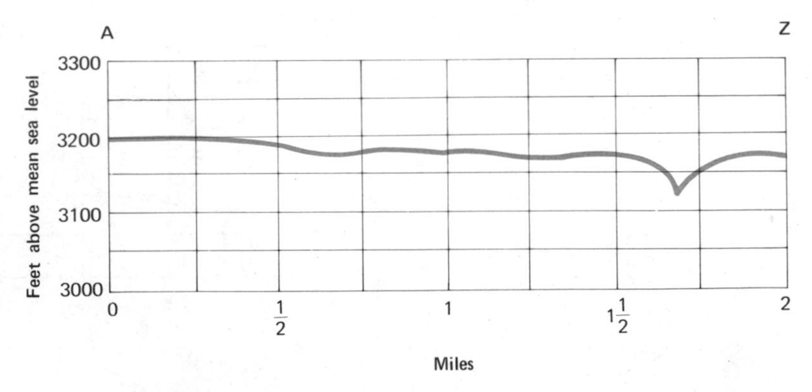

Wind and water erosion—level surface in semiarid climate

This map depicts a minor canyon cutting through part of the windswept plains of western Texas. A few small tributary streams appear immediately after storms and carve tiny V-shaped trenches in the sloping sides of the major stream valley. Otherwise, the terrain is extremely youthful, essentially level with little surface drainage. Such a landscape could not survive long in a rainy climate; it would soon succumb to river dissection. A second line of evidence for aridity is furnished by the shallow circular depressions on the upland. These are products of a long and self-reinforcing cycle: wind removes material during dry seasons; water fills the hollows after occasional storms; leaching weakens the structure of wet soils; bison and domestic livestock wallow in the water and mud; and trampled soil blows more readily in dry seasons and stays wetter in wet seasons than nearby soils on slightly higher grassy plains.

Snaggletooth Quadrangle, California

latitude: 34° 30′
longitude: 114° 37½′
interval: 50 feet

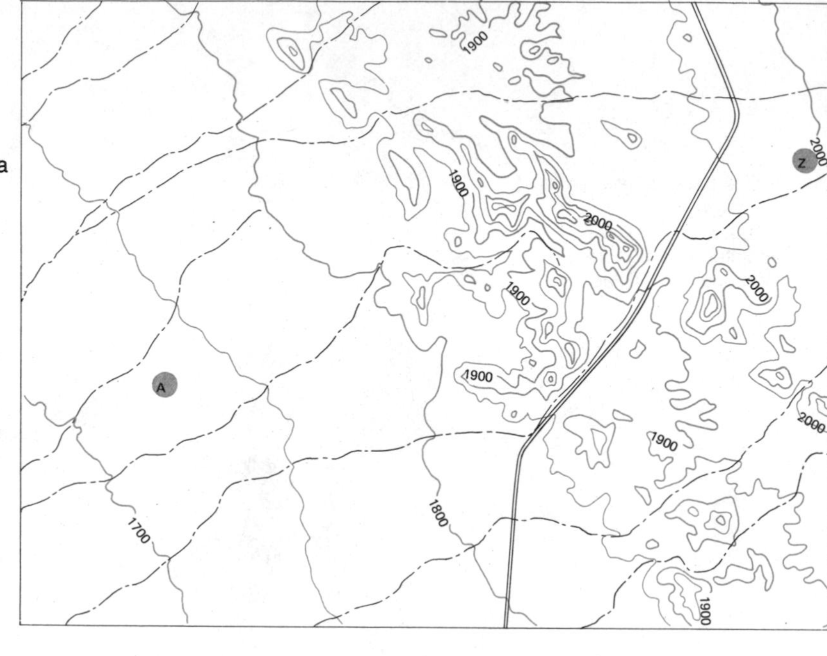

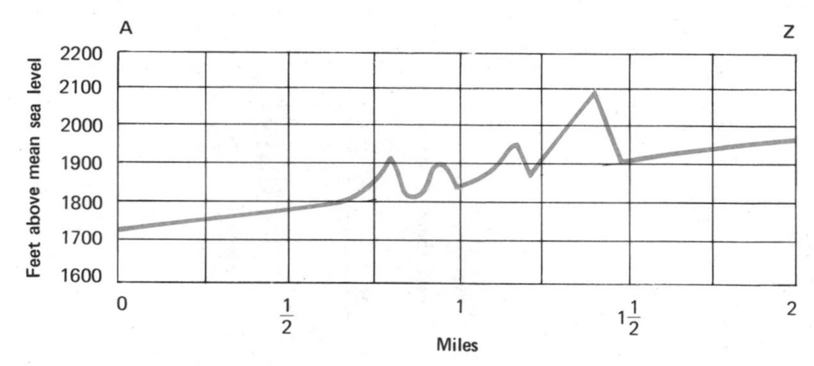

Stream deposition—pediment in arid climate

The cluster of steep hills on this map is slowly being buried in river deposits. The mountains are in a very dry part of California. Intermittent streams form sheet floods, which last for only a brief time after the infrequent storms. The water sinks into unconsolidated debris when it reaches the margins of the rocky core of the mountains. As the streams shrink and slow down, they drop their load of material. In time, a mountain in an arid region will be shattered by weathering, carried away by occasional flash floods, and spread out as a broad sheet of debris that gradually slopes away from the original rocky core. The hills on this map are outliers of a lofty range to the northeast. They protrude above the smooth slope of flood-planed rock and flood-deposited debris from the nearby mountains. A major highway shares the gap of one of the rivers through the outlier range.

Orange Cliffs 4 S W Quadrangle, Utah

latitude: 38° 0′
longitude: 110° 7½′
interval: 200 feet

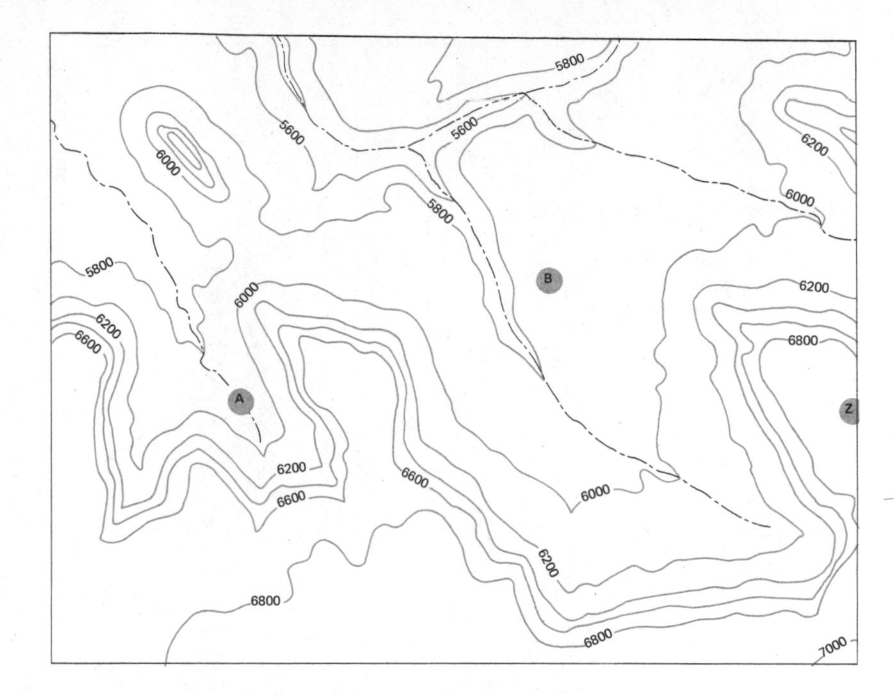

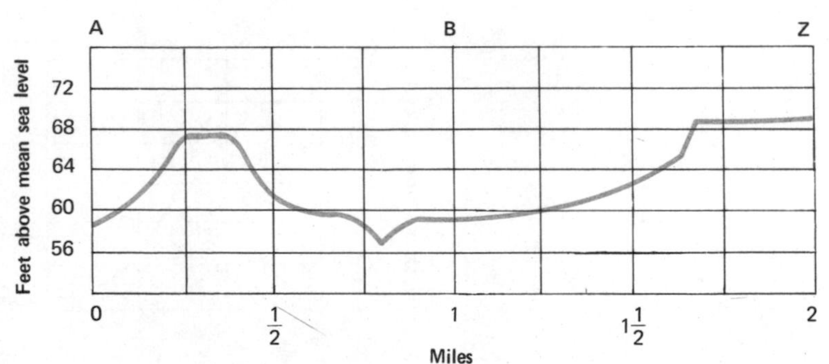

Stream erosion—plateau in arid climate

The profile shows the characteristic angularity of the landforms that a desert environment creates on horizontal rock layers. A resistant layer of rock forms a flat upland plateau about 6700 feet above sea level. Gentle slopes of rockfall debris spread out like aprons around the bases of the steep cliffs at the edge of the plateau. On the northern part of the map, the rivers have begun to cut valleys in another level bench about 5900 feet above sea level. The isolated hill in the northwestern part of the map is actually a former extension of the plateau between points A and B. Removal of intervening material by erosion left the hill, a monument to the previous extent of the upland. In time, the cliff will retreat still farther to the south and the valley in the low bench will be widened as the rivers carry the fractured remains of the plateaus away.

Many Glacier Quadrangle, Montana

latitude: 48° 45′
longitude: 113° 37½′
interval: 400 feet

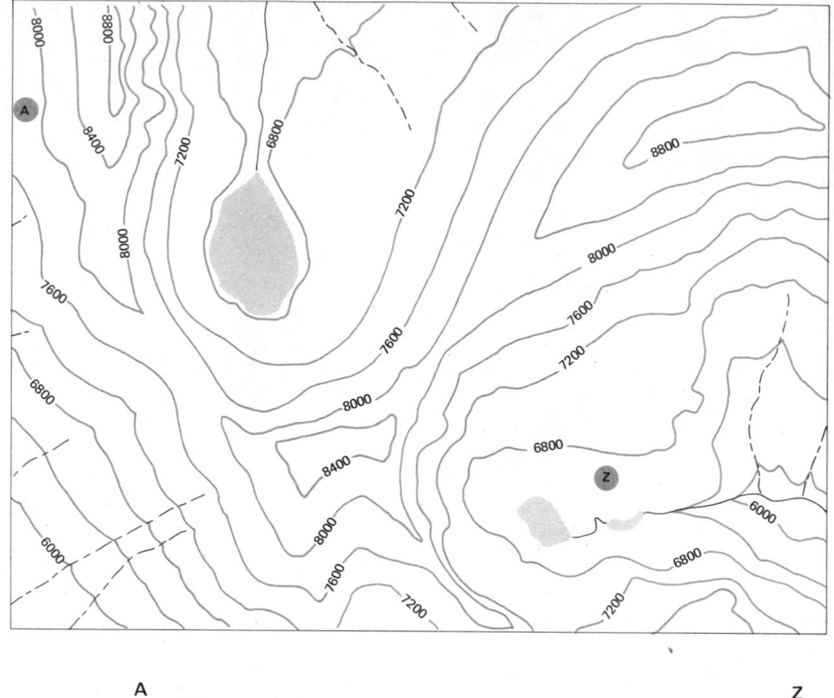

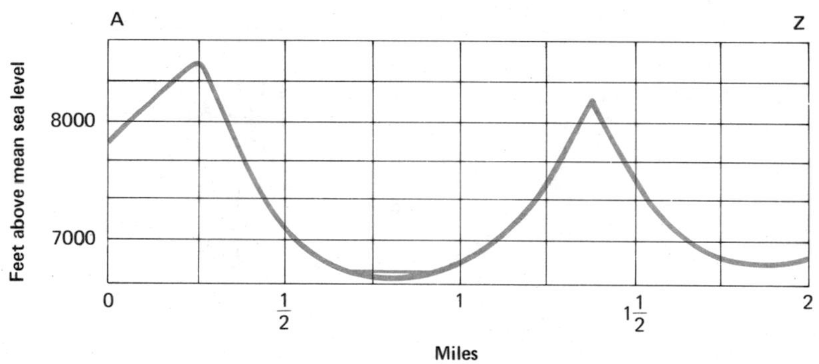

Ice scouring—horn and U-shaped valleys

Steep slopes around lakes near the crest of a sharp ridge are clues to the origin of the landforms on this map. There are two distinct bowl-shaped valleys on the north and east sides of the 8400-foot peak near the south edge of the map. These depressions are cirques, consequences of former valley glaciers. The profile shows the classic "U" shape created when a glacier tries to fit its bulk into the tight confines of a former river valley. The great mass of ice bearing down on the bottom of a narrow valley scours away material from both sides. Ridges between glacial valleys are often very sharp. There was no glacier in the valley beyond the left hand ridge on the profile, thus the ridge is not so symmetrically pointed as the one to the right on the profile. The smooth slope in the southwest corner of the map is also unmarked by glacial activity.

Skaneateles Quadrangle,
New York

latitude: 42° 52½′
longitude: 76° 22½′
interval: 20 feet

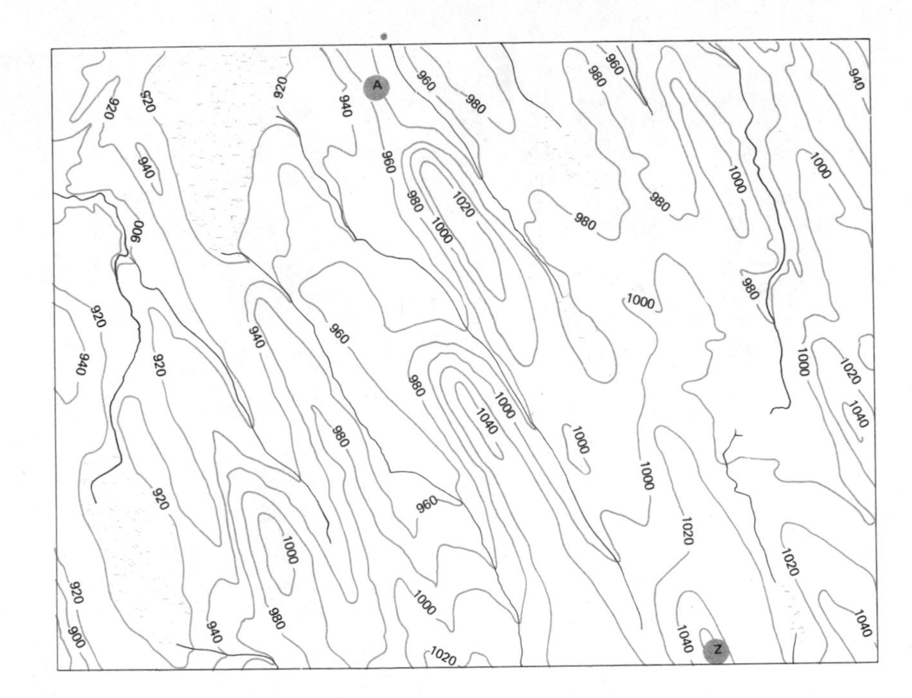

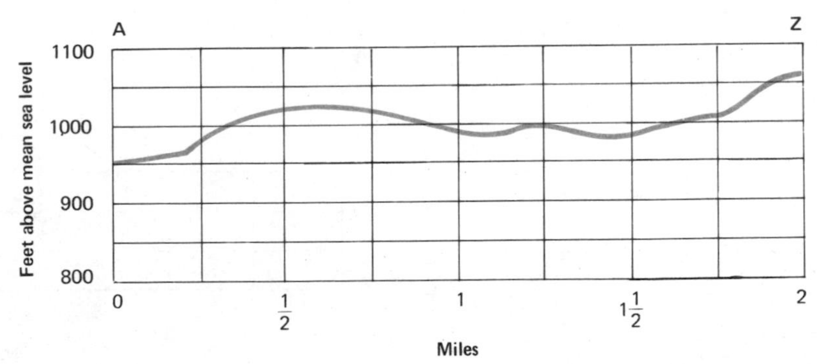

Ice deposition—drumlins and deranged drainage

Terrain that was shaped by a slowly moving and melting sheet of ice often contains swamps and aligned oval hills of loose material. The alignment of hills and valleys indicates that a former glacier advanced from the northwest, gouged out shallow valleys, and shaped the hills into smoothly rounded ridges. The left half of the profile shows a classic drumlin with a steep slope on the side from which the glacier came. Modern rivers occupy valleys between the elongated hills, so that the drainage system has a pronounced orientation from northwest to southeast, but the rivers have not had enough time since the Ice Age to deepen their valleys and drain all of the swampland. Aligned deposits do not occur where glaciers are vigorous and able to carry much material. Likewise, orientation of material is not apparent where glaciers stop, melt away, and abruptly drop their loads of debris.

Floodwood Lake Quadrangle, Minnesota

latitude: 47° 7½′
longitude: 93° 0′
interval: 20 feet

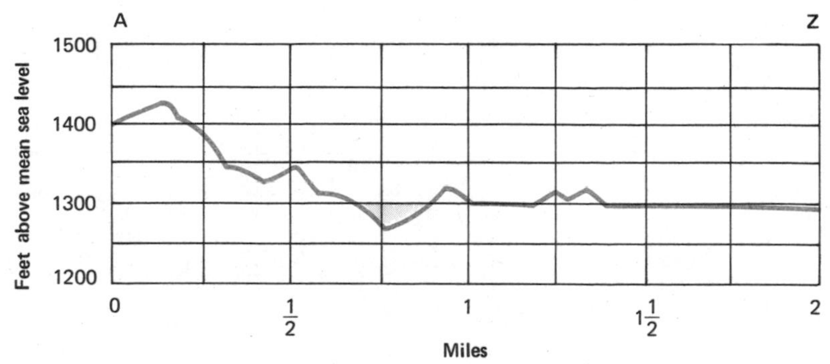

Ice and water deposition—moraine and outwash plain

Vast sheets of ice once covered most of the United States north of the Missouri and Ohio Rivers. The continental glaciers left their mark on the landscape in the form of irregular arcs of hills (the terminal moraines of various glacial advances), gently rolling plains of glacial till (ground moraines where stagnant ice melted and dropped its loads of debris), and flat plains of sorted debris (the beds of temporary meltwater rivers and lakes). The hills and lakes in the northwest corner of this map are part of a terminal moraine; the swampy plain on the east side of the map is the exposed bottom of a former lake that occupied the area between the ice margin and another moraine to the south. Iowa is not as swampy as Minnesota because the glaciers left Iowa long ago, and postglacial processes have had time to reshape the landscape and obliterate many of the distinctive glacial features.

Spotted Horse Valley Quadrangle, Nebraska

latitude: 41° 22½′
longitude: 101° 30′
interval: 50 feet

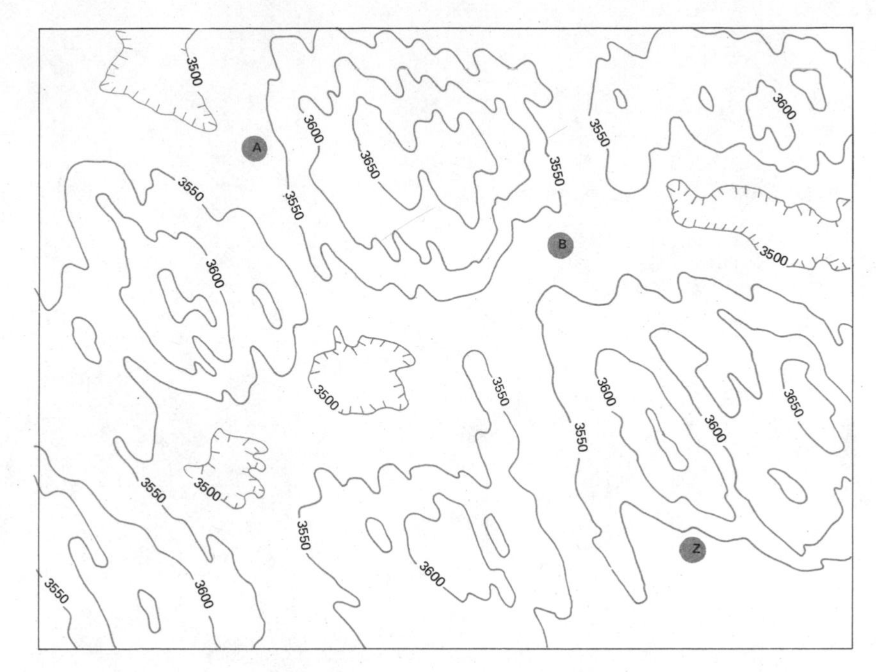

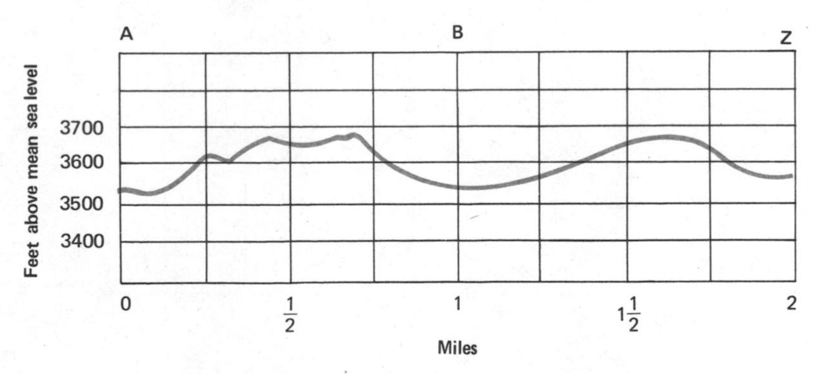

Wind deposition—aligned dunes in semiarid climate

Grassy hills of sand cover a sizeable part of west-central Nebraska and smaller areas in many other states. Coarse soil can absorb nearly all of the rain that falls, even in the most severe storms. Wind becomes a significant shaper of the terrain, because rivers have very little chance to form. The hills are not organized around treelike river systems, nor do they have the random arrangement of a glacial moraine. Rather, there appear to be two general patterns to the contours of the land. A map of a large area reveals a general alignment of hills in a series of irregular east-west ridges, like waves rolling onto a beach. The detailed map shows that the individual dunes are elongated and smoothed by winds from the northwest. The profile from A to B illustrates the irregular shape of a dune when crossed at an angle to the prevailing wind, whereas B to Z shows the smooth profile of a dune parallel to the wind.

Horse Cave Quadrangle, Kentucky

latitude: 37° 7½′
longitude: 85° 52½′
interval: 50 feet

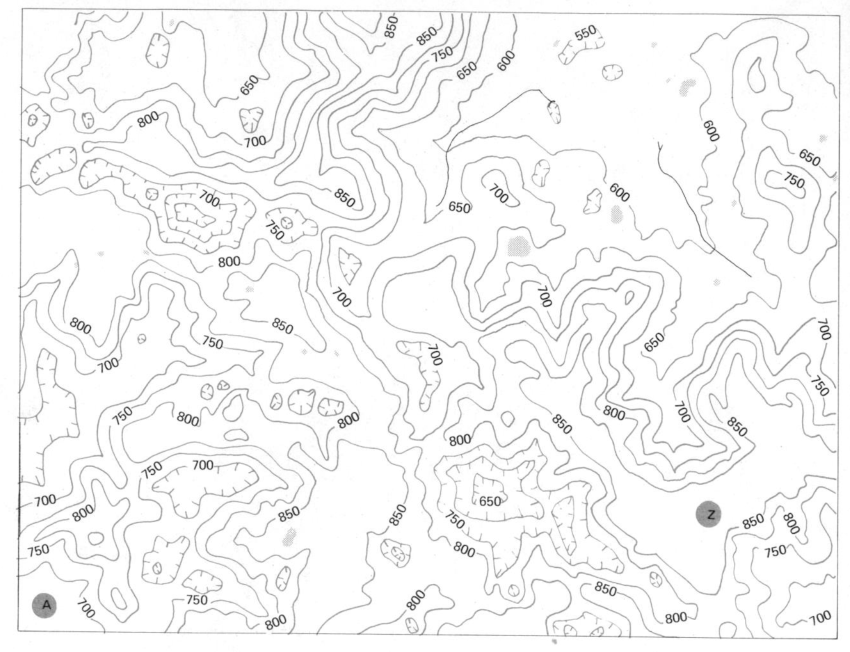

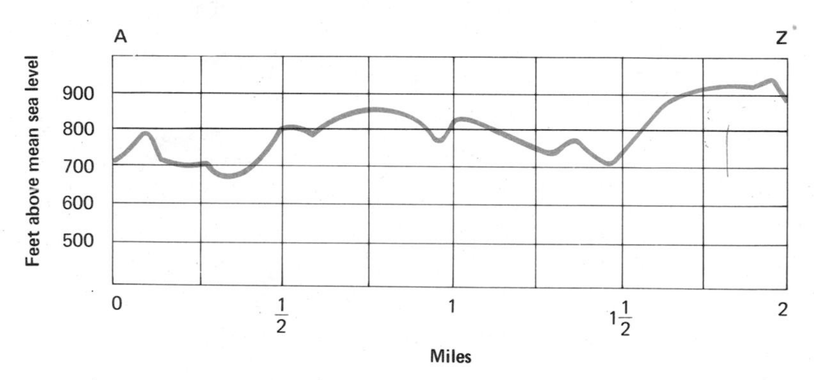

Rock solution—sinkholes and underground streams on limestone

The rock beneath this area is riddled with cracks and caves. Water moving through underground passageways slowly erodes the soluble limestone and enlarges the caverns. The caves eventually become so large that they cannot support their ceilings. This map shows a maze of unconnected valleys and sinkholes, a typical product of occasional cavern collapse. Two streams enter the underground drainage network after a short trip across the surface of the earth. Other V-shaped valleys may have supported surface streams in the past, but cave enlargement beneath the surface has created passageways for water to sink into the underground drainage system sooner than it did formerly. Scattered ponds indicate places where the floors of sinkholes are below the water table or are lined with enough fine-textured material, so that water does not seep through the bottom rapidly enough to empty the ponds.

Lava Butte Quadrangle, Oregon

latitude: 43° 52½′
longitude: 121° 15′
interval: 50 feet

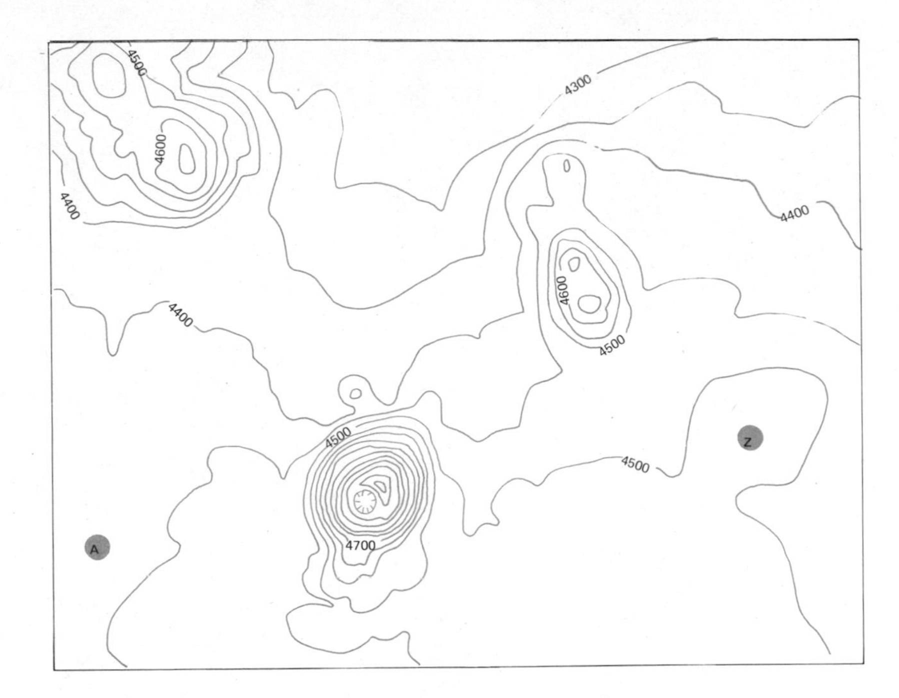

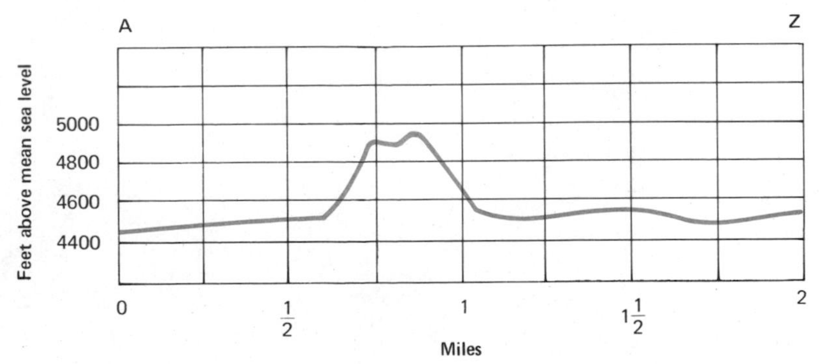

Volcanic eruption—cinder cones and lava flows

Some rocks are so hard or so recently exposed that the landforming work of water, ice, or wind is not apparent on a topographic map. Several different periods of volcanic activity are responsible for this cluster of cinder cones and associated flows of lava. The pair of cones in the northwest corner of the map are partly covered by vegetation, a tribute to centuries of soil formation and biotic succession. The tall cone to the south (also shown on the profile) is flanked by a large and irregularly lobed sheet of lava that extends to the base of the northern pair of cones. A lack of plant cover on the lava indicates that it is more recent or more resistant to weathering than the material on the north and east sides of the map. The volcanoes are part of a long chain of similar cones that are arranged along a weakness in the earth's crust; the weakness, in turn, is a result of Oregon's position near the leading edge of a moving crustal plate.

EPILOGUE

Earth Systems

It's the only world we have; we better take care of it.

Lowary, U.S. Forest Service

The interdependence of climate, soil, vegetation, rivers, and landforms has been emphasized many times throughout this book. The climatic pattern gives an underlying unity to other environmental patterns, because the weather is the major "landshaper," "rivermaker," "soilformer," and "plant-grower." Attributes of location—latitude, continentality, elevation—are the major controls of climate. For this reason, knowledge of the location of a place enables a person to make reasonably accurate general descriptions of the soil and its productivity, the natural vegetation and probable agricultural crops, the weather characteristics and their hydrological implications, and the typical shapes of the land surfaces. These physical features are an important part of the value of the area for human life (see Appendix).

Knowledge of local characteristics is necessary for a precise analysis of a local ecological system. The type of parent material affects the texture of the soil, which in turn influences the fertility of the soil, the type of vegetation, and the hydrographs of the streams. Certain terrain features—elevation, aspect, slope—affect the local microclimate and thus influence soil and vegetation. At the same time, plant cover and soil material can alter the heat and moisture characteristics of an area, have a pronounced effect on streamflow, and thus indirectly modify the erosional prowess, sediment load, and ecological characteristics of the rivers.

This list of interrelationships could be expanded considerably, but the main point is clear: the environment is a complex mechanism whose parts are meshed in many ways, not all of them fully understood. There is no such thing as altering only one part of an ecosystem (Table 18-1).

The very complexity of the physical world has led to a number of different ways of approaching the study of environment. This book has examined air movements, geologic structures, energy budgets, flow patterns, water balances, nutrient cycles, spatial arrangements, and chronological sequences. These and many other approaches have been tried with different kinds of success in different environments. There is no one way to describe an environment; dissimilar purposes imply different procedures. A farmer sees aspects of the world that an engineer misses, and prospectors, artists, hikers, and ecologists add their unique perspectives. Nobody can see the total environment, but everyone must become aware that it is much larger than one observer's corner of it. Without such awareness, humanity will continue down an already well-worn path of local "control" with unforeseen consequences to larger systems. History and landscape are littered with unanticipated environmental repercussions.

However, two things set the modern situation apart from any in the past. First, man's technical ability to alter the physical environment has never been greater than it is now. Second, and more important, an expanding world population decreases the margin for error. People can no longer get up and walk away from a mistake; there is nowhere else to go. Perhaps, when the full implications of that fact sink in, people will seek understanding, act with care, and recognize that different kinds of behavior are appropriate in different parts of the world.

1. Interconnections, where a cause can have an effect far away—e.g., clearing a hillside field decreases infiltration, increases runoff, raises the level of the river during flood but lowers it between storms, gives sediment and nutrients to the water, and thus causes floods, sedimentation, and eutrophication downstream (Chapters 8, 10, 12, 13, 14).
2. Leverage points, where a small cause can have a much larger effect—e.g., plowing a field near a shore lowers the albedo of the ground, raises surface temperatures, makes the air less stable, lowers air pressures over the land, causes winds to blow onshore, and thus increases local precipitation (Chapters 1, 2, 3).
3. Feedback loops, where an effect can reinforce or cancel its cause—e.g., putting a glacier on the land increases the albedo of the surface, reduces its temperature, and thus favors further growth of the glacier (Chapters 1, 16); raising the temperature of a surface increases the loss of energy by outgoing radiation and turbulent transfer and thus lowers the temperature (Chapter 1).
4. Nonlinearities, where doubling a cause does not double the effect—e.g., adding five units of food to a lake increases fish populations by stimulating their growth, but putting in another five units kills the fish by lowering the oxygen content of the water (Chapters 11 and 13).
5. Lag times and storages, where a cause can have a delayed effect—e.g., preventing or putting out all fires in a fire environment leads to a slow buildup of nutrients in the biomass and standing litter that ultimately produces a catastrophic fire (Chapter 12).

TABLE 18-1 INNATE CHARACTERISTICS OF ENVIRONMENTAL SYSTEMS

Environmental System in Which Change Is Made	**Years Needed for System to Adjust**
Lower troposphere, small streams	.01
Upper troposphere, large streams, microbe populations, soil water	.1
Lower stratosphere, seasonal ponds, insect populations, shallow groundwater	1
Upper stratosphere, herbivore populations, ocean surface, responsive ecosystems, valley glaciers	10
Deep groundwater, midenergy ecosystems, large lakes, steep slopes, carnivore populations	100
Deep ocean basins, low energy ecosystems, ice caps	1000
Crustal movements	?

APPENDIX TO EPILOGUE

Generalized Geographic Correlation of Major Environmental Processes and Features

Climate	"Summer" Weather	"Winter" Weather	Average Latitude	Continental Location	Soil-forming Processes
Instability	Hot wet	Hot wet	0°	West, center	Ferralization
Instability/ subsidence	Hot wet	Hot dry	15°	All across	Calcification
Subsidence	Hot dry	Warm dry	30°	West, center	Salinization
Subsidence/ marine frontal	Hot dry	Cool wet	35°	West	Variable
Marine frontal	Warm wet	Cool wet	45°	West	Podsolization
Marine frontal/ inversion	Cool wet	Cold dry	60°	West	Podsolization
Inversion	Cold dry	Cold dry	80°	All across	None
Inversion/ rainshadow frontal	Warm dry	Cold dry	50°	Center	Calcification
Inversion/ typical frontal	Warm wet	Cold dry	50°	East	Podsolization
Rainshadow frontal	Hot dry	Cold dry	45°	Center	Calcification
Typical frontal	Hot wet	Cold wet	45°	East	Podsolization
Typical frontal/ instability	Hot wet	Cold wet	35°	East	Podsolization

Major Soils	Natural Vegetation	Nutrient Storage Area	Landforming Processes	Typical Landforms	Agricultural Use
Oxisol, ultisol	Selva	Biomass	Weathering, soil flowage	Gentle slopes, rapids	Slash-and-burn, tree crops
Alfisol, ultisol	Savanna	Soil	Sheetfloods, soil hardening	Plateaus, prominences	Cattle, grains
Aridisol, entisol	Desert	Soil	Wind, floods, mass movement	Pavements, monuments	Little except in oases
Alfisol, entisol	Chaparral	Soil	Rivers, mass movement	Rugged valleys, mudflow scars	Tree crops, vegetables
Inceptisol, spodosol	Forest	Biomass	Trees, rivers, glaciers	Rolling hills, alpine mountains	Trees, tree crops
Spodosol, inceptisol	Taiga	Litter	Rivers, glaciers	Alpine mountains, rolling hills, lakes	Trees
Entisol, inceptisol	Tundra	Soil	Soil flowing, frost action	Permafrost features	Reindeer
Mollisol, inceptisol	Grassland	Soil	Rivers, wind, mass movement	Rolling hills, badlands	Cattle
Spodosol, inceptisol	Taiga	Litter	Rivers, past glaciers	Rolling hills, lakes	Trees
Mollisol, aridisol	Grassland	Soil	Wind, rivers	Pediments, badlands	Grains, cattle
Alfisol, ultisol	Hardwoods	Biomass	Trees, rivers	Rolling hills, treefall pits	Mixed farming
Ultisol, alfisol	Hardwoods	Biomass	Rivers, trees	Rolling hills, treefall pits	Tree crops, vegetables

INDEX

Boldface numbers refer to marginal definitions.

ablation **351**
absorbing slope 323
abrasion **148,** 148
abyssal **283**
acidity 162
actinomycetes **167**
adiabatic chart 36
adiabatic process **32**
aeration **186, 201**
aerobic **180**
agent of erosion **332**
air composition 43
air conditioning 20–21
air density 32
air instability 30–47
air mass **67,** 67
air pollution 39–42
air pressure 31–36, 51–54, 68
air profile 36
air stability **31,** 30–47, 49
albedo **11,** 11
Alberta Low 87–88
alfisol 222–223
alluvial fan **342,** 342
alluvium **341**
alpine glacier **345,** 345
aluminum 169
altitude zonation **237,** 237
American Association of State Highway Officials (AASHO) 204
amino acids **188**
amorphous state **139**
anaerobic **180, 274, 278**
anaerobic bacteria **190,** 190
anaerobic organism **167**
analemma 27
angle of repose **322**
angular blocky 209
animal **185**
animal energy budget 22–24
annular drainage 325
anoxic **180, 274**
anthracite coal 156
anticline **127**
aquatic biome 289–293
aquiclude 307
aquifer 307
architecture 20

Arctic barrens 255
Arctic grassland 255
arête **345**
aridosol 216–217
arroyo **340**
asthenosphere **115**
atmosphere 10
atmospheric radiation 14–15
automobile 21
autotroph **244**
avalanche **323,** 338
avalanche chute **338,** 338
azonal soil **207**
Azotobacter 188

back (backing) **82,** 82–83
backwash **358**
badland 380
bajada **339**
balloon profile 69
balloon sounding **35**
bankfull stage **312, 322**
bar **163**
barchan **352**
barometer **51, 82,** 82
barrier island **360**
bars 360–361
basalt 154
base flow **309,** 309
baselevel **327,** 327–328
base saturation **162,** 162
basin **127**
basin drainage 325
bauxite 153
beach 358–360
bearing strength **203,** 203
bench 343, 356–357, 391
benthic **283**
Bermuda High 88
biochemical oxygen demand (BOD) 273–287
biogeochemical cycle 257–270, **266**
biological control **186**
biomass 258, **259**
biosequence **196,** 196, **207**
biotic succession 281
bituminous coal 155
blocky 171

BOD 273–282, **277**
bog **281,** 281
boreal forest 254
bornhardt **334**
boron **175**
boulder 158
boulder-clay **349**
boundary layer **43, 106**
Bowen's Reaction Series 139
brick 155
broadleaf evergreen forest 248
buffer **186,** 186
bunch grassland 252

cactus forest 250
calcareous material **142**
calcification 205
calcite 153
calcium 169, 174
calorie **12**
canyon **340**
capillarity 168–169
capillary rise 304
carbon 183–184
carbon dioxide 14
carbon-14 132
carbonic acid **145,** 145, **149**
carbon/nitrogen ration 183–184
carnivore **24, 244,** 244–246
carp 274
cataract **341**
chaparral 251, 267, **267,** 366
chemical sediment 155
chemical weathering **146**
chert **143,** 155
chlorine **175**
chronosequence **196**
cinder 154
cindercone **125,** 125–126, 384–385, 401
cirque **345,** 345, 396
cirrus cloud **72**
clastic sediment **142**
clay 147, 158, 159, 164
clear-air turbulence 36
cleavage plane **149**
climate **43, 93**
climate graph 110

climax **239,** 239
climosequence **196**
closed system **25,** 257–260, **261**
cloud 10, 37–38
cloud cover 70
clutch size 243
C/N ratio **184**
coal 143, 144
coalescing alluvial fans 342
coastal energy budget 17–19
cobble 158
coefficient of friction **322**
cold current 287
cold desert 255
cold front 73
colluvium **201**
Colorado Low 89
columnar 171, 209
compartment **257**
compartmental modeling **257,** 257–263
compass bearing 388 (389)
compensation **263**
competitive exclusion **237,** 237, **241**
complex climate 58–60, **59**
component of gravity **322,** 323
concrete 155
condensation 32
condensation nucleus **261**
conduction **15,** 15
conductivity **106, 172,** 172
confidence limit **101**
conglomerate **142,** 155
coniferous forest 254
continental drift 121
continental energy budget 17–19
continental glacier 344,374
continentality **95**
continental shelf **119, 286**
contingent probability **101**
contour line 387
contour map 387
contour strip cropping 199
convection **15,** 51, **52**
convergence **54**
cooling degree-day 106
copper 175
coral reef 290–292
cordillera **115**
Coriolis effect **55,** 55, 61
corrosion 203
counter radiation **14**
cover index 315
crumblike 171
crust **115,** 115–123
crustal plate **138**
crystal **137,** 137–139, 152
crystal lattice **159**
cultural eutrophication **282**
cumulonimbus cloud **37,** 37
cumulus cloud **37,** 37
curve number 315
cutoff 78
cycle of erosion 327–329
cyclone **62, 75**
cyclone belt **84**
cyclonic storm 75

Davisian Cycle **327**
daylength 28
decay 259
deciduous forest 253, **266**
deciduous tree **184**
decomposer 177-190
deep blue sky 82
deepening of pressure **76**
deep-weathering profile **377**
deflation **351**
deflocculation **169**
degree day **104**
delta **341,** 359
dendritic drainage 325, **373**
denitrification **190,** 190
density **32, 71**
depth-yield graph 307–308
desert 250, 267, 364
desert climate 97
desert energy budget 11–15
desert pavement **351,** 351
desert water budget 298
design storm **103,** 103
destabilization **39**
dew point **32,** 48
dew point depression 69
diagnostic horizon **205**
diastrophism **126**
diffuse insolation **10**
dirt 192
discharge rate 304, **305,** 305
dissociation **161**
dissolved oxygen **273,** 273–282
ditching 201
divergence **54, 71**
diversity **241,** 241–242, **278**
doline **149**
dolomite **143,** 155
dome **127**
drainage divide **328, 379**
drainage pattern 325, 389, 390
Driftless Area **377**
drumlin 374, 375, **375,** 397
dry adiabatic rate **32**
dry-bulb temperature 48
dry farming **354**
dune 399
dust bowl 353, 354
dust cap 353–356
dynamic equilibrium **270**

earth measurement 29
earthquake 119–120, 128
earth radiation 14, 16
earth science 2
easterly wave **47**
Eastern Low 89
ecosystem **257**
edge organism **246**
effective surface **322**
emphysema **39**
end moraine **348,** 348, **375**
energy **8, 320,** 320
energy budget 8–25, **10**
entisol 210–211
entropy **23**
epeirogeny **129**
epipedon 206
equator 5
equatorial forest 248, 266
equatorial low 56
equilibrium **259,** 259, 269, 270
erosion **146,** 186, **323**
escarpment 374, 376
esker **349**
estuary 286–287
euphotic zone **283,** 283
eurythermic **236**
eutrophication **282,** 280–283
evaporation 10
exchange capacity **161,** 162, 183
exchange process 160–161
exfoliation **334,** 334
exotic river **354**
extinct volcano **385**
extracellular enzyme **178**
extrusion **124**

fallout 259
fallow **200**
fault 128, 132
feedback loop 405
feldspar 136, 137–138, 145, 149, 153
felsic **138**
felsite 154
ferralization 205
fertilizer 263–265
field capacity **164,** 164, **304**
fire environment 3
first law of thermodynamics **320**
flocculation **169,** 169
flood crest **309,** 309
flood multiplier graph 314
floodplain **300, 341,** 392
flood-routing graph 312–313
flood stage 311–312, **312,** 331
floor plan 21
fluorescence 156
flushing **169**
flushing time **282, 306**
fog **57,** 57
folding 127–128, 372
food chain 24, 244–246
footing **203**
forest soil 185, 196
forest water budget 298
fossil **116,** 116, 131
fracture 156
friction **320,** 320
front **67,** 71–75
frontal climate 84, 94, 96
frostbite **107**
frost-heaving **202**
frost penetration 202
frost-wedging **147,** 147

gabbro 154
galena 153
game theory **99**
garrigue 251
gas 8
gauging station **298,** 298, 305
genetic classification **205**
geographical cycle of erosion 327–329
geography 3
geologic time 130–131, 134
geology 2
geomorphology 2
geosystem nutrient cycle 261
geothermal energy **15**
glacial flow **354**
glacial lake 350
glacial melt water 379
glacial till 398
glacier 344–349, **345**
glass 139
gley layer **171**
global energy budget 10
globe 5
gneiss 141, 144, 155
gradational force **123**
graded stream 324–326, **326,** 330
gradient **379**
granite 136, 141, 144, 147, 149, 154
granular 171, 209
grassland 252
grassland climate 97
grassland soil 185
gravity 31, 32, 34, 146, 321–325
gravity movement 304
greenhouse effect **14**
green sunset 83
grid 5
ground moraine **348,** 348
groundwater discharge 306
groundwater graph 307–308
growing season **99**
growth form **235**
gypsum **143,** 155

habitat **236**
Hadley Cell 56–58, **57,** 63
half-life 132–133
halite 137, **143,** 155
hanging valley **347**
hardness 152–153, **330**
hardwood 253
harvest 263–265
headland **357**
heat 8, **10**
heat equator 58
heathland 255
heating 20–21
heating degree-days 104–105
heat island **22**
hematite 153
herbivore **23, 244,** 244–246
hierarchy 243
highland climate 95
high zonal index 78
histosol 228–229
homoiotherm **276**
horn **346,** 396
hot forest climate 97
humification 205
humus **178,** 178, **186,** 186–187
hurricane **60,** 60–62
hurricane track **62**
hybrid climate 58–60, **59**
hydrochloric acid 152
hydrogen ions 162
hydrogen sulfide 274
hydrologic cycle **296,** 296
hydrologic equation 298, 301
hydrology 2
hydrolysis **147,** 147
hypothetical continent 99

Ice Age 344
ice cap 368
ice climate 97
ice scour 396
ice sheet **345**
igneous rock **137,** 137–141, 154
immobilization **184**
inceptisol 212–213
incoming radiation 13–15
index cycle **78,** 78
index fossil 131
Indian Summer High 90
infiltration 186, 301–307, **302**
infiltrometer **304**
initiating slope 323
insolation 9
instability climate 46–47, 94, 96
intensification of pressure **76**
interception **302,** 302
interconnection 405
interface **136**
interflow **306**
intergrade **207**
interlobate moraine **375**
intermittent stream **381**
intertropical convergence **58**
intertropical convergence zone **47**
intrusion **124,** 132
inversion **44**
inversion climate 43–46, 94, 96
ion **160**
iron 175
iron compound 167

irrigation 169
isoline **18,** 18
isostasy **128,** 128–129, 328–329

jet stream 63, 78, **79, 84**
joint **149**

kaolinite clay **147, 204**
karst terrain **149,** 149
Koeppen climate classification 97
k-selection **241**

lacustrine plain **376,** 376
lagoon **360**
lag time 309, 405
lake 278–283
lakebed 374, 376
laminar flow **275,** 275
lamination 131
land breeze 52, **53**
land capability class **198,** 198–202, 230
landmark 4
langley **13**
lapse **31**
latent heat **10, 32**
latitude 5
lava **124,** 124–125, **139,** 154
lava flow 401
Law of the Minimum **232,** 232–233
leaching **161,** 161
leeward **352**
legume 188
Leibig 233
length of day 28
lens **382**
leverage point 405
L-factors 318
lichen **238,** 238–239
lignite 155
limestone **143,** 143, 144, 149, 155
liming **162**
limnology 2
liquid 8
lithologic correlation 131
lithosequence **196**
lithosphere **115**
litter 258, **259**
littoral **286**
load-discharge ratio **331**
loam **150,** 158, 164
lobe **374,** 374
local baselevel **329,** 329
location 4–6
loess 192
longshore transport **358,** 358
long wave **78**
longwave radiation 14–15
loose soil 186
low pressure area 61
low zonal index 78
luster 156

magma **124,** 124, **137,** 137–139, 154
magnesium 170, 174
magnetic orientation **118,** 118
magnetic reversal **118,** 118
maintaining slope 323
manganese 175
mangrove **288,** 292
mantle **115,** 115, 119
marble 156
mare's tail 82
marine terrace **356**
maritime climate 95
maritime effect **17**
maritime variant 84–85
marsh **281,** 281
maquis 251
mass 115
massive 171, 209
mass movement **321,** 321–325
mature stream 328
maximum entropy **25**
meander **343,** 392
mechanical weathering **146**
medial moraine **346,** 346
Mediterranean climate **60**
Mediterranean scrub 251
megalopolis **298**
melting 10
meridian 5
mesotropic 281
metamorphic rock **140,** 140–141, 144, 155–156
meteorite **320,** 320
meteorology 2
mica 136, 145, 153
microbe **177**
microclimate 238
mid-latitude broadleaf forest 253
mid-latitude grassland 252, 263, 368
mid-latitude forest 266, 366
mid-ocean ridge **118**
millibar **31**
mineral **136,** 153
mineralization **184**
mixing layer **34,** 34
moisture capacity of air 33
mollisol 218–219
molybdenum 175
monomolecular film **275**
monsoon **53,** 53
Monte Carlo simulation 80
montmorillonite clay **204**
monument 148, **351,** 382
moor 255
moraine 347–349, 374–376, 398
morphogenetic region **334,** 363–369
mortar 155
mounds 361–362
mountain energy budget 19
mountain glacier 385
muck **180,** 180
mudflow 336
mycelia **170, 186**

natural levee **341**
natural selection **236,** 236
neap tide **288**
needleleaf forest 254, 267
net primary productivity **177**
nevé **345**
niche **241**
nickpoint **330**
nimbostratus 38
nitrification **189,** 189
Nitrobacter 189
nitrogen 174, 183–184
nitrogen cycle 188–189
nitrogen fixation **189**
nitrogen-fixing bacteria 188
Nitrosomonas 189
nonlinearity 405
normal stage 331
Northwest Low 90
numerical method **79**
nutrient cycles 257–270
nutrient deficiency 174
nutrient sink **283**
nutrient storage 158–159

obsidian **139,** 154
occluded front **74**
occlusion **74**
ocean current 285–286
ocean energy budget 11–12
oceanography 2
ocean productivity 282–283
offshore bar **360,** 360
oil shale 143
old age stream 328
oligotrophic 281
olivine 138, 139, 153
open system **25,** 261–263, **262**
optimum **232,** 232
organic matter 178–209
organic sediment 155
orogeny **129**
orographic rain **38**
outcrop **238**
outgoing radiation 13–15
outlier 374, 376
outwash **349,** 349–350, 398
overland runoff **302,** 302
oxidation **147,** 147, **196**
oxidized iron **167**
oxisol 226–227
oxygen 273–287
oxygen recharge **274,** 274–282
oxygen sag **274,** 274–282
ozone 42, 43

paddy **182**
paleosol **193, 198**
pampas 252
parallel 5
particulate load **354**
paternoster lake **346**
pavement 350–351
peat 155, **180**
pebble 158
ped **169**
pedestal rock 351
pediment **339,** 339, 394
pedology 2
pegmatite 154
pelagic **283**
peneplain **328**
perched water table **166,** 166
percolation **161**
perennial **185**
permafrost **147, 202, 341,** 341
permeability **166,** 203
petroleum 143
pH **162,** 162
phosphorus 174
photochemical smog **21**
photosynthesis 22–23, 275
physical geography 2–3
pinnate drainage 325
pioneer 241
pioneer species **239**
pit 361–362
pitted outwash **350**
plain 341–343
plant energy budget 22–24
plant nutrient 174
plant-wedging **148,** 148
plastic soil **198**
plate tectonics **123**
platy 171, 209
Pleistocene **350**
plinthite **172**
plunging fold **372**
plutonic rock **139**
podzolization 205
poikilotherm **276**
polar anticyclone **44**
polar coordinates 4
polarity **118**
pond 281–283, 292–293
population control 242–243
porosity **161**
porphyry 154
positive feedback **274**
potassium 170, 174
potassium-40 133
potential evapotranspiration **108,** 109
ppm **273**
prairie 252, **266**
prairie soil 196
precipitation **296,** 302
pressure gradient **68**
pressure tendency **76**
prevailing wind 63–65
primary efficiency **23**
primary producer **244,** 244–246
primary production **23, 232,** 287
primary succession **240**
prime meridian 5
prismatic 171, 209
profile 388
P/R ratio **242**
pseudo-adiabatic chart 49
psychrometer 48
psychometric graph 48
pumice 154
pure climate **59,** 59, **94**
P-wave **116**
pyrite 153
pyroxene 138, 139, 153

quartz 136, 138, **145,** 145, 147, 149, 153
quartzite 144, 156
quickflow 309, **310**

radial drainage 325
radiation **12**
radioactive dating 132
rainshadow **58,** 85–86, **86, 267**
rainshadow climate 95
rat-tail maggot 274
recharge area **308**
recovery time **270**
rectangular coordinates 4–5
recurrence interval **102,** 102–103
red sky 83
reduced iron **167**
reef **288,** 288, 290–292
regolith **150**
rejuvenation **329**
relative humidity 33, 48
residency time **42**
respiration 275
retreating escarpment **376**
Rhizobium 188
rhyolite 154
ridge 118–119
ridge and valley 127, 372
rift **118**
rift zone **119**
Ring of Fire 119
river regime 310–311
roche moutonnée **346**
rock **136**
rockfall 334
roof 20
rotation speed 29
row crop **200**
r-selection **241**
runoff 301–307

salinization **169,** 205
saltation **352**

salt marshes 286–287, **288,** 292
sand 158, 164
sand dunes 352–353
sandstone **142,** 143, 144, 149, 155, 373
Santa Ana High 91
savanna 249, **266,** 364
schist 141, 156
sclerophyll forest 251
scour 345–347
scrub desert 250
scrub woodland 249
sea breeze **52,** 52
sea cliff **356**
seamount **119**
seasonal forest 249
seasons 26
sea stack **357**
secondary succession **240**
second law of thermodynamics **320**
sediment **141,** 154–155
sedimentary rock 141–144, 154
seif **352**
seismograph **115**
selva 248, **266,** 364
semiarid 366
septic drainage field **202**
sere **239**
sesquioxide **147**
Seventh Approximation 205–207
sewage treatment 279
S-factors 318
shale **142,** 144, 154
sheet wash **304**
shelter belt 354, 355
shelves 286–287
shield volcano **125,** 125
shortwave radiation 14–15
shrink-swell potential 203–204
shrub desert 250
sial **121**
sialic material (sial) **138**
silicate clay **145**
silt 158
siltstone **142,** 143
silvaturbate **338**
sima **121**
simatic material (sima) **138**
single-grain soil 171
sink **149, 262,** 400
sky, color of 10
slate 156
slope stability 204
sludgeworm 274
slump **336,** 337
snow 93
snow forest climate 97
sodium 170
soil **192,** 192
soil aggregates 169, 170
soil creep **336,** 336
soil drainage 167, **187**
soil-forming processes 205
soil horizon **184**
soil moisture discharge 305–306
soil moisture tension 162–164, **163,** 168–169, 304
soil orders 207–229
soil profile **184,** 211–229
soil sequence 196–197
soil series 196–197
soil structure 169–172
soil texture 158, 183–183
solid 8
solifluction **336,** 337
source **262**
source-forced transfer **262**
spacing of homes 21
specific gravity 156
specific heat 12, **17**
specific surface **160,** 160
spillway **377**
spodosol 220–221
spring tide **288**
stability 69
stable air **32**
stage **309**
stage-discharge rating **311,** 311
stage-frequency graph 310–311, **311**
station model **67**
statistical probability **99,** 99
Stefan-Boltzmann Law **13**
stemflow **302,** 302
stenothermic **236**
steppe 252, **266**
storage 405
stormflow 309, **310**
storm-rating **103,** 103
storm runoff graph 316
strand line **358**
stratigraphic column **117**
stratosphere **42**
stratus cloud **38,** 38
streak 156
stream deposition 392
stream erosion 389, 390
stream hydrographs 309–310
stream order 325
stream profile 326
stream zones 277–279
streets 21
striation **346**
stripcropping **198**
structural aggregates 186
structural dome 390
subalpine forest 254
subangular blocky 209
subarctic 254
subduction **140**
subduction zone **119**
subsidence 56
subsidence climate 94, 96
subtropical **54**
subtropical desert 57
subtropical high **56,** 56–57
succession 238–242
succulent desert 250
sulfur 175
sun angle 27
sun radiation 14, 16
superimposed stream **373**
superposition 130
surface detention **304**
surface front **72**
swamp **281,** 281
swash 358
S-wave **116**
syncline 127
synergism 234, 277
synoptic map 67
system 25
system linkage 269

taiga 254, **267**
talus slope **334,** 335
tar sand 143
tectonic force **123**
temperature 8, 68
temperature anomaly **81**
temperature gradient **18**
Temperature-Humidity Index **107,** 107
temperature map 17–18
temporary baselevel **32,** 32
terrace 343
terrestrial biome 247–255
territoriality 243

textural diagram 158
texture 156
thermal equator **58**
thermal equilibrium **12, 15**
thorn forest 249
Thornthwaite climate classification 108
Thornthwaite method 297
throughfall **302,** 302
thunderhead **37**
tidal zone 292
tide 288
tile 181
tiling 201
till **349**
till plain **348**
tilth **171**
tolerance 232–236
tolerance limit **232**
topographic map 387–401
toposequence **196,** 196
trade wind **63,** 63
transfer **259**
transparency 152
transpiration **17, 298,** 298
treefall 338
trellis drainage 325
trench 119
trophic level **24**
tropic **54**
tropical depression **60**
tropical grassland 249
tropical nutrient cycle 267
tropical rainforest 248
tropical savanna 266
tropical storm **60**
tropopause **42**
troposphere **42**
trout 274
truncated fold **372,** 372
tsunami **125**
tundra 255, **268,** 268, 368
tundra climate 97
turbulent flow **274,** 274
turbulent transfer **15,** 15, **34**
typhoon **60**

ultisol 224–225
ultraviolet radiation **42,** 42
unconformity 131
underfit stream **377**
Unified classification 204
United States Department of Agriculture 205
United States Rectilinear Survey 5,193
Universal Soil Loss Equation 316–317
unstable air **34**
upper-air inversion **39,** 39
uptake 258
upwelling **286,** 286, 287
uranium 133, 238
U-shaped valley 346, 396

valley 339–341
valley glacier 345–346
valley inversion **40**
valley wind 83
varving 131
veer **82,** 82–83
ventifact **351**
vertisol 214–215
viscosity **137**
volcano **119,** 119–121, 123–124
V-shaped valley 339–341, 389, 390, 391, 393

warm current 287
warm forest climate 97
warm front 72
water budget 296–307
water molecule 163
water pollution 273–282
watershed **298,** 298, 305
water table **165,** 165–167
water vapor 14
wave **284,** 284
wave cut 356–357
wave refraction 357
weather 43, **93**
weather analog **80**
weathering **141,** 145–151
weather map 68
weir 299
westerly wind 63, **64**
wet adiabatic rate **32**
wet-bulb temperature 48
wetted perimeter 326
white pine 299
Wien's Law **14**
wilderness 246
willow scrub 255
wilting point **164,** 164
wind chill **106,** 106–107
wind direction 68
wind rose 64–65, **65**
wind stress 235

youthful stream 327, 393

zenith 28
zenith angle 28
zinc 175
zonal soil **205**